MOLECULAR ASPECTS OF ANTI-CANCER DRUG ACTION

TOPICS IN MOLECULAR
AND STRUCTURAL BIOLOGY

General Editors:

Watson Fuller
(University of Keele)
and
Stephen Neidle
(King's College
University of London)

Previous published books in this series:

Topics in Nucleic Acid Structure Part 1 Edited by Stephen Neidle
Topics in Nucleic Acid Structure Part 2 Edited by Stephen Neidle

MOLECULAR ASPECTS OF ANTI-CANCER DRUG ACTION

Edited by

STEPHEN NEIDLE

Cancer Research Campaign
Biomolecular Structure Research Group
King's College University of London

and

MICHAEL J. WARING

Department of Pharmacology
University of Cambridge Medical School

First published 1983 by
The Scientific and Medical Division
THE MACMILLAN PRESS LIMITED
London and Basingstoke
Companies and representatives
throughout the world

ISBN 0 333 31556 1

Typeset by
RDL Artset Ltd, Sutton, Surrey

Printed in Great Britain by The Pitman Press, Bath

Contents

The Contributors

B. C. Baguley
Cancer Research Laboratory
University of Auckland
School of Medicine
Auckland
New Zealand

C. R. Beddell
Wellcome Research Laboratories
Langley Court
Beckenham
Kent BR3 3BS
UK

E. Bliss
University of Oxford
Nuffield Department of Pathology
John Radcliffe Hospital
Oxford OX3 9DU
UK

J. R. Brown
Department of Pharmaceutical Chemistry
Faculty of Pharmaceutical Sciences
Sunderland Polytechnic
Sunderland SR1 3SD
UK

B. F. Cain (the late)
Cancer Research Laboratory
University of Auckland
School of Medicine
Auckland
New Zealand

T. A. Connors
Medical Research Council Toxicology
Unit
Woodmansterne Road
Carshalton
Surrey SM5 6JB
UK

W. A. Denny
Cancer Research Laboratory
University of Auckland
School of Medicine
Auckland
New Zealand

K. R. Fox
University of Cambridge
Department of Pharmacology
Medical School
Hills Road
Cambridge CB2 2QD
UK

K. W. Kohn
Laboratory of Molecular Pharmacology
Division of Cancer Treatment
National Cancer Institute
National Institute of Health
Bethesda
Maryland 20205
USA

J. W. Lown
Department of Chemistry
University of Alberta
Edmonton T6G 2G2
Alberta
Canada

S. Neidle
Cancer Research Campaign
Biomolecular Structure Research Group
Department of Biophysics
King's College
26–29 Drury Lane
London WC2B 5RL
UK

M. P. Pera, Jr.
Institute of Cancer Research
Royal Cancer Hospital
Pollards Wood Research Station
Nightingales Lane
Chalfont St. Giles
Bucks HP8 4SP
UK

L. F. Povirk
Department of Pharmacology
Harvard Medical School
Boston
Massachusetts 02115
USA

J. J. Roberts
    Institute of Cancer Research
    Royal Cancer Hospital
    Pollards Wood Research Station
    Nightingales Lane
    Chalfont St. Giles
    Bucks HP8 4SP
    UK

B. Roth
    Burroughs Wellcome Co.
    3030 Cornwallis Road
    Research Triangle Park
    North Carolina 27709
    USA

M. R. Sanderson
    Gorlaeus Laboratories
    The State University of Leiden
    2300 RA Leiden
    The Netherlands

H. S. Schwartz
    Department of Experimental Therapeutics
    and Grace Cancer Drug Center
    Roswell Park Memorial Institute
    New York State Department of Health
    665 Elm Street
    Buffalo
    New York 14263
    USA

M. J. Waring
    University of Cambridge
    Department of Pharmacology
    Medical School
    Hills Road
    Cambridge CB2 2QD
    UK

D. E. V. Wilman
    Department of Biochemical Pharmacology
    Institute of Cancer Research
    Royal Cancer Hospital
    Clifton Avenue
    Belmont
    Sutton
    Surrey SM2 5PX
    UK

Preface

The idea that understanding the molecular bases of drug action will lead to major advances in medical practice is not new. It has been hallowed by the award of prestigious prizes. It has become enshrined in the thinking of a generation of scientists and physicians. Few people are brave (or foolhardy) enough to doubt the enormous potential of fundamental studies as a guide to drug design. This book represents an attempt to summarise certain areas of endeavour which provide a timely assessment of progress: it is unashamedly wedded to the concept that molecular and structural studies will point the way ahead.

A glance at the list of contents will reveal that it is in no sense comprehensive. On the contrary, in selecting chapter titles and authors we were guided first and foremost by considerations of topicality, particularly as regards molecular interpretations. Even so, it is easy to point to yawning gaps in our coverage of the subject and we hope no-one will feel offended if there is no mention of his or her favourite drug or modality of cancer treatment. With much regret we have tended to avoid discussion of clinical applications/applicability though we are conscious of the vitally important goal of transporting the insights and innovations of the laboratory into the clinic. Perhaps that aspect alone is worthy to form the subject of another volume.

There is an obvious preoccupation with matters concerning drug–nucleic acid interactions, if only because they are so amenable to investigation at the molecular level. But this is not intended to imply that drugs have a single site-specific mode of action or that DNA is the only important receptor for anti-cancer drugs. Indeed, we have deliberately encouraged several authors to adopt the role of "devil's advocate" and to examine critically the possibility of significant effects mediated by antitumour drugs at levels other than that of nucleic acids. It is clear that for many useful compounds the very concept of a primary site of action must be held open to question.

It may appear that the treatment of antimetabolites is particularly deficient, granted their prominent place in various cancer treatment regimes. This is chiefly due to our insistence on limiting attention to topics where a substantial body of information at the molecular level is available for the receptor — realistically a crystal structure from which precise molecular structural data can be derived. Such information is now available for dihydrofolate reductase, which therefore represents the outstanding example of a well-characterised antimetabolite receptor.

Until two or three years ago it was natural to regard the double-helical ten-fold B-form as the final word on DNA structure, and therefore the best-defined receptor for anticancer drug action. However, recent developments outlined in the first two volumes of this series reveal that we must discard the notion that DNA is a monotonously repeating biopolymer. Sequence-dependent variation of the local conformation of DNA is now a serious concept which presents an alluring challenge to those who would seek molecular interpretations of drug action.

Inevitably considerations of space have forced us to neglect a number of interesting areas where molecular insights are currently being applied to develop new drugs with promising clinical potential. We can only hope that the principles which emerge from the existing chapters will prove to be of sufficient generality to serve as pointers to developments in related fields.

Lastly, we hope the book may prove useful to physicians and others whose principal concern is with the use of drugs in the clinic to treat cancer, as well as to the more obvious audience of laboratory research workers. Our underlying belief remains that fundamental studies cannot be neglected in assessing future priorities: indeed we are prepared to argue that they represent the cornerstone of the subject, and this attitude (perhaps better described as an article of faith) is implicit in the message of several contributions. In this claim we make no boast for originality, only a driving concern to make best use of molecular and structural biology in the service of medicine.

Stephen Neidle
Michael Waring

1

Antitumour Acridines

William A. Denny, Bruce C. Baguley, Bruce F. Cain* and Michael J. Waring

INTRODUCTION

The first use of acridine compounds in clinical medicine occurred before the turn of the century when the chrysaniline derivatives (1; R_1, R_2 = H or CH_3), by-products of aniline dye manufacture, were found weakly active against malaria (Mannaberg, 1897). Acridine derivatives did not come into widespread use, however, until the First World War, with the discovery of the local anti-bacterial action of acriflavine and proflavine (2; R = H). The fact that commercial acriflavine was a mixture of euflavine (3) and proflavine hydrochloride was not recognised until 1934. The drugs received widespread use until the Second World War, finally being displaced as local antibacterials by another acridine derivative, 9-aminoacridine (aminacrine) (4), which afforded less acidic solutions (Albert, 1966).

Meanwhile, the early suggestions of antimalarial activity of acridine derivatives were followed up. The exemplary activity of mepacrine (atebrin, quinacrine) (5; R = H; X = CH) was discovered in the 1930s, and the drug came into wide-spread use during the Second World War as a prophylactic antimalarial. The clinical success of mepacrine stimulated interest in this class of compound, and hundreds of analogues were produced and tested. Only two of these, amino-acrichine (5; R = NH_2, X = CH) and azacrine (5; R = H; X = N) have been used clinically to any great extent (Albert, 1966), and these have structures very similar to mepacrine.

Thus the main impetus to develop clinically useful acridine derivatives began before 1900 with the production of antibacterials, and continued with the

*Deceased January 1981

Molecular Aspects of Anti-cancer Drug Action, ed. Neidle & Waring
0333-315561/83/001-034

development of antimalarial and antitrypanosomal agents. Although the antineo-plastic potential of these compounds was also explored (see below) minimal activity was found. Since 1960, however, the main development of biologically active acridine derivatives has been the successful elaboration of clinically useful antitumour drugs.

SIMPLE AMINOACRIDINES

It has long been known that acridine derivatives, particularly aminoacridines, bind preferentially to nucleic acids in living cells, and acridine orange (*6*) was introduced in the 1940s as a biological stain (Bukatsh and Haitingen, 1940;

Strugger, 1948). It was later found that this compound differentiated DNA (green fluorescence) from RNA (red fluorescence) in living cells (Smiles and Taylor, 1957; Stockinger, 1958). Some work has been done on the use of acridine orange staining to detect malignant cells in cancer screening procedures (Riva and Turner, 1962; Sherif, 1964). Aminoacridines were known to accumulate in the tumours of animals (Ackerman *et al.*, 1968), and many of the antimalarial candidates have been tested for antitumour activity, but with minimal results (Lewis and Goland, 1948). Acriflavine was shown to inhibit the growth of trans-plantable animal tumours (Lettré, 1941), and to induce partial responses when used to treat some human tumours (Blumenthal, 1931). These studies in the late 1920s significantly preceded the development of chemotherapy as a major treat-ment modality (usually dated from the use of nitrogen mustard derivatives immediately following the Second World War), but consistent clinical results were not found (Marsh and Simpson, 1927). More recently, euflavine (*3*) and acridine yellow (*2*; R = CH₃) have been shown to inhibit the growth of Ehrlich ascites tumours (Schummenfelder *et al.*, 1959). Much clinical work has been carried out with mepacrine, mainly to control serous effusions associated with metastatic malignant disease (Gellhorn *et al.*, 1961; Rochlin *et al.*, 1964; Stiksa *et al.*, 1979), but local cytostatic effects have been observed (Ultmann *et al.*, 1963).

ALKYLATING ACRIDINES (THE ICR COMPOUNDS)

The effectiveness of alkylating agents such as the simple nitrogen mustards in treating cancer led to numerous efforts to attach these agents to suitable carrier molecules. Noting that the cellular target of the alkylating agents was nucleic acids, and that acridine derivatives were reported to localise in the nuclei of cells, one group at the Institute for Cancer Research in Philadelphia tested a number of substituted acridines as potential carriers of alkylating functions, particularly N and S mustards (Peck *et al.*, 1959). Using the mepacrine chromophore they found that compounds bearing either a single (*7*) or *bis*chloroethyl mustard group had essentially equal potency as antitumour agents. Similar compounds possessing quinoline chromophores were found to be active only if the side-chain carried a bifunctional mustard (Peck *et al.*, 1961). They suggested that the potent antitumour and mutagenic activities of the acridine monofunctional mustards were due to the high affinity of the acridine chromophore for DNA; the compounds thus exhibiting a modified type of bifunctionality for DNA binding (Peck *et al.*, 1966; Preston *et al.*, 1964).

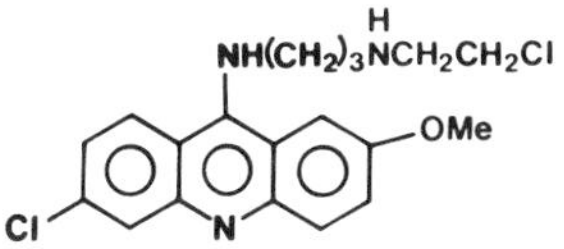

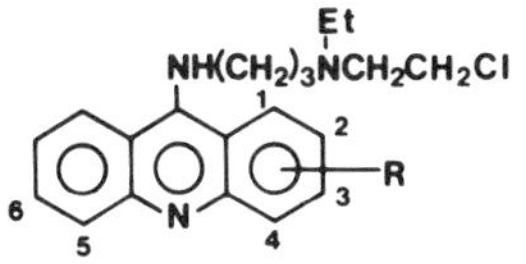

8	R =	2-OMe, 6-Cl
9		2-OnBu, 6-Cl
10		2-Ph, 6-Cl
11		1-aza-2-OMe, 6-Cl
12		H

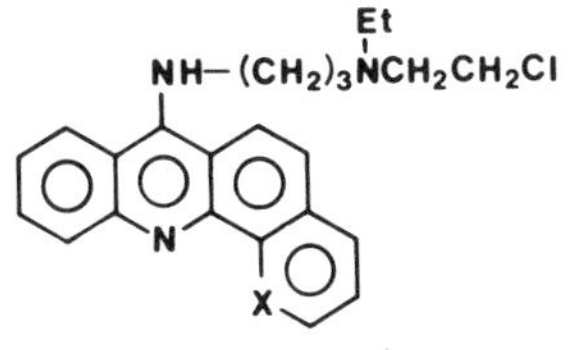

13	X = CH
14	X = N

A cursory structure–activity analysis of some of the published compounds, in the light of later work on the DNA-binding of substituted acridines, supports the view that the chromophore acts as a DNA-binding moiety. For the subset of compounds (*8–14*), bearing an identical side chain, potency against Ehrlich ascites tumour *in vivo* closely parallels the DNA-binding affinity of similarly-substituted acridine derivatives of the 9-anilinoacridine series (Baguley *et al.*, 1981*a*, *b*). Thus, attachment of bulky groups at C-2 to give (*9*) and (*10*), or introduction of an aza atom instead of C-1 (*11*) reduces potency; these changes also reduce DNA-binding of the acridine chromophore in the 9-anilinoacridine series. Conversely, the benz(*c*)acridine derivative (*13*) shows equal potency to the parent (*12*) in spite of a considerable increase in lipophilicity, and the pyridyl(*c*)acridine (*14*) is four times more potent than the parent. In the 9-anilinoacridine series these substitutions lead to greater DNA binding affinity.

In the ICR series of compounds, the tertiary amines such as (*8–14*) showed high antitumour activity, but low mutagenic activity against *Salmonella* species, as did a related series of sulphur mustards (Creech *et al.*, 1972). In contrast, the related secondary amine monofunctional mustards possess minimal antitumour activity but are exceptionally potent frameshift mutagens in bacteria, and have been widely used for studies of mutagenesis; ICR-191 (*7*) has become one of the standard agents for producing frameshift mutations in *Salmonella typhimurium* and *Escherichia coli*. This class of compound also proved highly carcinogenic (Peck *et al.*, 1976*a*), but compounds without the alkylating moiety showed no marked carcinogenicity or mutagenicity, even in sensitive systems (Peck *et al.*, 1976*b*).

Thus two structural features were found necessary for compounds of this class to exhibit high antitumour, mutagenic and carcinogenic activity (effects related to the ability of the compounds to interact with DNA): an alkylating group on the sidechain, and a highly structure-specific polyaromatic nucleus. It is interesting to observe the effect of previous work with antimalarial compounds on the development of this series. Although it was demonstrated quite early (Preston *et al.*, 1964) that acridine nuclei lacking the 6-Cl group were more active mutagens, it was the mepacrine analogue ICR-191 (*7*) that emerged as the most widely-used derivative. Finally, the development of this series of compounds as antitumour agents has been overshadowed by their utility as mutagens, and no member of the series has been considered for clinical development.

1-NITRO-9-AMINO (ALKYLAMINO) ACRIDINES

In the late 1960s Ledóchowski and coworkers began a systematic study of the antitumour properties of acridine derivatives bearing a charged aminoalkylamino group on the 9-position, and screened a large number of these derivatives against the sarcoma 180 solid tumour in mice (Radzikowski *et al.*, 1969*a*, *b*).

As previously noted by the ICR group using the Ehrlich ascites tumour system (Peck *et al.*, 1961; Peck *et al.*, 1966; Preston *et al.*, 1964), they found that the majority of the compounds were inactive, although some of the methoxylated acridines showed marginal antitumour activity. However, early in their investigations they noted the remarkable effect of a 1-nitro group placed on the acridine nucleus (Radzikowski *et al.*, 1969*b*), which resulted in a 100-fold increase in host toxicity compared to the unsubstituted acridine, and provided compounds (*15–18*) with marked activity in the S-180 tumour screen *in vivo*.

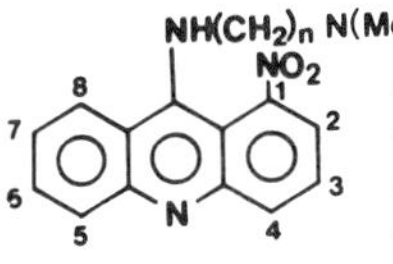

15	n = 2	
16	3	
17	4	
18	5	

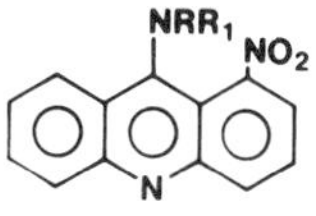

19	R = R_1 = H
20	= H, R_1 = Me
21	= R_1 = Me
22	= H, R_1 = C_6H_5

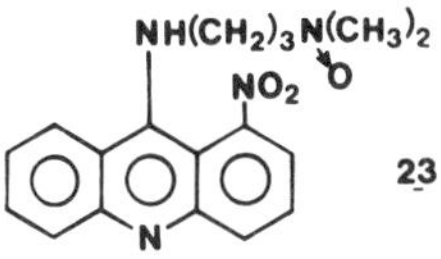

23

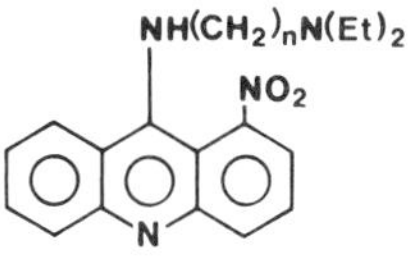

24	n = 3
25	n = 2

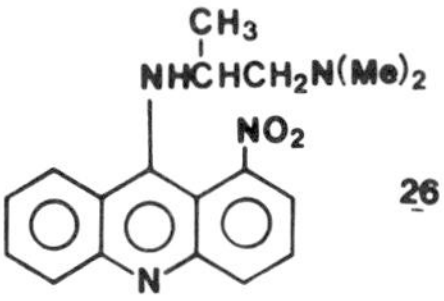

26

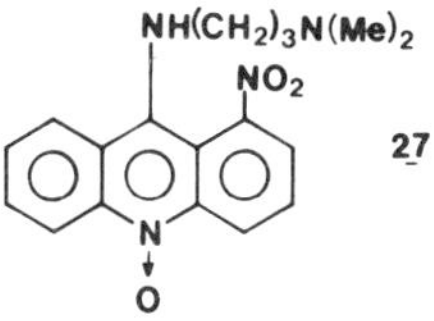

27

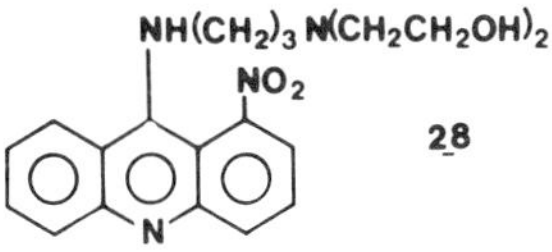

28

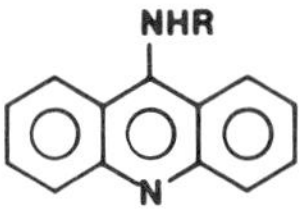

29

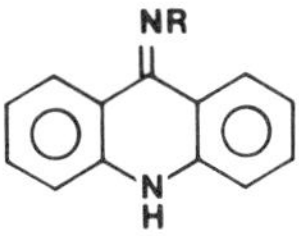

30

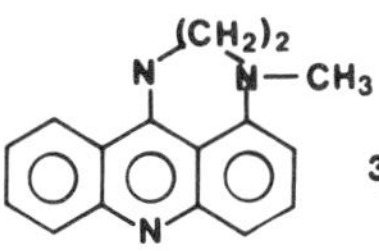

31

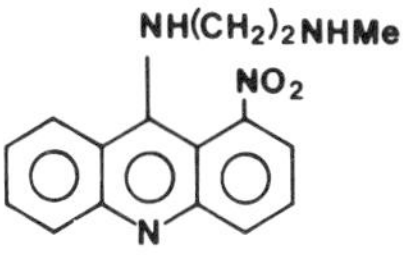

32

Optimal activity was associated with the C-3 and C-5 side-chains (for example, compounds *16* and *18*) (Hrabowska *et al*., 1970*a*), and with the dimethylamine moiety, although this could be replaced with piperidinyl, morpholino or even isopropylamine to give a secondary amine with retention of activity (Hrabowska *et al*., 1977). However, the diethyl analogue proved inactive (Hrabowska *et al*., 1970*a*).

Subsequent work confirmed that the 1-nitro group was essential for activity. Replacement by other groups that matched it for steric bulk (trifluoromethyl, dimethylamino, iodo) or electron-withdrawing capacity (chloro, bromo) led to inactive compounds (Hrabowska *et al*., 1971). The only other derivatives to show even slight activity were 3-nitro analogues but at 20-fold higher dose-levels than the corresponding 1-nitro compounds (Hrabowska *et al*., 1976). Further substitution of 1-nitro derivatives at various other positions around the acridine nucleus had only slight effects on both toxicity and antitumour activity (Hrabowska *et al*., 1970, 1971). *N*-oxide derivatives of either the acridine nitrogen, the tertiary side-chain nitrogen, or both nitrogens in compounds such as compound *16* were active, but at 10-fold higher doses than the parent compound, which is presumably released *in vivo* as the active principle (Hrabowska *et al*., 1976).

Other acridine derivatives with confirmed activity in the S-180 system are the 9-amino, 9-methylamino and 9-dimethylamino-1-nitro compounds (*19–21*), which lack the cationic side chain. Higher alkyl homologues of (*20*) proved inactive, possibly due to excessive lipophilic character (Hrabowska *et al*., 1970*a*). Interestingly, in view of the concurrent development of the 9-anilinoacridine antitumour agents (see below), the 1-nitro-9-anilino derivative (*22*) was found to be active in the S-180 system (Hrabowska *et al*., 1970*b*) but this observation does not appear to have been followed up.

Nitracrine

From the early examples of this series of compounds, 1-nitro-9-(3-dimethyl-aminopropylamino) acridine (*16*, originally coded C-283, marketed under the trade name Ledakrin, now known as nitracrine) and its dimethylamino *N*-oxide (*23*) were chosen for preclinical evaluation (Gieldanowski *et al*., 1972*a, b*). Both compounds were shown to have strong local irritant properties and hypotensive activity (Gieldanowski, 1972*a*). Nitracrine (*16*) was five times more potent than the *N*-oxide (*23*), but the latter compound was better tolerated. No marked renal or liver toxicity was noted for either compound, and it was concluded that they possessed relatively low toxicity in relation to other cytostatic agents (Gieldanowski *et al*., 1972*b*). The immunosuppressive properties of the compounds were studied (Gieldanowski *et al*., 1971), and the nonlinear dose–response effect of nitracrine in such studies was noted (Gieldanowski and Skowronska, 1979). Phase I and II trials of nitracrine were carried out in Poland on patients with advanced mammary carcinoma and ovarian carcinoma, and useful antitumour effects were noted (Kwaśniewska-Rokicińska *et al.*, 1973; Warwas *et al.*, 1977).

No hepatic or renal damage was seen at clinical dose levels, except in patients who already suffered dysfunction, nor was bone marrow depression observed (Bratkowska-Seniow *et al.*, 1974). A detailed study of the blood biochemistry of patients receiving nitracrine was carried out (Gerber *et al.*, 1972; Dobryszycka *et al.*, 1974).

Preclinical pharmacology and toxicology have been completed on a number of other 1-nitroacridine derivatives, including the close homologues of nitracrine (*18*), (*24*) and (*25*) (Kowalcysk-Bronisz *et al.*, 1975*a, b*: Gieldanowski *et al.*, 1980*a, b*) and (*26*) and the *N*-oxide derivative (*27*) (Blaszcyk *et al.*, 1979*a, b*), 1-nitro-9-dimethylaminoacridine (*21*) and the dihydroxy derivative (*28*) (Gieldanowski *et al.*, 1980*a, b*). The nitracrine alkyl homologues (*18, 24, 25* and *28*) produced levels of biological activity and side-effects so similar to those of nitracrine itself that it was not considered useful to introduce them into clinical trials. The analogue (*26*), almost alone among nitracrine derivatives, showed some sedative activity, and it was postulated that this might ameliorate the usually severe emetic side-effect found with nitracrine (Blaszczyk *et al.*, 1979*a, b*). However no clinical data have yet been reported. The *N*-oxide (*27*) was tested in the hope that it would show a reduced local irritant effect but this was not the case (Blaszczyk *et al.*, 1979*a, b*), and the compound also showed a higher level of renal toxicity than nitracrine. The monobasic derivative (*21*) also possessed the typical local irritant action of the dibasic nitroacridines, but had the lowest potency of all the compounds considered (Pelczarska *et al.*, 1974*a, b*) and was not recommended for clinical evaluation (Gieldanowski *et al.*, 1980*a, b*). On the basis of the clinical results gathered in Poland, nitracrine was evaluated by the American National Cancer Institute in 1975 (as NSC 247561), but was eventually dropped due to toxicity in clinical trials (Schepartz *et al.*, 1981).

Mechanism of action of nitracrine

Nitracrine and derivatives are known to bind tightly to nucleic acids. The use of 9-dialkylaminoalkylaminoacridines for the staining of human metaphase chromosomes has been described (Liman *et al.*, 1975; Comings *et al.*, 1978). Derivatives with electron-donating groups on the acridine rings and a $-(CH_2)_2-$ link between the N atoms of the side-chain showed the most intense fluorescence (Liman *et al.*, 1975). Both nitracrine (*16*) and its 2-nitro analogue form reversible intercalative complexes with DNA as evidenced by their ability to remove and reverse the supercoiling of closed circular duplex DNA (Filipski *et al.*, 1977; Gniazdowski *et al.*, 1979). In the absence of thiols, both compounds affected RNA synthesis in isolated rat liver nuclei only very slightly. However, in the presence of sulphydryl compounds such as mercaptoethanol or dithiothreitol, nitracrine but not the 2-nitro analogue produced significant inhibition of RNA synthesis. While both nitro compounds showed greater mutagenic activity than the unsubstituted parent compound against various *Salmonella typhimurium* strains, nitracrine (*16*) proved more potent than the 2-nitro analogue by a factor of more than 10^4 (Kalinowska and Chorazy, 1980). With L1210 cells *in vivo*, only nitracrine caused observable ultrastructural changes (margination of chromatin, nuclear segregation) (Filipski *et al.*, 1977).

These effects are attributed to the ability of 1-nitro derivatives to form covalent links to DNA, both *in vitro* (Gniazdowski *et al.*, 1975; Slaska *et al.*, 1979; Konopa *et al.*, 1976) and *in vivo* (Filipski *et al.*, 1977; Gniazdowski *et al.*, 1979). DNA isolated from drug-treated cells has been shown to contain bound drug molecules which are not removed by precipitation with trichloroacetic acid, solvent extraction, or other procedures which would dissociate noncovalently bound complexes (Konopa *et al.*, 1976; Gniazdowski *et al.*, 1979). Up to one drug molecule per thousand nucleotides can be bound in this way, and the phenomenon can be detected using drug which is radioactively labelled either in the acridine or in the alkylamino side-chain moiety. The modified DNA displays characteristics of reversible denaturation and renaturation, suggesting the formation of cross-links, but only if the cells have been exposed to biologically active 1-nitro isomers such as nitracrine itself (Filipski *et al.*, 1977, Gniazdowski *et al.*, 1979). The presumption that this form of inactivation of DNA operates *in vivo* is strengthened by the finding that nuclei isolated from hepatocytes of drug-treated rats are inactive in RNA synthesis in an *in vitro* system but recover activity following supplementation with exogenous DNA (Gniazdowski *et al.*, 1979).

In vitro in the presence of thiol compounds, nitracrine binds irreversibly to both native and denatured DNA, apparently with a certain preference for the latter and some specificity for deoxyguanosyl residues (Gniazdowski *et al.*, 1981). The resulting complexes show a decrease in template activity for transcription by purified RNA polymerase which is directly proportional to the amount of drug covalently bound (Gniazdowski *et al.*, 1979, 1981). Only very small numbers of bound drug molecules are needed to produce significant inhibition of RNA synthesis: 50% impairment of activity can be detected with as little as one drug molecule bound per 2,000 nucleotides, and with one molecule bound per 200 nucleotides the amount of RNA synthesised falls to 10% of the control (Gniazdowski *et al.*, 1979, 1982). By contrast, the complex remains able to bind the enzyme when up to 20 molecules per thousand nucleotides are bound. Features of the altered template activity of the nitracrine–DNA complexes are a lower number of initiated RNA chains and a shorter average chain length, attributable to prevention of movement of the polymerase enzyme along the template (Slaska *et al.*, 1979). The thiol-dependence of these effects is significant. In the absence of thiol compounds the level of inhibition of RNA polymerase activity produced by nitracrine and other 1-nitro derivatives is surprisingly low, a good deal lower than that produced by 2-, 3-, or 4-nitro isomers (Gniazdowski *et al.*, 1982). This may be attributed to structural distortion of the acridine ring produced by steric hindrance between the 1-nitro substituent and the alkylamino side-chain, which provokes a butterfly-like puckering of the tricyclic ring system as is evident in the crystal structure of nitracrine (figure 1). Such distortion would be expected to impede intercalation of the acridine moiety according to

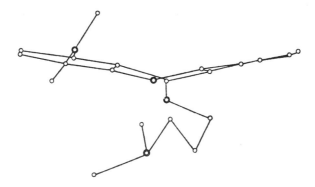

Figure 1   Molecular conformation of nitracrine (Ledakrin, C-283) as determined by X-ray diffraction analysis of the crystalline monoiodide. Nitrogen atoms are represented by circles in heavy outline; carbon and oxygen are shown as lightly-drawn circles. Hydrogen atoms are omitted. The direction of view is slightly to the left of a line passing through C(9) and N(10) of the acridine moiety. Note the 60° displacement of the nitro group (at left) from co-planarity with the aromatic ring system and the butterfly-like inclination of the two halves of the acridine chromophore. (From Dauter *et al.*, 1975).

the classical scheme (Gale *et al.*, 1981), but should not occur for isomers bearing the nitro group at other positions of the acridine ring. In support of these ideas Gniazdowski *et al.* (1979, 1982) have reported that the inhibitory effect of halogen-substituted analogues diminishes as the size of the substituent at position 1 increases, and that the anomalously low inhibitory activity of 1-nitro compounds is comparable with that seen with tetrahydroacridine derivatives in which another of the six-membered rings is puckered. Perhaps more importantly, the great enhancement of inhibitory activity toward RNA polymerase provoked by thiols is limited to the 1-nitro derivatives. Indeed, a good correlation exists between the effects of different derivatives on cell growth and the property of thiol-dependent enhancement of their inhibitory activity in the RNA polymerase system *in vitro*. All indications are to the effect that formation of irreversible (covalent) complexes with DNA both *in vitro* and within the cell reflects the same properties of nitro-acridine drugs as are responsible for their biological activity (Gniazdowski *et al.*, 1979, 1982). The structure–activity spectrum of the 9-alkylaminoalkyl acridines is suspiciously narrow, compounds without a 1-nitro group showing at best only marginal antitumour activity.

The detailed mechanism of the reaction between nitracrine and DNA is not clear. 9-Alkylamino acridines, like 9-aminoacridine itself, are generally accepted to exist in the amino form (for example, compound *29*), but the addition of a strongly electron-attracting group to acridine predisposes the compounds to adopt the imino form (*30*) (Skonieczny, 1980). The 1-nitro acridines such as nitracrine (*16*) also appear to adopt the iminoacridan form, both in solution (Skonieczny, 1980) and in the crystal state (Dauter *et al.*, 1975). The crystal structure of nitracrine (figure 1) clearly shows an iminoacridan form, with a C9-N11 bond length of 1.313Å, considerably shorter than that for other 9-amino and alkylamino acridines (1.34–1.36Å) (Dauter *et al.*, 1975). This explains the hydrolysis kinetics of the 1-nitro acridines, where the rate decreases with rising pH (as the more hydrolysis-resistant iminoacridan form is adopted), compared to hydrolysis kinetics of the 3-nitro derivatives, which increase with rising pH (Skonieczny, 1980). The crystal structure also shows that the crowded C-9 geometry resulting from the iminoacridan form forces the 1-nitro group 60° out of plane with the acridine ring. The resulting lack of conjugation accounts for the unusually high reactivity of this group, which in turn may be responsible for the high biological activity of the 1-nitroacridines. Nevertheless, the exact pathway leading to covalent bond formation of nitracrine derivatives with DNA remains to be elucidated. The facile reduction of the 1-nitro group *in vivo* has been demonstrated, and products of reduction such as the 1-hydroxylamino compound detected (Pawlak and Konopa, 1979). However, it is not certain that the reductive pathway is involved in the process of covalent bonding to DNA. The 1-nitro group is easily replaced by nucleophiles either intramolecularly, for example to give *31* from *32*, (Skonieczny *et al.*, 1978), or intermolecularly by sulphydryl groups (Weltrowski and Ledóchowski, 1978), and this may well be the important process.

9-ANILINOACRIDINES

As mentioned in the previous section, derivatives of 9-anilinoacridine had been synthesised in the late 1960s. The 1'-dimethylamino derivative (*33*; R = $N(CH_3)_2$; see numbering system which is used throughout this chapter) was tested in the S-180 system and found to be inactive (Radzikowski *et al.*, 1967), although it had been independently shown to be active against L1210 leukaemia (Goldin *et al.*, 1966). More recently, Zhang *et al.* (1980) have independently reported antitumour activity of an anilinoacridine. However, it has been the work of Cain and coworkers in Auckland, dating from 1972, which has demonstrated the broad spectrum of biological activities of 9-anilinoacridines, and the outstanding experimental antitumour activity of many of the analogues.

Anilinoacridines were initially synthesised in an extension of work with *bis*quaternary ammonium heterocyclic antitumour compounds (Cain *et al.*, 1971). Early studies of these linear, doubly charged molecules, which are thought to bind in the minor groove of the DNA double helix (Cain *et al.*, 1969; Denny *et al.*, 1979; Baguley, 1982) produced symmetrical molecules (for example, compound *34*) which were highly active against L1210 leukaemia. These compounds exhibited a lethal long-term (30–60 day) host toxicity (Atwell and Cain, 1973), which was minimised in a further series of asymmetric antitumour com-

33

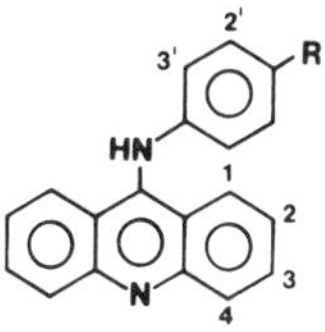

34 CH₃CH₂ / CH₂CH₃ ... NHCO ... NHCO ... CONH ... CONH ...

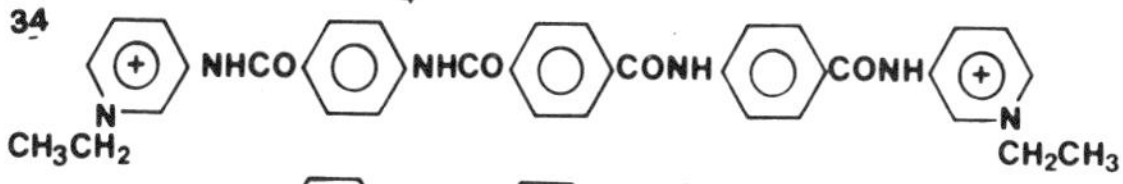

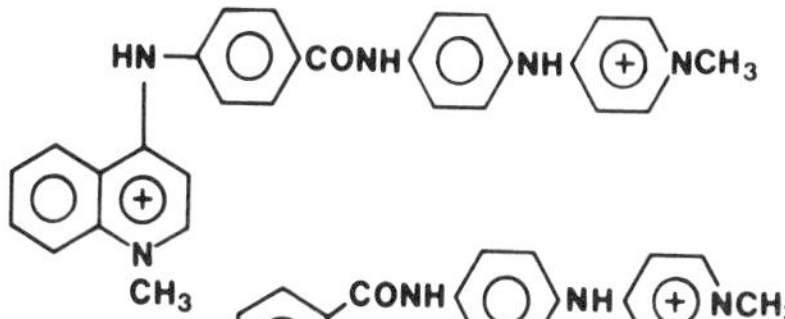

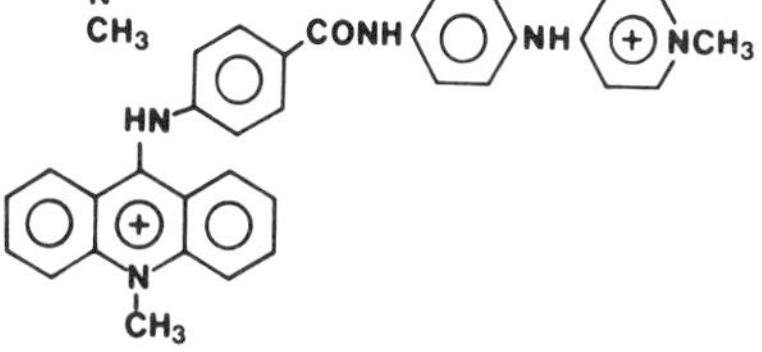

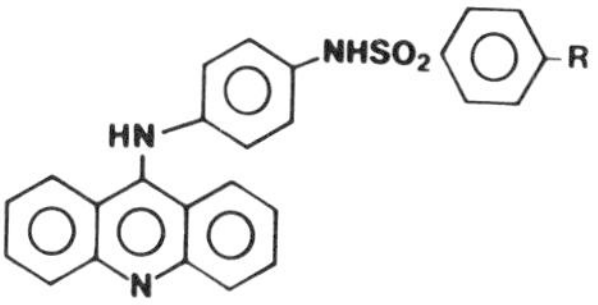

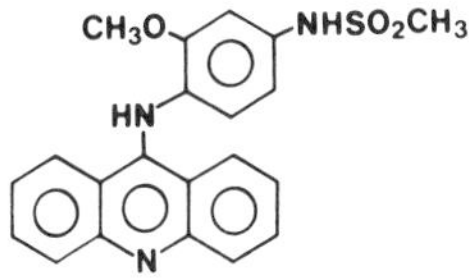

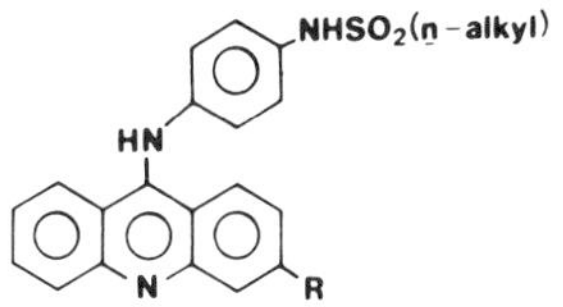

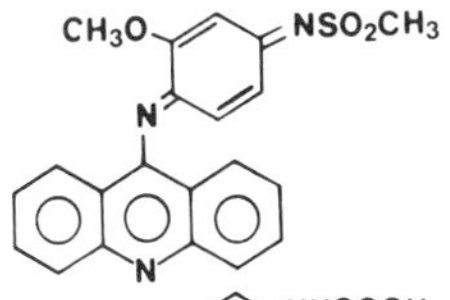

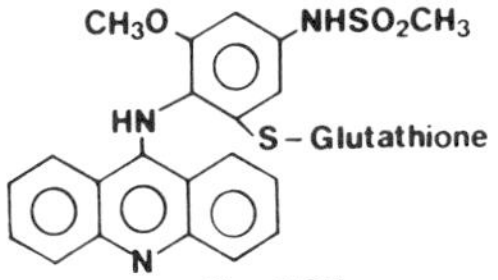

35 CH₃ ... HN ... CONH ... NH ... NCH₃

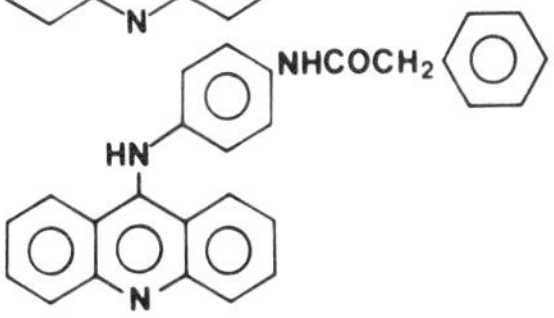

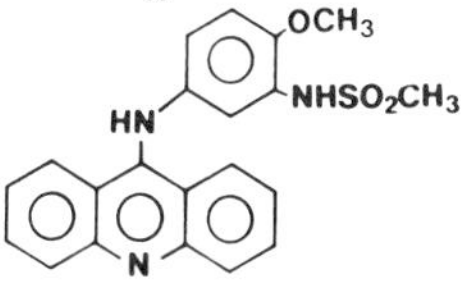

36 CH₃ ... HN ... CONH ... NH ... NCH₃

37 ... NHSO₂ ... R ... HN

38 CH₃O, NHSO₂CH₃ ... HN

39 NHSO₂(n-alkyl) ... HN ... R

40 CH₃O, NSO₂CH₃

41 CH₃O, NHSO₂CH₃ ... HN ... S – Glutathione

42 NHCOCH₂ ... HN

43 OCH₃ ... NHSO₂CH₃ ... HN

44 NH ... NHSO₂CH₃ ... CH₃

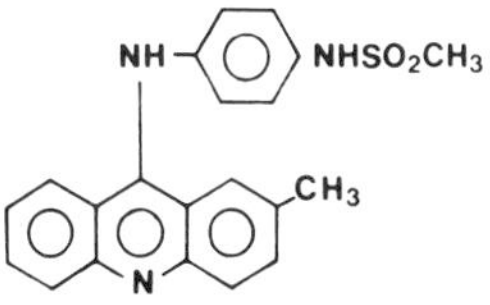

pounds possessing an amino-substituted *N*-methylquinoline at one end of the molecule (*35*). A logical extension to *N*-methylacridine derivatives produced a further series of highly active agents (*36*) (Cain *et al.*, 1971). Later studies with one member of this series (Braithwaite and Baguley, 1980) showed that it unwound closed circular duplex DNA, suggesting that it bound by intercalation (Waring, 1970; Gale *et al.*, 1981), in contrast to the N-methylquinolinium compounds. Further work on the biological activity of these compounds showed that neither the alkyl quaternising function on the acridine nor the second cationic group at the other end of the molecule were necessary, initiating the development of the simpler 9-anilinoacridine derivatives based on (*33*) (Cain *et al.*, 1971).

Two unusual features of the 9-anilinoacridines as antitumour compounds are the very wide variation of structure that can be tolerated while retaining biological activity (over 700 derivatives with antitumour activity are known), and, among the active derivatives, the extreme range of activity and potency afforded by small changes in structure. Thus, in the 1'-substituted 9-anilinoacridine series activity *in vivo* ranges from curative at a dose of 4.5 mg per kg body weight (*37*; R = NH$_2$) to inactive and nontoxic at a dose of 500 mg per kg (Cain *et al.*, 1975). This behaviour is reflected in tissue culture, where growth inhibitory concentration ranges from 0.3 nM (*37*; R = NH$_2$) to 17,000 nM (*33*; R = CN) (Baguley and Nash, 1981). The wide range of activity *in vivo* seen with monosubstituted derivatives prompted synthesis of further polysubstituted compounds; placement of a 3'-methoxy group on the anilino ring usually provided much more potent derivatives, active at low concentrations, one such being the clinically-useful derivative m-AMSA (amsacrine) (*38*) (Cain and Atwell, 1974).

Cain and coworkers espoused a rational basis for drug design, and have published a number of studies considering the effects on biological activity of the overall lipophilic–hydrophilic balance of derivatives, and of the electronic and steric effects of substituent groups (Cain *et al.*, 1975).

Effect of lipophilicity

A number of alkanesulphonamide homologues of 4'-(9-acridinylamino) methanesulphonanilide (AMSA) (*39*; R = H) and acridine monosubstituted variants have been prepared. In both the parent and 3-acetamido series, (*39*; R = NHCOCH$_3$) antitumour activity (measured by a life extension assay with L1210 leukaemia) was parabolically related to drug lipophilicity (Cain *et al.*, 1974). The latter property was measured by relative mobility (R_m) in a thin layer chromatographic system. The values so obtained were shown to correlate linearly with measured octanol-water log P values (log P = 2.00 R_m + 0.51; r = 0.99), and with log P values calculated by the fragment constant technique (Denny *et al.*, 1982). Optimal biological activity was shown by compounds with lipophilic properties close to that of the parent AMSA (*33*; R = NHSO$_2$CH$_3$). This relationship was used to identify other derivatives whose activity was greater than that expected on the basis of their lipophilicity alone (Cain *et al.*, 1975). In this way the enhanced activity of the 3-halo m-AMSA derivatives, in particular the 3-bromo derivative, was noted.

Effect of group electronic properties

X-ray crystallographic studies of 9-anilinoacridine derivatives (Hall *et al.*, 1974; Karle *et al.*, 1980) show a shortened C9-anilino nitrogen bond, consistent with an appreciable degree of double bond character. In agreement with this, there is a close correlation between the electronic properties (sigma-*para* values) of groups at the 1'-position of the anilino ring and acridine pK_a values, suggesting excellent transmission of electronic effects between the rings (Denny *et al.*, 1978; Baguley *et al.*, 1981*a*). Thus the 9-anilinoacridine framework allows transmission of the electronic effects of groups placed anywhere on the molecule with the primary effect being modification of the pK_a of the acridine nitrogen. For 193 derivatives of 9-anilino-acridine, measured pK_a values can be related to substituent sigma values in a typical Hammett relationship (Denny *et al.*, 1982). Nevertheless, the relationship between the electronic properties of substituents and antitumour activity was found to be complex. While electron-donating substituents at the 1'-position of the anilino ring in general provided active compounds, the most preferred substituents (for example acetamido, methanesulphonamide, methyl carbamate) are characterised by small or zero Hammett sigma values. Variation of substituents at the 2'- and 3'-positions of the anilino ring produced a pattern of activity consistent with the hypothesis that the 6'-position of the anilino ring requires a relatively high electron density (Cain *et al.*, 1975). There was a nonlinear dependence on Hammett sigma values for groups at the 3'-position with the highly electronegative amino substituent proving less active than a methyl or methoxy group. This lower activity may be due, partly or wholly, to the ease of oxidation of amino-bearing derivatives.

For a series of acridine-substituted derivatives of *m*-AMSA, antitumour dose potency (expressed as a logarithmic function) was positively linearly correlated with the pK_a values of the acridine base (Denny *et al.*, 1979), which were in turn highly correlated with Hammett sigma values of the acridine substituents (sigma-*para* for 3-substituents and sigma-*meta* for 2- and 4-substituents) (Baguley *et al.*, 1981*b*; Denny *et al.*, 1982).

The interpretation of these results is complicated by the observation that 9-anilinoacridines undergo nucleophilic attack by thiols to produce inactive species, this reaction occurring at a significant rate under physiological conditions (Cain *et al.*, 1976; Wilson *et al.*, 1977). The rate of this reaction with thiol is a function of acridine pK_a (Denny *et al.*, 1979; Baguley *et al.*, 1981*b*). The half life of *m*-AMSA in the presence of fresh mouse blood at 37°C is approximately 23 minutes, the products being protein adducts linked by a thiol group to the C9 position of the acridine and the released sidechain. Since thiolysis also occurs in tissue culture medium (Khan *et al.*, 1980; Cysyk *et al.*, 1979), this reaction could be important in determining activity both *in vivo* and *in vitro*. More recent studies (Khan *et al.*, 1980; Shoemaker *et al.*, 1982) suggest alternative mechanisms for degradation of *m*-AMSA by mammalian liver extracts, in which the anilino ring is oxidised to a quinoneimine (*40*), then conjugated with glutathione to produce (*41*) as a major metabolite.

Effect of group steric properties

Addition of bulky substituents at various positions on the 9-anilinoacridine framework leads to reductions in biological activity. This phenomenon is not due merely to increased lipophilicity, as was clearly shown in an extensive quantitative structure-activity relationship (QSAR) study of the antileukaemic activity of more than five hundred 9-anilinoacridine derivatives (Denny *et al.*, 1982). This showed that, when the effects of lipophilic and electronic properties of substituent groups on biological activity were allowed for, the steric effects of the group played a dominant role.

No bulk tolerance was evident for the 1- and 2-positions of the acridine ring, and the larger the groups in these positions the more dystherapeutic they proved. Small groups in the 3-position had a beneficial effect on activity, but only up to a limited size (compare Cain *et al.*, 1975). In contrast, 3,5-disubstitution was unfavourable. Only the 1'- and 4-positions appeared relatively free from steric restraint.

These findings are fully consistent with the hypothesis that the biologically-important receptor for 9-anilinoacridines is DNA, with the acridine chromophore intercalated between the base-pairs and the anilino ring lying in the minor groove of the double helix (Wilson *et al.*, 1981; Denny *et al.*, 1982). In this model (figure 2) the 1- and 2-positions of the acridine ring lie close to the sugar-phosphate backbone of the DNA. There is limited steric freedom around the 3-position, and considerable tolerance to the presence of bulky groups at the 1'- and 4-positions without distortion of the binding geometry.

Importance of DNA-binding for activity

Studies by Wilson (1973) first demonstrated that *m*-AMSA (*38*) binds to double-stranded DNA, and Waring (1976) showed that *m*-AMSA and three related 9-anilinoacridines bind to DNA by intercalation. A number of DNA binding studies have now appeared (for example Baguley and Falkenhaug, 1978; Gormley *et al.*, 1978; Wilson *et al.*, 1981*a*; Hudecz *et al.*, 1981). A method for determination of DNA binding constants based on the displacement of a fluorescent intercalating probe molecule, ethidium, gives results similar to those obtained by spectrophotometry and equilibrium dialysis (Wilson *et al.*, 1981*a*), and has been applied to measure the DNA binding of a large series of 9-anilinoacridines (Baguley *et al.*, 1981*a, b*).

Structure-activity studies for different series of 9-anilinoacridines have shown that DNA binding is closely correlated with antitumour activity, both *in vivo* (Robertson *et al.*, 1980; Cain *et al.*, 1980; Baguley *et al.*, 1981*b*) and in leukaemia cell cultures (Baguley and Nash, 1981; Ferguson and Baguley, 1981*b*), and is the statistically dominant parameter. These results provide additional evidence that DNA is the target for the cytotoxic action of these compounds.

The availability of DNA binding data for a series of 9-anilinoacridines has allowed confirmation of the predictions of a pure QSAR approach discussed

Figure 2   Space-filling (CPK) model of an intercalated complex between *m*-AMSA (*38*) and DNA. To facilitate identification of the drug molecule its hydrogen atoms have been marked with crosses. The acridine chromophore is partly obscured by the anilinosulphonamide side-chain which extends into the narrow groove of the helix. Models can also be built having the side-chain located in the wide groove. The conformation of the drug molecule is based on that of the parent compound AMSA (*39*; R = H) in the crystalline state as determined by Hall *et al*. (1974). Note the proximity of the $-OCH_3$ substituent of the drug molecule to one of the DNA base-pairs, effectively restricting further intercalative penetration of the acridine ring system.

above by a study of experimentally observed changes in DNA binding. Thus, substitution at the 1- and 2- positions of the acridine ring causes reduction of DNA binding with all groups studied except 2-amino (Baguley *et al*., 1981*b*; Ferguson and Baguley, 1981*a* and unpublished data). Bulky 3-substituents such as 3-methyl-sulphone and 3-methanesulphonamide decrease binding. Substitution by bulky groups at the 4-position is in general not accompanied by a decrease in DNA binding (Baguley *et al*., 1981*b*).

Substitution of the 1′-position with a *para*-aminobenzenesulphonamide group (*37*; R = NH_2) does not diminish DNA binding compared to compounds with smaller substituents such as methyl, methoxy and even hydrogen (Baguley *et al.*, 1981*a*). The bulk tolerance at these positions has been put to advantage in the design of a variety of compounds containing hydrophilic (Cain *et al.*, 1977), cationic (Atwell *et al.*, 1977), and anionic (Denny *et al.*, 1977; Denny and Cain, 1978) substituents.

Examination of the biological activity of these derivatives revealed complex relationships. For small series of closely-related derivatives, antitumour activity both *in vivo* (Robertson *et al.*, 1980) and in culture (Baguley and Nash, 1981) is positively related to the size of the 1′-substituent. This contrasts with bacterio-phage T4 inhibition (Robertson *et al.*, 1980) and ability to induce respiratory-deficient *petite* mutants in yeast (Ferguson and Baguley, 1981*b*) where activity is negatively correlated with size of substituent. Large differences in antileukaemic activity can be induced with only small changes in substituents; thus the benzene-sulphonamide derivative (*37*; R = H) is highly active, whereas the benzylic amide (*42*) is inactive (Atwell *et al.*, 1972).

The effects of acridine ring substituents on activity depend not only on the nature of the 1′-substituent but also on the pattern of acridine substitution. For neutral or cationic 1′-substituents, 3-amino or 3-nitro groups usually enhance activity (Atwell *et al.*, 1977). (It is possible that the 3-nitro group is reduced *in vivo*). With anionic 1′-substituents, on the other hand, 3-amino or 3-nitro groups reduce or abolish activity (Denny *et al.*, 1978). These results have been interpreted to mean that different 1′-groups dictate slightly different DNA-binding modes, providing altered structure–activity relationships for the different series.

While the above results provide reasonable explanations for the broad structure–activity relationships of the 9-anilinoacridines, they demonstrate that DNA binding and lipophilic character are not the only determinants of antitumour activity. Furthermore, substitution at the 2′- or 3′-positions can substantially modify activity, a dramatic example being the isomeric compounds *o*-AMSA (*43*) and *m*-AMSA (*38*), respectively the 2′- and 3′-methoxy derivatives. *o*-AMSA is completely inactive in L1210 tests (Cain *et al.*, 1975), and although its 3-amino analogue shows some activity *in vivo*, it is 100-fold less potent than the corres-ponding *m*-AMSA analogue (unpublished data).

The structure of AMSA in the crystalline state shows the plane of the anilino ring to be almost orthogonal to the plane of the acridine chromophore, and in-clined to one side above the 1-position (Hall *et al.*, 1974). In the currently-favoured model where the drug is intercalated into double-stranded DNA (see figure 2) the anilino ring is in the minor groove with the 1′-substituent pointing tangentially away from the helix. It is therefore in a position to interact with a second macromolecule, such as a protein. Such a ternary complex could mediate the biological effects of the anilinoacridines. This model would explain the key role of the 1′-substituent in determining the nature and extent of biological activity. It would also explain why the addition of sterically demanding 3′-substituents, or of acridine substituents which potentially modify intercalation geometry, greatly influence biological activity. A ternary complex could form either by recognition of the DNA-drug complex by a protein, or by combination of a protein-drug complex with DNA.

The clinical antitumour drug m-AMSA (amsacrine)

m-AMSA (*38*) was first reported in 1974 as an anilinoacridine derivative showing both high activity against the L1210 leukaemia (as measured by a life extension assay) and high dose potency (Cain and Atwell, 1974). It also showed antiviral (Byrd, 1977), immunosuppressive (Baguley *et al.*, 1974), and mutagenic activity (Baguley *et al.*, 1978; Ferguson and Denny, 1979). It proved active against a spectrum of experimental tumours (Cain and Atwell, 1974; Corbett *et al.*, 1977; Oseicka *et al.*, 1977; Sordillo *et al.*, 1980; Goldin *et al.*, 1981), and following considerable preclinical testing at the National Cancer Institute in the United States, entered Phase I clinical trials in 1978. Successful results in initial trials led to Phase II trials where good activity was demonstrated towards acute leukaemia (Legha *et al.*, 1978, 1980) and malignant lymphoma (Cabanillas *et al.*, 1981). A series of Phase II trials against a number of solid tumours provided, on the whole, negative results (Schneider *et al.*, 1980; Casper *et al.*, 1980; Sordillo *et al.*, 1980; Carroll *et al.*, 1980; Ferraro *et al.*, 1981; Yap *et al.*, 1981). Nevertheless *m*-AMSA, particularly in combination with other cytotoxic drugs (Arlin *et al.*, 1980; Laster *et al.*, 1980) remains an extremely promising drug in the treatment of leukaemia and malignant lymphoma. Its efficacy in the management of these diseases may be explained, at least in part, by the more aggressive treatment schedules adopted for these diffuse neoplasms as opposed to solid tumours (Arlin *et al.*, 1980). This must surely be taken into account when assessing the narrow antitumour spectrum of *m*-AMSA in man. The goal in the therapy of acute leukaemia is bone marrow aplasia, a condition which one specifically wishes to avoid in treating solid tumours.

The relative sensitivities to *m*-AMSA of cycling and non-cycling cells *in vitro* have been compared in several studies. Non-dividing precursors of antibody-forming cells were less sensitive to *m*-AMSA than were dividing precursors (Baguley *et al.*, 1974). Similar results have been reported for CHO cells: non-cycling cells arrested in G-1-phase by isoleucine deprivation were 2-4 times more resistant to low concentrations of *m*-AMSA than were cycling cells, but at higher concentrations this differential disappeared (Tobey *et al.*, 1978). When a non-cycling state was induced by growth of CHO or V79 Chinese hamster cells into plateau phase, the marked resistance of non-cycling cells to *m*-AMSA was again apparent (Wilson *et al.*, 1981*b*). However, in this instance the differential between cycling and non-cycling cells was maintained at high drug concentrations, as was also observed in a similar comparison of growth-arrested and cycling human colon carcinoma (LoVo) cells (Drewinko *et al.*, 1981). V79 cells in the interior of multicellular spheroids, which provide a model for the non-cycling compartment in solid tumours, have also been shown to be highly resistant to *m*-AMSA relative to the high-growth-fraction population near the spheroid surface (Wilson *et al.*, 1981). It is thus likely that the insensitivity of non-cycling cells limits the utility of *m*-AMSA in the treatment of solid tumours.

m-AMSA can act at all or most stages of the cell division cycle. A single exposure to *m*-AMSA for 15 minutes inactivated 96% of precursors to antibody-forming cells, whereas in a tritiated thymidine "suicide" experiment where only S-phase cells are killed, 75% were inactivated (Baguley *et al.*, 1974). More detailed studies using CHO cells showed that killing efficiency varied throughout the cell cycle, reaching a maximum during S phase (Wilson and Whitmore, 1981). The pattern of differential sensitivity to *m*-AMSA complemented that of X-ray irradiation, suggesting that combination of irradiation and *m*-AMSA could provide an effective treatment (Wilson and Whitmore, 1981).

m-AMSA probably acts on cells by causing DNA strand breaks (Furlong *et al.*, 1978; Ralph, 1980; Zwelling *et al.*, 1980). DNA breakage could in turn result in the appearance of sister chromatid exchanges (Crossen, 1979) and chromosomal aberrations in mitotic cells (Tobey *et al.*, 1978; Baguley *et al.*, 1978). Chromosomal aberrations are most evident in cells which are exposed in the G-2 phase of the cell cycle (Deaven *et al.*, 1978). Chromosomal breaks could readily lead to direct inactivation of genes, translocation processes, or to loss of genetic material after cell division by production of chromosomal fragments lacking a centromere.

DNA breakage could be initiated by recognition of a DNA-*m*-AMSA complex by a repair nuclease. This model is consistent with that discussed in the previous section, based on structure-activity considerations, in which a drug–DNA–enzyme complex mediates the cytotoxic effects of *m*-AMSA. The question then arises as to what extent the molecular characteristics of that complex can be held to account for the antitumour activity of *m*-AMSA and, in particular, its selectivity. We can only speculate as to the involvement of the protein moiety in a ternary complex, but the nature of *m*-AMSA-DNA interaction is relatively well defined.

All studies are in agreement that the binding of the drug occurs by a process of intercalation and that it is not an unusually strong interaction; indeed the binding constant for natural DNAs is consistently reported to be lower than that for proflavine or other aminoacridines under comparable conditions (Waring, 1976; Gormley *et al.*, 1978; Wilson *et al.*, 1981; Hudecz *et al.*, 1981). Likewise, the ability of *m*-AMSA (and other anilinoacridines) to stabilise the double helix against heat-denaturation is conspicuously lower than that of proflavine or 9-aminoacridine (Gormley *et al.*, 1978; Wilson *et al.*, 1981). Plainly the antitumour efficacy and dose-potency of *m*-AMSA cannot be attributed to a high affinity for DNA *per se*. Actually, the binding constant for the inactive positional isomer *o*-AMSA (*43*) is 2 to 3-fold greater than that for *m*-AMSA, illustrating quite well the lack of correlation between gross DNA-binding affinity and antitumour activity which sometimes occurs with closely-related compounds (Wilson *et al.*, 1981). It is clear from comparative binding studies that the methanesulphon-anilide ring present in AMSA drugs does not of itself interfere with binding to double-stranded DNA, in agreement with earlier studies which established that sterically demanding substituents at position 9 of the acridine nucleus do not prevent DNA binding (Blake and Peacocke, 1968; Wilson *et al.*, 1981). This is consistent with the model illustrated in figure 2 where the methanesulphonanilide moiety of *m*-AMSA is located in the minor groove of B-DNA [that is, orientated in a fashion analogous to that accepted for ethidium (Gale *et al.*, 1981)] without interfering with intercalation of the acridine nucleus.

A 2'- substituent such as the $-OCH_3$ group present in *o*-AMSA (*43*) is not likely to impede this interaction either, but with the methoxy substituent in position 3' as in *m*-AMSA some steric hindrance would be expected to arise, limiting the extent to which the acridine nucleus can be made to overlap with the base-pairs. Doubtless this steric hindrance to intercalation accounts for the lower association constant of *m*-AMSA as well as its relative inability to stabilise double-stranded DNA against thermal denaturation. Other AMSA derivatives with sterically demanding 3'-substituents (CH_3, $NHSO_2 CH_3$, NO_2, Cl, OH) also display reduced affinity for double-helical DNA (Baguley and Nash, 1981; Baguley *et al.*, 1981*a*). Steric factors may also account for the weak binding of 2-methyl-AMSA (*44*) (Wilson *et al.*, 1981) since accommodation of the 2-methyl group would probably require a change in the orientation of the acridine nucleus within the intercalation site, thus bringing the 3' position of the methane-sulphonanilide ring into close contact with base-pairs in the minor groove. Such interactions could again impede optimal ring overlap. Support for this hypo-thesis can be drawn from the observation that the unwinding angle of inter-calated 2-methyl-AMSA (*44*) is significantly lower than that for AMSA (*33*; R = $NHSO_2 CH_3$), *m*-AMSA (*38*), or *o*-AMSA (*43*) (Waring, 1976), consistent with an altered acridine ring orientation.

If the broad quantitative features of *m*-AMSA-DNA interaction cannot explain its antitumour effectiveness, what other clues may point to the molecu-lar basis for its selective action? Several groups have considered the possibility of selective binding to particular nucleotide sequences in DNA, but simple equilib-rium dialysis experiments (Gormley *et al.*, 1978) or measurement of circular dichroism spectra (Hudecz *et al.*, 1981) failed to reveal significant differences in the binding of *m*-AMSA to a limited number of polynucleotides. However, using an ethidium-displacement fluorescence assay and a variety of synthetic poly-nucleotides, Wilson *et al.* (1981) reported association constants for binding sites of different sequence varying over a tenfold range. As with other intercalating drugs (Gale *et al.*, 1981) a marked preference for alternating purine-pyrimidine polymers over polypurine·polypyrimidine DNAs was noted. Interestingly, the logarithmic mean of association constants determined for synthetic polynucleo-tides was found to lie close to the value observed for calf thymus DNA, as if the value measured for the natural DNA might represent a mean averaged over the microscopic constants for all the sequences contained within it (Wilson *et al.*, 1981).

One feature of the interaction between *m*-AMSA and nucleic acids which is striking is its conspicuous ability to discriminate sensitively between native DNA, denatured DNA and RNA (figure 3). Its binding constants for the latter two polymers are respectively 5- and 25-fold lower than its association constant for native DNA (Wilson *et al.*, 1981). This property appears to be a common feature of drugs in the AMSA series and is substantially more marked than the preferences found with simpler aminoacridines. Other studies concur with the conclusion that *m*-AMSA is strongly selective for DNA over RNA (Hudecz *et al.*, 1981). A possible consequence of this selectivity may be efficient binding to

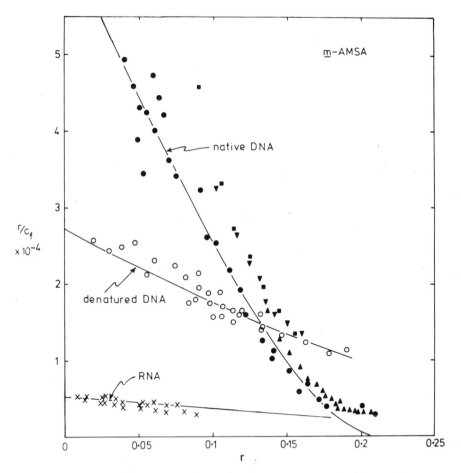

Figure 3   Scatchard plots for the interaction of *m*-AMSA (*38*) with nucleic acids at ionic strength 0.01 (from Wilson *et al.*, 1981). Data were obtained by equilibrium dialysis (●, native calf thymus DNA; ○, heat-denatured DNA; ×, ribosomal RNA) or spectrophotometry (■, ▼, ▲, native DNA). The symbol r denotes drug molecules bound per nucleotide of the polymer; C_f is the free ligand concentration. The curves are theoretical, computed to represent a best fit of the equilibrium dialysis data to the excluded-site model of binding to a linear lattice (for details see Wilson *et al.*, 1981).

DNA *in vivo* as a result of diminished binding to the considerable excess of RNA present in cells, which might otherwise serve as a large "silent" (or at any rate unproductive) sink for intracellular drug molecules.

In many ways *m*-AMSA resembles adriamycin, a DNA binding antibiotic (DiMarco *et al.*, 1975) in binding to DNA and causing DNA breakage (see Chapter 2). Like *m*-AMSA, adriamycin is active towards a variety of experimental tumours. Adriamycin has a broader spectrum of activity towards human tumours, and it is possible that further structural modification of *m*-AMSA could provide analogues with a broader spectrum of clinical antitumour activity but without the dose-limiting cardiomyopathy of adriamycin.

DIACRIDINES

Although several examples of diacridine compounds had been reported in the search for antimalarial drugs (Elslager *et al.*, 1968), the recent growth of interest in these compounds as potential antitumour agents (Fico *et al.*, 1977; Cain *et al.*, 1978; Chen *et al.*, 1978) and as probes for drug-DNA binding interactions (Wakelin *et al.*, 1978; Capelle *et al.*, 1979; Gaugain *et al.*, 1978*a*) has occurred for two main reasons. Considering the demonstrated relationship between DNA binding affinity and antitumour effectiveness for several series of intercalating drugs (Le Pecq *et al.*, 1974; Baguley *et al.*, 1981), diacridines capable of intercalating both chromophores into DNA were prepared in searches for compounds with the highest possible binding affinities (Le Pecq *et al.*, 1975; Canellakis *et al.*, 1976*a*; Gaugain *et al.*, 1978*b*). *Bis*-intercalating compounds were also considered to provide a way of increasing binding selectivity, both to the primary DNA

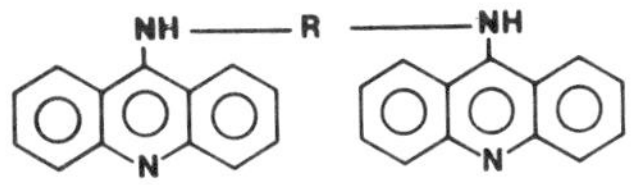

45	R = $(CH_2)_2$	50	R = $(CH_2)_7$
46	$(CH_2)_3$	51	$(CH_2)_8$
47	$(CH_2)_4$	52	$(CH_2)_9$
48	$(CH_2)_5$	53	$(CH_2)_{10}$
49	$(CH_2)_6$	54	$(CH_2)_{12}$

55 $(CH_2)_3NH(CH_2)_4NH(CH_2)_3$

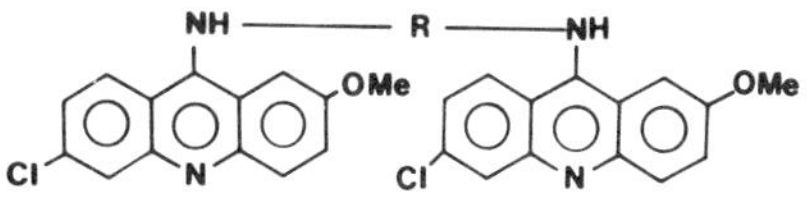

56	R = $(CH_2)_3NH(CH_2)_3$
57	$(CH_2)_3NH(CH_2)_4$
58	$(CH_2)_3NH(CH_2)_4NH(CH_2)_3$
59	$(CH_2)_6$

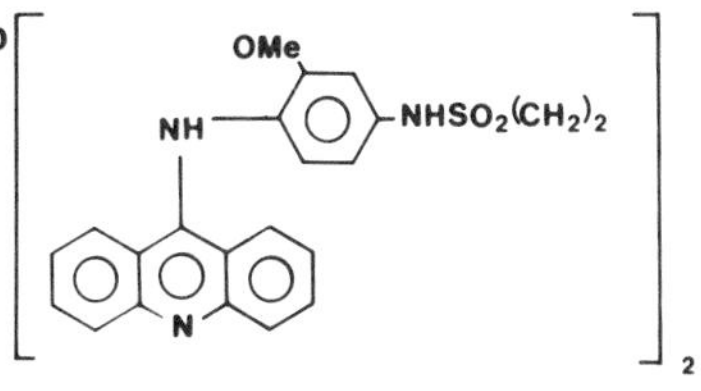

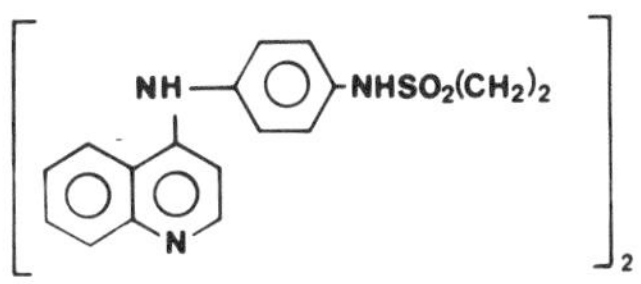

sequences by extending the binding site size of the molecule (Wakelin *et al.*, 1978) and topologically by expected enhanced binding to supercoiled DNA (Davidson, 1972; Cain *et al.*, 1978). Most work on the DNA binding of diacridines has been carried out with the symmetrical compounds (*45-55*). For a true *bis*-intercalating ligand, the total free energy of binding might be expected to equal the sum of the binding energies of the monomers, so that binding constants of the order of those observed for repressor proteins and DNA polymerase (10^{10} to 10^{12} M^{-1}) (Von Hippel and McGhee, 1972) might be attainable. Binding constants of this order have been reported for some diacridines (*56-58*) (Le Pecq *et al.*, 1975), but these compounds possess a linker chain containing functional groups capable of ionic and hydrogen bonding. The simple diacridines (*45-51*) display binding constants much smaller than the theoretical maximum (square of the binding constant of the monomer), suggesting that significant unfavourable entropic and stereochemical effects are involved (Capelle *et al.*, 1979; Wakelin *et al.*, 1979). A detailed investigation of the mode of binding of the homologous series of diacridines (*45-54*) has been carried out (Wakelin *et al.*, 1978) using three types of analysis; sedimentation of closed circular duplex DNA, helix extension of short rod-like DNA fragments, and electric dichroism. All of the derivatives were shown to be intercalators by virtue of their ability to remove and reverse the supercoiling of circular DNA, but they fell into two distinct groups as far as their equivalence points (inversely related to their unwinding angles; Waring, 1970) were concerned (figure 4). The monomeric compounds 9-aminoacridine and 9-methylamino-acridine, together with the C-2, C-3 and C-4 diacridines (*45-47*) yielded values typical for simple (monofunctional) intercalators, whereas the C-6 to C-12 compounds (*49-54*) exhibited twice the unwinding angle, consistent with *bis*-intercalation. The C-5 compound (*48*) showed intermediate behaviour. These results were confirmed by the helix extension studies. The electric dichroism studies indicated little stereochemical strain for the compounds with chains C-7 and longer, with the angles between the chromophores and the base pair planes being essentially zero for these compounds. A value of $10°$ for the average angle between the planes of the chromophores and the base pairs for the C-6 compound (*49*) indicates some degree of distortion from the (assumed) ideal *bis*-intercalation geometry, although the compound clearly *bis*-intercalates by the other criteria. These studies (Wakelin *et al.*, 1978) delineate the minimum length of linker chain (8.8Å) necessary for *bis*-intercalation to occur. The result is of some interest, as the minimum length of linker needed for a *bis*-intercalator to span two base pairs, in agreement with the "neighbouring site exclusion" principle (Von Hippel and McGhee, 1972) is 10.2Å (figure 5). The neighbour-exclusion model was favoured by Le Pecq *et al.* (1975) as a result of their studies with the mepacrine dimers (*56-58*). Compounds (*57*) and (*58*), having chain lengths of 16.1 and 11.2Å respectively, bound to DNA by *bis*-intercalation. The removal of one $-CH_2-$ group from (*57*) to provide (*56*), with a chain length of 9.9Å, led to a monointercalator, in agreement with the exclusion model. However, the *bis*-intercalating C-6 diacridine

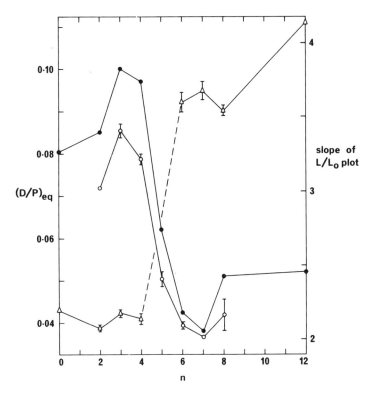

Figure 4 Variation of the helix unwinding angle and the helix extension associated with intercalative binding of dimeric acridine compounds (45–54) to DNA. Each diacridine molecule contained two 9-aminoacridine chromophores connected through their amino substituents by a methylene linker chain of varying length $-(CH_2)_n-$. The left-hand ordinate records the ligand/nucleotide ratio at equivalence (relaxation of the supercoiling) for PM2 DNA determined by two different methods: filled circles show results from sedimentation experiments where the equivalence points are not corrected for the drug–DNA binding equilibrium; open circles represent the results from viscometric measurements performed as described by Waring and Henley (1975) and are true equivalence binding ratios whose standard deviations are indicated by the error bars. For the C2 and C7 compounds, the errors were smaller than the symbol plotted. Triangular symbols represent the results of viscometric experiments to determine changes in the contour length of sonicated rod-like fragments of calf thymus DNA (right-hand ordinate). They are expressed as the slope of plots of L/L_0 versus D/P, where L_0 is the contour length of control DNA and L is the contour length at the level of binding given by the drug/nucleotide ratio D/P. For an ideal process of simple (monofunctional) intercalation by a tightly-binding ligand the expected slope is 2.0; for an ideal process of *bis*-intercalation it should be close to 4.0. Standard deviations are indicated. The results plotted for $n = 0$ were obtained with 9-methylamino-acridine (From Wakelin *et al.*, 1978).

(49) with a chain length of only 8.8Å must violate the neighbour-exclusion model (provided normal DNA geometry is retained) and bind with its two chromophores on either side of a single base pair. It has been suggested (Gaugain *et al.*, 1978*a*) that the excluded-site rule is a thermodynamic limitation applying

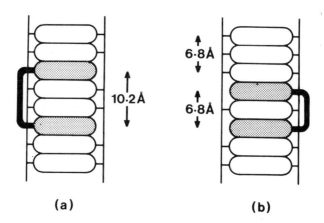

(a)   **(b)**

Figure 5   Schematic illustration of bifunctional intercalation by diacridines subject to the constraint of neighbouring site-exclusion (a) or in violation of that principle (b). The intercalated chromophores (shaded) and the DNA base-pairs are shown in edgewise projection as discs 3.4Å thick, with the sugar-phosphate backbones reduced to vertical lines for clarity; the helical twist of the DNA is neglected. The chain of atoms linking the chromophoric moieties is shown as a heavy line; its length effectively determines whether the separation between chromophores is sufficient to allow sandwiching of two base-pairs (a) or one (b).

to monofunctional intercalators of relatively low binding affinity, which may be violated by derivatives with very high binding affinity.

Later studies with a series of diacridines substituted in the chromophores (Wright *et al.*, 1980) showed that the ability of this class of compound to *bis*-intercalate depends more on the acridine substitution pattern than the nature of the linker chain. Modification of the unsubstituted *bis*-intercalating C-6 diacridine (*49*) gives (*59*), which was found to be capable only of monofunctional intercalation. Generally, substitution with groups close to the long axis of the acridine chromophore produced greater restrictions on *bis*-intercalative ability than groups placed close to the minor axis (Wright *et al.*, 1980). Alterations in the linker chain led to less marked changes in *bis*-intercalative potential. The dimeric anilinoacridine (*60*) was found to be a *bis*-intercalating agent, with an unwinding angle of 41° (Cain *et al.*, 1978). It is interesting that the size of the chromophore also plays a critical role in determining *bis*-intercalative ability; *bis*quinoline compounds such as (*61*) appear to behave as monofunctional intercalators (Cain *et al.*, 1978; Wright *et al.*, 1980).

The kinetics of binding of diacridines to DNA are of special interest for two reasons. Firstly, for molecules of very high binding affinity, accurate measurement of binding constants (above 10^8 M^{-1}) by equilibrium methods becomes difficult, and such determinations are best made by a comparison of the appropriate kinetic parameters (off-rate and on-rate) (Capelle *et al.*, 1979). It has also been demonstrated for actinomycin derivatives (Müller and Crothers, 1968; Bittman and Blau, 1975) that slow off-rates from the DNA, rather than merely high binding affinity, correlate better with high biological activity.

The diacridines, whose dissociation is envisaged as involving coordinate disengagement of both chromophores from the DNA, might be expected to display slow off-rates (Canellakis *et al.*, 1976; Capelle *et al.*, 1979). A study of the kinetics of binding of several diacridine derivatives confirms these ideas (Capelle *et al.*, 1979). Association rate constants for the monomeric acridine species (*4*), the monofunctional intercalating diacridine (*56*) and the *bis*-intercalating diacridines (*57, 58*) were all of the same order ($3 \times 10^7 \, M^{-1} \, s^{-1}$), suggesting a diffusion-controlled process in all cases. Dissociation rate constants for the diacridines, measured by exchange of the ligands between DNAs of different base composition, were of the order of $10^{-3} \, s^{-1}$, giving rise to kinetically determined binding constants of the order of $10^{10} \, M^{-1}$ for these compounds.

One interesting result of the above study was the finding that the reactivity of diacridines towards DNA was reduced at *p*H values above 7.0; under these conditions unstacking of the compound from its solution conformation prior to DNA binding appears to be the rate-determining process. The folded solution conformation is quite different from the crystal conformation as determined by X-ray diffraction, where the linker chain is in an extended position (Courseille *et al.*, 1977*a, b*).

Several diacridines have been shown to be potent inhibitors of the growth of L1210, P388 and HeLa cells in culture (Canellakis and Bellantone, 1976). They also inhibit RNA synthesis, mainly by preventing chain initiation by RNA polymerase (Canellakis and Bellantone, 1976; Canellakis *et al.*, 1976*b*; Sarris *et al.*, 1977). In this regard the diacridines act similarly though more efficiently than the simple acridines, but differently from actinomycin D which primarily inhibits chain elongation (Sarris *et al.*, 1977). For a homologous series of diacridines the potency of inhibition increased with chain length, plateauing from the C-8 to C-12 compounds (*51–54*) corresponding to *bis*-intercalated derivatives.

A considerable number of acridine-substituted diacridines have been synthesised and tested for antitumour activity *in vivo* and in culture (P388 leukaemia), and for their ability to inhibit RNA and DNA synthesis (Fico *et al.*, 1977; Chen *et al.*, 1978). No direct correlation was observed between activity *in vivo* and *in vitro*, although no attempt was made to allow for varying lipophilic/hydrophilic balance; a parameter shown to have considerable importance in similar correlations with other series of antitumour compounds. A significant inverse correlation was noted between the antitumour activity *in vivo* (% ILS values) and the ability of the compounds to accelerate the rate of agglutination of concanavalin A-treated cells. However, the strength of this correlation rests heavily on the inclusion of a number of compounds (25% of the total data base) whose %ILS values lie below the range (25% and up) properly accepted as defining significant activity *in vivo* (Geran *et al.*, 1972). The hypotheses which have been put forward (Fico *et al.*, 1977; Chen *et al.*, 1978) on the basis of the correlation need corroborating. Significant levels of activity *in vivo* (against L1210 and P388 leukaemia) were seen for a number of *bis*-AMSA compounds, including (*60*) (Cain *et al.*, 1978). From earlier work with non-intercalating *bis*-

cationic AMSA derivatives (Atwell *et al*., 1976), the dimeric compounds could be regarded as excessively lipophilic, but the addition of hydrophilic functionality at positions acceptable in the monomers abolished rather than enhanced activity. For the unsubstituted *bis*-AMSA compounds, maximum biological effectiveness did not correlate with measures of DNA-binding, such as ability to displace ethidium from polynucleotides, or unwinding angles for closed circular DNA (Cain *et al*., 1978). It was concluded that constraints on the intercalating nuclei alter with chain length, complicating the derivation of structure–activity relationships.

One diacridine *bis*-intercalator (*49* as NSC 219733) was evaluated for clinical trial (Goldin *et al*., 1981; Beltagy *et al*., 1980), but was dropped when the soluble form of the drug displayed CNS toxicity and only marginal activity (Schepartz *et al*., 1981). The *bis*-intercalating compounds as a class have not yielded the exceptional antitumour activity *in vivo* expected, given the extensive theoretical reasoning underlying their development (Cain *et al*., 1978; Canellakis *et al*., 1976*a*, Le Pecq *et al*., 1975). These *bis*-cationic compounds may be expected to have pharmacokinetic properties that limit their effectiveness against remote tumour sites *in vivo*. However, they do not exhibit the very high levels of potency *in vitro*, or in simple tests *in vivo*, that might be expected from their extremely high DNA binding affinities and slow dissociation kinetics (Capelle *et al*., 1979). In this regard, it should be noted that it is erroneous to regard these compounds as molecular staples attaching to the DNA (Canellakis *et al*., 1976*a*). While it is true that one expects both chromophores to have to come free almost simultaneously for the drug to dissociate from the DNA, the flexible nature of the linker chain will allow most of the diacridines to move quite efficiently along the DNA in a stepwise fashion without dissociation occurring. Thus the average residence time of these chromophores at a particular binding site, a property recently shown to be correlated with antitumour activity *in vivo* for a large number of intercalating drugs (Feigon *et al*., 1982), might not be anything like as long as inferred from their very high binding affinities. An obvious step in the development of *bis*-intercalating antitumour agents would be preparation of compounds with relatively rigid groups linking the two chromophores, to provide a true "molecular staple". The useful antitumour activity of the quinoxaline antibiotic echinomycin (Chapter 5) which does possess such rigidity, is notable considering the small noncationic quinoxaline chromophores which mediate its intercalative binding with DNA.

ANTITUMOUR ACRIDONES

The major end products of the metabolism of 9-substituted acridines such as nitracrine (*16*) and *m*-AMSA (*38*) *in vivo* are the corresponding acridones (Hashimoto *et al*., 1949), the formation of which is probably mediated by thiol attack at C-9, followed by hydrolysis (Wilson *et al*., 1977; Cysyk *et al*., 1977). Noting this, Ledóchowski and coworkers prepared a limited series of methoxy- and nitroacridone and thioacridone derivatives possessing a dialkylaminoalkyl side-chain attached to the N-10 position (Radzikowski *et al.*, 1971). Compounds of this general type (for example, compounds *62, 63*) are known to have convulsant properties, and to concentrate in the nucleolus of cells in the central nervous system following intravenous injection (Mayer and Bain, 1956). Only the 1-nitroacridone derivatives were shown to have antitumour activity in the S-180 system *in vivo*, and two compounds (*64, 65*) have undergone extensive preclinical testing (Patkowski *et al.*, 1973*a, b*; Pelczarska *et al.*, 1974*a, b*). The pattern of biological activity found for both of these compounds was surprisingly similar to that of nitracrine (*16*), given the quite different chemical structures of the acridone derivatives. Both compounds were acutely toxic in mice and rats at doses below 1 mg per kg, but at therapeutic doses there was little evidence of hepatotoxicity. The severe local irritant action noted for the acridine derivatives, which constitutes a severe clinical limitation in the use of nitracrine (Kwaśniewska-Rokicińska *et al.*, 1973), was still present in the acridone derivatives. Although the dimethyl-propylamino compound (*65*) was recommended as worthy of clinical trial (Patkowski *et al.*, 1973*a, b*) no clinical data have yet been reported.

A number of acridone alkaloids occur naturally, but only acronycine (acronine) (*66*; R = CH$_3$) has been noted for its antitumour activity. This alkaloid is readily obtainable by extraction of the bark of the Australian scrub ash *Acronychia baueri* Schott (Hughes *et al.*, 1948; McDonald and Robertson, 1966; Svoboda *et al.*, 1966; Govindachari *et al.*, 1966), and several total syntheses have been reported (Beck *et al.*, 1968; Hlubucek *et al.*, 1970; Blechert *et al.*, 1978).

Acronycine has been reported to have broad-spectrum antitumour activity in experimental animals, although it was inactive *in vivo* against some of the standard screening systems such as the L1210 leukaemia and Ehrlich carcinoma in mice, and the Walker 256 carcinosarcoma in rats (Svoboda *et al.*, 1966; Liska, 1972). Other workers have claimed low activity in the latter two test systems, but the activity is at best marginal, as it is also against the solid form of S-180 *in vivo* (Schneider *et al.*, 1972).

A number of analogues of acronycine have been made and tested for antitumour activity (Svoboda, 1966; Schneider *et al.*, 1972), but only the derivative (*66*; R = (CH$_2$)$_2$ N(CH$_3$)$_2$), bearing a charged dimethylaminoethyl side chain, is reported to have activity *in vivo* (Schneider *et al.*, 1972). Acronycine is thought to act by inhibition of RNA synthesis. In studies carried out using cultured leukaemia L5178 cells, inhibition of RNA synthesis and cessation of cell growth was seen from 0.5 μg ml^{-1} (Gort *et al.*, 1971). DNA synthesis was also inhibited at higher drug concentrations, but this was not a prerequisite for inhibition of cell growth. The metabolism of acronycine *in vivo* has been studied in several species, including man (Sullivan *et al.*, 1970). The major metabolites were the 9- and 11-hydroxy compounds; although both *N*- and *O*-dealkylation products were detected they were minor metabolites. The major metabolite, 9-hydroxyacronycine, is also the main product of microbiological hydroxylation of acronycine (Betts *et al.*, 1974; Brannon *et al.*, 1974). It shows no activity even in tumour systems where acronycine is active, and thus this metabolism probably results in detoxification *in vivo*.

Acronycine received extensive study with a view to clinical trial, but the study was not pursued since adequate therapeutic effects in animal models could not be demonstrated (Suffness and Douros, 1979; Finkelstein *et al.*, 1975). Attempts have been made to overcome the extreme insolubility of acronycine by the formation of *O*-acyl quaternary salts (such as compound *67*). These compounds are 2 to 5 orders of magnitude more water-soluble than acronycine itself (Repta *et al.*, 1977) and with half-lives under simulated *in vivo* conditions of approximately five minutes they may serve as suitable latentiated forms of acronycine.

Recent work (Wright *et al.*, 1980) on *bis*-cationic derivatives of acridone itself (such as compound *68*) shows that acridone can be an acceptable chromophore for DNA intercalation. The authors suggest that this chromophore may be better suited to the development of polyfunctional agents, for it lacks the marked non-specific affinity of charged acridines for macromolecules in general, and may permit the construction of more selectively-binding compounds. Thus, despite the apparent lack of success of acronycine, it is likely that interest in antitumour acridones will continue.

CONCLUSIONS

The broad spectrum of biological activities shown by the different types of acridines discussed here attest the importance of this class of compound in experimental and clinical medicine. These activities are almost certainly due to the high binding affinity of acridines for macromolecules, particularly nucleic acids: acridines such as 9-aminoacridine *(4)*, proflavine *(2*; R = H) and acridine orange *(6)* being classical DNA-intercalating agents. The recent success of the acridine derivatives nitracrine *(16)* and *m*-AMSA (amsacrine) *(38)* as clinical antitumour drugs, and the relative synthetic accessibility of the acridine system, provide incentives for the further development of acridines as antineoplastic agents.

ACKNOWLEDGEMENTS

The authors wish to thank members of the Cancer Research Laboratory for helpful discussions, and particularly Rosalee Nash for preparing the diagrams. William A. Denny and Bruce C. Baguley thank the Auckland Division, Cancer Society of New Zealand, and the Medical Research Council of New Zealand for support. Michael J. Waring acknowledges the support of the Medical Research Council of Great Britain and the Cancer Research Campaign.

REFERENCES

Ackerman, N. B., Haldorsen, D. K., Tendick, F. H. and Elslager, E. F. (1968). *J. med. Chem.*, **11**, 315

Albert, A. (1966). *The Acridines*, 2nd edn, Edward Arnold, London

Arlin, Z. A., Sklaruff, R. B., Gee, T. S., Kempin, S. J., Howard, J., Clarkson, B. D. and Young, C. W. (1980). *Cancer Res.*, **40**, 3304

Atwell, G. J. and Cain, B. F. (1973). *J. med. Chem.*, **16**, 673.

Atwell, G. J., Cain, B. F. and Denny, W. A. (1977). *J. med. Chem.*, **20**, 1128

Atwell, G. J., Cain, B. F. and Seelye, R. N. (1972). *J. med. Chem.*, **15**, 611

Baguley, B. C. (1982). *Molec. cell. Biochem.*, **43**, 167

Baguley, B. C., Denny, W. A., Atwell, G. J. and Cain, B. F. (1981a). *J. med. Chem.*, **24**, 520

Baguley, B. C., Denny, W. A., Atwell, G. J. and Cain, B. F. (1981b). *J. med. Chem.*, **24**, 170

Baguley, B. C. and Falkenhaug, E. M. (1978). *Nucleic Acids Res.*, **5**, 161

Baguley, B. C., Falkenhaug, E. M., Rastrick, J. M. and Marbrook, J. (1974). *Eur. J. Cancer*, **10**, 169

Baguley, B. C. and Nash, R. (1981). *Eur. J. Cancer*, **17**, 671

Baguley, B. C., Wilson, W. R., Ferguson, L. R. and Cain, B. F. (1978). In: *Current Chemotherapy Proc. 10th Int. Congress of Chemotherapy* (ed. W. Siegenthaler and R. Luthy) International Society of Chemotherapy, vol 2, 1210

Bauer, W. and Vinograd J. (1970). *J. molec. Biol.*, **47**, 419

Beck, J. R., Kwok, R., Booher, R. N., Brown, A. C., Patterson, L. F., Pronc, P., Rockey, B. and Pohland, A. (1968). *J. Am. Chem. Soc.*, **90**, 4706

Beltagy, Y. A., Waugh, W. and Repta, A. J. (1980). *Drug Dev. ind. Pharmac.*, 6, 411
Betts, R. E., Walters, D. E., Rosazza, J. P. (1974). *J. med. Chem.*, 17, 599
Bittman, R. and Blau, L. (1975). *Biochemistry*, 14, 2138
Blake, A. and Peacocke, A. R. (1968). *Biopolymers*, 6, 1225
Blaszczyk, B., Gieldanowski, J., Patkowski, J., Kowalczyk-Bronisz, H. (1979). *Archvm Immun. Ther. exp.*, 27, 749, 765
Blechert, S., Fichter, K. E., Winterfeldt, E. (1978). *Chem. Berichte*, 111, 439
Blumenthal, S. (1931). *Beitr. klin. Chim.*, 154, 50
Braithwaite, A. W. and Baguley, B. C. (1980). *Biochemistry*, 19, 1101
Brannon, D. R., Horton, D. R. and Svoboda, G. H. (1974). *J. med. Chem.*, 17, 653
Bratkowska-Seniow, B., Ozyhar-Hajpel, H., Runge, I. and Wahl-Mungenska, M. (1974). *Archvm Immun. Ther. exp.*, 22, 441
Bukatsh, P. and Haitingen, M. (1940). *Protoplasma*, 34, 515
Byrd, D. M. (1977). *Ann. N.Y. Acad. Sci.*, 284, 463
Cabanillas, F., Legha, S. S., Bodey, G. P. and Freireich, E. J. (1981). *Blood*, 57, 614
Cain, B. F. and Atwell, G. J. (1974). *Eur. J. Cancer*, 10, 539
Cain, B. F., Atwell, G. J. and Denny, W. A. (1975). *J. med. Chem.*, 18, 1110
Cain, B. F., Atwell, G. J. and Denny, W. A. (1977). *J. med. Chem.*, 20, 987
Cain, B. F., Atwell, G. J. and Seelye, R. N. (1969). *J. med. Chem.*, 12, 199 .
Cain, B. F., Atwell, G. J. and Seelye, R. N. (1971). *J. med. Chem.*, 14, 311
Cain, B. F., Baguley, B. C., Denny, W. A. and Atwell, G. J. (1980). In: *Clinical Pharmacology and Therapeutics. Proc. First World Conference* (ed. P. Turner) Macmillan, London, p. 436
Cain, B. F., Baguley, B. C. and Denny, W. A. (1978). *J. med. Chem.*, 21, 658
Cain, B. F., Wilson, W. R. and Baguley, B. C. (1976). *Molec. Pharmac.*, 12, 1027
Canellakis, E. S. and Bellantone, R. A. (1976). *Biochim. biophys. Acta*, 418, 289
Canellakis, E. S., Fico, R. M., Sarris, R. H. and Shaw, Y. H. (1976a). *Biochem Pharmac.*, 25, 231
Canellakis, E. S., Shaw, Y. H., Hanners, W. E. and Schwartz, R. A. (1976b). *Biochim. biophys. Acta*, 418, 277
Capelle, N., Barbet, J., Dessen, P., Blanquet, S., Roques, B. P. and Le Pecq, J-B. (1979). *Biochemistry*, 18, 3354
Carroll, D. S., Kemeny, N., Lynch, G. and Woodcock, T. (1980). *Cancer Treat. Rep.*, 64, 1149
Casper, E. S., Gralla, R. J. and Kemeny, N. E. (1980). *Cancer Treat. Rep.*, 64, 345
Chen, T-K., Fico, R. M. and Canellakis, E. S. (1978). *J. med. Chem.*, 21, 868
Comings, D. E., Liman, J., Ledóchowski, A. and Tsou, K. C. (1978). *Expl cell. Res.*, 117, 451
Corbett, T. H., Griswold, D. P. Jr, Roberts, B. J., Peckham, J. C. and Schabel, F. M. Jr (1977). *Cancer*, 40, 2660
Courseille, C., Leroy, F., Hospital, M. and Barbet, J. (1977a). *Acta. crystallogr.*, B33, 1565
Courseille, C., Leroy, F., Busetta, B. and Barbet, J. (1977b). *Acta. crystallogr.*, B33, 1570
Creech, H. J., Preston, R. K., Peck, R. M. and O'Connell, A. P. (1972). *J. med. Chem.*, 15, 739
Crossen, P. E. (1979). *Mutat. Res.*, 68, 295
Cysyk, R. L., Shoemaker, G. J. and Adamson, R. H. (1977). *Drug Metab Disp.*, 5, 579
Cysyk, R. L., Shoemaker, D. A., Ayers, D. C. and Adamson, R. H. (1978). *Pharmacology*, 16, 206
Dauter, Z., Bogucka-Ledóchowska, M., Hempel, A., Ledóchowski, A. and Kosturkiewicz, Z. (1975). *Roczniki. Chem.*, 49, 859
Davidson, N. (1972). *J. molec. Biol.*, 66, 307
Deaven, L. L., Oka, M. S. and Tobey, R. A. (1978). *J. natn. Cancer Inst.*, 60, 1155
Denny, W. A., Atwell, G. J., Baguley, B. C. and Cain, B. F. (1979). *J. med. Chem.*, 22, 134
Denny, W. A., Atwell, G. J. and Cain, B. F. (1977). *J. med. Chem.*, 20, 1242
Denny, W. A., Atwell, G. J. and Cain, B. F. (1978). *J. med. Chem.*, 21, 5
Denny, W. A., Atwell, G. J. and Cain, B. F. (1979). *J. med. Chem.*, 22, 1453
Denny, W. A. and Cain, B. F. (1978). *J. med. Chem.*, 21, 430
Denny, W. A., Cain, B. F., Atwell, G. J., Hansch, C., Panthananickal, A. and Leo, A. (1982). *J. med. Chem.*, 25, 276

DiMarco, A., Arcamone, F. and Zunino, F. (1975). In *Antibiotics III* (ed. J. W. Corcoran and F. E. Hahn), Springer, Berlin, p. 101

Dobryszycka, W., Wozniak, M., Gerber, J. and Jablonski, K. (1974). *Archvm Immun. Ther. exp.*, 22, 251

Drewinko, B., Patchen, M., Yang, L-Y. and Barlogie, B. (1981). *Cancer Res.*, 41, 2328

Elslager, E. F., Short, F. W. and Tendick, F. H. (1968). *J. heterocyclic Chem.*, 5, 599

Feigon, J. F., Denny, W. A., Leupin, W. and Kearns, D. R. (1982). *Biochemistry* (submitted)

Ferguson, L. R. and Baguley, B. C. (1981a). *Mutat. Res.*, 82, 37

Ferguson, L. R. and Baguley, B. C. (1981b). *Mutat. Res.*, 90, 41

Ferguson, L. R. and Denny, W. A. (1979). *J. med. Chem.*, 22, 251

Ferraro, J. A., Horton, J., Weismann, C., Ruckdeschel, J. C., O'Donnell, M. and Ludlum, D. (1981). *Cancer Treat. Rep.*, 65, 345.

Fico, R. M., Chen, T-K. and Canellakis, E. S. (1977). *Science*, 198, 53

Filipski, J., Marczyński, B., Sadzinska, L., Chalupko, G. and Chorazy, M. (1977). *Biochim. biophys. Acta.*, 478, 33

Finkelstein, T. Z., Tittle, K., Meshnik, R. and Weiner, J. (1975). *Cancer Chemother. Rep.*, 59, 975

Furlong, N., Burr, S. J., Brown, T., Chavez, F. and Hurlbert, R. B. (1978). *Cancer Res.*, 38, 1329

Gale, E. F., Cundliffe, E., Reynolds, P. E., Richmond, M. H. and Waring, M. J. (1981). *The Molecular Basis of Antibiotic Action*, 2nd edn, Wiley, London, pp. 258–401

Gaugain, B., Barbet, J., Oberlin, R., Roques, B. P. and Le Pecq, J-B. (1978a). *Biochemistry*, 17, 5078

Gaugain, B., Barbet, J., Oberlin, R., Roques, B. P. and Le Pecq, J-B. (1978b). *Biochemistry*, 17, 5071

Gellhorn, A., Zeidenweber, J., Ultmann, J. and Hirschberg, E. (1961). *Dis. Chest.*, 39, 165

Geran, R. I., Greensberg, N. H., McDonald, M. M., Schumacher, A. M. and Abbott, B. J. (1972). *Cancer Chemother. Rep.*, 3, 1

Gerber, J., Jabloński, K., Osada, J. and Dobryszycka, W. (1972). *Archvm Immun. Ther. exp.*, 20, 139

Gieldanowski, J., Patkowski, J., Szaga, B. and Teodoroczyk, J. (1972a). *Archvm Immun. Ther. exp.*, 20, 399; (1972b). *ibid* 419

Gieldanowski, J. and Skowrońska, J. (1979). *Archvm Immun. Ther. exp.*, 27, 315

Gieldanowski, J., Szaga, B., Patkowski, J. and Pelczarska, A. (1971). *Archvm Immun. Ther. exp.*, 19, 465

Gieldanowski, J., Wieczorek, Z., Kowalczyk-Bronisz, S. H., Kowalewska, D., Blaszczyk, B., Skibiński, G. and Zimecki, M. (1980a). *Archvm Immun. Ther. exp.*, 28, 735; (1980b) *ibid* 755

Gniazdowski, M., Ciesielska, E. and Szmigiero, L. (1981). *Chem-biol. Interact.*, 34, 355

Gniazdowski, M., Filipski, J. and Chorazy, M. (1979). In *Antibiotics V/Part 2* (ed. F. E. Hahn), Springer-Verlag, Berlin, p. 275

Gniazdowski, M., Szmigiero, L., Slaska, K., Jaros-Kaminska, B. and Ciesielska, E. (1975). *Molec. Pharmac.*, 11, 310

Gniazdowski, M., Szmigiero, L. and Wilmanska, D. (1982). *Cancer Lett.* (submitted)

Goldin, A., Serpick, A. A. and Mantel, N. (1966). *Cancer Chemother. Rep.*, 50, 173

Goldin, A., Venditti, M. M., McDonald, J. S., Muggia, F. M., Henney, J. E. and DeVita, V. T. Jr (1981). *Eur. J. Cancer*, 17, 129

Gormley, P. E., Sethi, V. S. and Cysyk, R. L. (1978). *Cancer Res.*, 38, 1300

Gort, P. W., Dunn, B. P. and Beer, C. T. (1971). *J. cell. Physiol.*, 78, 127

Govindachari, T. R., Pai, B. P. and Subramanian, P. S. (1966). *Tetrahedron*, 22, 3245

Hall, D., Swann, D. A. and Waters, T. N. M. (1974). *J. Chem. Soc. Perkin* II, 1334

Hashimoto, J., Ito, R. and Kitaura, S. (1949). *J. Japan Biochem. Soc.*, 21, 27

Hlubucek, J., Ritchie, E. and Taylor, W. C. (1970). *Aust. J. Chem.*, 23, 1881

Hrabowska, M., Ledóchowski, A. and Onoszko, K. (1976). *Archvm Immun. Ther. exp.*, 24, 249

Hrabowska, M., Ledóchowski, A. and Onoszko, K. (1977). *Archvm Immun. Ther. exp.*, 25, 253

Hrabowska, M., Ledóchowski, A., Stefańska, B. and Onoszko, K. (1970a). *Archvm Immun. Ther. exp.*, 18, 557

Hrabowska, M., Ledóchowski, A., Horowska, B., Bogucka, M. and Onoszka, K. (1970*b*). *Archvm Immun. Ther. exp.*, 18, 230

Hrabowska, M., Ledóchowski, A., Horowska, B., Konopa, J. and Onoszka, K. (1971). *Archvm Immun. Ther. exp.*, 19, 879

Hudecz, F., Kajtár, J. and Szekerke, M. (1981). *Nucleic Acids Res.*, 9, 6959

Hughes, G. K., Lahey, F. N., Price, J. R. and Webb, L. J. (1948). *Nature, Lond.*, 162, 223

Kalinowska, E. and Chorazy, M. (1980). *Mutat. Res.*, 78, 7

Karle, J. M., Cysyk, R. L. and Karle, I. L. (1980). *Acta. crystallogr.*, B36, 3012

Khan, M. N., Soloway, A. H., Cysyk, R. L. and Malspeis, L. (1980). *Proc. Am. Ass. Cancer Res.*, 21, 306

Konopa, J., Koldej, K., Pawlak, J. W. (1976). *Chem.-biol. Interact.*, 13, 99

Kowalcysk-Bronisz, S. H., Blaszcyk, B., Gieldanowski, J. and Pelczarska, A. (1975*a*). *Archvm Immun. Ther. exp.*, 23, 391; (1975*b*). *ibid* 415

Kwaśniewska-Rokicińska, C., Swiecki, J., Wieczorkiewicz, A. (1973).*Archvm Immun. Ther. exp.*, 21, 863

Laster, W. R. Jr, Witt, M. H. and Schabel, F. M. Jr (1980). *Proc. Am. Ass. Cancer Res.*, 21, 271

Legha, S. S., Gutterman, J. U., Hall, S. W., Benjamin, R. S., Burgess, M. A., Valdivieso, M. and Bodey, G. P. (1978). *Cancer Res.*, 38, 3712

Legha, S. S., Keating, M. J., Zander, A. R., McCredie, K. B., Bodey, G. P. and Freireich, E. J. (1980). *Ann. internal Med.*, 93 (Part 1), 17

Le Pecq, J-B., Le Bret, M., Barbet, J. and Roques, B. P. (1975). *Proc. natn. Acad. Sci. U.S.A.*, 72, 2915

Le Pecq, J-B., Xuong, N-D., Gosse, C. and Paoletti, C. (1974). *Proc. natn. Acad. Sci. U.S.A.*, 71, 5078

Lettré, H. (1941). *Z. physiol. Chem.*, 271, 192

Lewis, M. R. and Goland, P. P. (1948). *J. Am. med. Sci.*, 215, 282

Liman, J., Babinska, M., Ledóchowski, A. (1975). *Expl cell. Res.*, 92, 299

Liska, K. J. (1972). *J. med. Chem.*, 15, 1177

Mannaberg, J. (1897). *Arch. klin. Med.*, 59, 185

Marsh, M. C. and Simpson, B. T. (1927). *J. Cancer Res.*, 11, 416

Mayer, S. E. and Bain, J. A. (1956). *J. Pharmac. Exp. Ther.*, 118, 17

McDonald, P. L. and Robertson, A. V. (1966). *Aust. J. Chem.*, 19, 275

Müller, W. and Crothers, D. M. (1968). *J. molec. Biol.*, 35, 251

Oseicka, R., Houchens, D. P., Goldin, A. and Johnson, R. K. (1977). *Cancer*, 40, 2640

Patkowski, J., Kowalczyk-Bronisz, S. H., Szaga, B., Teodorczyk, J. and Gieldanowski, J. (1973*a*). *Archvm Immun. Ther. exp.*, 21, 775; (1973*b*). *ibid* 797

Pawlak, J. W. and Konopa, J. (1979). *Biochem. Pharmac.*, 28, 3391

Peck, R. M., O'Connell, A. P. and Creech, H. J. (1966). *J. med. Chem.*, 9, 217

Peck, R. M., Preston, R. K. and Creech, H. J. (1959).*J. Am. chem. Soc.*, 81, 3984

Peck, R. M., Preston, R. K. and Creech, H. J. (1961).*J. org. Chem.*, 26, 3409

Peck, R. M., Tan, T. K. and Peck, E. B. (1976*a*). *Cancer Res.*, 36, 2423

Peck, R. M., Tan, T. K. and Peck, E. B. (1976*b*).*J. med. Chem.*, 19, 1422

Pelczarska, A., Kowalczyk-Bronisz, S. H., Blaszcyzk, B., Gieldanowski, J. and Patkowski, J. (1974*a*). *Archvm Immun. Ther. exp.*, 22, 823; (1974*b*). *ibid* 843

Preston, R. K., Peck, R. M., Breuninger, E. R., Miller, A. J. and Creech, H. J. (1964).*J. med. Chem.*, 7, 471

Radzikowski, C., Ledóchowski, Z., Ledóchowska, A., Peryt, J., Hrabowska, M., Konopa, J., Balk, M. and Jereczek, E. (1967). *Archvm Immun. Ther. exp.*, 15, 233

Radzikowkski, C., Ledóchowski, A., Hrabowska, M., Morowska, B., Stefańska, B., Konopa, J. and Morowska, E. (1969*a*). *Archvm Immun. Ther. exp.*, 17, 86

Radzikowski, C., Ledóchowski, A., Hrabowska, M., Stefańska, B., Horowska, B., Konopa, J., Morowska, E. and Urbanska, M. (1969*b*). *Archvm Immun. Ther. exp.*, 17, 99

Radzikowski, C., Wysocka-Skrzela, B., Hrabowska, M., Konopa, J., Jereczek-Marawska, E., Onoszko, K. and Ledóchowski, A. (1971). *Archvm Immun. Ther. exp.*, 19, 219

Ralph, R. K. (1980). *Eur. J. Cancer*, 16, 595

Repta, A. J., Dimmock, J. R., Kreilgard, B. and Kaminski, J. J. (1977). *J. pharm. Sci.*, 66, 1501

Riva, H. L. and Turner, T. R. (1962). *Obstet. Gynecol.*, **20**, 451

Robertson, I. G. C., Denny, W. A. and Baguley, B. C. (1980). *Eur. J. Cancer*, **16**, 1133

Rochlin, D. B., Smart, C. R., Wagner, D. E. and Silva, A. R. M. (1964). *Surgery Gynaecol. Obstet.*, **118**, 991

Rozencweig, M., Von Hoff, D. D., Cysyk, R. L. and Muggia, F. M. (1979). *Cancer Chemother. Pharmac.*, **3**, 135

Sarris, A. H., Niles, E. G. and Canellakis, E. S. (1977). *Biochim. biophys. Acta.*, **474**, 268

Schepartz, S., Venditti, J. M., Plowman, J. and Wolpert, M. K. "Report on the Program of the DCT, NCI" (1981): quoted by Skipper, H. E., in *"Some Thoughts on Screening for New Anti-cancer Drugs — Past, Present and Future"* Booklet No. 5 (1981), Southern Research Institute, Birmingham, Alabama

Schneider, J., Evans, E. L., Grundberg, E. and Fryer, R. I. (1972). *J. med. Chem.*, **15**, 266

Schneider, R. J., Woodcock, T. M. and Yagolla, A. (1980). *Cancer Treat. Rep.*, **64**, 183

Schummenfelder, N., Wessel, W., Nessel, E. (1959). *Z. Krebsforsch.*, **63**, 129

Sherif, M. (1964). *Nature, Lond.*, **204**, 390

Shoemaker, D., Padmanabhan, S., Cysyk, R., Bhat, H. B. and Malspeis, L. (1982). *Drug Metab. Disp.*, **10**, 35

Skonieczny, S. (1980). *Heterocycles*, **14**, 985

Skonieczny, S., Organiak, A., Snarska, A., Kunikowska, A., Nowak, S. and Ledóchowski, A. (1978). *Pol. J. Chem.*, **52**, 2125

Slaska, K., Szmigiero, L., Jaros-Kaminska, B., Ciesielska, E. and Gniazdowski, M. (1979). *Molec. Pharmac.*, **16**, 287

Smiles, J. and Taylor, A. E. R. (1957). *Nature, Lond.*, **179**, 306

Sordillo, P. P., Helson, L. and Lesser, M. (1980). *Cancer clin. Trials*, **3**, 385

Sordillo, P. P., Magill, G. B., Gralla, R. J. and Golbey, R. B. (1980). *Cancer Treat. Rep.*, **64**, 1129

Stiksa, G., Korsgaard, R. and Simonsson, B. G. (1979). *Scand. J. respir. Dis.*, **60**, 197

Stockinger, L. (1958). *Z. Naturforsch.*, **136**, 407

Strugger, S. (1948). *Can. J. Res.*, **26E**, 229

Suffness, M. and Douros, J. (1979). *Meth. Cancer Res.*, **21**, 73

Sullivan, H. R., Billings, R. E., Occolowitz, J. L., Boaz, H. E., Marshall, F. J. and McMahon, R. F. (1970). *J. med. Chem.*, **13**, 904

Svoboda, G. H. (1966). *Lloydia*, **29**, 206

Svoboda, G. H., Poore, G. A., Simpson, P. J. and Boder, G. B. (1966). *J. pharm. Sci.*, **55**, 758

Tobey, R. A., Deavan, L. L. and Oka, M. S. (1978). *J. natn. Cancer Inst.*, **60**, 1147

Ultmann, J., Gellhorn, A., Osnos, K. and Hirschberg, E. (1963). *Cancer*, **16**, 283

Von Hippel, P. H. and McGhee, J. D. (1972). *A. Rev. Biochem.*, **41**, 231

Wakelin, L. P. G., Creasy, T. S. and Waring, M. J. (1979). *FEBS Lett.*, **104**, 261

Wakelin, L. P. G., Romanos, M., Chen, T-K., Glaubiger, D., Canellakis, E. S. and Waring, M. J. (1978). *Biochemistry*, **17**, 5057

Waring, M. J. (1970). *J. molec. Biol.*, **54**, 247

Waring, M. J. (1976). *Eur. J. Cancer*, **12**, 995

Waring, M. J. and Henley, S. M. (1975). *Nucleic Acids Res.*, **2**, 567

Warwas, M., Narezewska, B. and Dobryszycka, W. (1977). *Archvm Immuno. Ther. exp.*, **25**, 235

Weltrowski, M. and Ledóchowski, A. (1978). *Pol. J. Chem.*, **52**, 215

Wilson, W. R. (1973). *Chemistry in New Zealand*, **37**, 148

Wilson, W. R., Baguley, B. C., Wakelin, L. P. G. and Waring, M. J. (1981*a*). *Molec. Pharmac.*, **20**, 404

Wilson, W. R., Cain, B. F. and Baguley, B. C. (1977). *Chem-biol. Interact.*, **18**, 613

Wilson, W. R., Giesbrecht, J. L., Hill, R. P. and Whitmore, G. F. (1981*b*). *Cancer Res.*, **41**, 2809

Wilson, W. R. and Whitmore (1981). *Radiat. Res.*, **87**, 121

Wilson, W. R., Whitmore, G. F. and Hill, R. P. (1981). *Cancer Res.*, **41**, 2817

Wright, R. G. McR., Wakelin, L. P. G., Fieldes, A., Acheson, R. M. and Waring, M. J. (1980). *Biochemistry*, **19**, 5825

Yap, B. S., Plager, C., Benjamin, R. S., Murphy, W. K., Legha, S. S. and Bodey, G. P. (1981). *Cancer Treat. Rep.*, **65**, 341

Zhang, B-X., Xie, K-Q. and Liu, J-Y. (1980). *Acta pharmaceutica sinica*, **15**, 40

Zwelling, L. A., Michaels, S., Erikson, L. C., Ungerleider, R. S., Nichols, M. and Kohn, K. W. (1981). *Biochemistry*, **20**, 6553

2

The Interactions of Daunomycin Adriamycin with Nucleic Acids

S. Neidle and M. R. Sanderson

INTRODUCTION

Daunomycin (NSC 82151, daunorubicin) (figure 1) and Adriamycin (NSC 123127, doxorubicin) are anthracycline anticancer antibiotics isolated from strains of *Streptomyces peucetius.* Both drugs are in current clinical use, the former being active mainly against acute lymphocytic leukaemia. Adriamycin has an exceptionally wide spectrum of antitumour activity (Goldin *et al*, 1981; Arcamone, 1981), having clinical applications in the treatment of a number of solid tumours as well as acute leukaemias. Both drugs produce severe dose-dependent cumulative cardiomyopathy in a high percentage of patients. Thus there has been continuing effort in many laboratories to develop analogues with

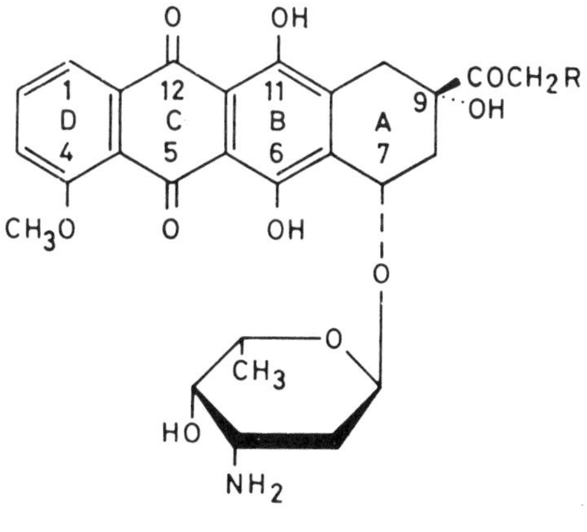

Figure 1   The molecular structures of daunomycin (R-H) and Adriamycin (R-OH).

Molecular Aspects of Anti-cancer Drug Action, ed. Neidle & Waring
0333–315561/83/035–055

reduced toxicity, and an improved spectrum of activity; to date well over 500 daunomycin derivatives have been examined. The supposition that DNA is the major cellular receptor for the drug's action has provided rationalisations for many of these endeavours. (See Chapter 3 of this volume and the numerous reviews available for further information on the relationships between daunomycin modification, DNA binding and biological activity (Arcamone, 1978, 1981; Arlandini, Vigevani and Arcamone, 1980; Brown, 1978; Di Marco and Arcamone, 1975; Henry, 1976; Neidle, 1978, 1979).) This chapter is concerned with the interactions of the parent antibiotics with nucleic acids, an area which has undergone extensive development since it was last reviewed in detail (Neidle, 1978).-

BIOLOGICAL ACTION

Daunomycin and Adriamycin inhibit the *in vitro* growth of both normal and cancer cell lines, and as a consequence, nuclear damage is characteristically observed (Di Marco, 1975). Chromosomal and genetic aberrations of several types are also produced by the drugs. The nuclear concentrations of both drugs in cultured fibroblasts are very similar, in spite of their known differences in cellular uptake (Noel *et al.*, 1978).

A number of studies have demonstrated that both RNA and DNA synthesis are inhibited by the drugs (for example, Zunino *et al.*, 1974; Meriwether and Bachur, 1972). A fundamental finding (Zunino *et al.*, 1974) was that for *Escherichia coli* DNA-dependent RNA polymerase the inhibition is dependent upon polymerase concentration, and therefore is consistent with direct drug-template (DNA) interaction.

Several studies have shown that daunomycin and Adriamycin bind to both cellular DNA and RNA. For example, rat liver cells have been found to accumulate significant binding of labelled Adriamycin to RNA (Sinha and Sik, 1980), as have cultured Yoshida sarcoma cells (Garner and Fox, 1981).

Binding to chromatin

Interactions of daunomycin and adriamycin with chromatin, in general parallel results obtained with isolated DNA (see below). Both Zunino *et al.* (1980) and Sabeur *et al.* (1979) report only a slight reduction in the apparent binding constant compared to that found for calf thymus DNA. The former study also showed that the two drugs produce quite distinct Scatchard plots for chromatin binding; the daunomycin one is significantly nonlinear at high drug: nucleotide ratios, in contrast to the linearity found with Adriamycin in this region of the plot. Zunino *et al.* (1980) suggest that this finding may be relevant to other evidence from the effect of ionic concentration on binding of the drugs to DNA, which indicates that Adriamycin binding is less sensitive to changes in DNA conformation and structure. They further relate these findings to considerations of the interactions at the molecular level.

An interesting recent observation has been made by Waldes and Center (1981), who found a marked drug-induced compaction of chromatin following incubation with Adriamycin, which was not produced by ethidium bromide (another established intercalating agent). The drug treatment also resulted in a marked resistance to digestion by micrococcal nuclease, suggesting that the nuclease sensitive sites (such as linker regions) are involved in this compaction. Since the nuclease-resistance behaviour is also not paralleled by ethidium, it is suggested that Adriamycin intercalation into DNA is not responsible for these effects, with cross-linking of DNA strands being advanced as a possible alternative binding mode.

DNA strand scission

It is now well established that the anthracycline drugs may be enzymatically reduced to free-radical species which can bind covalently to DNA and subsequently produce strand breakage. There is some evidence that the mechanisms involved in these processes do not necessarily require the drugs to be intercalated into DNA (Berlin and Haseltine, 1981). Free-radical strand breakages are discussed in detail in Chapter 9, and will not be further described here.

There have been a number of reports that drug intercalation itself produces a marked propensity for single-strand enzymatic cleavage. Thus, several studies (Lee and Byfield, 1976; Center, 1979) have shown that Adriamycin induces strand breaks, which can be produced by nuclease attack following incubation of DNA with drug. Center (1979) found that this behaviour was displayed by the intercalators Adriamycin, daunomycin and ethidium, although surprisingly, no discernible effect was found for actinomycin D. Non-intercalating DNA-binding antitumour drugs such as *cis*diaminodichloroplatinum, produced negative results.

The effect of Adriamycin on the DNA of L1210 leukaemia cells has been extensively explored by the method of alkaline elution (Ross, *et al.*, 1978, 1979; Ross and Bradley, 1981; Zwelling *et al.*, 1981). A variety of intercalating agents have been examined in addition to Adriamycin and daunomycin, differing widely in detailed molecular structure. These are ellipticine, actinomycin D, ethidium bromide, lucanthone and the acridine antitumour agent *m*-AMSA (Chapter 1). In all cases, DNA single-strand breaks are induced; these breaks are characterised by cross-link association with a cellular protein. It was also established that these lesions and cross-linkings are only produced by intercalators.

It is a plausible hypothesis that the process of intercalation itself results in a perturbation of the DNA structure at, and adjacent to the intercalation site (Berman *et al.*, 1978). This discontinuity and strain in the helix could be recognised by repair or topoisomerase enzymes. Most recently evidence has been presented (Ross and Bradley, 1981) that many of these single-strand breaks are actually double-strand ones, thus lending support to the suggestion that the scissions are produced by DNA topoisomerases. These would serve to relieve the conformational strain produced by intercalation. The analysis by Zwelling *et al.* (1981) shows that the relationship between single and double-stranded breaks is actually a complex one, with different intercalating drugs producing wide variations in the ratios of the two types of break.

The strand scissions produced do not seem to be free radical-dependent, since many of the intercalators examined, such as ellipticine or *m*-AMSA, do not possess the molecular features necessary for radical formation. Thus, the phenomenon appears to be a fundamental property of intercalating drugs when binding to DNA *in vivo*. It is not known how or indeed whether it relates to the cytotoxic and anti-cancer properties of daunomycin and Adriamycin.

INTERACTIONS WITH ISOLATED NUCLEIC ACIDS

Binding to DNA

Determination of equilibrium affinity constants

Daunomycin and Adriamycin bind reversibly to DNA, in a manner typical of other intercalating molecules. Spectroscopic methods have been widely used to characterise the interactions, which produce spectral shifts in both the visible and ultraviolet regions of the drug's spectra (Calendi *et al.*, 1965; Porumb, 1978; Zunino *et al.*, 1972; Chaires *et al.*, 1982*a*). The quenching of the drug's intrinsic fluorescence upon binding to DNA has also been monitored (Zunino *et al.*, 1979; Plumbridge and Brown, 1978). Scatchard plot analysis has been extensively employed to obtain association constants (K) and number of binding sites per nucleotide (n). The deficiences of this method, as applied to the anthracyclines have been pointed out (Plumbridge *et al.*, 1978; Bryn and Dolch, 1978; Chaires *et al.*, 1982*a*), and nonlinear regression analysis as well as neighbour exclusion models of the binding data have been used to obtain values for K and n. Table 1 presents data from a number of studies, some of which have employed other experimental approaches such as equilibrium dialysis. The recently-devised method of solvent-enhanced partition analysis (Krugh *et al.*, 1981) has been used to measure daunomycin binding to DNA at very low drug levels, where the interaction was found to be cooperative. At higher drug levels, used in the studies reported in table 1, the intercalative binding follows a neighbour exclusion model. The intercalative binding has an important electrostatic component, as indicated by its dependence on ionic strength. This has been interpreted in terms of interactions between the protonated amine group on the daunosamine sugar, and the negatively-charged phosphate oxygen atoms of the DNA backbone (Zunino *et al.*, 1977). The studies referenced in table 1, were performed on calf thymus DNA, unless otherwise stated. In general, the origin of the DNA has been found to have little effect on the final binding parameters. Thus, no difference in K or n has been found using DNA isolated from human leukaemic sources compared to that from either normal patients, or from calf thymus (Gray and Phillips, 1976). Studies with PM2 DNA (Zunino *et al.*, 1980)

Table 1 Apparent association constants (K), in $M^{-1} \times 10^6$, and number of binding sites per nucleotide (n)

K	n	Method of anlaysis	Reference
For daunomycin			
2.6	0.16	Equilibrium dialysis	Zunino *et al.*, 1972
3.3	0.18	Spectrophotometric	Zunino *et al.*, 1972
6.8	0.17	Spectrophotometric	Gabbay *et al.*, 1976
9.3	0.17	Solvent distribution	Gabbay *et al.*, 1976
2.1	0.16	Equilibrium dialysis	Huang and Phillips, 1977
7.2	0.16	Spectrophotometric	Gray and Phillips, 1976
1.27*	0.20	Difference spectroscopy	Byrn and Dolch, 1978
4.8	0.24	Spectrophotometric	Wiesehahn *et al.*, 1981
3.3	0.18	Spectrofluoremetry	Zunino *et al.*, 1979
1.3[†]	0.20	Spectrophotometric	Plumbridge and Brown, 1977
0.31[‡]	0.18	Spectrophotometric	Schneider *et al.*, 1979
0.14[§]	0.16	Spectrophotometric	Schneider *et al.*, 1979
0.18[‖]	0.18	Spectrophotometric	Schneider *et al.*, 1979
0.70	0.14	Spectrophotometric, fluorescence, equilibrium dialysis	Chaires *et al.*, 1982*a*
For Adriamycin			
2.8	0.20	Equilibrium dialysis	Zunino *et al.*, 1972
2.3	0.19	Spectrophotometric	Zunino *et al.*, 1972
2.04*	0.20	Difference spectroscopy	Byrn and Dolch, 1978
0.40	0.16	Equilibrium dialysis	Arlandini, Vigevani and Arcamone, 1980
4.2	0.19	Spectrofluorometry	Zunino, Di Marco and Zaccara, 1979
1.89[†]	0.19	Spectrophotometric	Plumbridge and Brown, 1977
0.54[‡]	0.15	Spectrophotometric	Schneider *et al.*, 1979
0.19[§]	0.19	Spectrophotometric	Schneider *et al.*, 1979
0.23[‖]	0.16	Spectrophotometric	Schneider *et al.*, 1979

*Using computerised curve-fitting to obtain K and n.
[†]Using nonlinear regression analysis.
[‡]With phosphate buffered saline.
[§]With 10% newborn calf serum.
[‖]With 10% fetal calf serum.

have shown that both drugs bind somewhat more strongly to the closed-circular form of this DNA than to its nicked linear-form; in the case of daunomycin, an apparent K of $6.7 \times 10^6 M^{-1}$ was found for the former, and $3.6 \times 10^6 M^{-1}$ for the latter. The number of bonding sites (0.13) calculated for the closed-circular form is abnormally low (compare table 1), which possibly suggests that the precise DNA conformation at, and adjacent to the binding sites, is of importance in determining binding affinity. The average n value of about 0.17 for the drugs corresponds to a maximal binding of one drug molecule per six nucleotides (three base pairs). The variation of K values listed in table 1 can probably be accounted for by minor differences in experimental conditions, such as buffer type, ionic strength (for which the binding dependence is well documented (Porumb, 1978; Zunino *et al.*, 1980)), and extent of DNA deproteinisation. Whatever the precise values of K found, the majority of studies find a small though significant increase in the affinity of Adriamycin to DNA compared to that of daunomycin. Under conditions which approximate cells in culture (Schneider *et al.*, 1979), K for both drugs decreases by an order of magnitude.

Changes in DNA properties

Anthracycline drug-DNA interactions have frequently been monitored by hydrodynamic methods. Typical changes observed in DNA include an increase in viscosity and a decrease in sedimentation coefficient. (Zunino, 1971; Kersten *et al.*, 1966; Zunino *et al.*, 1972). These are due to increases in the length of DNA molecules, consistent with stiffening of rod-like character and a decrease in the mass per unit length of the molecules. All of these changes are characteristic of intercalation.

The melting point (T_m) of double-stranded DNA is the point at which the thermal denaturation helix $\rightarrow$ coil transition takes place. This temperature is higher after binding intercalating molecules, and the change (ΔT_m) characterises the enhanced stabilisation of the DNA double helix produced by the inter-calation. (Although a high ΔT_m value does not definitively mean that a particular DNA-binding agent is an intercalator). ΔT_m values are straightforward to measure, and are not subject to the complications surrounding the determination of affinity constants. Thus, they have been widely used as a simple indicator of DNA binding for daunomycin and Adriamycin analogues (Henry, 1976, Di Marco and Arcamone, 1975; Zunino *et al.*, 1972; Zunino *et al.*, 1979; Plumbridge and Brown, 1979; Neidle, 1978, 1979). Reported ΔT_m values for daunomycin range from 13.4°C (Zunino *et al.*, 1972) to 16.8°C (Henry, 1976); those for Adriamycin are typically 1.0 to 1.5°C higher. The slightly increased DNA stabilisation produced by Adriamycin thus parallels its higher affinity constant. However, comparisons should be interpreted with caution in view of the dependence of T_m on ionic strength and the drug: DNA ratio, as well as the fact that T_m values do not measure strength of binding.

Both drugs unwind closed-circular duplex DNA in the characteristic manner of intercalators (Waring, 1970), and both produce an unwinding angle of about 11° (Gale *et al.*, 1981), relative to the 26° unwinding angle of ethidium bromide. Such angles are exceptionally low for intercalating drugs, and a more recent study (Fritzsche *et al.*, 1982) has resulted in an angle of 15°, suggesting that the actual molecular geometry of the daunomycin and Adriamycin intercalative site is quite distinct from that for, say, ethidium.

Transient electric dichroism studies (Fritzsche *et al.*, 1982) have indicated that the long axis of the daunomycin molecule is perpendicular to the DNA helix axis and that the short axis of the drug is twisted by 25°.

Interactions with other nucleic acids

Base and sequence dependence

There has been an early indication that the strength of the anthracycline-DNA interaction increases with increasing C + G content of the DNA (Kersten *et al.*, 1966). However, this has been discounted by measurements of the degree of inhibition of DNA-dependent RNA polymerase (Ward *et al.*, 1965), and ΔT_m measurements on different DNAs (Barthelemy-Clavey *et al.*, 1973). However, melting transition studies of various synthetic polymers have shown increasing stabilisation with increasing A+T content (Zunino *et al.*, 1974; Phillips *et al.*, 1978), even though the apparent binding constants calculated (Phillips *et al.*, 1978) failed to reveal these differences. Furthermore, this study showed that the alternating polynucleotide poly (dA-dT)·poly (dA-dT) not only had a T_m with daunomycin some 9° higher than that of the non-alternating poly d(A)·poly d(T), but had a significantly higher number of binding sites. The report of higher K values for poly d(G)·poly d(C) compared to poly d(A)· poly d(T) for Adriamycin binding (Tsou and Yip, 1986) is in conflict with these results, although surprisingly not with data on the dissociation of polynucleotide-daunomycin complexes (Grant and Phillips, 1979). A recent study (Chaires *et al.*, 1982*a*) has indicated that (C+G) rich DNA shows enhanced affinity for daunomycin.

It is well established that DNA structure and conformation are dependent on polymer sequence (see for example, Arnott, 1981), and in the present context it is especially noteworthy that poly d(A)·poly d(T) is exceptional in not having been observed to date in the classical B-form structure accessible to all other sequences. B-DNA is generally considered to be the form most likely to be present under physiological conditions, even though there may be differences in the microstructural conformation around a particular residue or sequence. Thus, in general terms, the differences in binding discussed above are doubtless resultants of structural differences between the various polynucleotides. It is at present however, not possible to relate one to another.

Several of the structurally simple intercalating agents such as ethidium and proflavine are known to bind preferentially at pyrimidine-3',5'-purine sites (Krugh and Reinhardt, 1975; Patel and Canuel, 1977). Although there is as yet no directly comparable solution evidence for this, the daunomycin-hexanucleotide crystal complex (Quigley *et al.*, 1980) has the drug intercalating at CG sites rather than any others in the d(CGTACG) sequence of the hexamer.

It has been established that daunomycin binds to DNA even at high ionic strengths (Huang and Phillips, 1977; Zunino *et al.*, 1980). Comparable experiments have not been reported with the alternating polynucleotide poly (dG-dC)·poly (dG-dC), which has recently been found to undergo a profound structural transition under conditions of high ionic strength (Pohl and Jovin, 1972). This transition is from a standard right-handed double helix to a left-handed one (Z-DNA) with novel and unexpected features (Wang *et al.*, 1980).

Of especial relevance to this discussion is the location of the base pairs in **Z-DNA**, which are situated on the exterior of the helix, in contrast to their relatively buried position in **B-DNA**. Thus, the base pairs in **Z-DNA** would be expected to be especially sensitive to external influences, including perhaps those of intercalating drugs.

Interactions with ribonucleic acids

There is good evidence that daunomycin and Adriamycin do not bind to double-stranded ribonucleic acids (Plumbridge and Brown, 1977, 1979) by intercalative processes, in contrast to simpler intercalators such as acridines and ethidium. Double-stranded RNAs are exclusively found in the A-form (Arnott, 1981), and consequently do not have the conformational variability shown by deoxypolynucleotides. It is possible that since the anthracyclines can actually bind to deoxypolymers that are not in the B-form (see above), the important determinants of binding are not only the initial nucleic acid structure, but are also the degrees of conformational flexibility inherent in it necessary to accommodate the incoming intercalative molecule. (RNAs are conformationally rigid due primarily to the presence of 2′ hydroxyl groups producing steric hindrance to movement.) The binding of daunomycin to transfer RNA has been reported, with an apparent K of $1.2 \times 10^8 M^{-1}$ (Shafer, 1977). No information is as yet available on the localisation of these interactions. This question is especially intriguing in view of the fact that transfer RNAs have highly compact tertiary structures and rigidised helical stems. Whether it is more likely that the drug molecules (about three per transfer RNA) bind in clefts and folds of the tertiary structure, or classical intercalation occurs, remains a matter for speculation.

NMR Studies

Proton (^{1}H) NMR investigations of anthracycline-nucleic acid interactions are not as extensive as similar studies for other intercalating drugs such as proflavine. Figure 2 shows that, in contrast to proflavine, daunomycin has non-exchangeable protons only as the extremities of the chromophore − these are the ones which can be monitored by ^{1}H NMR in D_2O. Therefore, it is not

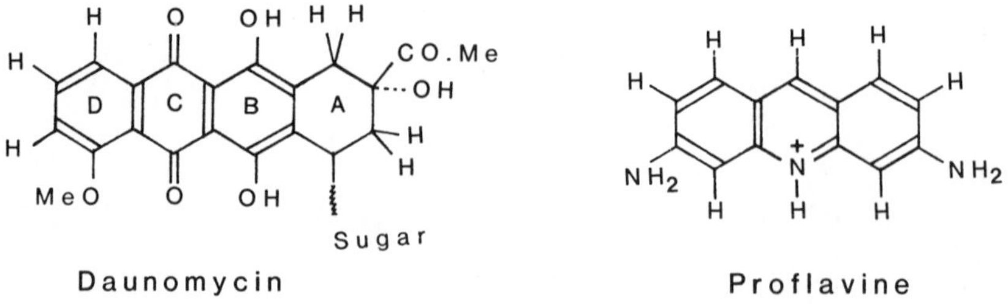

Figure 2 The proton distribution in daunomycin and proflavine.

straightforward to use this technique as a sensitive probe of the shielding environment (and indirectly the geometry) of the daunomycin molecule, following intercalation.

Assignment of the proton spectrum for daunomycin itself in chloroform has been made (Arcomone *et al.*, 1986; Iwamoto *et al.*, Bhacca, 1968), and its pronounced self-aggregation in aqueous solution has been monitored (Barthelemy-Clavey *et al.*, 1974), although the association constant determined ($700 M^{-1}$) differs significantly from that subsequently found ($6,000$–$7,000 M^{-1}$) in a circular dichroism study (Martin, 1980). A recent thorough analysis using absorbance, sedimentation and NMR techniques (Chaires *et al.*, 1982*b*) has shown that aggregation beyond the dimer level is extensive at the concentrations employed in NMR experiments.

Interactions with oligonucleotides
The assumption that suitable oligonucleotides can model nucleic acids themselves has been a major factor in the study of a number of intercalating drugs, by a variety of techniques. Such an approach has been followed largely because of the enhanced possibility of exhaustively analysing the conformational and structural features of oligonucleotide complexes, compared to polynucleotide ones, whether by NMR or X-ray or theoretical methods. This is undoubtedly true in principle; however, it has to be borne in mind that such systems do not necessarily accurately model polynucleotide systems, and moreover that features observed in a model may be peculiar to that system.

The interaction of daunomycin with dinucleotides has been studied by 100 MHz and 360 MHz proton NMR (Nuss *et al.*, 1980). The chemical shifts of all the drug protons were monitored at varying nucleotide: drug ratios, and some measure of interaction was observed with all the dinucleotides examined (pdApdT, pdTpdA, pdGpdC, pdCpdG and pdCpdC). At the temperature used (40°C) none of these are base-paired. It was concluded that in all cases the complexes formed have 1:1 stoicheoimetry. The H1, H2 and H3 aromatic protons on ring D displayed the largest upfield shifts on binding, with the pdGpdC complex consistently showing the greatest effect. Evidence was also presented for a conformational change involving the sugar group, as well as for the ring A conformation of the drug without nucleotide being as shown in figure 3b. This data was used to suggest a tentative model for the interaction involving a hydrogen bond between the 9-hydroxyl group on ring A, and the phosphate group at the intercalation site. However, since these experiments were performed at a temperature at which the dinucleotides could not form Watson-Crick base-paired duplex dimers, the results of this study do not obviously relate to double-stranded nucleic acids.

The interaction of the tetranucleotide d(GpCpGpC) with daunomycin has been studied in relation to its helix-coil transition (Patel, 1979*a*), by high-resolution proton and phosphorus NMR. The temperature-dependent chemical shifts for guanine H8, cytosine H5 and H6 and sugar H1′ protons, were

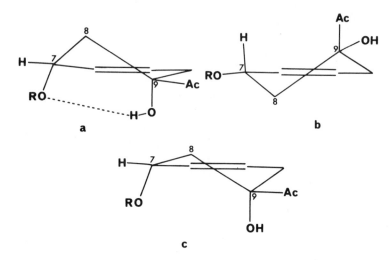

Figure 3 The conformation of the daunomycin A ring; (a) as found in the crystal structure of the pyridine adduct (Neidle and Taylor, 1977): (b) as suggested by Henry (1976); (c) as found in the N-bromoacetyl derivative (Angiuli *et al.*, 1971). Reproduced by permission from *Topics in Antibiotic Chemistry*, Vol. 2 (1978) ed. P.G. Sammes, Ellis Horwood Ltd., Chichester.

examined for the tetranucleotide itself. These enabled duplex formation in solution to be followed; it was concluded that the resulting helical fragment has B-DNA type base pair overlap. Complex formation with the drug was monitored by examining both chromophore and sugar proton resonances. Those on ring D (figure 3) undergo small shifts upfield ($\simeq 0.15$ p.p.m.) whereas the sugar protons do not show significant shifts. The drug binding imparts extra stabilisation to the d(GpCpGpC) duplex, for which $t_{1/2}$, the midpoint of the melting curve rises by at least 40°C from the 40°C value for the tetramer alone. It was inferred that the intercalative binding observed is consistent with a model having ring D not being overlapped by base pairs, since the ring-current effect induced by these base pairs is very small compared with that seen with other intercalators such as proflavine (Patel, 1979b). Typically, large upfield shifts are observed for these molecules, indicative of extensive basepair-intercalator association and overlap.

A thorough proton NMR study has been made of the interactions between the hexanucleotide d(pTpA)$_3$ and daunomycin (Phillips and Roberts, 1980). In general, since AT base pairs are less stable than GC ones, there was considerable evidence of fraying of both terminal and internal base pairs. Daunomycin binding imparted a large degree of stabilisation to the duplex, with $t_{1/2}$ rising by 21°C for a drug: duplex ratio of 0.75. As with the tetranucleotide discussed above, surprisingly small upfield shifts were noted for the daunomycin aromatic protons; in contrast, the aromeric H1' daunomycin sugar proton shifted upfield by 0.22 p.p.m. This latter effect was tentatively attributed to ring-current shielding effects.

Interactions with polynucleotides

The complex between daunomycin and poly (dA-dT) · poly (dA-dT) has been extensively investigated (Patel and Canuel, 1978; Patel *et al.*, 1981). This polynucleotide is especially suited for proton NMR studies since it gives signals with narrow line widths. As with the oligomers, small (0.1–0.25 p.p.m.) resonance shifts were observed for the ring D daunomycin protons, and a slightly larger one for the anomeric H1′ sugar proton. Patel *et al.* also found the following: (1) Base pairing observed in the polynucleotide alone, was intact in the complex. (2) The phenolic protons (figure 2) of the anthracycline which for the drug alone are exchangeable in aqueous solution, have a reduced exchange rate when daunomycin binds to the polynucleotide. This is attributed to intramolecular carbonyl-hydroxyl hydrogen bonding and to shielding from solvent after intercalation. It is thus concluded that rings B and C (figure 2) extensively overlap with base pairs at the site of intercalation. (3) The ^{31}P NMR spectrum of the complex suggests site-specific intercalation, although it was not possible to distinguish between d(ApT) and d(TpA) sites.

Patel and Canuel (1978) also reported parallel helix-coil thermal transition studies on daunomycin binding to synthetic DNAs containing bulky halogen substituents protruding into the major groove of the helix. The T_m values found were similar to those for the unsubstituted DNA, indicating that either the drug intercalates via the minor groove, or via the major groove without steric hindrance with the bulky substituent. It was also observed that a daunomycin derivative with the amino group on the sugar replaced by the more bulky dimethylglycine group had a markedly reduced DNA binding affinity, presumably due to a decrease in electrostatic interaction between the amino group and the phosphate backbone (Gabbay *et al.*, 1976; Zunino *et al.*, 1972, 1974, 1977).

Geometric and structural information derived from the NMR data has been summarised and compared with X-ray structural results (Patel *et al.*, 1981). We defer details of this comparison to a subsequent section.

STRUCTURAL STUDIES

The conformational properties of daunomycin and Adriamycin

The molecular structure of *N*-bromoacetyl daunomycin determined by X-ray crystallographic analysis (Angiuli *et al.*, 1971), has been employed in model-building studies of the DNA-drug complex (Pigram *et al.*, 1972). The half-chair conformation of ring A, shown in figure 3c, has the daunosamine sugar ring axially substituted, and the C13 side-chain equatorial. Similar arrangements have been found in the more accurate crystal structures of daunomycin itself, both as a pyridine (Neidle and Taylor, 1977) (figure 4), and as a butanol adduct (Courseille *et al.*, 1979), and in 4-hydroxy daunomycin (carcinomycin) (Von Dreele and Einck, 1977). In all these structures, ring D has the pucker

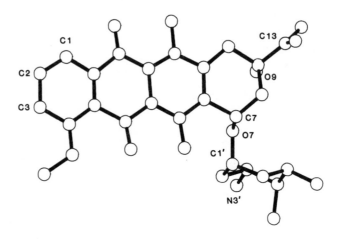

Figure 4 The daunomycin molecular structure, as found in its pyridine adduct (Neidle and Taylor, 1977).

shown in figure 3a, with an O7 . . . . . . O9 intramolecular hydrogen bond. Remarkably, in all the structures, the overall molecular structure is very similar, with the relative orientation of sugar to chromophore being conserved (table 2).

In view of the importance of the molecular structure of daunomycin for models of its DNA complex, several groups have performed conformational calculations on the drug. An analysis varying only the two glycosidic angles around the C7–O7 and O7–C1′ bonds (Neidle and Taylor, 1979), found that the minimum-energy state indeed corresponded to the crystal structures. A subsequent study of Adriamycin (Nakata and Hopfinger, 1980a), however, found a quite distinct global minimum conformation (table 2), which, it is suggested, is stabilised by a hydrogen bond between the hydroxyl proton on O6, and the sugar ring oxygen atom. These controversal findings have not been supported by two more recent studies. A complete molecular mechanics optimisation of the daunomycin structure (Brown *et al.*, 1982), is consistent with the X-ray-derived conformation and with the earlier calculations (Neidle and Taylor, 1979). This study has also found that the alternative ring A pucker (figure 3b) is energetically as least as favourable as that found in the crystal structures, in agreement with an NMR study in solution (Nuss *et al.*, 1980). Detailed analysis of the hydrogen-bonding in the daunomycin molecule (Islam and Neidle, 1983), suggests that the intramolecular quinone-hydroxyl hydrogen bonds are strong and unlikely to be severed.

Fibre diffraction studies of a daunomycin-DNA complex

X-ray diffraction patterns have been obtained from oriented fibres of this complex (Pigram *et al.*, 1972). It was not possible to analyse these patterns *ab initio* in terms of molecular structure in the same manner as single crystal

Table 2   Features of the low-energy conformation of daunomycin

	Torsion angles*		ΔE in Kcal mol^{-1}	
	ϕ_1	ϕ_2		
In the crystalline state				
Daunomycin as pyridine adduct	125°	292°		Neidle and Taylor, 1977
Carminomycin	117°	290°		Von Dreele and Einck, 1977
N-Bromoacetyl daunomycin	102°	281°		Angiuli et al., 1971
Daunomycin in hexanucleotide complex	91°	282°		Quigley et al., 1980
From calculations				
Daunomycin	93°	290°	0	Neidle and Taylor, 1979
Adriamycin (a)	77°	288°	0	Nakata and Hopfinger, 1980a
(b)	259°	257°	1.5	Nakata and Hopfinger, 1980a
Daunomycin (a)	79°	288°	0	Nakata and Hopfinger, 1980a
(b)	258°	257°	1.8	Nakata and Hopfinger, 1980a
Daunomycin (a)	95°	288°	0.9	Brown et al., 1982
(b)	258°	259°	10.8	Brown et al., 1982

*ϕ_1, ϕ_2 are around the C7–O7 and O7–C1′ bonds respectively.

structure determinations; however, systematic examination of various types of trial models enabled a plausible structure to be proposed for the complex, which was in accord with the fibre patterns and much of the binding data.

The resulting model for the daunomycin-DNA complex has the drug intercalated in from the major groove of the double helix. The unwinding angle at the binding site was judged to be 12°, in surprisingly good agreement with later estimates from unwinding experiments. The positioning of the sugar group in the major groove is of some interest; in accord with the binding requirement for a considerable electrostatic component, the positively-charged nitrogen atom is in close contact with an anionic phosphate-oxygen atom two base-pairs distant. An additional element of stabilisation is imparted by a hydrogen bond involving O9 and an adjacent-site phosphate group.

This model has commanded widespread acceptance since it explains much of the binding data for daunomycin and Adriamycin derivatives (Neidle, 1978, 1979). Thus, the total binding site spans three base pairs, in good agreement

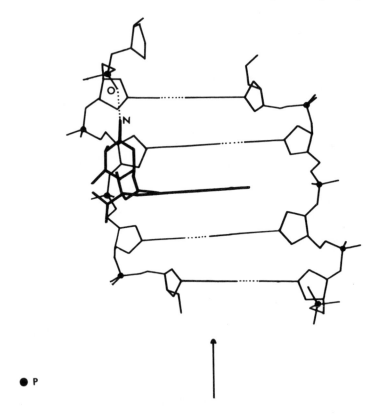

Figure 5 The modified major-groove binding DNA-daunomycin model, with the drug molecule drawn in bold outline. Reproduced by permission from *Topics in Antibiotic Chemistry*, Vol. 2 (1978) ed. P.G. Sammes, Ellis Horwood, Chichester.

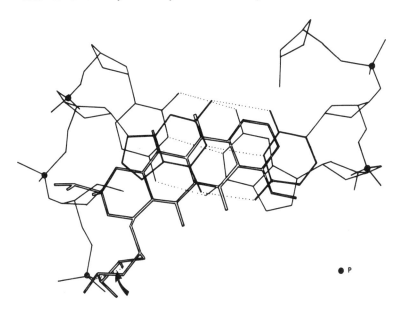

Figure 6 A view of the model in figure 5, observed looking onto the base-pair planes. The arrow indicates the hydrogen bond between the sugar amine and a phosphate oxygen atom. Reproduced by permission from *Topics in Antibiotic Chemistry*, volume 2 (1978) ed. P.G. Sammes, Ellis Horwood, Chichester.

with the finding from solution data of ~0.17 binding sites per nucleotide (table 1). The electrostatic interaction of the drug with the backbone accords with the requirement, again from solution evidence, of a considerable electrostatic component in the interaction. The Pigram-Fuller-Hamilton model has had minor modification (Neidle, 1977, 1978), with the substitution of the daunomycin molecular structure (Neidle and Taylor, 1977) for the less accurate N-bromoacetyl one (Angiuli *et al.*, 1971). This modified model is illustrated in figures 5 and 6, which show that the long axis of the drug chromophore is roughly aligned with the base pairs. There is relatively little overlap of the chromophore as a whole with adjacent base pairs, with rings B and C of the drug having the most base-pair interaction. An additional hydrogen-bonding interaction has also been suggested for Adriamycin, involving the C14 hydroxyl group and a phosphate oxygen atom at the interaction site.

Theoretical model-building

A molecular model for the Adriamycin-DNA complex has been suggested (Henry, 1976) on the basis of the alternative ring A conformation (figure 3b) and constructed by manual model-building. Intercalation is again via the major groove, with a maximum number of hydrogen-bonded contacts being made. Thus, contacts are made with a total of three phosphate groups, one by the protonated amino sugar, and two by the 4' and 9 hydroxyl group. A measure of specificity is proposed by means of a hydrogen bond between the 14-hydroxyl group and a N7 atom of an adjacent-site purine base. This model, in spite of its almost totally hypothetical nature and relative lack of detail, has been influential and useful in the rational search for active daunomycin analogues (Henry, 1976).

A theoretical molecular mechanics energy study has been made of the interactions between Adriamycin and base-paired, self-complementary dinucleoside monophosphates (Nakata and Hopfinger, 1980b). Conformations for these nucleic acid models were taken either from an ethidium-5-iodo-CpG crystal-structure analysis (Jain *et al.*, 1977), or from a previous theoretical analysis of intercalation into dinucleosides (Malhotra and Hopfinger, 1980). Sequences examined were the ribodinucleosides TpC, CpG and UpA, base-paired with their complementary strands. Both major and minor groove intercalation was examined, as well as differences in drug conformation (Nakata and Hopfinger, 1980a; Neidle and Taylor, 1979). A consistent finding was that interaction from the minor-groove direction produces an energetically more stable agreement, between 1 and 30 kcal mol^{-1} less (depending on the sequence analysed and the precise drug conformation utilised), compared to major-groove models. Interestingly, this study, although agreeing with both NMR (Patel *et al.*, 1981) and X-ray crystallographic data (Quigley *et al.*, 1981), for the minor-groove preference, suggests subsidiary interaction between the daunomycin sugar amine and a nucleotide backbone phosphate group. This interaction has not been observed in the crystal structure.

Theoretical model-building studies have not been reported for anthracylines binding to deoxyoligonucleotide sequences, which would have enhanced relevance to the structure and dynamics of DNA-drug complexes. Such studies, in conjunction with other physico-chemical data could be of promise in the rational design of new daunomycin and Adriamycin analogues.

Structure of a daunomycin-oligonucleotide complex

A complex between daunomycin and the hexanucleoside pentaphosphate d(CpGpTpApCpG) has been crystallised, and its structure determined by X-ray crystallography (Quigley *et al.*, 1980; Rich *et al.*, 1981). This analysis has provided the first atomic-level visualisation of anthracycline drug-nucleic acid interaction, and is therefore discussed in some detail.

Two drug molecules are bound per self-complementary oligonucleotide duplex. Each drug is intercalated between the terminal CG base pairs (figure 7). Thus, in spite of the choice of four types possible at sites (CG, GT, TA or AC), the drug has preferentially bound to only one. This evidence for a pyrimidine-3', 5'-purine sequence preference at the intercalation site accords with that for several other intercalators, as described previously.

A striking feature of the complex is that the drug molecules show minor groove intercalation (figure 8), in contrast to the other molecular models

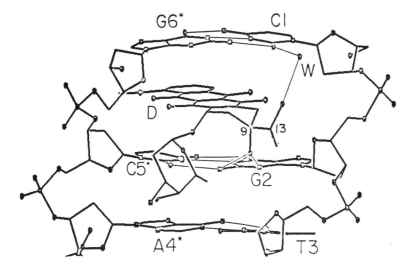

Figure 7   View of daunomycin (D) intercalated into d(CpGpTpApCpG.) Intermolecular contacts are shown. Reproduced with permission from Quigley *et al.* (1980).

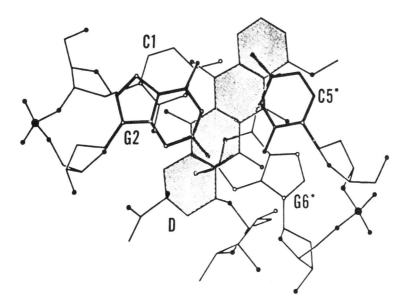

Figure 8 View of the daunomycin-hexanucleotide complex perpendicular to the base plane. The drug chromophore is shaded. Reproduced with permission from Quigley *et al.*, (1980).

proposed. The amino sugar fits snugly into the minor groove; this position excludes any interaction between the charged amino group and backbone phosphates. The daunomycin chromophore lies skew to the CG base pairs, and it is notable that only rings A and D protrude out from them, in complete agreement with the NMR observations (Patel *et al.*, 1981). The B and C ring overlap with base pairs is small, although the hydroxyquinone oxygen atoms on each side of the drug molecule clearly play a stabilising role (figure 8), by being stacked with the G2 and G6 guanine bases.

In overall terms, the d(CpGpTpApCpG) sequence has a right-handed DNA conformation. The base pairs at the intercalation sites are inclined to each other at 36°. This base turn angle (Berman *et al.*, 1978) is the same as that for B-DNA itself, although the detailed aspects of the oligonucleotide conformation do not fit into any of the classical DNA structural types. The zero unwinding at the actual intercalation site is balanced by an 8° unwinding of adjacent residues. Effects of steric compression and extension have been observed in the backbone, which suggest that the daunomycin molecule has induced markedly asymmetric structural changes. It is of course possible that this asymmetry is a result of the drug binding at the termini of the helix fragment, rather than in a region with balancing numbers of base pairs on either side. Although atomic coordinates for the structure are not available as yet, it is of interest to compare the overall features with several accurate molecular structures of drug-dinucleoside and deoxydinucleoside phosphate complexes (Neidle *et al.*, 1977; Shieh *et al.*, 1980; Jain *et al.*, 1977; Neidle, 1981). All these structures have very similar backbone conformations, in spite of binding different intercalators. These are characterised by resembling A form DNA, a feature not discernable in the daunomycin-hexanucleoside complex. However, the pattern of changes in nucleic acid helicity as measured by glycosidic angles round the intercalation site, is broadly similar in all of these structures. It remains to be established whether the dinucleoside-drug structures or the hexamer-daunomycin one (with its asymmetry in number of residues around the binding site), more accurately model the geometry of polynucleotide-drug intercalation.

The daunomycin-hexamer structure reveals several specific hydrogen-bonding interactions which serve to provide additional stabilisation for the binding. There is a doublet of hydrogen bonds involving the hydroxyl oxygen atom O9, and the N2 and N3 atoms of the guanine residue in the first strand of the duplex. The C13 side-chain is anchored via a water molecule to the O2 atom of the cytosine adjacent to the above guanine. It is pointed out that these interactions are not completely specific for particular residues; adenine could replace guanine, albeit with a loss of one hydrogen bond. As with other models, substitution of the Adriamycin molecule could well result in an additional hydrogen-bonding stabilisation.

Perhaps the least understandable feature of this otherwise illuminating model is its inability to account for the role of the charged amino group in DNA binding (and in biological activity). It is perhaps plausible that this group acts as an initial site for the exterior of the DNA molecule to recognise, and thus to bring the intercalative grouping into the appropriate orientation for its binding. Quigley *et al.*, (1980) do suggest that the amino group may serve to recognise RNA polymerase as it progresses along DNA, and thus block its progress. The question of whether the amine group position in the structure is merely a special aspect of its particular environment in the hexanucleotide complex, or is a general feature of daunomycin-DNA structure, will only be resolved by further X-ray structure analyses of relevant drug-oligonucleotide complexes. The positioning of the sugar group in the minor groove of DNA provides a likely explanation for the lack of intercalation with double-stranded RNA, since the minor groove in RNA is very shallow and would be unable to accommodate (and hence interact with) the bulky sugar group.

Correlations between solution and crystallographic data

The high-resolution proton NMR study of the daunomycin-poly (dA-dT)· poly (dA-dT) complex (Patel *et al.*, 1981) has enabled a direct comparison of aspects of binding-site geometry to be made between the solution and crystallographic complexes. The overlap of daunomycin rings B and C with adjacent base pairs, and the non-overlap of ring D have been found to be very similar with both techniques. The crystallographically-observed perpendicular position of the chromophore with respect to the base-pair hydrogen bonds is likewise in agreement with the NMR data. Although the NMR solution data is suggestive of an overall position for the sugar group, details for this (as well as for the other subsidiary interactions indicated from the crystal structure) must await more refined NMR analyses. Nuclear Overhauser enhancement studies, which can be used to obtain inter-atomic distances, should be of especial importance in resolving these issues.

CONCLUSIONS

The many studies reviewed in this chapter, taken as a whole, provide highly detailed information on many aspects of the interactions of daunomycin and Adriamycin with nucleic acids. The subject may well be soon at the stage where rational design of analogues with enhanced and more selective DNA-binding properties can be achieved, by utilisation of this large body of data.

ACKNOWLEDGEMENTS

We are grateful to the Cancer Research Campaign for a Career Development Award (to S.N.) and to the Science and Engineering Research Council for a NATO Fellowship (to M.R.S.). Professors P.A. Kollman and A. Rich are thanked for the provision of material prior to publication.

REFERENCES

Angiuli, R., Foresti, E., Riva Di Sanserverino, L., Issacs, N. W., Kennard, O., Motherwell, W. D. S., Wampler, D. L. and Arcamone, F. (1971). *Nature new Biol.*, **234**, 78

Arcamone, F. (1978). In *Topics in Antibiotic Chemistry*, Vol. 2 (ed. P. G. Sammes), Ellis Horwood, Chichester, p. 89

Arcamone, F. (1981). In *Doxorubicin, Anticancer Antibiotics*, Academic Press, New York

Arcamone, F., Cassinelli, G., Franceshi, G., Orezzi, P. and Mondelli, R. (1968). *Tetrahedron Lett.*, **30**, 3353

Arlandini, E., Vigevani, A. and Arcamone, F. (1980). *Farmaco, Sci.*, **35**, 65

Arnott, S. (1981). In *Topics in Nucleic Acid Structure* (ed. S. Neidle), Macmillan Press, London, p. 65

Barthelemy-Clavey, V., Maurizot, J.-C., Dimicoli, J. C. and Sicard, P. (1974). *FEBS Lett.*, **46**, 5

Barthelemy-Clavey, V., Maurizot, J.-C. and Sicard, P. J. (1973). *Biochimie*, **55**, 859

Berlin, V. and Haseltine, W. A. (1981). *J. biol. Chem.*, **256**, 4747

Berman, H. M., Neidle, S. and Stodola, R. K. (1978). *Proc. natn. Acad. Sci. U.S.A.*, **72**, 828

Brown, J. R., (1978). *Prog. med. Chem.*, **15**, 125

Brown, S. C., Kollman, P. A. and Weiner, P. K. (1982). *Biochim. biophys. Acta*, **717**, 49

Byrn, S. R. and Dolch, G. D. (1978). *J. pharm. Sci.*, **67**, 688

Calendi, E., Di Marco, A., Reggiani, M., Scarpinato, B., and Valentini, L. (1965). *Biochim. biophys. Acta.* **103**, 25

Center, M. S. (1979). *Biochem. biophys. Res. Commun.*, **89**, 1231

Chaires, J. B., Dattagupta, N. and Crothers, D. M. (1982*a*). *Biochemistry*, **21**, 3933

Chaires, J. B., Dattagupta, N. and Crothers, D. M. (1982*b*). *Biochemsitry*, **21**, 3927

Courseille, C., Busetta, B., Geoffre, S. and Hospital, M. (1979). *Acta crystallogr.*, B35, 764

Di Marco, A. (1975). *Cancer Chemotherapy Rep.*, 6, 91

Di Marco, A. and Arcamone, F. (1975). *Arnz. Forschung*, **25**, 368

Fritzsche, H., Triebel, H., Chaires, J. B., Dattagupta, N. and Crothers, D. M. (1982). *Biochemistry*, **21**, 3940

Gabbay, E. J., Grier, D., Fingerle, R. E., Reiner, R., Levy, R., Pearce, S. W. and Wilson, W. D. (1976). *Biochemistry*, **15**, 2062

Gale, E. F., Cundliffe, E., Reynolds, P. E., Richmond, M. H. and Waring, M. J. (1981). *The Molecular Basis of Antibiotic Action* 2nd edn John Wiley, London

Garner, A. and Fox, B. W. (1981). *Chem.-biol. Interactions*, **36**, 189

Goldin, A., Venditti, J. M., MacDonald, J. S., Muggian, F. M., Henny, J. E. and deVitta, V. T. Jr. (1981). *Eur. J. Cancer*, **17**, 129

Grant, M. and Phillips, D. R. (1979). *Molec. Pharmac.*, **16**, 357

Gray, P. J. and Phillips, D. R. (1976). *Eur. J. Cancer*, **12**, 237

Henry, D. W. (1976). *Am. chem. Soc. Symp. Ser.*, **30**, 15

Huang, Y. M. and Phillips, D. R. (1977). *Biophys. Chem.* 6, 363

Islam, S. A. and Neidle, S. (1983). *Acta crystallogr.*, B39, 114

Iwamoto, R. H., Lim, P. and Bhacca, N. S. (1968). *Tetrahedron Lett.*, 3891

Jain, S. C., Tsai, C. C. and Sobell, H. M. (1977). *J. molec. Biol.*, **114**, 317

Kersten, W., Kersten, H. and Szybalski, W. (1966). *Biochemistry*, **5**, 236

Krugh, T. R. and Reinhardt, C. G. (1975). *J. molec. Biol.*, **97**, 133

Krugh, T. R., Winkle, S. A. and Graves, D. E. (1981). *Biochem. biophys. Res. Commun.* **98**, 317

Lee, Y. C. and Byfield, J. E. (1976). *J. natn. Cancer Inst.*, **57**, 221

Lown, J. W. and Chen, H.-H. (1981). *Can. J. Chem.*, **59**, 3212

Malhotra, D. and Hopfinger, A. J. (1980). *Nucleic Acids Res.*, 8 , 5289

Martin, S. R. (1980). *Biopolymers*, **19**, 713

Meriwether, W. D. and Bachur, N. R. (1972). *Cancer Res.*, **32**, 1137

Nakata, Y. and Hopfinger, A. J. (1980*a*). *FEBS Lett.*, **117**, 259

Nakata, Y. and Hopfinger, A. J. (1980*b*). *Biochem. biophys. Res. Commun.*, **95**, 583

Neidle, S. (1977). *Cancer Treat. Rep.*, **61**, 928

Neidle, S. (1978). In *Topics in Antibiotic Chemistry*, Vol. 2 (ed. P. G. Sammes), Ellis Horwood, Chichester, p. 230

Neidle, S. (1979). *Prog. med. Chem.*, **16**, 151

Neidle, S. (1981). In *Topics in Nucleic Acid Structure* (ed. S. Neidle), Macmillan Press, London, p. 177

Neidle, S., Achari, A., Taylor, G. L., Berman, H. M., Carrell, H. L., Glusker, J. P. and Stallings, W. C. (1977). *Nature, Lond.*, **269**, 304

Neidle, S. and Taylor, G. (1977). *Biochim. biophys. Acta.*, **479**, 450

Neidle, S. and Taylor, G. (1979). *FEBS Lett.*, **107**, 348

Noël, G., Peterson, C., Trouet, A. and Tulkens, P. (1978). *Eur. J. Cancer*, **14**, 363

Nuss, M. E., James, T. L., Apple, M. A. and Kollman, P. A. (1980). *Biochim. biophys. Acta.*, **609**, 136

Patel, D. J. (1979a). *Biopolymers*, **18**, 553

Patel, D. J. (1979b). *Acc. Chem., Res.*, **12**, 118

Patel, D. J. and Canuel, L. L. (1977). *Proc. natn. Acad. Sci. U.S.A.*, **74**, 2624

Patel, D. J. and Canuel, L. L. (1978). *Eur. J. Biochem.*, **90**, 247

Patel, D. J., Kozlowski, S. A. and Rice, J. A. (1981). *Proc. natn. Acad. Sci. U.S.A.*, **78**, 3333

Phillips, D. R., Di Marco, A. and Zunino, F. (1978). *Eur. J. Biochem.*, **85**, 487

Phillips, D. R. and Roberts, G. C. K. (1980). *Biochemistry*, **19**, 4795

Pigram, W. J., Fuller, W. and Hamilton, L. D. (1972). *Nature new Biol.*, **235**, 17

Plumbridge, T. W., Aarons, L. J. and Brown, J. R. (1978). *J. pharm. Pharmac.*, **30**, 69

Plumbridge, T. W. and Brown, J. R. (1977). *Biochim. biophys. Acta.*, **479**, 441

Plumbridge, T. W. and Brown, J. R. (1978). *Biochem. Pharmac.*, **27**, 1881

Plumbridge, T. W. and Brown, J. R. (1979). *Biochim. biophys. Acta.*, **563**, 181

Pohl, F. M. and Jovin, T. M. (1972). *J. molec. Biol.*, **67**, 375

Porumb, H. (1978). *Prog. biophys. molec. Biol.*, **34**, 175

Quigley, G. J., Wang, A. H.-J., Ughetto, G., van der Marel, G., van Boom, J. H. and Rich, A. (1980). *Proc. natn. Acad. Sci. U.S.A.*, **77**, 7204

Rich, A. Quigley, G. J. and Wang, A. H.-J. (1981). In *Biomolecular Stereodynamics* Vol. 1 (ed. R. H. Sarma), Adenine Press, New York, p. 25

Ross, W. E. and Bradley, M. O. (1981). *Biochim. biophys. Acta.*, **654**, 129

Ross, W. E., Glaubiger, D. and Kohn, K. W. (1979). *Biochim. biophys. Acta*, **562**, 41

Ross, W. E., Glaubiger, D. L. and Kohn, K. W. (1978). *Biochim biophys Acta.*, **519**, 23

Sabeur, G., Genert, D. and Aubel-Sadron, G. (1979). *Biochem. biophys. Res. Commun.*, **88**, 722

Schneider, Y.-J., Baurain, R., Zenebergh, A. and Trouet, A. (1979). *Cancer Chemother, Pharmac.*, **2**, 7

Shafer, R. H. (1977). *Biochem. Pharmac.*, **26**, 1729

Shieh, H.-S., Berman, H. M., Dabrow, M. and Neidle, S. (1980). *Nucleic Acids Res.*, **8**, 85

Sinha, B. K. and Sik, R. H. (1980). *Biochem. Pharmac.*, **29**, 1867

Tsou, K. C. and Yip, K. F. (1976). *Cancer Res.*, **36**, 3367

Von Dreele, R. B. and Einck, J. J. (1977). *Acta crystallogr.*, B**33**, 3283

Waldes, H. and Center, M. S. (1981). *Biochem. biophy. Res. Commun.*, **98**, 95

Wang, A. H.-J., Quigley, G. J., Kolpak, F. J., Crawford, J. L., van Boom, J. H., van der Marel, G. and Rich, A. (1980). *Nature, Lond.*, **282**, 680

Ward, D. C., Reich, E. and Goldberg, I. H. (1965). *Science*, **149**, 1259

Waring, M. J. (1970). *J. molec. Biol.*, **54**, 247

Wiesehahn, G., Varga, J. M. and Hearst, J. E. (1981). *Nature, Lond.*, **292**, 467

Zunino, F. (1971). *FEBS Lett.*, **18**, 249

Zunino, F., Di Marco, A. and Zaccara, A. (1979). *Chem.-biol. Interactions*, **24**, 217

Zunino, F., Di Marco, A., Zaccara, A. and Gambetta, R. A. (1980). *Biochim. biophys. Acta.*, **607**, 206

Zunino, F., Di Marco, A., Zaccara, A. and Luoni, G. (1974). *Chem.-biol. Interactions*, **9**, 25

Zunino, F., Gambetta, R., Di Marco, A., Velcich, A., Zaccara, A., Quadrifoglio, F. and Crescenzi, V. (1977). *Biochim. biophys. Acta*, **476**, 38

Zunino, F., Gambetta, R., Di Marco, A. and Zaccara, A. (1972). *Biochim. biophys. Acta.*, **277**, 489

Zwelling, L. A., Michaels, S., Erickson, L. C., Ungerleider, R. S., Nichols, M. and Kohn, K. W. (1981). *Biochemistry*, **20**, 6553

3
New Natural, Semisynthetic and Synthetic Anthracycline Drugs

J. R. Brown

INTRODUCTION

The anthracycline antibiotics Adriamycin (1) and daunomycin (2) are important drugs in cancer chemotherapy, despite their inhibitory effects on rapidly-dividing tissues (which leads to alopecia, stomatitis and myelosuppression) and their specific toxicity to the heart (Blum and Carter, 1974; Davis and Davis, 1979; Carter, 1980; Wiernik, 1980; Olson et al., 1981; Young et al., 1981). The latter is dose-limiting and is a major restriction in the use of these drugs: cardiotoxicity is rare below cumulative doses of 450 mg m^2 but the likelihood of incidence of cardiotoxicity increases markedly above a cumulative dose of 550 mg m^2 for Adriamycin and 600 mg m^2 for daunomycin. Daunomycin has clinical uses in the induction of remission in acute lymphocytic leukaemia and in acute non-lymphocytic leukaemia, however, it is not the first-choice drug for the former and the possibility of dose-related cardiotoxicity generally precludes its continued use for maintenance of remission. By contrast, Adriamycin is used, in combination therapy to treat a wide spectrum of solid tumours. It has a better therapeutic index than daunomycin and is one of the most valuable of current antitumour drugs. Note that this difference in clinical usefulness results only from a minor difference in chemical structure. On the basis of these observations on toxicity and clinical utility (Carter, 1980; Young et al., 1981) we can surmise that the full potential of the anthracyclines has probably not yet been realised. Not surprisingly then, there has been a vigorous search for improved analogues, and this has spanned the 15 years since Adriamycin was isolated as a product of a mutant strain of the daunomycin-producing organism (Arcamone et al., 1969).

Molecular Aspects of Anti-cancer Drug Action, ed. Neidle & Waring
0333–315561/83/057–092

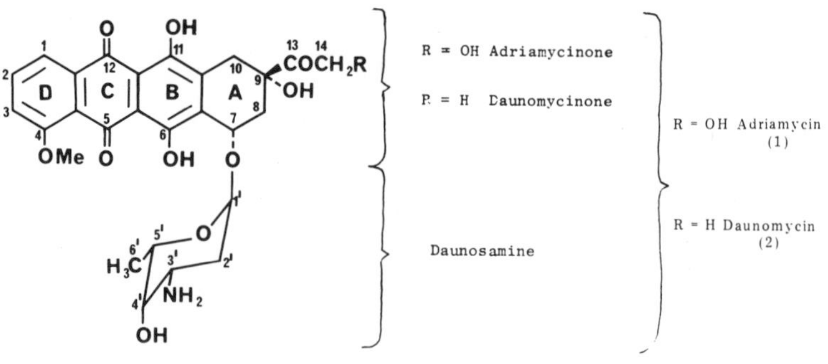

Most of the work on the development of anthracyclines has been of an empirical nature. One approach has been to search for novel 'natural' anthracyclines produced by newly-isolated microorganisms or by strains of anthracycline-producing organisms derived via exposure to some form of mutagenic treatment. A second approach has been to chemically modify known anthracyclines or to couple either a synthesised alternative aglycone to daunosamine, or a synthesised alternative sugar to a nature aglycone (all of these will be classed here as semisyntheses). The final approach has been to attempt to mimic the molecular biological actions of Adriamycin and daunomycin in simplified, totally synthetic compounds. Each of these approaches has led to important new analogues. This work has been well-reviewed previously (Arcamone, 1978; Brown, 1978; Strauss, 1978; Henry, 1979; Remers, 1979; Wiley, 1979; Crooke and Reich, 1980; Myers, 1980; Nettleton *et al.*, 1980; Arcamone, 1981; Crooke, 1981; Crooke *et al.*, 1981; Di Marco, 1981; Du Vernay, 1981) but the pace of research in this field is so rapid and the breadth of research so wide that a high frequency of appearance of reviews is called for. For example, this chapter contains the first full review of the development of substituted anthraquinones modelled on Adriamycin, yet already two members of this group are undergoing clinical trial.

Adriamycin and daunomycin have long been known to have an effect on nuclear nucleic acid, and this is regarded as one of their major cytotoxic effects, if not the major effect (for reviews see Brown, 1978; Arcamone, 1978, 1981). Consequently, most *in vitro* studies of analogues have involved examination of their interaction with DNA or their inhibition of cell-free nucleic acid synthesis. Most cell culture studies have considered uptake of drug, and/or cellular localisation of drug and/or effect on cellular nucleic acid synthesis. Some *in vivo* studies have given evidence of the effect on the nucleus. From consideration of all these studies there is a composite picture emerging of the relationship between structural changes in the anthracyclines and differences in molecular action. However, the natural anthracyclines are such an extensive group that, before considering these molecular actions of natural and semisynthetic anthracyclines, their biogenetic origins will be briefly outlined to clarify their structural interrelationships.

STRUCTURAL RELATIONSHIPS OF NATURAL ANTHRACYCLINES

Adriamycin and daunomycin are glycosides of the sugar daunosamine, and the aglycones adriamycinone and daunomycinone respectively. These aglycones are polyketides derived from a propionate starter unit and 9 malonate extender units *via* the postulated intermediate compound (3) in figure 1 (Casey *et al.*,

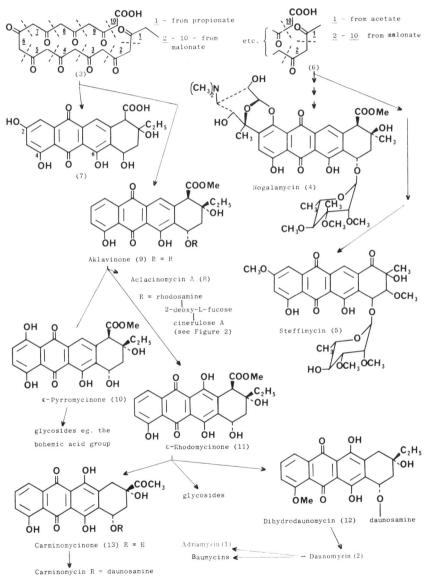

Figure 1 Biogenetic interrelationships of anthracyclines (Blumauerova *et al.*, 1979*a*; Yoshimoto *et al.*, 1980*a,b*)

1978; Shaw *et al.*, 1979). The fact that there is an atypical three-carbon (pro-pionate) starter unit suggests that compounds will also exist having the two-carbon (acetate) starter which is more usual for polyketides, and indeed the ring skeletons of anthracyclinones such as nogalarol and steffimycinone (the aglycones of nogalamycin (4) and steffimycins, for example steffimycin (5), respectively) have been shown to be formed in this manner and we can postulate the intermediate (6) (figure 1) (Wiley *et al.*, 1978). (Note that in nogalarol, the terminal acidic group of the polyketide precursor (6) is retained as the methyl ester; this is common with anthracyclines).

If we consider cyclisation of (3) (figure 1) to give the tetrahydronaphthacene-quinone ring system of anthracyclines, then the oxidation to the quinone structure will be facile and the simplest expected product would be (7) with oxygenation at the 2, 4 and 6 positions of the tricyclic anthraquinone unit (for simplicity, all compounds will be numbered in the manner used for Adriamycin (1) above). The methyl ester of (7) has in fact been isolated from a blocked mutant of an organism which produces aclacinomycin A (8) (Oki *et al.*, 1981a). Almost invariably however, the 2-oxygen function is lost (the only other known exception where it is retained is steffimycinone): hence the first product from cyclisation of intermediate (3) is normally the compound termed aklavinone (9). The current understanding of anthracycline biogenesis is shown in figure 1. Aklavinone can be glycosylated to give, for example, aclacinomycin A (8) (glycosylation occurs sequentially; Oki *et al.*, 1980), or can be further oxygenated at either the 1-position, to give ε-pyrromycinone (10), or at the 4-position to give ε-rhodomycinone (11). These may then be glycosylated sequentially such that anthracyclines with mono- di- or oligosaccharide chains are formed: generally the primary sugar in these chains is an amino-sugar, usually rhodosamine (figures 1 and 2). There can therefore be a multiplicity of anthracyclines produced by any one organism, the compounds differing in their aglycone and/or their sugar unit(s) (Oki *et al.*, 1979; Pandey and Toussaint, 1980).

Other variations occur; for example ε-rhodomycinone (11) can be converted to daunomycin (2) via dihydrodaunomycin (12) (McGuire, 1980; Yoshimoto *et al.*, 1980a): if 4-*O*-methylation does not occur then carminomycin is pro-duced, via carminomycinone (13), or if 14-hydroxylation additionally occurs then Adriamycin (1) is formed (Oki *et al.*, 1981b). Daunomycin can be further converted to the *N*-substituted compounds, the baumycins (Oki *et al.*, 1981c). Many permutations can therefore occur and it is not surprising to find that compounds such as 13-deoxydaunomycin (14) (Oki *et al.*, 1981c) and 11-deoxy-daunomycin (15) (Cassinelli *et al.*, 1980a) have been isolated. In some cases where the terminal carboxylic acid function is lost, 10-oxygenation occurs to give a 10-keto group (as in steffimycinone) or more frequently, a 10-hydroxy group for example γ-rhodomycinone (16). 7-Deoxy- compounds have been found and with these glycosidation will occur at the 10-position, for example in roseorubins A (17) and B (18) (Matsuzawa *et al.*, 1979). With 7,10-dihydroxy

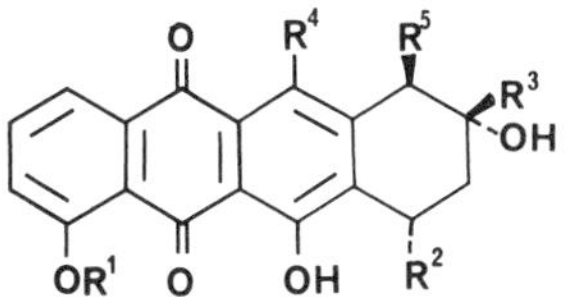

	R^1	R^2	R^3	R^4	R^5
13-Deoxydaunorubicin (14)	CH_3	0-daunosamine	CH_2CH_3	OH	H
11-Deoxydaunorubicin (15)	CH_3	0-daunosamine	$COCH_3$	H	H
γ-Rhodomycinone (16)	H	H	CH_2CH_3	OH	OH
Roseorubicin A (17)	H	H	CH_2CH_3	OH	0-(rhodosamine)$_2$-deoxy-L-fucose-(rhodinose)$_2$
Roseorubicin B (18)	H	H	CH_2CH_3	OH	0-(rhodosamine)$_2$

compounds, glycosidation can occur at the 7, or the 7 and 10 positions (Brockmann and Greve, 1975).

There is consequently a wide range of possible natural anthracyclines; it is only possible to review here the diversity of structures which occur; a full review of all natural anthracyclines is beyond the scope of this chapter, but has recently been covered by Remers (1979) and Arcamone (1981). Undoubtedly we will see a continuing discovery of novel members of the group. In addition to isolation of new compounds, attempts have been made to obtain new derivatives by feeding natural anthracyclines to microorganisms in the hope of obtaining novel transformations. Commonly however, reductive cleavage to 7-deoxyaglycones occurs (along with reduction to 13-dihydro derivatives in the case of Adriamycin, daunomycin and carminomycin) (Blumauerova *et al.*, 1979*b*; Komiyami *et al.*, 1979; Rueckert *et al.*, 1979). Perhaps greater success will be achieved by use of blocked mutants, for example feeding of a range of aglycones to an aclacino-mycin-producing organism which cannot produce aglycones has yielded novel trisaccharides containing the typical aclacinomycin A trisaccharide unit (figure 2) (Matsuzawa *et al.*, 1980; Oki *et al.*, 1980). This may be developed further in the future by hybridisation of *Streptomyces* spp (Ihn *et al.*, 1980; Schlegel and Fleck, 1980; Schlegel *et al.*, 1980).

Running parallel to studies on isolation of new anthracyclines, there has been an intense effort directed at development of improved semisynthetic analogues. Discussion of the molecular actions of these natural and semisynthetic anthra-cyclines will be grouped here according to the oxygenation pattern of the anthraquinone unit of the aglycone of the natural antibiotic. We have seen why, on biogenetic grounds, we can regard 4,6-dihydroxy compounds as the 'basic' anthracyclines: these will be discussed first. Reference will be made to the known molecular actions of Adriamycin and daunomycin (the most studied anthracyclines) where relevant.

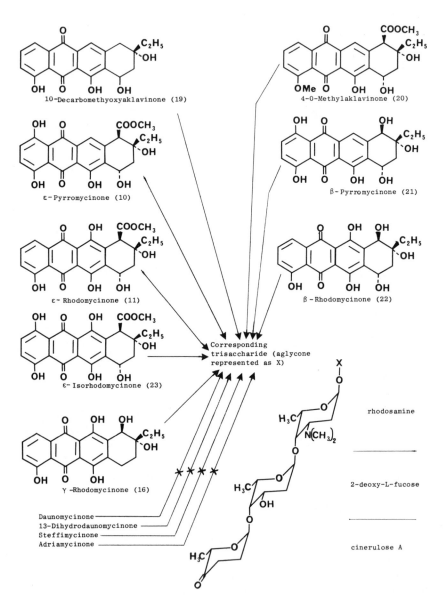

Figure 2 Derivation of novel natural anthracyclines using blocked mutant organisms (Matsuzawa *et al.*, 1980; Oki *et al.*, 1980)

MOLECULAR ACTIONS OF NATURAL AND SEMISYNTHETIC ANTHRA-CYCLINES

4,6-Dihydroxy compounds

We can regard the 'basic' anthracycline structure as that containing a 9-ethyl group (since the biogenetic starter unit is propionate) hence the base compound in this group is aklavinone (9). The main aklavinone glycosides are the aclacino-mycins which are a large group of related compounds isolated from *Strepto-myces galilaeus* M144-M1 (Oki *et al.*, 1977). Most contain the 4,6-dihydroxy system (although some are 1,4,6-trihydroxylated) and they differ in the sugar unit at C7. Sugars are added sequentially to give mono-, di- and tri-saccharide derivatives. In the latter the terminal sugar may undergo one or more oxidative steps to give for example, aclacinomycin A (8) (Yoshimoto *et al.*, 1979; Oki *et al.*, 1980). Of these compounds, aclacinomycin A (8) is the most active against L1210 leukaemia (Hori *et al.*, 1977). It is equivalent to, or slightly less active than, Adriamycin, but is less cardiotoxic (Hori *et al.*, 1977), as shown by electron microscopy of the myocardial cells of animals treated with the drug (Hori *et al.*, 1977; Wakabayashi *et al.*, 1980). In addition to its cytotoxic action, aclacinomycin A selectively depresses natural suppressor cells hence giving an enhancement of the body's own immune system; this occurs to a greater extent than with Adriamycin (Ishizuka *et al.*, 1981). Aclacinomycin A is therefore a promising alternative to it.

The uptake of aclacinomycin A into L1210 and P388 cells is more rapid than with Adriamycin and daunomycin (Egorin *et al.*, 1979*a*; Seeber *et al.*, 1980). This difference in rate of cell uptake is reflected in the fact that there is no difference in apparent rate of uptake in Adriamycin-resistant cells (Seeber, 1980; Seeber *et al.*, 1980) in contrast to Adriamycin itself where, due to increased extrusion from the cell, apparent uptake is reduced since uptake is too slow to compensate for increased efflux. There is conflicting evidence on the localisation of this drug. Assay of drug extracted from nuclei of treated cells suggested that 78% of the drug is associated with the nucleus compared to 87% for Adriamycin (Seeber *et al.*, 1980), whereas fluorescence microscopy suggests that aclacino-mycin is located mainly in the cytoplasm, and Adriamycin in the nucleus (Egorin *et al.*, 1979*a*). Note, however, that aclacinomycin A fluorescence is quenched by DNA (Du Vernay *et al.*, 1979*b*) so this may give misleading con-clusions in fluorescence microscopy. There is no doubt that aclacinomycin A has an effect on DNA function since in Chinese hamster ovary cells treated with low concentrations of drug, there is initially a slowing of the traverse through the G2 and M phases and eventually the G1 phase. At high concentrations there is also a slowing of S phase progression and cells accumulate in the G2 and M phases (Traganos *et al.*, 1981). Aclacinomycin A causes chromosome aberrations and swelling of the nuclei, and the nucleoli reduce in size (Shuin *et al.*, 1981). Consistent with this is the finding that the RNA content of the cells is reduced by up to 50% (Traganos *et al.*, 1981). The evidence suggests that aclacinomycin

A blocks cell replication by interfering with the synthesis of an RNA required for maturation of chromatin (Kajiwara *et al.*, 1979).

Whereas Adriamycin and daunomycin inhibit DNA and RNA synthesis to approximately the same degree, aclacinomycin A preferentially inhibits RNA synthesis (Oki, 1977). This is true for both *in vitro* and *in vivo* systems (Yamaki *et al.*, 1978). For example for the latter case, HeLa cells exposed to low concentrations of drug show up to 40% inhibition of uracil incorporation whilst showing no inhibition of thymidine incorporation although as concentration of drug is increased then the latter is also inhibited (Shuin *et al.*, 1981). In fact, the specificity is for nucleolar RNA synthesis. Aclacinomycin A inhibits this in Novikoff hepatoma cells at a 170-fold lower concentration than that required to inhibit DNA synthesis (Du Vernay *et al.*, 1979*a*). There may be some sequence selectivity with this drug, since it is effective when poly(dA·dT) is the template, whereas it is not active with poly(dI·dC) (Yamaki *et al.*, 1978).

The inhibitory effects of aclacinomycin A on nucleic acid synthesis can be reversed by addition of further DNA but not by increasing the concentration of enzyme (Yamaki *et al.*, 1978; Misumi *et al.*, 1979) suggesting that aclacinomycin A acts by binding to DNA as do Adriamycin and daunomycin (see the previous chapter). It is assumed that the aglycone intercalates into the DNA helix, and the trisaccharide chain then lies in either the minor or major groove. In accord with this postulated DNA binding mechanism, aclacinomycin A is found to undergo bathochromic and hypochromic shifts in its absorption spectrum in the visible region, and shows a reduction in fluorescence intensity on binding to DNA (Misumi *et al.*, 1979); the changes are typical of intercalation of anthracyclines (Plumbridge *et al.*, 1978; Plumbridge and Brown, 1978, 1979*a*). Equilibrium dialysis studies of the binding of ^{14}C-aclacinomycin A to calf thymus DNA gave values of $K = 1.2 \times 10^6 \, M^{-1}$ and $n = 0.17$ (from a Scatchard plot) (Misumi *et al.*, 1979) these values being typical of anthracyclines which intercalate (Plumbridge and Brown, 1979*b*): again poly(dA·dT) was preferred to poly(dI·dC) (Misumi *et al.*, 1979). Fluorescence titration yielded values of $K = 2.48 \times 10^6 \, M^{-1}$ and $n = 0.18$ for binding to calf thymus DNA (from a Scatchard plot), template specificity again being observed – poly(dA·dT) being preferred to poly(dG·dC) (Du Vernay *et al.*, 1979*b*). By contrast, the affinity constant for binding to denatured DNA is about two orders of magnitude lower ($K = 3.5 \times 10^4 \, M^{-1}$) suggesting that the double-helical structure is required (Misumi *et al.*, 1979); this is again consistent with intercalation as the mode of interaction. This is borne out further by studies on the thermal denaturation of DNA in presence of aclacinomycin A (Misumi *et al.*, 1979). Studies with covalently closed-circular PM-2 DNA show unwinding in the presence of the drug (Mong *et al.*, 1980).

Adriamycin and daunomycin are known to cause strand breaks in DNA (see Chapters 8 and 9). There is some debate over the relevance of these findings, since for example Adriamycin gives strand breaks in covalent closed circular DNA *in vitro* only in the presence of borohydride. Whatever the significance, this test can be used to compare aclacinomycin A with Adriamycin: again with the former no breaks are produced in the absence of borohydride. Breaks do occur in the presence of borohydride albeit at a lower incidence than with Adriamycin (Someya and Tanaka, 1979): this suggest an overall similarity in the actions of the two drugs. The effect of aclacinomycin is reduced in the presence of catalase, superoxide dismutase and free radical scavengers (Someya and Tanaka, 1979) so is undoubtedly mediated via reduction of drug, followed by reoxidation via formation of active oxygen species. This is consistent with the finding that in cell-free systems, both aclacinomycin A and Adriamycin are reduced to free radicals by NADPH cytochrome P-450 reductase (Bachur *et al.*, 1979). Since the DNA-nicking effect is weaker with aclacinomycin A than with Adriamycin, it is tempting to speculate that this effect has some relationship to cardiotoxicity.

We can conclude that the mechanism of action of aclacinomycin A shows many similarities to that of Adriamycin. Since aclacinomycin A is the most active compound of this group, most studies have been with it. To date the 10-decarboxymethyl-, the 7-*R* epimer of the 10-decarboxymethyl- and the 10-epimer derivatives of aclacinomycin A have all been shown to have reduced antitumour activity whereas the 4-*O*-methyl- and 6-*O*-methyl-derivatives retain activity but are less potent (Tanaka *et al.*, 1980; 1981). Loss of the terminal L-cinerulose sugar unit of aclacinomycin A produces a reduction in activity: removal of the terminal α-L-cinerulose-2-deoxy-L-fucose two-sugar unit, to give 1-deoxypyrromycin (24), gives a dramatic reduction in activity (Hori *et al.*, 1977). Other natural anthracyclines where the terminal sugar is modified are similar in activity to aclacinomycin A, but those not containing a rhodosamine unit or where the sugar is desmethylrhodosamine are inactive (Hori *et al.*, 1977). It appears that a rhodosamine-2-deoxysugar disaccharide unit at C7 is required for activity. Since Adriamycin contains only one sugar (daunosamine, *bis*desmethylrhodosamine) this begs the question; what would be the activity of a compound comprising daunomycinone and the trisaccharide unit of aclacinomycin A?

1,4,6-TRIHYDROXY COMPOUNDS

Two major groups of anthracyclines of this type are the 'operatic' anthracyclines which contain the ε-pyrromycinone (10) aglycone, and nogalamycin (4) and its derivatives based on the aglycone nogalarol.

'Operatic' anthracyclines

A mixture of anthracyclines isolated from *Actinosporangium* sp. C36145 was termed the bohemic acid complex (after La Bohème; with new anthracycline constituents being named after characters in this opera). Three components were already known compounds, namely pyrromycin (25) cinerubin A (26) and cinerubin B (27). Additionally the deoxyfucose and (deoxyfucose)$_2$ derivatives of pyrromycin were isolated and named as musettamycin (28) and marcellomycin (29) respectively (Nettleton *et al.*, 1980); rudolphomycin (30) with a modified terminal sugar (probably biosynthesised from marcellomycin) was also found. Preparative high-pressure liquid chromatography of marcellomycin revealed the minor constituents alcindoromycin (31) the mono-desmethyl-analogue of marcellomycin) collinemycin (32) and minimycin (33), the latter two being the 10-epimers of musettamycin and marcellomycin respectively (epimerisation probably occurring in solution) (Nettleton *et al.*, 1980). There is hence a range of known ε-pyrromycinone glycosides; additional to those mentioned above are the rhodirubins A and B (Kitamura *et al.*, 1977).

In 1978, Daskal *et al.*, showed that marcellomycin (29) and Adriamycin gave nucleolar lesions in Novikoff hepatoma cells in cell culture. With the latter there was evidence of fibrillar centre formation and redistribution of the granular component to give segregation of the nucleolus. This suggests that there is generalised inhibition of nucleolar function. By contrast, marcellomycin caused nucleolar segregation at low doses and at high doses, microspherules were formed in the fibrillar component suggesting a more specific effect on some aspect of nucleolar function. Marcellomycin did not cause formation of fibrillar centres. Hence there appears to be some difference in the effects of Adriamycin and marcellomycin on the nucleolus.

In line with the properties of aclacinomycin described above, it was found that marcellomycin and musettamycin inhibit RNA synthesis at a lower concentration than that required to inhibit DNA synthesis (Crooke *et al.*, 1978). The IC_{50} values for these compounds for DNA synthesis in Novikoff hepatoma cells

	R^1	R^2	R^3	R^4
1-Deoxypyrromycin (24)	H	$COOCH_3$	H	rhodosamine
ε-Pyrromycinone (10)	OH	$COOCH_3$	H	H
Pyrromycin (25)	OH	$COOCH_3$	H	rhodosamine
Cinerubin A (26)	OH	$COOCH_3$	H	rhodosamine-2-deoxy-L-fucose-cinerulose A
Cinerubin B (27)	OH	$COOCH_3$	H	rhodosamine-cinerulose B
Musettamycin (28)	OH	$COOCH_3$	H	rhodosamine-2-deoxy-L-fucose
Marcellomycin (29)	OH	$COOCH_3$	H	rhodosamine-(2-deoxy-L-fucose)$_2$
Rudolphomycin (30)	OH	$COOCH_3$	H	rhodosamine-2-deoxy-L-fucose-rednose
Alcindoromycin (31)	OH	$COOCH_3$	H	desmethylrhodosamine-(2-deoxy-L-fucose)$_2$
Collinemycin (32)	OH	H	$COOCH_3$	rhodosamine-2-deoxy-L-fucose
Minimycin (33)	OH	H	$COOCH_3$	rhodosamine-(2-deoxy-L-fucose)$_2$

were seven times higher than those for RNA synthesis compared to less than two times higher in the case of pyrromycin (Crooke *et al.*, 1978). This increase in selectivity against RNA synthesis as the number of sugars is increased is supported by the 10- and 7.5-fold higher activity of cinerubins A and B against RNA synthesis compared to DNA synthesis in an *in vitro* test system (Henry, 1976). It was already known that nucleolar RNA synthesis is more sensitive to Adriamycin than is extranucleolar RNA synthesis, so the effect of the operatic anthracyclines on nucleolar RNA synthesis was studied (Crooke *et al.*, 1978). The ratios of IC_{50} values for inhibition of DNA and nucleolar preribosomal RNA synthesis were 1.02, 0.93, 170, 240, 714 and 1,256 respectively for Adriamycin, pyrromycin, aclacinomycin A, rudolphomycin, musettamycin and marcellomycin (Crooke *et al.*, 1980; Du Vernay *et al.*, 1979a). Selectivity of inhibition of nucleolar preribosomal RNA synthesis therefore increases as the number of deoxyfucose units increases from 0 to 2. The addition of rednosamine to musettamycin (28) to give rudolphomycin (30) gives a reduction, however, so both the number of sugar residues *and* their structure are important.

There are thus two classes of anthracyclines. Those designated class I (Crooke *et al.*, 1978) include Adriamycin and pyrromycin (25), which require approximately equivalent concentrations to inhibit DNA and RNA synthesis. By contrast, class II anthracyclines require between 150- and 1,500-fold lower concentration to inhibit nucleolar preribosomal RNA synthesis than that required to inhibit DNA synthesis; aclacinomycin A and the di- and trisaccharide operatic anthracyclines fall in this group. Some further information has been derived from analysis of the 10-descarbomethoxy derivatives of marcellomycin and rudolphomycin (Du Vernay *et al.*, 1979a). The IC_{50} values for inhibition of nucleolar RNA synthesis were 14 nM for marcellomycin but 2.6 μM for its 10-descarbomethoxy analogues; and 300 nM for rudolphomycin yet 9 μM for its 10-descarbomethoxy analogue. There was not this disparity in the effects on inhibition of thymidine and uridine incorporation. This suggests the 10-substituent is important in determining the selectivity for nucleolar RNA synthesis and, since the 10-descarbomethoxy analogues are less cytotoxic, that this inhibition of nucleolar RNA synthesis is related to cytotoxic action.

There does not appear to be a difference in the overall mechanism of binding of these compounds to DNA compared with the binding of Adriamycin. Pyrromycin (25), musettamycin (28) and marcellomycin (29) all give large hypochromic and bathochromic spectral shifts with an isosbestic point on binding to DNA (Morris and Brown, 1978). Their fluorescence is quenched in the presence of DNA, there is an increase in polarisation of fluorescence and their phenolic groups cannot be ionised when drug is bound to DNA (Morris and Brown, 1978). All these effects are typical of anthracycline intercalation into DNA. Using a fluorescence titration method and data analysis by Scatchard plot, Du Vernay *et al.* (1979b,c) found K and n values for binding to calf thymus DNA for Adriamycin of 3.67×10^6 M^{-1} and 0.15; for pyrromycin of 0.98×10^6 M^{-1} and 0.38; for musettamycin of 2.21×10^6 M^{-1} and 0.20; for marcellomycin of 5.03×10^6 M^{-1} and 0.19; for 10-descarboxymethylmarcellomycin of 2.14×10^6 M^{-1} and 0.224; for rudolphomycin of 1.98×10^6 M^{-1} and 0.22; for 10-descarboxymethylrudolphomycin of 1.42×10^6 M^{-1} and 0.321. There is thus a correlation for the class II anthracyclines, between the affinity of binding and the relative selectivity of effect against nucleolar RNA synthesis. There is no sequence-specificity shown by class II anthracyclines up to a level of 100% A+T content when affinity of binding increases. By contrast, Adriamycin binding affinity decreases as G+C content decreases (Du Vernay *et al.*, 1979*b*). Du Vernay *et al.* (1979*b*) have suggested that it may bind to spacer or promoter regions rich in A and T residues.

There are therefore some pointers emerging on the structural basis for differences between the cellular effects of anthracyclines. Whilst showing no difference in binding to naked DNA in an *in vitro* system and whilst being as active as class I anthracyclines against DNA synthesis in cell culture, class II anthracyclines are two or more orders of magnitude more active against nucleolar RNA synthesis. The major determinant of this activity is the possession of a rhodosamine-deoxyfucose substituent at C7, a further deoxy sugar increasing the selectivity. The 10-COOCH$_3$ is required for optimal activity and the configuration at C10 must be the 'natural' one (Nettleton *et al.*, 1980). The aglycone is of relatively less importance. Comparison of pyrromycin (25) with its 1-deoxy analogue aklavine, or cinerubins A and B with aclacinomycins A and B (compounds containing the same sugar units but different aglycones) shows that whilst potency is marginally lower for 4,6-dihydroxy compounds than with 1,4,6-trihydroxy compounds, they are more active at the optimal dose. Marcellomycin (29) is the compound which so far shows greatest selectivity for nucleolar RNA synthesis; it would be informative to evaluate its 1-hydroxy analogue in this test system.

Nogalamycin and its derivatives

Nogalamycin (4) is unique among anthracyclines in that the amino-sugar has its glycosidic linkage with the 1- (not the 7-) hydroxyl group, and there is a C-C linkage to C2. It was isolated from *Streptomyces nogalater* var. nogalater in the mid 1960s and the structure was published in 1977 (Wiley *et al.*, 1977*a*). The stereochemistry is not yet fully resolved. It is however proven that the C7 and C9 oxygens project on opposite sides of the cyclohexene ring, that the amino-sugar projects on the same face as the nogalose group and that the C10 substituent is on the opposite face. The most likely of the two possible structures is that shown as (4) (C7*S*, C9*S*, C10*R*, aminosugar α-L-glucopyranose) (Wiley, 1979). More definitive X-ray crystallography studies are needed to resolve the stereochemistry before models of any nogalamycin/DNA interaction can be proposed. Although nogalamycin was shown to have antitumour activity in both *in vitro* and *in vivo* systems, it is poorly soluble and has a poor therapeutic index (it gives for example renal toxicity, pulmonary thrombosis and venous occlusion) hence it was not developed as an antitumour agent.

Nogalamycin was shown to inhibit nucleic acid synthesis and it is significant in view of the previous discussion, that it was shown to inhibit nucleolar RNA synthesis to a higher degree than chromosomal RNA synthesis (Arrighi, 1967). It is a more potent inhibitor of sarcoma 180 RNA synthesis than is daunomycin: both compounds inhibit protein synthesis (Chowdhury and Neogy, 1980). A number of studies have shown that the aromatic ring system intercalates into DNA (Kersten *et al.*, 1966; Marciani *et al.*, 1977; Brown, 1978; Plumbridge and Brown, 1979*c*). On binding to DNA, the drug undergoes hypochromic and bathochromic shifts with an isosbestic point, it cannot be reduced polarographically and dichroism studies showed that the plane of the ring is perpendicular to the helix axis. In presence of drug, DNA solutions show increased viscosity, the DNA is more stable to thermal denaturation and it shows a decreased buoyant density and decreased sedimentation rate; kinks in DNA are removed, indicating stretching, and supercoiled DNA is unwound. The binding parameters determined by a spectrophotometric titration method ($K = 1.432 \times 10^6 \text{ M}^{-1}$, $n = 0.14$; Plumbridge and Brown, 1979*c*) are similar to those for Adriamycin and daunomycin. It is not clear whether nogalamycin is base-specific. Some studies suggest that affinity for DNA increases as A + T content increases (for example Bhuyan and Smith, 1965; Marciano *et al.*, 1977), others that affinity increases as G + C content increases (Kersten *et al.*, 1966) and yet others suggest that binding is not affected by base composition (Chowdhury *et al.*, 1978). A comparison of binding of nogalamycin to DNA and to chromatin by a spectral titration method (Chatterji *et al.*, 1980) showed that both the affinity of binding and the number of binding sites were reduced with chromatin compared to DNA (DNA, $K = 3.55 \times 10^6$, $n = 0.15$: chromatin, $K = 1.48 \times 10^6$, $n = 0.07$). The number of binding sites were also reduced in presence of f_1 histone, which indicates that blocking of the major groove leads to reduced binding. Although the evidence clearly indicates that the chromophore intercalates into DNA, model-building studies (Brown and Neidle, unpublished data) suggest that neither of the two possible stereochemically different forms of nogalamycin can intercalate with the long axis almost parallel to the base pairs (as in the classical anthracycline intercalation model; see chapter 2). However it can be accommodated if it 'spears' the helix.

The clinical utility of other anthracyclines prompted a re-evaluation of nogalamycin and its derivatives (table 1). Some derivatives had in fact undergone some evaluation during the initial studies on nogalamycin. Nogalarol (34) and nogalarene (35) had been isolated as acid degradation products; 7-*O*-methylnogalarol (36) and 7-deoxynogalarol (37) were prepared during structure elucidation studies (Wiley *et al.*, 1977*b*). Nogalarene is only a weak inhibitor of nucleic acid synthesis, and the other derivatives are all less active than nogalamycin as inhibitors of nucleic acid synthesis: the most active of the derivatives is 7-*O*-methylnogalarol (36) (Bhuyan & Reusser, 1970). All inhibit RNA synthesis at about 50% the concentration required for inhibition of DNA synthesis. The poor activity of nogalarene (35) is consistent with it giving none of the effects typical

Table 1 Evaluation of DNA-binding of nogalamycin and analogues.

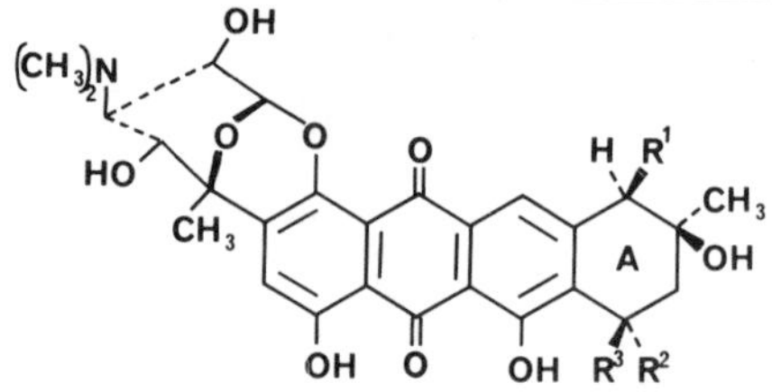

	R^1	R^2	R^3
Nogalamycin (4)	COOCH$_3$	nogalose	H
dis-Nogalarol (34)	COOCH$_3$	OH	H
Nogalarene (35) (7,9-deoxy, ring A aromatic)			
7-dis-O-Methylnogalarol (36)	COOCH$_3$	OCH$_3$	H
7-Deoxynogalarol (37)	COOCH$_3$	H	H
dis-Nogamycin (38)	H	nogalose	H
7-dis-O-Methylnogarol (39)	H	OCH$_3$	H
7-con-O-Methylnogalarol (40)	COOCH$_3$	H	OCH$_3$
7-con-O-Methylnogarol (41)	H	H	OCH$_3$

of intercalation (no shift in λ_{max}, no decrease in absorbance and no lack of spectral shift on basification, in the presence of DNA) (Plumbridge & Brown, 1979c). In contrast, 7-deoxynogalarol (37) was shown to undergo a 8 nm shift in λ_{max} and 25% decrease in absorbance on binding to DNA (compared to a 13 nm shift and 28% decrease with nogalamycin) and the DNA is stabilised to thermal denaturation by 6.8°C (drug to DNA ratio of 1:10, 0.003M Tris, 0.018M NaCl; nogalamycin gives an increase in Tm of DNA of 17.2°C under these conditions). (Plumbridge and Brown, 1979c). The binding constant is less than 10^5 M^{-1}. Any model for binding of nogalamycin must therefore allow for a dramatic decrease in affinity of binding on removal of the nogalose residue. Nogalamycin (4) must have a reinforced intercalative binding with close contact of the sugars within the DNA grooves.

A major stimulus to research on nogalamycin analogues was the finding that the 10-descarboxymethyl derivative (38) (which is formed by ready decarboxylation of nogalamycinic acid on mild base hydrolysis) has improved *in vivo* activity with an increase in life span of 67%; compound (39), the 10-descarboxymethyl derivative of 7-O-methylnogalarol, was even more active with an increase in life span of 155% compared to 40% for nogalamycin (Wiley *et al.*, 1977b). Here we need to clarify the trivial nomenclature used in the literature. Compounds lacking the nogalose residue are termed *nogalarols* those lacking the nogalose and carboxymethyl residues are termed *nogarols*: compounds with the

7 and 9 oxygens projecting on opposite sides of the cyclohexene ring are termed *dis* compounds, and the 7-epimers are termed *con* compounds. The aforementioned (36) is therefore 7-*dis*-*O*-methylnogalarol. During preparation of this compound by methanolysis some of the 7-*con*-*O*-methylnogalarol (40) is also formed. Similarly methanolysis of the 10-descarboxymethylnogalamycin (38) (termed *dis*-nogamycin; the -*al*- of nogalamycin is deleted since it is the 10-descarboxymethyl compound) gives both 7-*dis*-*O*-methylnogarol (39) and 7-*con*-*O*-methylnogarol (41) (7-OMEN). 7-*O*-Alkylnoga*lar*ols have higher activity against P388 leukaemia than 7-*O*-alkylnoga*lar*ols, and the 7-*con* epimers are more active than the 7-*dis* epimers (Wiley 1979). The most active compound is 7-OMEN which is equivalent in activity to Adriamycin, but has the advantages of being orally active and possibly less cardiotoxic. It may be relevant that neither aclacinomycin A nor 7-OMEN reduces cardiac guanyl cyclase activity in contrast to Adriamycin (Levey *et al.*, 1980). Paradoxically, comparison of the effects on DNA and RNA synthesis show 7-OMEN to be a poor inhibitor (Li *et al.*, 1979). Yet the cytotoxic effect of 7-OMEN against L1210 leukaemia is greater than that of nogalamycin (Li *et al.*, 1979). This was confirmed by examination of the cell kill kinetics for Chinese hamster ovary, L1210 and B16 melanoma cells (Bhuyan *et al.*, 1981*a*). Uptake into these cells was greater and more rapid than for Adriamycin (Bhuyan *et al.*, 1981*b*) but based on intracellular concentration, the latter is more lethal. Like aclacinomycin, musettamycin and marcellomycin, 7-OMEN was found to be localised in the cytoplasm of L1210 cultured cells (using fluorescence microscopy). All the other nogalamycin compounds evaluated (nogalamycin, *dis*-nogamycin, and 7-*dis*-*O*-methylnogarol) were found to be located in the cytoplasm (Egorin *et al.*, 1979*b*, 1980*a*). The different compounds have differing effects on the cell cycle (Bhuyan *et al.*, 1980); for example nogalamycin is most effective in the S phase, 7-*con*-*O*-methylnogalarol is equally active in all phases, and 7-OMEN most effective in the early G1, S and G2 phases.

Results from studies on DNA-binding of nogalamycin analogues parallel the effects on DNA and RNA synthesis rather than the cytotoxic effects. For example whilst nogalamycin undergoes marked spectral shifts in the presence of DNA, 7-OMEN shows no marked shift in λ_{max} and no isosbestic point on increasing the concentration of DNA, also the phenolic groups can be ionised at drug/DNA ratios at which the phenolic groups of nogalamycin are not ionisable due to sequestration within the DNA double helix (Hardman and Brown, unpublished results).

Results of DNA-binding experiments indicate that affinity for DNA reduces dramatically on removal of the nogalose residue (as discussed above) and even further reduces on loss of the 10-COOCH$_3$ group. The *dis* epimers have a higher affinity for DNA than the *con* epimers and there is some evidence that compounds containing nogalose are A+T specific (Richardson *et al.*, 1981). Scatchard plots of results from spectral titration were 'anomalous and curved' (Krueger *et al.*, 1981), giving binding parameters (K and n) as follows: nogalamycin (4),

$7 \times 10^6 \, \mathrm{M}^{-1}, 0.17; dis$-nogamycin (38), $2 \times 10^6 \, \mathrm{M}^{-1}, 0.20; dis\text{-}O$-methylnogalarol (36), $1 \times 10^6 \, \mathrm{M}^{-1}, 0.18; dis\text{-}O$-methylnogarol (39), $0.6 \times 10^6 \, \mathrm{M}^{-1}, 0.2; 7\text{-}con\text{-}O$-methylnogarol (7-OMEN, 41), $0.7 \times 10^6 \, \mathrm{M}^{-1}, 0.17$. Hence DNA-binding activity parallels cytotoxicity except for the *7-con-O*-alkyl compounds. However, 7-OMEN is well taken up into cells, and a displacement assay shows that there is some interaction of 7-OMEN with DNA (Richardson *et al.*, 1981). Furthermore all compounds tested by pulse radiolysis (nogalamycin, *dis*-nogamycin, *7-dis-O*-methylnogalarol, *7-dis-O*-methylnogarol and the *con*-epimers of the latter two compounds) showed lower rate constants for interaction with e_{aq}^- in the presence of DNA than in its absence (Krueger *et al.*, 1981). The significance of this is questionable, however, since this effect was seen even at drug:DNA ratios of > 1. Perhaps more significant is the finding that single-strand breaks in DNA are more abundant with 7-OMEN than with its *dis* epimer and no damage is detectable with nogalamycin (Petzold *et al.*, 1979).

2,4,6-Trihydroxy compounds

Steffimycin (5) was isolated from *Streptomyces steffiburgensis* in 1967 (Bergy and Reusser, 1967) and its 4'-*O*-methyl analogue (termed steffimycin B) from *S. elgretius* in 1974 (Brodasky and Reusser, 1974). Both compounds were found to inhibit nucleic acid synthesis *in vitro* but whereas steffimycin gives 50% inhibition of RNA synthesis (*Escherichia coli* RNA polymerase) with salmon sperm DNA template at 0.4 μM, steffimycin B has no effect even at 100 μM (it does however inhibit if poly (dA·dT) is a template). By contrast steffimycin only inhibits DNA synthesis 15–28% at 350 μM whereas steffimycin B causes 60–70% inhibition at 20 μM (Reusser, 1967, 1975). These differences are not reflected in their DNA binding properties: both compounds give bathochromic shifts in λ_{max}, with an isosbestic point on binding to DNA and stabilise the DNA helix to thermal denaturation (Reusser, 1967, 1975; Dall'Acqua *et al.*, 1979). There is a slightly greater stabilisation with steffimycin B and this is reflected in the binding parameters derived by calculation of free concentration and fraction bound from a spectral titration method, followed by analysis by the McGhee and Von Hippel method (steffimycin, $K = 7.7 \times 10^4 \mathrm{M}^{-1}$, $n = 0.066$; steffimycin B, $K = 9.03 \times 10^4 \, \mathrm{M}^{-1}$, $n = 0.0981$). Flow dichroism studies showed that the chromophore lies parallel to the plane of the base pairs (Dall'Acqua *et al.*, 1979). Taken as a whole the data indicate an intercalative mode of binding. Studies with synthetic nucleotides (Dall'Acqua *et al.*, 1979) suggest that drug has a preference for A-T sequences. Taken in conjunction with the results from the nucleic acid synthesis inhibition studies we can conjecture that perhaps base sequence specificity (as opposed to base pair specificity) may be an important determinant in the effects of these (and other?) anthracyclines on nucleic acid synthesis.

4,6,11-Trihydroxy compounds

The 'fundamental' anthracyclines in this group are those based on ϵ-rhodomycinone (11): however the most significant are Adriamycin and daunomycin, and their analogues and derivatives.

ϵ-Rhodomycinone derivatives
Glycosides containing ϵ-rhodomycinone were studied by Brockmann's group in the 1950's during structure elucidation studies on anthracyclines (see Remers, 1979; Arcamone, 1981). None came into clinical use, but due to the utility of Adriamycin, the duanosamine glycoside (42) of ϵ-rhodomycinone has been prepared (Smith *et al.*, 1978); it stabilises DNA to thermal denaturation to a lesser extent than Adriamycin ($\Delta T_m = 7.6\,^{\circ}C$ compared to $13.4\,^{\circ}C$). It is only a quarter as potent as Adriamycin as an inhibitor of DNA and RNA synthesis in cultured L1210 cells. Compound (42) was also prepared by Essery and Doyle (1980) as was its 10-epimer (43) and daunosaminylpyrromycin (44): the latter has lower experimental antitumour activity than pyrromycin (26).

Adriamycin and daunomycin analogues and derivatives
Here, reference will be made principally to work since 1978, (for work previous to this see for example, Brown, 1978).

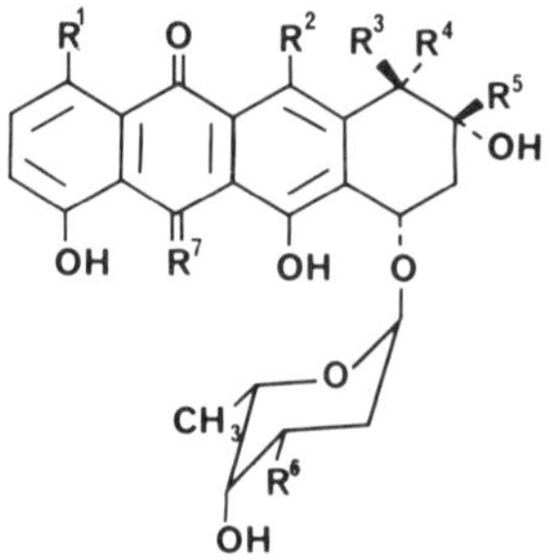

	R^1	R^2	R^3	R^4	R^5	R^6	R^7
(42)	H	OH	COOCH$_3$	H	C$_2$H$_5$	NH$_2$	O
(43)	H	OH	H	COOCH$_3$	C$_2$H$_5$	NH$_2$	O
(11)	OH	H	COOCH$_3$	H	C$_2$H$_5$	NH$_2$	O
(45) AD32	H	OH	H	H	COCH$_2$OCO(CH$_2$)$_4$CH$_3$	NHCOCF$_3$	O
(46)	H	OH	H	H	COCH$_3$	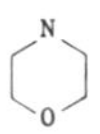 N (morpholino)	O
5-Iminoadriamycin (50)	H	OH	H	H	COCH$_2$OH	NH$_2$	N
Rubidazone (51)	H	OH	H	H	C(CH$_3$)=NNHCOC$_6$H$_5$	NH$_2$	O

Modified sugars. One of the earliest noted inconsistencies in molecular actions
of anthracyclines was that AD32 (the *N*-trifluoroacetyl-14-valerate derivative of
Adriamycin) (45) does not bind to DNA and is localised mainly (75%) in the
cytoplasm yet still is cytotoxic, albeit ten times less potent than its parent. It is
more rapidly taken up into cells than is Adriamycin (Seeber *et al.*, 1980); despite
this, it has a less damaging effect on macrophages (Facchinetti *et al.*, 1978) so
may have a lesser effect on the immune system than its parent. It is also less
cardiotoxic. One drawback, its lack of solubility, has recently been overcome by
Israel *et al.* (1981) with the development of a *N*-trifluoroacetyl-14-hemiadipate
analogue. Lack of binding to DNA (evidenced by circular dichroism, thermal
denaturation and exhaustive dialysis of the DNA-drug complex), and lack of
inhibition of DNA polymerases and reverse transcriptase yet retention of cytotoxic activity has also been shown for the other *N*-acyl compounds, *N*-octanoyl,
N-dodecanoyl and *N*-isonicotinoyl derivatives of daunomycin (Aszalos *et al.*,
1979). Neither AD32 nor these *N*-acyl derivatives are metabolised back to the
parent amine in significant amounts (Israel *et al.*, 1978; Aszalos *et al.*, 1979) so
activity is probably not mediated via the parent drug.

Despite its lack of intercalative binding to DNA, AD32 blocks the cell cycle
in the G2 phase, causes chromosome aberrations and inhibits cellular nucleic
acid synthesis, as does Adriamycin (Krishan *et al.*, 1981), though it is less potent
than Adriamycin. It causes single strand breaks in DNA which are associated
with protein cross-links (Chapter 4) (Kanter and Schwartz, 1979; Schwartz and
Kanter, 1979; Brox *et al.*, 1980; Gowans *et al.*, 1980; Potmesil *et al.*, 1981);
the protein may be a topoisomerase. This effect on DNA could explain the
ability of AD32 to inhibit synthesis of nuclear ribosomal and heterogenous RNA
(Wilson *et al.*, 1980) in L1210 cells. Both AD32 and Adriamycin inhibit chicken
myeloblastosis RNA polymerase II (Chuang *et al.*, 1980). The fact that this
enzyme uses denatured DNA could be used as evidence to support a hypothesis
that it is not the intercalative interaction of anthracyclines which is biologically
important. One interesting observation is that although AD32 inhibits thymidine
incorporation by cells in cell culture, if cells are preloaded at 0°C with thymidine
(when DNA synthesis does not occur), then on raising the temperature to 37°C,
thymidine incorporation is not inhibited (Chuang *et al.*, 1981) suggesting inhibition of transport rather than incorporation. AD32 (and by inference other
N-acyl compounds) thus does affect DNA structure although it does not intercalate into DNA; the actual biologically significant action is not yet clear.

N-amino acid and *N*-peptide derivatives can be regarded as a special class of
acyl derivatives. These have been prepared as candidate drugs and as potential
prodrugs to generate the parent on lysosomal cleavage. These derivatives are
more lipophilic and have a lower pK_a than the parent compounds but are less
rapidly taken up into cells (Baurain *et al.*, 1980, Masquelier *et al.*, 1980), the
rate of uptake varying, with L-leucyl derivatives being among the most rapidly
taken up.

These derivatives bind to DNA less avidly than the parent compounds, spectral titration showing that affinity is decreased about threefold for an amino acid and about tenfold for a dipeptide derivative (Masquelier *et al.*, 1980), and the lower binding is reflected in a decreased degree of fluorescence quenching on addition of DNA compared to the parent compounds (Levin and Sela, 1980) (in the latter study the basic lysyl-daunomycin showed the strongest interaction with DNA). Consistent with this lowered DNA binding they only poorly stain the nucleus (Levin and Sela, 1981) and only give a low induction of prophage λ – a screen for DNA damage (Anderson *et al.*, 1980a). However, lysosomal cleavage does occur, being particularly effective for L-leucyl-derivatives. For example L-leucyl-daunomycin is 75% metabolised to daunomycin in four hours and totally cleared within 24 hours (Masquelier *et al.*, 1980). Hence the compounds are cytotoxic, acting as prodrugs; the L-leucyl derivatives were found to be less cardiotoxic and haematotoxic than their parents (Jaenke *et al.*, 1980). A conflicting report (Sela and Levin, 1981) showed that the L-leucyl derivatives are not demonstrably more effective than the parent compounds in experimental tumour systems but basic amino acid derivatives were more effective, optimum activity being shown by the diaminobutyryl derivative.

Whilst acylation results in reduced affinity for DNA, N-alkylation gives retention of this activity, possibly due to the retention of basicity at the nitrogen atom (Tong *et al.*, 1979a, Acton and Mosher, 1981; Henry *et al.*, 1981). For example N,N-diethyldaunomycin shows an ED_{50} for inhibition of DNA synthesis in L1210 cells of 0.67 μM, and a value of 0.13 μM for RNA synthesis inhibition (0.66 μM and 0.33 μM respectively for daunomycin) and produces a similar increase in DNA melting temperature compared to daunomycin. Increasing the alkyl chain length gives a reduction in affinity. Interestingly, two compounds (N,N-dimethyldaunomycin, and N,N-pentamethyleneadriamycin) show properties resembling Class II anthracyclines in that the ratios of ED_{50} DNA synthesis to ED_{50} RNA synthesis are between 10 and 20 compared to 2 to 3 for the parent compounds. As would be expected from their DNA-binding properties, the N-dimethyl derivatives have been shown to be localised in cell nuclei (Egorin *et al.*, 1980b). Gordon and Kashiware (1981) showed that these N-alkyl derivatives cause DNA strand scission and this was found to correlate with oxygen consumption in a NADPH fortified rat liver microsome system: for example N,N-dimethyldaunomycin and N,N-dimethyladriamycin have 80–100% of the activity of Adriamycin, N-cyclohexyldaunomycin 20–70%, and N-benzyldaunomycin and N,N-dibenzyldaunomycin <25%. Whereas N-benzyldaunomycin is presumed to intercalate (ΔT_{m} = 10.2°C compared to 11.2° for daunomycin), the N,N-dibenzyldaunomycin derivative does not (ΔT_{m} = 1.35°C). The same trend is found in the Adriamycin series. Yet these compounds show good cytotoxic properties. There is an apparent lack of correlation between antitumour effect and DNA intercalating ability with 3′-deamino-3′-(3-morpholinoyl)-daunomycin (46) which although highly potent against P388 produces half the ΔT_m of daunomycin and shows a lower frequency of strand scission of covalently closed circular DNA than does daunomycin (Acton and Mosher, 1981).

There is not an absolute necessity for an amino group at the 3'-position: 3'-deamino-3'-hydroxy and 3'-deamino-3'-acetoxy-4'-O-acetyl analogues are active in the P388 system though much less potent than their parent (Fuchs *et al.*, 1979*a*; Horton and Turner, 1979; Horton *et al.*, 1981; El Khadem and Swartz, 1981). If the 3'-amino group is present it should preferably be in the 'natural' configuration: epimerisation gives inactivity in *in vivo* systems (Fuchs *et al.*, 1979*b*) and in the aforementioned prophage λ system (Anderson *et al.*, 1980*b*) and binding to DNA is weaker. Scatchard plots of data from fluorescence quenching experiments (Zunino *et al.*, 1977) give $K = 0.86 \times 10^6 \, M^{-1}$ and $n = 0.215$ for 3'-*epi*-daunomycin compared to $K = 3.3 \times 10^6 \, M^{-1}$ and $n = 0.176$ for daunomycin. This reduction is too great to be accounted for by the change in pK_a on epimerisation at the 3' position. Equilibrium dialysis studies (Arlandini *et al.*, 1980) confirm this reduced affinity.

Modifications at other positions of the daunosamine units have been examined. Changes can be made at the 4'-position without loss of activity; for example, 4'-O-methyl-derivatives are equiactive compared to the parent compounds (Cassinelli *et al.*, 1979) as are the 4'-O-tetrahydropyranyl compounds (Umezawa *et al.*, 1979). The equilibrium dialysis study mentioned above (Arlandini *et al.*, 1980) gave binding constants for 4'-O-methyladriamycin and for 4'-O-methyl-daunomycin comparable to that for Adriamycin. 4'-*epi* compounds retain activity in the prophage λ system (Anderson *et al.*, 1980*b*) and the fluorescence titration study quoted above for the 3'-*epi* compound (Zunino *et al.*, 1977) gave values of $K = 2.0 \times 10^6 \, M^{-1}$ and $n = 0.185$. This apparent slight reduction in binding strength compared to daunomycin can be accounted for by its slightly lower pK_a, hence resulting in a decrease in per cent ionisation at the pH used, rather than a decrease in binding affinity (DiMarco *et al.*, 1977). This equivalent binding also pertains for Adriamycin and its 4'-*epi*-derivative. 2'-Hydroxy derivatives are equivalent to the parent compounds in DNA binding ability. This does not apply to 6'-hydroxy-derivatives which have a reduced affinity over and above that due merely to differences in pK_a (Di Marco *et al.*, 1977). Deletion of the 4'-hydroxy group also leads to a reduction in affinity for DNA (Plumbridge and Brown, 1979*b*).

There has been little reported work on derivatives at C7 other than hexose glycosides, undoubtedly due to the poor activity of previously-reported compounds (for example, see Brown, 1978). A study of the β-alanine ester (47), the 2-aminoethylether (48) and the 2-aminoethylthioether (49) of daunomycinone

(47) R = $OCOCH_2CH_2NH_2$

(48) R = $OCH_2CH_2NH_2$

(49) R = $SCH_2CH_2NH_2$

showed that these stabilised the DNA helix (ΔT_m values of 1.0°C, 8.2°C and 6.3°C respectively compared to 11.2°C for daunomycin under the same conditions) and that they inhibited DNA and RNA synthesis in cultured L1210 cells. The ether (48) was as active as daunomycin (Acton *et al.*, 1979). Hence DNA binding activity is reduced but not abolished, suggesting further studies may be worthwhile. This is substantiated by the finding that esters with 3-aminocyclo-hexane-carboxylic acid show *in vivo* activity against L1210 leukaemia (Londos-Gagliardi *et al.*, 1980), and affinity constants for DNA binding equivalent to those for daunomycin using an electrochemical method (Molinier-Jumel *et al.*, 1978).

Modified substitution in the tetracyclic ring system. Most modifications have been at C1-4 and C6-11, and only one alteration of the quinone unit has been reported to date. This produced the compounds 5-iminoadriamycin (50) and 5-iminodaunomycin (Tong *et al.*, 1979*b*; Acton and Tong, 1981). The latter is as active as daunomycin against P388 leukaemia cells but is slightly less potent. Surprisingly 5-iminoadriamycin whilst as active as its parent, is only one sixth as potent. All four compounds are equivalent in their effects on DNA and RNA synthesis in cultured L1210 cells but the 5-imino derivatives stabilise the DNA double helix less than the parents (ΔT_m values of 6.25, 6.9, 11.2 and 13.4°C respectively for 5-iminodaunomycin, 5-iminoadriamycin, daunomycin and Adriamycin). Perhaps a significant feature is that the 5-imino compounds give rise to a much lower oxygen consumption by liver microsomes than do the parents (Acton and Tong, 1981) in line with the fact that they are more difficult to reduce polarographically and to reoxidise, leading to a lower nicking of DNA in the presence of borohydride than with the parents (Lown *et al.*, 1979*a*). This might well be related to their reported lower mutagenicity and cardiotoxicity. Comparison of the effects of Adriamycin and its 5-iminoderivative on heart cells of rats treated intravenously with drug (Peters *et al.*, 1980) showed that the former gave irreversible nucleolar segregation into fibrillar and granular sections, wheras there were only slight effects with the latter.

Other reported modifications of the dihydroxy-quinone system have been methylation or deletion of one or more hydroxyl functions. 6,11-di-*O*-Methyl-daunomycin has only a weak, if any, stabilising effect on DNA; with a ΔT_m = 0.6°C and a DNA-binding parameter a hundred-fold less than that of dauno-mycin (Zunino *et al.*, 1977; 1979). Monomethylation at the 6-*O*- or 11-*O*-position gives a marginally less drastic reduction in binding affinity and in addition cell uptake is reduced (Zunino *et al.*, 1979). Nevertheless 6-*O*-methyl-daunomycin shows activity *in vivo* against L1210 (Forenza *et al.*, 1981). A similar drastic loss of DNA binding ability is seen on 11-*O*-methylation of carminomycin (the 4-*O*-demethyl analogues of daunomycin). This compound has a much reduced antitumour activity (Essery and Doyle, 1979). It shows only weak inhibition of thymidine incorporation, and produces a lower degree of fluorescence quenching than does carminomycin on addition of DNA. A

> 5 fold concentration is required to achieve the same degree of unwinding of PM2 DNA that carminomycin produces, which in turn has weaker DNA binding ability than Adriamycin (Du Vernay *et al.*, 1980). Nevertheless, this 11-ether and the 6-*O*-methyl isomer are active against P388 cells (Masi *et al.*, 1980). The effects of 6- and 11-*O*-methylation on DNA-binding are probably steric rather than electronic since 11-deoxy compounds show antitumour activity (Cassinelli *et al.*, 1980a,b; Tatsuta and Takeuchi, 1980). The 4-demethoxy-11-deoxy-analogues of Adriamycin and daunomycin show anti-tumour activity, cyto-toxicity to cultured L1210 cells and inhibition of DNA and RNA synthesis, comparable to the parent compounds (Umezawa *et al.*, 1980). NMR studies of the interaction of 11-deoxydaunomycin (15) with poly (dA · dT) have shown that the chromophore intercalates into the polynucleotide helix (Patel *et al.*, 1981; see Chapter 2). Whilst discussion the 6 and 11 hydroxyl groups mention should be made of quelamycin, an orally-absorbed 3:1 ferric chelate of Adria-mycin which can pass the blood-brain barrier; it probably exists in the plasma for several hours after administration (Gosalvez *et al.*, 1978; May *et al.*, 1980a, 1980b). Chelation to ferric ions probably occur via the 6-OH, 11-OH and 3'-NH_2 groups.

With respect to changes in ring D, it has already been mentioned that carmino-mycin has weaker DNA-binding ability than Adriamycin (Du Vernay *et al.*, 1980b). It causes nucleolar segregation but a sixfold higher dose is required than for Adriamycin (Merski *et al.*, 1979) and the unwinding angle produced in PM-2 DNA is less than that for Adriamycin (Pachter *et al.*, 1980). Other 4-*O*-alkyl compounds retain activity though potency is reduced (Bernardi *et al.*, 1979), whereas 4-*O*-demethyl compounds have increased potency and show equivalent DNA-binding compared to the parent compounds (Plumbridge and Brown, 1978b; Zunino *et al.*, 1979). Substitution at the 2,3 or 1,4 positions with Me or Cl, in 4-demethoxydaunomycin gives some reduction in activity (Di Marco *et al.*, 1978).

The stereochemistry at the C7 position in ring A is an important determinant of the interaction with DNA, since a comparison of the epimers of 9-deacetyl-9-dehydroxy-4-demethoxydaunomycin showed that only the 7*S* epimer can intercalate: in the presence of DNA, the 7-*R* epimer showed no spectral shifts and no decrease in absorbance. Its phenolic groups could still be ionised and there was no significant increase in its degree of fluorescence polarisation when irradiated with polarised light. There was also no significant increase in the melting temperature of DNA in the presence of this compound (Plumbridge and Brown, 1979b). Substitution with O-Me at the 10 position gives retention of activity against P388 cells when the stereochemistry is *R* (as in other 10-substituted anthracyclines) but the *S*-epimer is inactive (Penco *et al.*, 1979). Removal of the 9-OH group gives active compounds (Penco *et al.*, 1978) which show only marginally weaker DNA binding than the parent drug (Arlandini *et al.*, 1980) but 7-deoxy-9-*O*-α-daunosaminyldaunomycinone has a much lower DNA affinity (Arlandini *et al.*, 1980). 9-*O*-Methylation or replacement of the

9-OH with 9-CH_3 gives inactivity against P388 cells (Giardino *et al.*, 1980; Zunino *et al.*, 1981). 9,10-Anhydrodaunomycin is inactive even though it still binds to DNA, with about the same binding constant as daunomycin, using equilibrium dialysis (Arlandini *et al.*, 1980). Replacement of the 9-acetyl group with a hydroxymethyl group gives retention of DNA binding activity (Arlandini *et al.*, 1980), but replacement with formyl or bulky acyl substituents or ethyl gives a reduction in DNA binding affinity (Smith *et al.*, 1978, 1979). Briefly turning to 13-carbonyl derivatives, the benzoylhydrazone rubidazone (51) is a daunomycin prodrug reportedly less cardiotoxic, which is active in the G1 and S phases (Maral, 1979) and affects DNA replication (Sartiano *et al.*, 1979). A series of derivatives substituted in the phenyl ring was prepared by Tong *et al.* (1978); the DNA binding ability of daunomycin derivatives was judged to be equivalent to that of daunomycin, using measurements of ΔT_m and inhibition of nucleic acid synthesis. Cleavage to daunomycin rapidly occurs. The Adriamycin derivatives have lowered DNA affinities. Whilst differing little in cytotoxicity, the compounds differed in cardiotoxicity and a quantitative structure-activity relationship (QSAR) study showed electron-withdrawing groups gave decreased cardiotoxicity. *Bis*(hydrazones) have been prepared (Henry and Tong, 1978) as potential *bis*-intercalating agents but hydrolysis is rapid (Henry, 1979) although for example the succinylbishydrazone derivative gives a high degree of stabilisation of DNA to thermal denaturation, suggesting *bis*-intercalation.

Surprisingly, despite the vast amount of work on testing of analogues and derivatives, few QSAR studies have been reported. In addition to that of Tong *et al.* (1978) mentioned above, Fink *et al.* (1980) showed that activity increases with hydrophilicity and that generally structural changes producing increased activity result in cardiotoxicity, though separation may be achievable with electron-donating groups in ring D.

Attempts have been made to improve selectivity by preparation of daunomycin and Adriamycin-macromolecule complexes. These may be reversible (to slowly release drug) using liposomes (for example, Rahman *et al.*, 1980), albumin microspheres (Morimoto *et al.*, 1981) magnetic albumin microspheres (Widder *et al.*, 1981) or DNA-drug complexes (for example, Blanchard *et al.*, 1981; Nilsson *et al.*, 1981): in the latter case the daunomycin complex acts as a slow-release form whereas the Adriamycin complex at least in part undergoes endocytosis to give release of drug in lysosomes. Covalent complexes have been prepared using dextrans (Bernstein *et al.*, 1978; Hurwitz *et al.*, 1978) or proteins with potential targetting properties such as antibodies (Pagé *et al.*, 1981; Sela *et al.*, 1981; Suzuki *et al.*, 1981), lectins (Monsigny *et al.*, 1980), thyrotropin (Kaneka, 1981) and melanotropin (Wiesehahn *et al.*, 1981). A good demonstration of the effectiveness carriers can have in improving drug distribution is seen with trypanocidal properties of daunomycin. Although the drug is active *in vitro* against trypanosomes it is inactive *in vivo*. However, activity *in vivo* is retained when the drug is coupled to protein via a labile link (Williamson *et al.*, 1981; Brown *et al.*, 1982) but only when the link is labile, suggesting that the drug must be released before acting. Studies with labelled drug and labelled protein support this hypothesis (Hurwitz *et al.*, 1978), as does the finding that the melantropin–daunomycin conjugate does not itself interact with DNA (Wiesehahn *et al.*, 1981) (no shift in drug spectrum was observed in the presence of DNA and no unwinding of covalently closed-circular DNA was found). Drug is presumably released before exerting its effect. Most conjugates are linked via the amino sugar (either intact or after periodate oxidation); it would be informative to see if drug linked to protein via the 9-substituent could still intercalate.

γ-Rhodomycinone derivatives

γ-Rhodomycinone (16) has been found as the aglycone of iremycin (its 10-*O*-rhodosamine derivative) (*cf* roseorubicin B (18), which is the 10-*O*-(rhodos-amine)$_2$ analogue) (Ihn *et al.*, 1980). The interaction of iremycin with DNA has been studied by transient electric dichroism (Fritzsche and Dattagupta, 1980) and the results are consistent with an intercalative mode of interaction. There is a greater degree of unwinding and extension of the DNA than with daunomycin.

Iremycin is produced by *Streptomyces violaceus* subs. iremycetius an inter-species-recombinant of *S. hygroscopicus* (a macrolide-producing organism) and *S. violaceus* which produces the violamycins (Schlegel and Fleck, 1980; Schlegel *et al.*, 1980). The latter are a group of anthracyclines, individual members containing α_2-rhodomycinone (a 1,4,11-trihydroxy compound), β-rhodomycinone (22) or ε-isorhodomycinone (23) as aglycone in glycosidic linkage with one or more of rhodosamine, 2-deoxyfucose and rhodinose. The structures of individual members of the group have not yet been fully elucidated. One constituent, violamycin B1, has been shown to unwind covalently closed-circular DNA and the binding of ethidium to DNA is reduced in its presence (Popa *et al.*, 1981) indicative of intercalation. The drug undergoes hypochromic and bathochromic shifts and the phenolic groups cannot be ionised, when bound to DNA; the fluorescence is quenched on binding to DNA and the fluorescence titration method yielded values of $K = 4.8 \times 10^6 \text{M}^{-1}$ and $n = 0.27$. There is a higher degree of thermal stabilisation of DNA than with daunomycin (Löber *et al.*, 1980). An estimate of the unwinding angle (covalently closed-circular plasmid pBR313 DNA) gave a value of 10–11° (Triebel *et al.*, 1980) which is similar to that found for daunomycin.

As might be predicted therefore, the substitution pattern at positions 1,4,6 and 11 in anthracyclines has no effect on the ability of the compound to intercalate, though it was seen earlier that this only applies if there is no steric interaction. The configuration at position 7 is, however, crucial. The presence of a 10-carbomethoxy group in the 'correct' configuration and an appropriate di- or trisaccaride unit is necessary to confer selectivity against nucleolar RNA synthesis.

Affinity for naked DNA *in vitro* does not always correlate with cytotoxicity and so cannot be taken as a fool-proof indicator of *in vivo* properties. Nevertheless it does provide a starting point for rationalisation of the action of these compounds. There is a need for more information on the interaction of anthracyclines with chromatin and it is still not clear whether it is interaction with DNA, some consequence of this interaction or some other effect which is the primary cytotoxic event.

SYNTHETIC ANTHRACYCLINE ANALOGUES

Anthracyclines can be viewed as substituted anthraquinones, so an analysis of the antitumour effects of anthraquinones could be worthwhile particularly since (in addition to their long-recognised laxative effect) some anthraquinones and/or their glycosides have antibacterial and antipsoriatic activity (Anton and Haag-Berrurier, 1980; Friedman, 1980). Certain hydroxy- nitro- and amino-anthra-quinones are known to be mutagenic (Brown and Brown, 1976; Brown and Dietrich, 1979; Venturini and Tamaro, 1979), and some anthraquinones have been found to undergo spectral shifts in presence of DNA (Swanbeck, 1966). The related anthrone dithranol is widely used for treatment of psoriasis because of its cytotoxic effects (Krebs and Schaltegger, 1965). However, on the basis of routine screening at NCI, Driscoll (1974) commented ". . . the most noteworthy observation concerning the anthraquinones is the relative lack of activity. . .". Of 379 tested, only 2-bromo- and 1,2-dihydroxyanthraquinone showed *in vivo* activity and then only at high dose and in only one tumour system.

The fact that it is the anthraquinone unit of anthracyclines which intercalates into DNA (Patel and Canuel, 1978; Phillips and Roberts, 1980; Quigley *et al.*, 1980; Patel *et al.*, 1981; and previous chapter) suggests it should be possible to derive anthraquinones which mimic anthracyclines. In 1971, Müller *et al.*, stated

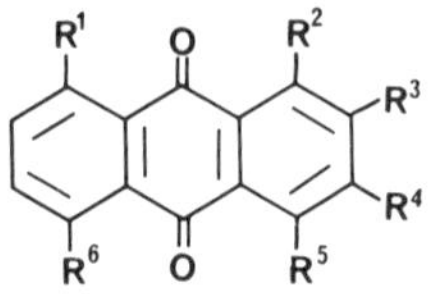

(52) $R^2 = R^5 = OH$ $R^3 = CH_2NMe_2$ $R^4 = (CH_2)_2NMe_2$ $R^1 = R^6 = H$
(53) $R^2 = R^5$ NHCH(CH_3)(CH_2)_3NEt_2 $R^1 = R^3 = R^4 = R^6 = H$
(54) $R^2 = NH(CH_2)_3NMe_2$ $R^5 = OH$ $R^1 = R^3 = R^4 = R^6 = H$
(55) $R^2 = NHpC_6H_5CH_2NMe_2$ $R^5 = OH$ $R^1 = R^3 = R^4 = R^6 = H$
(56) $R^2 = R^5 = NH(CH_2)_2NHCH_2CH_2OH$ $R^1 = R^3 = R^4 = R^6 = H$
(57) $R^2 = R^5 = NH(CH_2)_2NHCH_2CH_2OH$ $R^1 = R^6 = OH$ $R^3 = R^4 = H$
(58) $R^2 = NH(CH_2)_2NHCH_2CH_2OH$ $R^5 = OH$ $R^1 = R^3 = R^4 = R^6 = H$
(60) $R^2 = R^5 = NH(CH_2)_2NHEt$ $R^1 = R^3 = R^4 = R^6 = H$
(61) $R^1 = R^2 = NH(CH_2)_2NEt_2$ $R^3 = R^4 = R^5 = R^6 = H$

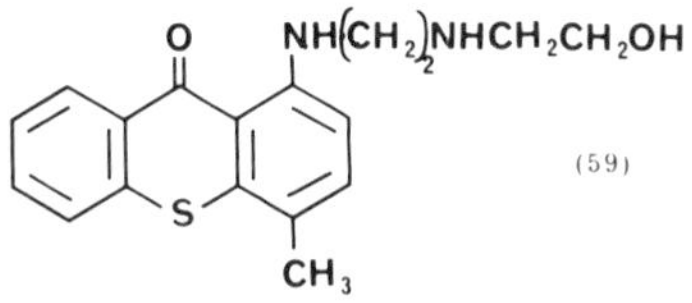

(59)

that substances such as compound (52) (modelled on daunomycin) interact with DNA, but no details were given. In 1975, Double and Brown reported a series of 1-substituted anthraquinones, such as (53), designed as intercalating agents incorporating essential features of anthracyclines. These compounds were found to undergo spectral shifts in the presence of DNA and to have affinity constants for the interaction with DNA of $0.5 - 4.2 \times 10^6 \, M^{-1}$. Although inhibitors of DNA and RNA synthesis in L1210 cells in culture (D. W. Henry, personal communication) they were inactive *in vivo*. Concurrently, Henry (1974) reported the related (54) as an inhibitor of nucleic acid synthesis and both this and the related (55) were found to stabilise DNA to thermal denaturation (ΔT_m 12.4°C and 10.6°C respectively when Adriamycin gives 17.8°C stabilisation) the former being marginally active against P388 *in vivo* (Henry, 1976). Full-aromatic tetracyclic analogues were also found to bind to DNA but primarily by stacking on the exterior of the helix (Double and Brown, 1976). One such derivative was found to be a weak inhibitor of DNA and RNA synthesis in cultured L1210 cells (Henry, 1976).

Following on from this, an evaluation of twenty-one 1,4-disubstituted anthraquinones at Lederle Research, Inc, where work on related antiprotozoal compounds was under way, (Fabio *et al.*, 1978) showed some anthraquinones such as (56) to be active *in vivo*, with compound (56) showing optimum activity (Murdock *et al.*, 1978, 1979). A major factor discovered here was that activity and potency were markedly enhanced by 5,8-hydroxylation: of 35 compounds tested, mitoxantrone (57) had optimal activity. Random screening of dyestuffs for antitumour activity had also uncovered the antitumour activity of (56) — prepared as a ball-point ink (Cheng *et al.*, 1981). In the late 1970s structural modifications based on the earlier work and on a hypothesis that a nitrogen and two oxygens in a fixed spatial relationship are essential for antileukemic activity, were carried out (Zee-Cheng and Cheng, 1978). Greatest activity was found with (57) (the only dihydroxy compound tested): a further study (Zee Cheng *et al.*, 1979) showed activity was retained but to a lower degree, in (58) and (59). From all these studies, it is apparent that activity is greatest where there is disubstitution on the anthraquinone ring, where there is a dimethylene spacer between the two nitrogen atoms, where the aliphatic amine is secondary, and where the substituent is hydroxyethyl.

Most studies have been with mitoxantrone (57) and its dedihydroxy analogue ametantrone (56), which shows activity equal to or superior to that of Adriamycin, is cross-resistant with it and possibly less cardiotoxic (Murdock *et al.*, 1978; Zee-Cheng *et al.*, 1978; Murdock *et al.*, 1979; Wallace *et al.*, 1979; Cheng *et al.*, 1979; Johnson *et al.*, 1979). It gives a greater number of survivors than does Adriamycin in animal tumour systems (Murray and Wallace, 1980) and it has a wide spectrum of activity in a human tumour-cloning system (Von Hoff *et al.*, 1981). Mitoxantrone was soon entered into phase I trials (see for example, Valdivieso *et al.*, 1981) and pharmacokinetic data are now accumulating (for example, Von Hoff *et al.*, 1980).

Mitoxantrone gives nuclear aberrations similar to those produced by Adriamycin. It is not cell-cycle specific and although it is more effective against cells in the growth phase, it is also active against non-dividing cells (Murray and Wallace, 1980). Both mitoxanthrone and ametantrone produce an accumulation of cells in the G2 phase (Evenson *et al.*, 1979 and 1980; Kimler, 1980; Traganos *et al.*, 1980; Kapuscinski *et al.*, 1981). In blocked cells, RNA content increases (Evenson, 1980; Traganos *et al.*, 1980) especially in the nucleolus (Kapuscinski *et al.*, 1981) and examination of the localisation of mitoxantrone in treated cells shows it is predominantly in the nucleus. It binds to DNA and RNA, and cells pretreated with DNAase and RNAase almost completely fail to bind drug (Kapuscinsi *et al.*, 1981). Mitoxantrone thus appears to mimic Adriamycin in its effects on the cell nucleus. Similarities in behaviour extend to lipid membrane binding properties (Yesair and Taylor, 1981). Comparing a series of related compounds, Uyeki *et al.* (1981) showed that mitoxantrone was more potent as an inhibitor of proliferation of Chinese hamster ovary cells than was Adriamycin. Both mitoxantrone and ametantrone cause sister-chromatid exchange and chromosome breaks. The latter was the more effective, consistent with its higher antitumour potency (Au *et al.*, 1978).

It appears that mitoxantrone (and by inference the other compounds) interact with DNA and this is supported by the finding that it stabilises DNA to thermal denaturation (ΔT_m = 15.9°C compared with a value for Adriamycin of 13.6°C) and that it is an inhibitor of RNA and DNA synthesis (Johnson *et al.*, 1979). Comparison of the DNA binding of 14 related compounds however showed that affinity for DNA (as measured by ΔT_m) is not a good predictor of antitumour effect; for example the 1,4-*bis*-ethylaminoethyl analogue (60) gave a higher ΔT_m (19.7°C) than with mitoxantrone but is much less active against all tumour systems evaluated (Johnson *et al.*, 1979). Comparison of a 1,8-*bis* substituted analogue with its 1,4-*bis* substituted counterpart showed the latter to be a more effective DNA-binding agent but comparison of two 1,5-*bis* substituted analogues with their 1,4-*bis* substituted analogues showed the stabilisation of DNA to thermal denaturation to be higher with the 1,5-compounds. They also required lower concentrations to give the same inhibition of nucleic acid synthesis than the 1,4-analogues: the 1,5-compounds however were totally inactive *in vivo* (Johnson *et al.*, 1979). Confirmation of intercalation as the mechanism of binding was given by studies on (53) and related compounds (Plumbridge *et al.*, 1980); when bound to DNA the compounds could not be reduced polarographically, they exhibited spectral shifts in the presence of DNA, and competitively displaced ethidium. Again stabilisation of the helix to thermal denaturation was highest with the 1,5-*bis* substituted compound (ΔT_m 10.0°C compared with 16.25°C for Adriamycin, and compared to 3.4–5.0°C for the other compounds). As with other intercalators, mitoxantrone has been shown to give protein-associated single strand breaks in DNA (Cohen *et al.*, 1980) and to unwind covalently closed-circular DNA (Kapuscinski *et al.*, 1981; Hardman and Brown, unpublished). The unwinding angle has been calculated as

26.5° (based on 26° for ethidium) and the binding parameters found to be: $K = 1.84 \times 10^6$ M^{-1}, and $n = 0.2$ (Kapuscinski *et al.*, 1981). Consistent with intercalation as the mechanism of binding, mitoxantrone displaces acridine orange (as monitored by the change in fluorescence polarisation of the acridine orange as it is displaced): this technique revealed that mitoxantrone shows no base specificity in binding (Richardson *et al.*, 1980).

Although the evidence indicates intercalation as the mechanism of DNA binding, it is not possible for the compound to intercalate with the long axis parallel to the long axis of the base pairs. It is necessary to resolve the nature of the interaction between such anthraquinones and DNA to facilitate further drug design in this series. As an initial study, Islam *et al.* (1982) have examined (61) by X-ray crystallography, have confirmed that it intercalates into DNA (it shows spectral shifts in the presence of DNA, displaces ethidium, unwinds covalently closed-circular DNA and gives a ΔT_m of 9.5° compared to 16.25 for Adriamycin) and have examined the interaction with a DNA fragment by computer-graphics modelling. The anthraquinone ring is almost coplanar, and the effective planar area is increased by hydrogen bonding of the NH hydrogen atoms to the quinone oxygen atoms. Intermolecular mechanics calculation of the fit into a d(CpG) self-complementary dimer showed it could enter only from the major groove and the minimum-energy position was similar to that assumed by proflavine. Protonation of the side-chain nitrogens enhanced the interaction particularly through interaction with O6 guanine atoms.

Mitoxantrone is thus a highly potent antitumour compound, it can intercalate into DNA and its action in many ways resembles that of Adriamycin. It is however a 1,4-disubstituted compound whereas Adriamycin can be regarded as a 1,4-dihydroxy-2,3-disubstituted anthraquinone: it may be profitable therefore to examine 2-substituted anthraquinones. The first anthraquinones modelled on Adriamycin were of this type, (Müller *et al.*, 1971). Furthermore *bis*-substituted anthraquinones such as (62) (modelled on tilorone) have antiviral activity (Grisar *et al.*, 1974; Sill *et al.*, 1974) and related compounds have antiprotozoal activity (Winkelman and Raether, 1979). Compounds modelled on Adriamycin include (63) (Henry, 1976) and (64) (Henry, 1979): the former is a weak inhibitor of RNA and DNA synthesis and only weakly stabilises DNA to thermal denaturation. The latter compound was found to be inactive against P388 tumours. Compounds such as (65) (Henry, 1976) (Double and Brown, 1976) do not bind to DNA but the analogues such as (66), whilst inactive *in vivo*, show some inhibition of RNA and DNA synthesis in cultured L1210 cells (Henry, 1976). The *N*-ethyl analogue of (66) has been shown to stabilise the DNA helix to thermal denaturation, (Knight, Hardman and Brown, unpublished results). However, no antitumour activity was shown. This was also the case for a series of similar compounds such as (67) (Bennett *et al.*, 1982): all these compounds gave effects typical of an intercalative interaction (spectral shifts occur, the phenolic group cannot be ionised in the presence of DNA, and there is a marked reduction in fluorescence and increase in fluorescence

(62) $R^2 = R^6 = SO_2NH(CH_2)_3NBu_2$ $R^1 = R^3 = R^4 = R^5 = R^7 = H$

(63) $R^3 = CH_2O\text{-daunosamine}$ $R^1 = R^2 = R^4 = R^5 = R^6 = R^7 = H$

(64) $R^1 = R^4 = OH$ $R^3 = CH_2O\text{-daunosamine}$ $R^5 = OCH_3$ $R^2 = R^6 = R^7 = H$

(65) $R^1 = R^4 = OH$ $R^3 = CH_2\text{-p-}C_6H_4NH_2$ $R^2 = R^5 = R^6 = R^7 = H$

(66) $R^1 = R^4 = OH$ $R^3 = NH(CH_2)_3NMe_2$ $R^2 = R^5 = R^6 = R^7 = H$

(67) $R^1 = R^4 = OH$ $R^3 = CH_2CONH(CH_2)_3NMe$ $R^2 = R^5 = R^6 = R^7 = H$

(68) $R^4 = OH$ $R^3 = CONH(CH_2)_2NEt_2$ $R^1 = R^2 = R^5 = R^6 = R^7 = H$

(69) $R^2 = CH_2OCONHCH_3$ $R^1 = R^3 = R^4 = R^5 = R^6 = R^7 = H$

polarisation when irradiated with polarised light). Values of ΔT_m ranging from 5 to 7°C were obtained and affinity constants for the binding were in the range 1.3–$5.3 \times 10^5 M^{-1}$. Monohydroxylated compounds such as (68) had reduced affinity and affinity was reduced even further on removal of the second hydroxyl group.

There is thus evidence that 1,4-dihydroxy-2-substituted anthraquinones can intercalate into DNA, and hence there is scope for further structural modification within this series in the search to convert DNA binding ability into cytotoxic activity *in vivo*.

As with anthracyclines, there is some doubt over the relevance of the redox properties of the substituted anthraquinones with respect to cytotoxicity and cardiotoxicity. There is a possibility of DNA damage mediated in some way by the (reduced) semiquinone, for example carminic acid does not bind to DNA but produces nicking of DNA in the presence of borohydride and even some nicking of DNA in its absence (Lown *et al.*, 1979*a*). The reversal of this effect by free radical scavengers, catalase and superoxide dismutase suggests involvement of OH· or O_2^{-}. *In vitro* studies with ametantrone and mitoxantrone, however, suggest a lower propensity for enzymatic activation to the semiquinone than with, for example, Adriamycin. Kharash and Novac (1981) showed that both compounds interact with FAD and FMN residues on NADPH cytochrome P450 reductase but are poor substrates for reduction. This is probably due to the more negative one-electron reduction potentials compared to the anthracycline antibiotics: for example the value for ametantrone is $-348 \, mV$ compared to $-292 \, mV$ and $-305 \, mV$ for Adriamycin and daunomycin respectively (Powis *et al.*, 1981; Svingen and Powis 1982). The full relevance of these differences awaits elucidation, although the *in vivo* lipid peroxidation caused by ametantrone is certainly less than that caused by Adriamycin (Patterson *et al.*, 1982).

Exploitation of the reduction properties might be possible in order to achieve selectivity. A series of anthraquinones with a 2-methylene group substituted with a good leaving group such as in (69) have been shown to be more effective versus chronically hypoxic cells than aerated cells (Lin *et al.*, 1981). The compounds were designed on the basis that tumour cells, particularly those in the interior of solid tumours, are hypoxic. Therefore a greater degree of reduction and subsequent generation of anthraquinone methide should be produced than in oxygenated cells. The value of the anthraquinones for the future is that since (on the whole) they are more accessible synthetically than the anthracycline antibiotics and thus there are greater opportunities to evaluate drug design hypotheses.

CONCLUSION

The vast amount of research on anthracyclines has generated several promising compounds for clinical evaluation. This research has also revealed the complexity of unravelling the molecular actions of these drugs. The DNA-intercalation hypothesis provides just a convenient starting point although how the substituents on ring A play a role in determining the selectivity against nucleolar RNA synthesis is not yet clear. The effects of these drugs on chromatin requires more detailed examination, particularly in view of the differential effects of marcellomycin and Adriamycin on organisation of chromatin. The relevance of the causation of breaks in DNA due to drug binding, and of redox-cycling of drug to cytotoxicity is not clear at the present time: an understanding of these events may resolve interpretation of the anomalous effects of 7-OMEN and AD32. Whatever the cellular cytotoxic event of anthracyclines, it appears on current evidence that anthraquinones such as mitoxantrone mimic the effects of Adriamycin and so provide a new structural group for further development and study.

Certainly the anthracyclines continue to be an active area of antitumour research, and hopefully will continue to be a fertile area. Having generated a series of compounds with marked cytotoxic activity, we should expect the research to turn increasingly towards design of methods to selectively target such drugs to tumour cells.

REFERENCES

Acton, E. M. and Mosher, C. W. (1981). *Proc. Am. Ass. Cancer Res.*, **22**, 225
Acton, E. M. and Tong, G. L. (1981). *J. med. Chem.*, **24**, 669
Acton, E. M., Tong, G. L., Mosher, C. W., Smith, T. H. and Henry, D. W. (1979). *J. Med. Chem.*, **22**, 922.
Anderson, W. A., Boivin, Monneret, C. and Pais, M. (1980b). *Mutat. Res.*, **72**, 341
Anderson, W. A., Moreau, P. L., Devoret, R. and Maral, R. (1980a). *Mutat. Res.*, **77**, 197
Anton, R. and Haag-Berrurier, M. (1980). *Pharmacology*, **20**, 104
Arcamone, F. (1978). In *Topics in Antibiotic Chemistry*, Vol. 2 (ed. P. Sammes), Ellis Horwood, Chichester, p. 89
Arcamone, F. (1981). *Doxorubicin, Anthracycline Antibiotics*, Academic Press, New York

Arcamone, F., Cassinelli, G., Fantini, G., Grein, A., Orezzi, P., Pol, C. and Spalla, C. (1969). *Biotechnol. Bioeng.*, **11**, 1101
Arlandini, F., Vigevani, A. and Arcamone, F. (1980). *Il. Farmaco. Ed. Sci.*, **35**, 65
Arrhigi, F. E. (1967). *J. cell. Physiol.*, **69**, 45
Aszalos, A., Macy, M. L., Sethi, V. S., Luc, V. and Kaliha, C. (1979). *Biochem. Pharmac.*, **28**, 335
Au, W. W., Butler, M. A., Matney, T. S. and Loo, T. L. (1981). *Cancer Res.*, **41**, 376
Bachur, N. R., Gordon, S. L., Gee, M. V. and Kon, H. (1979). *Proc. natn. Acad. Sci., U.S.A.*, **76**, 954
Baurain, R., Masquelier, M., Deprez-De Campaneere, D. and Trouet, A. (1980). *J. med. Chem.*, **23**, 1171
Bennett, S., Sharples, D. and Brown, J. R. (1982). *J. med. Chem.*, **25**, 369
Bergy, M. E. and Reusser, F. (1967). *Experientia.*, **23**, 254
Bernardi, L., Masi, P., Sapino, O., Suarato, A. and Arcamone, F. (1979). *Il. Farmaco Ed. Sci.*, **34**, 884
Bernstein, A., Hurwitz, E., Maron, R., Arnon, R., Sela, M. and Wilchek, M. (1978). *J. natn. Cancer Inst.*, **60**, 379
Bhuyan, B. K. and Smith, C. G. (1965). *Proc. natn. Acad. Sci. U.S.A.*, **54**, 566
Bhuyan, B. K., Blowers, C. L., Crampton, S. L. and Shugars, K. D. (1981*a*). *Cancer Res.*, **41**, 18
Bhuyan, B. K., Blowers, C. L. and Shugars, K. D. (1980). *Cancer Res.*, **40**, 3437
Bhuyan, B. K., McGovern, J. P. and Crampton, S. L. (1981*b*). *Cancer Res.*, **41**, 582
Bhuyan, B. K. and Reusser, F. (1970). *Cancer Res.*, **30**, 984
Blanchard, J. C., Schneider, Y-J, Baurain, R. and Trouet, A. (1981). *Eur. J. Cancer*, **17**, 297
Blum, R. M. and Carter, S. K. (1974). *Annls intern. Med.*, **88**, 168
Blumauerova, M., Kralovcova, E., Hostalek, Z. and Vanek, Z. (1979*a*). *Folia Microbiol.*, **24**, 128
Blumauerova, M., Kralovcova, E., Mateju, J., Jizba, J. and Vanek, Z. (1979*b*). *Folia Microbiol.*, **24**, 117
Brockmann, H. and Greve, H. (1975). *Tetrahedron Lett.*, 831
Brodasky, T. F. and Reusser, F. (1974). *J. Antibiotics*, **27**, 809
Brown, J. E., Brown, J. R. and Williamson, J. (1982). *J. Pharm. Pharmac.*, **34**, 236
Brown, J. P. and Brown, R. J. (1976). *Mutat. Res.*, **40**, 203
Brown, J. P. and Dietrich, P. S. (1979). *Mutat. Res.*, **66**, 9
Brown, J. R. (1978). *Prog. med. Chem.*, **15**, 125
Brox, L., Gowans, B. and Belch, A. (1980). *Can. J. Biochem.*, **58**, 720
Carter, S. K. (1980). *Cancer Chemother. Pharmac.*, **4**, 5
Casey, M. L., Paulick, R. C. and Whitlock, H. W. (1978). *J. org., Chem.*, **43**, 1627
Cassinelli, G., Ruggieri, D. and Arcamone, F. (1979). *J. med. Chem.*, **22**, 123
Cassinelli, G., Di Matteo, F., Forenza, S., Ripamonti, M. C., Rivola, G., Arcamone, F., Di Marco, A., Casazza, A. M., Soranzo, C. and Pratesi, G. (1980*a*). *Antibiotics*, **33**, 1468
Cassinelli, G., Forenza, S., Ripamonti, M. C. and Ruggiori, D. (1980*b*). *Eur. Pat. Appl.* 22, 515
Chatterji, H., Deb, J. K. and Neogy, R. J. (1980). *J. Biochem. Biophys.*, **17**, 421
Cheng, C. C., Zbinden, G. and Zee-Cheng, R. K-Y (1979). *J. Med. Chem.*, **68**, 393
Cheng, C. C., Zee-Cheng, R.K-Y, Narayanan, V. L., Ing, R. B. and Paull, K. D. (1981). *Trends Pharmac. Sci.*, 223
Chowdhury, K., Chowdhury, I., Biswas, N. and Neogy, R. K. (1978). *Ind. J. Biochem. Biophys.*, **15**, 373
Chowdhury, K. and Neogy, R. K. (1980). *Ind. J. Biochem. Biophys.*, **18**, 120
Chuang, L. F., Kawahata, R. T. and Chuang, R. Y. (1980). *FEBS Lett.*, **117**, 247
Chuang, L. F., Kawahata, R. T. and Chuang, R. Y. (1981). *Proc. Am. Ass. Cancer Res.*, **22**, 24
Cohen, L. F., Glaubinger, D. L., Kann, H. E. and Kohn, K. W. (1980). *Proc. Am. Ass. Cancer Res.*, **21**, 277
Crooke, S. T. (1981). In *Cancer Chemotherapy, Vol. III, Antineoplastic Agents.* (ed. S. T. Crooke and A. W. Prestayko), Academic Press, New York, p. 112

Crooke, S. T., Du Vernay, V. H., Galvan, L. and Prestayko, A. (1978). *Molec. Pharmac.*, **14**, 290

Crooke, S. T., Du Vernay, V. H. and Mong, S. (1981). In *Molecular Actions and Targets for Cancer Chemotherapeutic Agents* (ed. A. C. Sartorelli, J. S. Lazo and J. R. Bertino), Academic Press, New York, p. 137

Crooke, S. T. and Reich, S. D. (eds) (1980). *Anthracyclines, Current Status and New Developments*, Academic Press, New York

Dall'Aqua, F., Vedaldi, D. and Gennaro, A. (1979). *Chem.-biol. Interact.*, **25**, 59

Daskal, Y., Woodard, C., Crooke, S. T. and Busch, H. (1978). *Cancer Res.*, **38**, 467

Davis, H. L. and Davis, T. E. (1979). *Cancer Treat. Rep.*, **63**, 809

Di Marco, A. (1981). *Chemoterapia. Oncol.*, **4**, 5

Di Marco, A., Casazza, A. M., Dasdia, T., Necco, A., Pratesi, G., Rivolta, P., Velcich, A., Zaccara, A. and Zunino, F. (1977). *Chem. biol. Interact.*, **19**, 291

Di Marco, A., Casazza, A. M., Soranzo, C. and Pratesi, G. (1978). *Cancer Chemother. Pharmac.*, **1**, 249

Double, J. C. and Brown, J. R. (1975). *J. Pharm. Pharmac.*, **27**, 502

Double, J. C. and Brown, J. R. (1976). *J. Pharmac. Pharmac.*, **28**, 166

Driscoll, J. S., Hazard, G. F., Wood, H. B. and Goldin, A. (1974). *Cancer Chemother. Rep.*, **4**, 1

Du Vernay, V. H. (1981). In *Cancer Chemotherapy Vol. III Antineoplastic Agents* (ed. S. T. Crooke and A. W. Prestayko), Academic Press, New York, p. 233

Du Vernay, V. H., Essery, J. M., Doyle, T. W., Bradner, W. T. and Crooke, S. T. (1979a). *Molec. Pharmac.*, **15**, 341

Du Vernay, V. H., Pachter, J. A. and Crooke, S. T. (1979b). *Biochemistry*, **18**, 4024

Du Vernay, V. H., Pachter, J. A. and Crooke, S. T. (1979c). *Molec. Pharmac.*, **16**, 623

Du Vernay, V. H., Pachter, J. A. and Crooke, S. T. (1980). *Cancer Res.*, **40**, 387

Egorin, M. J., Clawson, R. E., Cohen, J. L., Ross, L. A. and Bachur, N. R. (1979b). *J. Pharmac., Exp. Ther.*, **210**, 229

Egorin, M. J., Clawson, R. E., Cohen, J. L., Ross, L. A. and Bachur, N. R. (1980a). *Cancer Res.*, **40**, 4669

Egorin, M. J., Clawson, R. E., Ross, L. A. and Bachur, N. R. (1980b). *Cancer Res.*, **40**, 1928

Egorin, M. J., Clawson, R. E., Ross, L. A., Schlossberger, N. M. and Bachur, N. R. (1979a). *Cancer Res.*, **39**, 4396

El Khadem, K. S. and Swartz, D. L. (1981). *J. med. Chem.*, **24**, 112

Essery, J. M. and Doyle, T. W. (1979). *J. Antibiotics*, **32**, 247

Essery, J. M. and Doyle, T. W. (1980). *Can. J. Chem.*, **58**, 1869

Evenson, D. P., Darzynkiewicz, Z., Staiano-Coico, L., Traganos, F. and Melamed., M. R. (1979). *Cancer Res.*, **39**, 2574

Evenson, D. P., Traganos, F., Darzynkiewicz, Z., Staiano-Coico, L. and Melamed, M. R. (1980). *J. natn. Cancer Inst.*, **64**, 857

Fabio, P. F., Fields, T. L., Lin, Y-I, Burden, E. J., Carvajal, S., Murdock, K. C. and Lang, S. A. Jr (1978). *J. med. Chem.*, **21**, 273

Facchinetti, R., Raz., A. and Goldman, R. (1978). *Cancer Res.*, **38**, 3944

Fink, S. I., Leo, A., Yamakawa, M., Hansch, C. and Quinn, F. R. (1980). *Il. Farmaco. Ed. Sci.*, **35**, 965

Forenza, S., Arcamone, F., Cassinelli, G. and Ripamonti, M. C. (1981). *Belg. Pat.*, 883, 759

Friedman, C. A. (1980). *Pharmacology*, **20**, 113

Fritzsche, H. and Dattagupta, N. (1980). *Studia Biophys.*, **81**, 77

Fuchs, E-F, Horton, D., Weckerle, W. and Winter, B. (1979b). *J. Antibiotics*, **32**, 223

Fuchs, E-F., Horton, D., Weckerle, W. and Winter-Mihaly, E. (1979a). *J. med. Chem.*, **22**, 406

Giardino, P., Vigevani, A., Bernardi, L. and Arcamone, F. (1980). *Gazz. Chim. Ital.*, **110**, 101

Grisar, J. M., Hickey, K. R., Fleming, R. W. and Mayer, G. D. (1974). *J. med. Chem.*, **17**, 890

Gordon, G. R. and Kashiware, D. (1981). *Proc. Am. Ass. Cancer. Res.*, **22**, 256

Gosalvez, M., Blanco, M. F., Vivero, C. and Valles, F. (1978). *Eur. J. Cancer*, **14**, 1185

Gowans, B., Belch, A. and Brox, L. (1980). *Proc. Am. Ass. Cancer Res.*, **21**, 282

Henry, D. W. (1974). *Cancer Chemother. Rep.* Part 2, **4**, 5

Henry, D. W. (1976). In *Cancer Chemotherapy* (ed. A. C. Sartorelli), American Chemical Society, Washington, p. 15

Henry, D. W. (1979). *Cancer Treat. Rep.*, **63**, 845

Henry, D. W., Smith, T. H., Tong, G. L. and Wu, H. Y. (1981). *U.S. Pat.*, **4**, 202, 967

Henry, D. W. and Tong, G. L. (1978). *U.S. Pat.*, **4**, 112, 217

Hori, S., Shirai, M., Hirano, S., Oki, T., Inui, T., Tsukagoshi, S., Ishizuka, M., Takeuchi, T. and Umezawa, H. (1977). *Gann*, **68**, 685

Horton, D., Priebe, W. and Turner, W. R. (1981). *Carbohydrate Res.*, **94**, 11

Horton, D. and Turner, W. R. (1979). *Carbohydrate Res.*, **77**, C8

Hurwitz, E., Maron, R., Arnon, R., Wilchek, M. and Sela, M. (1978). *Eur. J. Cancer*, **14**, 1213

Ihn, W., Schlegel, B., Fleck, W. F. and Sedmera, P. (1980). *J. Antibiot.*, **33**, 1457

Ishizuka, M., Takeuchi, T., Masuda, T., Fukasawa, S. and Umezawa, H. (1981). *J. Antibiotics*, **34**, 331

Islam, S. A., Neidle, S., Gandecha, B. M. and Brown, J. R. (1982). *Biochem. Pharmac.*, in press.

Israel, M., Pegg, W. J. and Wilkinson, P. M. (1978). *J. Pharmac., exp. Ther.*, **204**, 696

Israel, M., Potti, G., Anderson, L. B. and Khetarpal, V. K. (1981). *Proc. Am. Ass. Cancer Res.*, **22**, 252

Jaenke, R. S., Deprez-De Campeneere, D. and Trouet, A. (1980). *Cancer Res.*, **40**, 3530

Johnson, R. K., Zee-Cheng, R. K-Y, Lee, W. W., Acton, E. M., Henry, D. W. and Cheng, C. C. (1979). *Cancer Treat Rep.*, **63**, 425

Kajiwara, K., Rogers, A. E. and Mueller, G. C. (1979). *Proc. Amer. Ass. Cancer Res.*, **20**, 239

Kaneka, Y. (1981). *Horm. Metab. Res.*, **13**, 110

Kanter, P. M. and Schwartz, H. S. (1979). *Cancer Res.*, **39**, 448

Kapuscinski, J., Darzynkiewicz, Z., Traganos, F. and Melamed, M. R. (1981). *Biochem. Pharmac.*, **30**, 231

Kersten, W., Kersten, H. and Szybalski, W. (1966). *Biochemistry*, **5**, 236

Kharasch, E. D. and Novak, R. F. (1981). *Biochem. Pharmac.*, **30**, 2881

Kimler, B. F. (1980). *Cancer Res.*, **40**, 42

Kitamura, I., Shibamoto, N., Oki, T., Inui, T., Naganawa, H., Ishizuta, M., Masuda, T., Takeuchi, T. and Umezawa, H. (1977). *J. Antibiotics*, **30**, 616

Komiyama, T., Oki, T., Inui, T., Takeuchi, T. and Umezawa, H. (1979). *Gann*, **70**, 403

Krebs, A. and Schaltegger, H. (1965). *Experientia*, **21**, 128

Krishan, A., Dutt, K., Israel, M. and Ganepathi, R. (1981). *Cancer Res.*, **41**, 2745

Krueger, W. C., Pschigoda, L. M., Schpok, S. C. F., Moscowitz, A., McGovren, J. P., Neta, P., Merrit, M. V. and Li, Z. H. (1981). *Chem.-biol. Interact.*, **36**, 1

Levey, B. A., Ruiz, E., Rogerson, B., Lehotay, D. C. and Levey, G. S. (1980). *Cancer Treatment Rep.*, **64**, 1127

Levin, Y. and Sela, B. (1980). In *Proc. 11th Int. Conf. Chemother. & 19th Interscience Conf. on Antimicr. Agents & Chemother.* Vol. II, American Soc. for Microbiology, Washington

Levin, Y. and Sela, B. A. (1981). *FEBS Lett.*, **98**, 119

Li, L. H., Kuentzel, S. L., Murch, L. L. Pschigoda, L. M. and Krueger, W. C. (1979). *Cancer Res.*, **38**, 4816

Lin, T-S, Teicher, B. A. and Sartorelli, A. C. (1980). *J. med. Chem.*, **23**, 1237

Löber, G., Kalmer, R., Smekal, E., Balcarova, Z., Klein-Wächter, U., Raim, T., Popa, L., Hanschmann, M. and Koudelka, J. (1980). *Studia Biophys.*, **81**, 103

Londos-Gagliardi, D., Capri, M., Aubel-Sadron, G. and Maral, R. (1980). *Biochem. Biophys. Res. Commun.*, **97**, 397

Lown, J. W., Chen, H-H and Plambeck, J. A. (1979a). *Biochem. Pharmacol.*, **28**, 2563

Lown, J. W., Chen, H-H, Sin, S-K. and Plambeck, J. A. (1979b). *Bioorg. Chem.*, **8**, 17

Maral, R. (1979). *Cancer Chemother. Pharmac.*, **2**, 31

Marciani, S., Terbojevich, M., Vedaldi, D. and Rodighiero, G. (1977). *Il Farmaco Ed. Sci.*, **32**, 248

Masi, P., Suarato, A., Giardino, P., Iraci, G., Bernardi, L. and Arcamone, F. (1980). *Il.*

Farmaco, Ed. Sci., **35**, 347

Masquelier, M., Baurain, R. and Trouet, A. (1980). *J. med. Chem.*, **23**, 1166

Matsuzawa, Y., Yoshimoto, A., Oki, T., Inui, T., Takeuchi, T. and Umezawa, H. (1979). *J. Antibiot.*, **32**, 420

Matsuzawa, Y., Yoshimoto, A., Oki, T., Naganawa, H., Takeuchi, T., and Umezawa, H. (1980). *J. Antibiot*, **33**, 1341

May, P. M., Williams, G. K. and Williams, D. R. (1980*a*). *Eur. J. Cancer*, **16**, 1275

May, P. M., Williams, G. K. and Williams, D. R. (1980*b*). *Inorg. Chim. Acta*, **46**, 221

McGuire, J. C., Thomas, M. C., Stroshane, R. M., Hamilton, B. K. and White, R. J. (1980). *Antimicrob. Ag. Chemother.*, **18**, 454

Merski, J. A., Daskal, Y., Crooke, S. T. and Busch, H. (1979). *Cancer Res.*, **39**, 1239

Misumi, M., Yamaki, H., Akiyama, T. and Tanaka, N. (1979). *J. Antibiotics*, **32**, 48

Molinier-Jumel, C., Malfay, B., Reynaud, J. A., Aubel-Sadron, G. (1978). *Biochem. Biophys. Res. Commun.*, **84**, 441

Mong, S., Du Vernay, V. H. , Strong, J. E. and Crooke, S. T. (1980). *Molec. Pharmacol.*, **17**, 100

Monsigny, M., Kieda, C., Roche, A-C. and Dehnotte, F. (1980). *FEBS Lett.*, **119**, 181

Morimoto, Y., Sugibayashi, K. and Kato, Y. (1981). *Chem. Pharm. Bull.*, **29**, 1433

Morris, M. J. and Brown, J. R. (1978). Presented at the VIth International Symposium on Medicinal Chemistry, Brighton, 1978

Müller, W., Flügel, R. and Stein, C. (1971). *Liebigs Ann. Chem.*, **754**, 15

Murdock, K. C., Child, R. G., Fabio, P. F., Angier, R. B., Wallace, R. E., Durr, F. E. and Citarella, R. V. (1979). *J. med. Chem.*, **22**, 1024

Murdock, K. C., Wallace, R. E., Angier, R. B., Durr, F. E., Child, R. G., Citarella, R. V., Fabio, P. F. and Lang, S. A. (1978). Presented at the VIth International Symposium on Medicinal Chemistry, Brighton, England

Murray, E. F. and Wallace, R. E. (1980). In *Anthracyclines: Current Status and New Developments* (eds S. T. Crooke and S. D. Reich), Academic Press, New York

Myers, G. E. (1980). In *Cancer Chemotherapy, Annual 2* (ed. H. M. Pinedo), Elsevier, New York, p. 66

Nettleton, D. E., Balitz, D. M., Doyle, T. W., Bradner, W. T., Johnson, D. L., O'Herron, F. A., Schreiber, R. H., Coon, A. B., Mosely, J. E. and Myllymaki, R. W. (1980). *J. Nat. Prods.*, **43**, 242

Nilsson, S. O., Andersson, B., Eksborg, S., Beran, M. and Ehrsson, H. (1981). *Cancer Chemother. Pharmac.*, **5**, 261

Oki, T. (1977). *J. Antibiot.*, **30**, Suppl. 70

Oki, T., Kitamura, I., Matsuzawa, Y., Shibamoto, N., Ogasawara, T., Yoshimoto, A., Inui, T., Noganawa, H., Takeuchi, T. and Umezawa, H. (1979). *J. Antibiot.*, **32**, 801

Oki, T., Matsuzawa, Y. Kiyoshima, K., Yoshimoto, A., Naganawa, H., Takeuchi, T. and Umezawa, H. (1981*c*). *J. Antibiot.*, **34**, 783

Oki, T., Shibamoto, N., Matsuzawa, Y., Ogasawara, T., Yoshimoto, A., Kitamura, I., Inui, T., Naganawa, H., Takeuchi, T. and Umezawa, H. (1977). *J. Antibiot.*, **30**, 683

Oki, T., Takatsuki, Y., Tobe, H., Yoshimoto, A., Takeuchi, T. and Umezawa, H. (1981*b*). *J. Antibiot.*, **34**, 1229

Oki, T., Yoshimoto, A., Matsuzawa, Y., Takeuchi, T. and Umezawa, H. (1980). *J. Antibiot.*, **133**, 1331

Oki, T., Yoshimoto, A., Matsuzawa, Y., Takeuchi, T. and Umezawa, H. (1981*a*). *J. Antibiot.*, **34**, 916

Olson, R. D., Boerth, R. C., Gerber, J. G. and Nies, A. S. (1981). *Life Sci.*, **29**, 1393

Pachter, J. A., Du Vernay, V. H., Crooke, S. T. and Prestayko, A. W. (1980). *Proc. Am. Ass. Cancer Res.*, **21**, 24

Pagé, M., Belles-Isles, M. and Emond, J-P (1981). *Proc. Am. Ass. Cancer Res.*, **22**, 835

Pandey, R. C. and Toussaint, M. W. (1980). *J. Chromatogr.*, **198**, 407

Patel, D. J. and Canuel, L. L. (1978). *Eur. J. Biochem.*, **90**, 247

Patel, D. J., Kozlowski, S. A. and Rice, J. A. (1981). *Proc. natn., Acad., Sci., U.S.A.*, **78**, 3333

Patterson, L. H., Gandecha, B. M. and Brown, J. R. (1982). *Biochem. biophys. Res. Commun.* (in press)

Penco, S. Franchi, G. and Arcamone, F. (1978). *Belg. Pat.*, 876, 100

Penco, S., Gozzi, F., Vigevani, A., Ballabio, M. and Arcamone, F. (1979). *Heterocycles*, **13**, 281

Peters, J. H., Evans, M. J., Jenson, R. E. and Acton, E. M. (1980). *Cancer Chemother., Pharmac.*, **4**, 263

Petzold, G. L., Krueger, W. C., Wooden, J. M., Pschigoda, L. M. and Li, L. H. (1979). *Proc. Am. Ass., Cancer Res.*, **20**, 122

Phillips, D. R. and Roberts, G. C. K. (1980). *Biochemistry*, **19**, 4801

Plumbridge, T. W., Aarons, L. J. and Brown, J. R. (1978). *J. Pharm. Pharmac.*, **30**, 69

Plumbridge, T. W. and Brown, J. R. (1978). *Biochim. Biophys. Acta*, **479**, 441

Plumbridge, T. W. and Brown, J. R. (1979a). *Biochem. Pharmac.*, **27**, 1881

Plumbridge, T. W. and Brown, J. R. (1979b). *Biochem. Biophys. Acta*, **563**, 181

Plumbridge, T. W. and Brown, J. R. (1979c). *Biochem. Pharmac.*, **28**, 3231

Plumbridge, T. W., Knight, V., Patel, K. L. and Brown, J. R. (1980). *J. Pharm. Pharmac.*, **32**, 78

Popa, L. M., Repanovici, R. and Löber, G. (1981). *Studia Biophys.*, **83**, 113

Potmesil, M., Levin, M. and Silber, R. (1981). *Proc. Am. Ass. Cancer Res.*, **22**, 209

Powis, G., Svingen, B. A. and Appel, P. (1981). *Molec. Pharmac.*, **20**, 387

Quigley, G. J., Wang, A. H-J., Ughetto, G., van der Marel, G., van Boom, J. H. and Rich, A. (1980). *Proc. natn. Acad. Sci. U.S.A.*, **77**, 7204

Rahman, A., Kessler, A., More, N., Sikic, B., Rowden, G., Woolley, P. and Schein, P. S. (1980). *Cancer Res.*, **40**, 1532

Remers, W. A. (1979). In *The Chemistry of Antitumour Antibiotics*, Vol. 1, John Wiley, New York, 63

Rueckert, P. W., Wiley, P. F., McGovren, J. P. and Marshall, V. P. (1979). *J. Antibiot.*, **32**, 4283

Reusser, F. (1967). *J. Bact.*, **93**, 65

Reusser, F. (1975). *Biochim. Biophys. Acta*, **383**, 266

Richardson, C. L., Grant, A. D., Schpok, S. L., Krueger, W. C. and Li, L. H. (1981). *Cancer Research*, **41**, 2235

Richardson, C. L., Roboz, J. and Holland, J. F. (1980). *Res. Commun. Path. Pharmac.*, **27**, 497

Sartiano, G. P., Lynch, W. E. and Billington, W. D. (1979). *J. Antibiotics*, **32**, 1038

Schlegel, B. and Fleck, W. F. (1980). *Z. Allg. Microbiol.*, **20**, 527

Schlegel, B., Ihn, W. and Fleck, W. F. (1980). *Z. Allg. Microbiol.*, **20**, 531

Schwartz, H. S. and Kanter, P. M. (1979). *Cancer Treat. Rep.*, **63**, 821

Seeber, S. (1980). *Proc. Am. Ass., Cancer Res.*, **21**, 272

Seeber, S., Loth, H. and Crooke, S. T. (1980). *J. Cancer Res. clin. Oncol.*, **98**, 109

Sela, M., Arnon, R., Hurwitz, E., Maron, R. and Levy, R. (1981). *U.S. Pat.*, 4, 263, 279

Sela, B-A and Levin, Y. (1981). *Cancer Treat. Rep.*, **65**, 277

Shaw, G. J., Milne, G. W. A. and Minghetti, A. (1979). *Phytochemistry*, **18**, 178

Shuin, T., Moriyama, M., Nishimura, R., Takai, S. and Umeda, M. (1981). *Gann*, **72**, 179

Sill, A. D., Andrews, E. R., Sweet, F. W., Hoffman, J. W., Tierman, P. L., Grisar, J. M., Fleming, R. W. and Mayer, G. D. (1974). *J. med. Chem.*, **17**, 965

Smith, T. H., Fujiwara, A. N. and Henry, D. W. (1978). *J. med. Chem.*, **21**, 280

Smith, T. H., Fujiwara, A. N., Henry, D. W. (1979). *J. med. Chem.*, **22**, 40

Someya, A. and Tanaka, N. (1979). *J. Antibiotics*, **32**, 839

Strauss, D. G. (1978). *Folia Microbiol.*, **23**, 152

Suzuki, T., Sato, E., Goto, K., Katsurada, Y., Unno, K. and Takahashi, R. (1981). *Chem., Pharm., Bull.*, **29**, 844

Svingen, B. A. and Powis, G. (1981). *Arch. Biochem. Biophys.*, **209**, 119

Swanbeck, G. (1966). *Biochim. biophys. Acta*, **123**, 630

Tanaka, H., Yoshimoti, T., Shimauchi, Y., Matsushita, Y., Matsuzawa, Y., Oki, T. and Inui, T. (1981). *J. Antibiotics*, **34**, 905

Tanaka, H., Yoshioka, T., Shimauchi, Y., Matsuzawa, Y., Oki, T. and Inui, T. (1980). *J. Antibiotics*, **33**, 1323

Tatsuta, K. and Takeuchi, T. (1980). *J. Antibiotics*, **33**, 1581

Tong, G. L., Cory, H., Lee, W. W., Henry, D. W. and Zbinden, G. (1978). *J. med. Chem.*, **21**, 732

Tong, G. L., Henry, D. W. and Acton, E. M. (1979b). *J. med. Chem.*, **22**, 36

Tong, G. L., Wu, H. Y., Smith, T. H. and Henry, D. W. (1979a). *J. med. Chem.*, **22**, 912

Traganos, F., Evenson, D. P., Staiano-Coico, L., Darzynkiewicz, Z. and Melamed, M. R. (1980). *Cancer Res.*, **40**, 671

Traganos, F., Staiano-Coico, L., Darzynkiewicz, Z. and Melamed, M. R. (1981). *Cancer Res.*, **41**, 2728

Triebel, H., Reinert, K. E., Bär, H., Schütz, H. and Hartmann, M. (1980). *Studia Biophys.*, **81**, 79

Umezawa, H., Takahashi, T., Kinoshita, M., Naganawa, H., Masuda, T., Ishizuka, M., Tatsata, K. and Takeuchi, T. (1979). *J. Antibiotics*, **32**, 1082

Umezawa, H., Takahashi, Y., Kinoshita, M., Naganawa, H., Tatsuta, K. and Takeuchi, T. (1980). *J. Antibiotics*, **33**, 1581

Uyeki, E. M., Nishio, A., Wittek, P. J. and Cheng, C. C. (1981). *J. pharm. Sci.*, **70**, 1011

Valdivieso, M., Bedikian, A. Y., Burgess, M. A., Savaraj, N., Jeffers, W. B. and Bodey, G. P. (1981). *Cancer Treat. Rep.*, **65**, 841

Venturini, S. and Tamaro, M. (1979). *Mutat. Res.*, **68**, 307

Von Hoff, D. D., Myers, J. W., Kuhn, J., Sandbach, J. F., Pocelinko, R., Clark, G. and Coltman Jr., C. A. (1981). *Cancer Res.*, **41**, 3118

Von Hoff, D. D., Pollard, E., Kuhn, J., Murray, E. and Coltman Jr, C. A. (1980). *Cancer Res.*, **40**, 1516

Wakabayashi, T., Oki, T., Tone, N., Hirano, S. and Omori, K. (1980). *J. Electron Microsc.*, **29**, 106

Wallace, R. E., Murdock, K. C., Angier, R. B. and Durr, F. E. (1979). *Cancer Res.*, **39**, 1570

Widder, K. J., Morris, R. M., Poone, G., Howard, D. P., Senyei, A. E. (1981). *Proc. natn. Acad. Sci. U.S.A.*, **78**, 579

Wiernik, P. H. (1980). In *Anthracyclines Current Status and New Development* (ed. S. T. Crooke and A. W. Prestayko), Academic Press, New York, p. 273

Wiesehahn, G., Varga, J. M. and Hearst, J. E. (1981). *Nature, Lond.*, **292**, 467

Wiley, P. F. (1979). *J. Nat. Prod.*, **42**, 569

Wiley, P. F., Elrod, D. W. and Marshall, V. P. (1978). *J. org., Chem.*, **43**, 3457

Wiley, P. F., Johnson, J. L. and Houser, D. J. (1977a). *J. Antibiotics*, **30**, 628

Wiley, P. F., Kelly, R. B., Caron, E. L., Wiley, V. H., Johnson, J. H., MacKellar, F. A. and Mizsak, S. A. (1977b). *J. Amer. Chem., Soc.*, **99**, 542

Williamson, J., Scott-Finnigan, T. J., Hardman, M. A. and Brown, J. R. (1981). *Nature, Lond.*, **292**, 466

Winkelmann, E. and Raether, W. (1979). *Arzneimitellforschung*, **29**, 1504

Wilson, R. G., King, M., Lockwood, R. and McNeil, M. (1980). *Chem. biol. Interact.*, **32**, 331

Yamaki, H., Suzuki, H., Nishimura, T. and Tanaka, N. (1978). *J. Antibiot.*, **31**, 1149

Yesair, D. W. and Taylor, R. F. (1979). *Proc. Am. Ass. Cancer Res.*, **20**, 149

Yoshimoto, A., Oki, T., Takeuchi, T. and Umezawa, H. (1980a). *J. Antibiotics*, **33**, 1158

Yoshimoto, A., Oki, T. and Umezawa, T. (1980b). *J. Antibiotics*, **33**, 1199

Yoshimoto, A., Ogasawara, T., Kitamura, I., Oki, T. and Inui, T. (1979). *J. Antibiotics*, **32**, 472

Young, R. C., Ozols, R. F. and Myers, C. E. (1981). *New Engl. J. Med.*, **305**, 139

Zee-Cheng, R. K-Y. and Cheng, C. C. (1978). *J. med. Chem.*, **21**, 291

Zee-Cheng, R. K-Y., Podrebarac, E. G., Menon, C. S. and Cheng, C. C. (1979). *J. med. Chem.*, **22**, 501

Zunino, F., Casazza, A. M., Pratesi, G., Formelli, F. and Di Marco, A. (1981). *Biochem. Pharmac.*, **30**, 1856

Zunino, F., Di Marco, A. and Velcich, A. (1977). *Cancer Lett.*, **3**, 271

Zunino, F., Di Marco, A. and Zaccara, A. (1979). *Chem.-biol. Interact.*, **24**, 217

4

Mechanisms of Selective Cytotoxicity of Adriamycin, Daunomycin and Related Anthracyclines

Herbert S. Schwartz

"To treat disease methodically and effectively, the nature and actions of the living tissue, in both the healthy and morbid conditions, must be correctly appreciated; the effects, which the articles of materia medica are capable of exerting under both those conditions, must be known from accurate observations, and not until then can the practioner prescribe with any well-founded prospect of success."

Robley Dunglison
Philadelphia, October 15, 1839
In: *New Remedies*

[And. . .]

"In any case, the goal of cancer therapy is to reduce the rate of growth in such cells, and, because the lesion is heritable within the affected cell line, to obliterate all such offending cells. Since such cells, in virtually all respects, resemble the normal cells from which they arose, the challenge to specific chemotherapy will be apparent".

In: *Biology and the Future of Man*
(1970) Edited by Philip Handler
Oxford University Press

Molecular Aspects of Anti-cancer Drug Action, ed. Neidle & Waring
0333–315561/83/093–125   © The Contributors 1983

INTRODUCTION

For the most part, this chapter is a brief deliberation on potential mechanisms of cytotoxic selectivity of Adriamycin and daunomycin (figure 1) in tumour cells in animals and in culture. It is no more than just a few reflections because we are unable to define with assurance any mechanisms that determine selectivity of

ANTHRACYCLINE AMINOGLYCOSIDE DERIVATIVES		R_1	R_2	R_3
1	ADRIAMYCIN	OCH_3	CH_2OH	H
2	4'-epi ADRIAMYCIN	OCH_3	CH_2OH	H
3	CARMINOMYCIN	OH	CH_3	H
4	4-DEMETHOXYDAUNO-MYCIN	H	CH_3	H
5	DAUNOMYCIN	OCH_3	CH_3	H
6	ADRIAMYCIN-14-BENZOATE	OCH_3	CH_2OC-	H
7	ADRIAMYCIN-14-(1-NAPHTHALENEACETATE)	OCH_3	CH_2OC-CH_2	H
8	ADRIAMYCIN-14-OCTANOATE	OCH_3	$CH_2OC(CH_2)_6CH_3$	H
9	N-ACETYLDAUNO-MYCIN	OCH_3	CH_3	$C-CH_3$
10	N-TRIFLUOROACETYL-ADRIAMYCIN-14-VALERATE	OCH_3	$CH_2OC(CH_2)_3CH_3$	$C-CF_3$

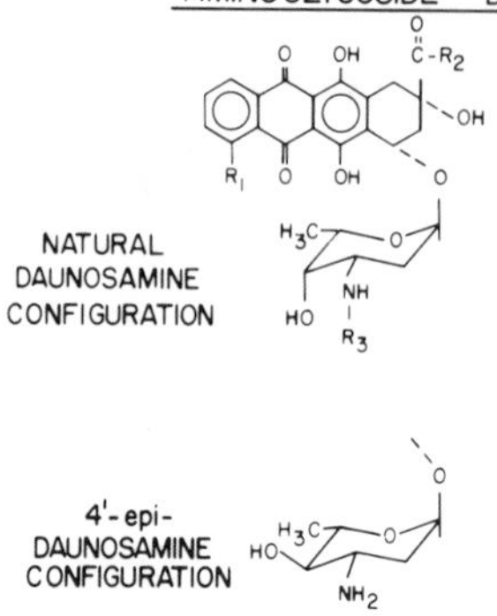

Figure 1   Structures of Adriamycin, daunomycin and some derivatives.

these drugs in any target, whether tumour or normal cells, although recent lines of evidence may point to types of irreversible lesions that correlate with cell sensitivity and thereby may contribute to selective cytotoxicity. As a pragmatic decision, confining the text primarily to animal tumours and to tumours in culture limits the scope of this review but I have taken the liberty to range somewhat afield of the anthracyclines to examine related mechanisms of cell death and some aspects of drug cytotoxicity that might contribute to or detract from their clinical utility.

Because of the limited nature of this review, the reader is referred to other sections in this volume for more detailed descriptions of new anthracyclines (Chapter 3) and of various aspects of drug activation to radicals and interactions with DNA (See Chapters 9, 10, 2). For other sources, Arcamone's *Doxorubicin* (1981) is as encyclopaedic as it is authoritative. Recent reviews on anthracyclines and other DNA-reactive agents include those by Remers (1979), Schwartz (1979), and Young *et al.* (1981). Proceedings of a number of symposia concerned primarily with anthracyclines have been published recently (Crooke and Reich, 1980; Bardos and Kalman, 1982; Muggia, 1982). The classic *Selective Toxicity* by Albert (1979) may be consulted for general background and for an approach to mechanisms of selectivity somewhat different from that presented here.

TARGET DEFINITION AND LIMITATIONS

Target cells of the anthracyclines include numerous experimental and human tumours. Those of the host organism include bone marrow, spleen, thymus, oral mucosa and gastrointestinal epithelium, hair follicles, liver, kidney, heart and skeletal systems, to varying degrees in different species (Maral *et al.*, 1967; Bertazzoli *et al.*, 1972; Philips *et al.*, 1975). Little is known about the biochemistry of most of the host lesions and, except for the cardiomyopathy, they will not be discussed further. Target tissues of Adriamycin and daunomycin also include those in which tumorigenesis occurs (Bertazolli *et al.*, 1971; Sternberg *et al.*, 1972; Marquardt *et al.*, 1976). Although all such host effects decrease efficacy of the anthracyclines, overall toxicity to the host is still less than that to at least some tumour targets which, obviously, is the reason that the agents are useful.

Differential immunosuppressive activities by the anthracyclines also make contributions to their selective antitumour activity, especially as known in animal systems (Schwartz and Grindey, 1973; Casazza *et al.*, 1975). It is difficult to separate in a quantitative way the amounts of selective antitumour activity attributable to the host and to the drug in most studies, although Casazza *et al.* (1975) have utilised the murine sarcoma virus (Moloney) to titrate host and tumour induction and regression responses with Adriamycin and daunomycin. In general, however, immunostimulation by tumours in allogeneic (or even putatively syngeneic) hosts is only vaguely understood. For the purposes of this review, then, it will be sufficient to note that tumour, host, and drug form a complex interactive triad. In a further simplification, it is assumed here that immunological effects are contributory to but not determinants of antitumour selectivity of the anthracyclines.

APPROACHES TO DETERMINING CYTOTOXIC MECHANISM

It is almost axiomatic that selectivity of the anthracyclines depends upon physical and chemical characteristics of the agents as a class and upon the special ways that they interact with target cells. As anticancer agents, the anthracyclines are expected to impose on tumour targets lesions that are inimical to survival of the target. Lethal actions, then, are interactions between the agent and the target cell at critical receptors and involve biochemical sequences that are or become irreversible. Accordingly, the stringency of structure–activity relationships determines the degree of drug selectivity, and the distribution of critical biochemical processes affected by the agent determines which cells in a population are targets. In this sense, both an agent's activity and a cell's response are probability functions that determine selectivity. It is from this point of view that the body of this review is written. Potentially lethal agents, moreover, generate cascades of different effects in target cells, and most such resultant effects may have dose-response and other correlations with growth inhibition and cytotoxicity. We

are faced consequently with the difficulty of knowing which receptor interactions, which processes, which effects are the determinants of selective cell
kill. There is no singular means of performing these evaluations, and we generally
tend to rely on intuition to select parameters likely to be important determinants. Interpretation is assisted, of course, when quantitative correlations are
available and when we conjure up schemes that are consistent with the correlations. In effect this is model building, and a few such reveries are described
below.

DNA-ASSOCIATED ACTIONS

The major effects of anthracyclines fall into broad categories with respect to
interactions with DNA. The first refers to structural changes related to the
binding, primarily intercalative and non-covalent, of the agents to duplex DNA
(See Chapter 2). The second includes the functional effects in which kinetics
of DNA-dependent enzymes may be altered. These are not necessarily mutually
exclusive. Nor is a third category, induced DNA damage, for which additional
mechanisms are invoked and so it is discussed separately. The objectives in this
chapter in discussions of binding and kinetic actions are only to evaluate their
respective roles as possible determinants of selectivity for these agents.

In considering DNA intercalation, there have been many attempts to correlate parameters of binding affinity with antitumour activities of anthracyclines
(see Henry, 1976). For native DNA, apparent binding constants of Adriamycin
and daunomycin and most derivatives with a free amino group are moderately
strong and fall within a rather narrow range (K_{dis} = 2-20 × 10^{-6}M) (Arlandini
et al., 1977; Arcamone, 1981). Similarly the number of binding sites (n; molar
drug: DNA-P) range from about 0.13 to 0.20. Antitumour activities against
L1210 and sarcoma 180 ascites cells by such compounds range from no effect to
more than   a doubling of host survival times at essentially molar-equivalent
doses. Comparing daunomycin (K_{dis} = 2.2 × 10^{-6}M; n = 0.16) with 4'-*epi*-
daunomycin (K_{dis} = 2.6 × 10^{-6}M; n = 0.15), survival times for mice bearing
sarcoma 180 ascites are increased 93% by daunomycin but only 20% by the
4'-*epi* derivative at optimal doses (1-1.1 mg per kg). Interestingly, the compound
with the lowest DNA affinity (20 × 10^{-6}M) in this series is the β-anomer of
4'-*epi*daunomycin and, at 1.3 and 5.7 mg per kg, survival times are increased
only 10 and 26% respectively, comparable to antitumour activity of the α-anomer
which has 8 times the affinity for DNA.

It appears from this brief example (and there are many others like it; for
examples see DiMarco, 1975; Chandra, 1975; Henry, 1976; DiMarco *et al.*,
1977) that correlations between physical binding to DNA and antitumour
selectivity are not consistently impressive. To take another example, *N*-acyl
derivatisation of Adriamycin and daunomycin (figure 1) generally results in
marked reductions or total loss of antitumour activity in experimental animals.
The lost activity has been associated with decreased ionic contributions from the

amino group to the strength of intercalative binding (DiMarco *et al.*, 1971; Pigram *et al.*, 1972). In considering such derivatives, it is well established that *N*-acetyldaunomycin has little or no activity (compared to daunomycin) against murine L1210 cells in mice and in culture, although it does have substantial activity against P388 (Henry, 1976). Yet the binding constant to DNA is 200-fold less than that of daunomycin while the number of binding sites is reduced three-fold (Zunino *et al.*, 1972). To take another, *N*-trifluoroacetyladriamycin (AD41) is rapidly formed by esterases that cleave the 14-valerate from AD32 (figure 1) in cells and plasma; neither AD32 nor AD41 binds to DNA by intercalation, yet the agents (AD32 or AD41 or both) inhibit thymidine incorporation, induce DNA damage, and inhibit growth of human CCRF-CEM cells (Kanter and Schwartz, 1979*a*,*b*). Furthermore, AD32 is reported to be an effective chemo-therapeutic drug against murine tumours in animals (Israel *et al.*, 1975; Krishan *et al.*, 1981). The overall picture from data available suggests that intercalation and the physical consequences to the structure of DNA undoubtedly contribute to some of the actions of the anthracyclines, but the correlations are either exceedingly subtle or very complex. Adriamycin and daunomycin bind to all duplex DNA, presumably in all cells into which the agents penetrate, so it is also difficult to see how intercalation *per se* contributes to selective cytotoxicity (even though the mode of binding may be more specific for one base sequence or another) without involving additional mechanisms, as those used in mathe-matical models (see below).

A second consequence of intercalation is that the moderately strong and reversible binding to DNA acts as a depot site within the target cell. In equili-brium studies Schneider *et al.* (1979) estimated that the proportions of bound: total daunomycin and Adriamycin depend on concentrations of the agent and DNA. Thus bound drug may be as much as 65%–85% of the total drug at con-centrations below the intercalating limits for isolated mammalian DNA. Kanter and Schwartz (1979*b*) assumed that the efflux of 4-demethoxydaunomycin (figure 1) from CCRF-CEM cells occurs in two phases. The first or α-phase is rapid ($t_{1/2} < 2$ hr) and probably represents loss of free or weakly bound agent. The second or β-phase clearly follows first order kinetics and the slow cellular depletion ($t_{1/2}$: 20–74 hr) is thought to be a consequence of release from binding sites on DNA or from other sites of moderately high affinity. Over an 8-fold range of 4-demethoxydaunomycin concentrations (0.03–0.25 μM), the ratio of bound:total drug ranges from 0.44 to 0.58 for intracellular concentrations at zero time after a 2-hour loading period. The constancy of the proportionally retained drug is in reasonable agreement with the estimates by Schneider *et al.* (1979) and with pharmacokinetics in man (Benjamin, 1977).

Bound intracellular drug levels correlate (with a correlation coefficient $R > 0.98$) empirically with inhibition of cell growth. Whereas it is possible to express growth inhibition in terms of drug retention during the β-phase for a single agent, there is no such obvious relationship in a series of anthracyclines incubated with CCRF-CEM cells under uniform conditions (Kanter and

Schwartz, 1979*b*). The reasons for this could be that rates of entry, activation and loss of each agent in the series are governed at different rates or by different mechanisms or both. There is reasonable evidence to suggest that anthracycline egress, for example, may be an energy-dependent process, and relative rates of loss for the different agents in drug sensitive and resistant cell populations may be functions of the specific energetics available to each population (Danø 1972, 1973; Danø, *et al.*, 1972; Skovsgaard, 1978). In addition, it is known that metabolic transformations occur in tumour cells at rates that cannot yet be incorporated into a model. Clearly such factors may be specifically different for each agent in the class and among different populations of tumour cells. Over a period of time, both binding and release of anthracyclines from DNA depots are presumably determined by extracellular, intracellular, and intracompart-mental free drug. Because free drug determines both sides of a depot function, loading and unloading, the net intercalated drug present at any time in a cell depends on the factors that regulate these functions — rates of cellular or com-partmental entry and egress, biotransformation, etc. From this it seems that these functions, along with specific biochemical actions, are likely to be among the important determinants of cytotoxic selectivity.

Inhibition of both RNA and DNA synthesis generally accompanies DNA intercalation. The inhibition may be more specific for one aspect of template function than for another (RNA versus DNA; synthesis of ribosomal versus transfer or messenger RNA; initiation versus elongation) but as a general rule the same type of specific effect is usually obtained in a broad spectrum of cells with effective levels of a given agent. Although evidence in support of these con-clusions is given elsewhere (Schwartz, 1979), the point of this rather sweeping generalisation is that inhibition of nucleic acid synthesis by intercalating agents may be specific but it is not selective; nevertheless it is likely to be one of the determinants. Generally, the anthracyclines are not equipotent inhibitors of either RNA or DNA synthesis (that is, precursor incorporation) even at the same intracellular concentrations in either sensitive or resistant cells (Danø *et al.*, 1972; Bachur *et al.*, 1976; Supino *et al.*, 1977; Kessel, 1979*a*), despite close affinity constants and binding site values (n) of the anthracyclines for DNA. Thus, precursor incorporation might reflect altered precursor pool sizes asso-cited with membrane changes, differing rates and products of drug metabolism, or different redox potentials, pK_a, and lipophilic properties of the parent drugs leading to distinctively different disposition of the agents within cells (Bachur *et al.*, 1976; Schwartz, 1976).

Because of the difficulties in interpreting inhibitions of nucleic synthesis by precursor incorporation, it has been valuable to utilise isolated DNA-dependent enzymes for mechanism studies (table 1). Before discussing the data shown in the table, it should be noted that the relative purity and composition of poly-merising enzymes (especially mammalian) and of activated DNA are critical variables that present significant hazards in cross comparison. This caveat not-withstanding, all of the DNA-dependent DNA and RNA polymerases in table 1 are inhibited by Adriamycin and daunomycin. The data furthermore indicate that DNA polymerase α, putatively the eukaryotic synthesising enzyme of S-phase replicating cells, is several-fold more sensitive to the drugs than is the β enzyme, presumably the DNA repair enzyme. Sartiano *et al.* (1979) suggest that this difference may underlie the cell cycle specificity of the anthracyclines. Similarly, it may also account for observations (Kanter and Schwartz, unpublished) that daunomycin-treated cells repair the DNA damage generated by X-irradiation (1,000 rad) within 2 hours, even while drug-induced DNA damage is increasing.

There are other points of importance in the data shown in table 1. Possibly the most striking is that the type of inhibition shown for the various enzymes are extraordinarily different, being competitive in the intermediary range of n (drug: DNA-P) from about 0.01 to 0.2; under these conditions template reversal is readily apparent. Noncompetitive-type kinetics are also obtained in the same range of n but template reversal is minimal. At lower n values (0.001-0.1) the inhibitions tend to be uncompetitive, and as n approaches high values the kinetics are mixed noncompetitive. At present there is no kinetic model which explains these differences but this may only reflect the varying affinities of the enzymes or drugs (or both) for different regions of available base sequences in activated DNA.

Synthetic polymers also display variable types of inhibition with partially purified calf thymus DNA polymerase α. The kinetics in figure 2 are obtained

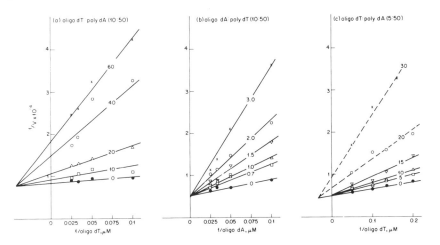

Figure 2 Lineweaver-Burk plots obtained with carminomycin from linear initial velocity of partially purified calf thymus DNA polymerase α with oligo $(dT)_{12-18}$:poly (dA) in panels (a) and (c) and with reverse template, oligo $(dA)_{12-18}$:poly (dT) in panel (b) in constant proportions of oligo:poly indicated in parenthesis. Drug concentrations (μM) are shown in the regressions. Fifteen-minute incorporations were carried out with 75 μM (0.2 μCi) ^{3}H-TTP (panels a and c) or ^{3}H-dATP (panel b) in assays (140 μl) containing 20 mM K-phosphate (pH 7.2), 1 mM dithiothreitol, 1 (b) or 6 (a,c) mM Mg acetate, bovine serum albumen (28 μg) and enzyme (specific activity 2,000–5,000 μ mg^{-1}) (Schwartz, unpublished).

Table 1 Inhibition of DNA-dependent polymerases by Adriamycin (Am) and daunomycin (Dm)

Enzyme Source (type)	Template	Approx. ID$_{50}$ (µM) Am	Dm	Inhibition type (n range)*	Ref.
DNA Polymerase					
E. coli (T4D, wild)	Activated DNA	80	100	Uncompetitive (0.01–0.05)	1, 2
		"Low"		Competitive (0.05–0.11)	
		"High"			
(L56, mutator)	Activated DNA	110	> 100		1
(L141, anti-mut.)	Activated DNA	30	40		1
E. coli (I)	Activated DNA	20			3–5
	Poly (dG):poly (dC)	6			5
	Poly (dAdT)	6			5
	Poly (dGdC)	7			5
	Poly (dA):poly (dT)	26			5
	Poly (dA):(dT)	70			5
M. luteus (I)	Activated DNA	22	26	Mixed noncompetitive (0.09–2.0)	3
M. lysodeikticus	Poly (dAdT)	4	4		6
Rat liver (high mw)	Activated DNA	36	38	Noncompetitive (0.02–0.27)	3
Rat liver (α)	Activated DNA		40		5
	Poly (dAdT)		36		5
	Poly (dA)–(dT)		> 110		5
Rat liver (α)	Activated DNA†	4	5		7
(β)	Activated (67 × K_m)	250	350		7
	Activated DNA^2	60	60		7
	Activated (67 × K_m)	400	400		7

Calf thymus (α)	Activated DNA	15‡	15‡	Competitive (0.1–0.6)	8
	Poly (dT):oligo (dA)	44	0.9		8
(β)	Activated DNA		65	Competitive (0.1–0.6)	8
	Poly (dT):oligo (dA)		25		8
Murine sarcoma virus transformed	Activated DNA	11	15	Competitive (0.02–0.06)	3
RNA Polymerase					
E. coli	Native DNA		10		4
E. coli	Native DNA	2			9
Chicken myeloblastosis (II)	Denatured DNA	2	5	Competitive	9
L1210 (II)	Native DNA	"Low"		Uncompetitive (0.001–0.08)	10
	Native DNA	"High"		Competitive (0.01–0.25)	10

*Drug:DNA-P ratio

†DNA concentration adjusted to K_m

‡K_i values

References: 1, Goodman et al. (1974); 2, Goodman et al. (1977); 3, Zunino et al. (1975); 4, Zunino et al. (1976); 5, Phillips et al. (1978); 6, Chandra (1975); 7, Sartiano et al. (1979); 8, Tanaka and Yoshida (1980); 9, Chuang and Chuang (1979); 10, Maniglia and Wilson (1981). References to earlier polymerase studies are listed by DiMarco (1975).

with the initiator, oligo $(dN)_{12-18}$, in constant molar proportion (10:50) to the template poly (dN') under conditions of linear rates of incorporation of dNTP. The inhibitor used is carminomycin (figure 1), one of the most potent anthracyclines. The reciprocal plots are calculated from incorporation rates for dNTP as a function of the oligo(dN) concentration which limits the duplex concentrations and provides the 3′-OH termini for chain elongation. The enzyme is processive, adding 13–18 (dN) without releasing the template, so that the calculated K_m is an apparent affinity constant, K_s. With oligo (dT):poly (dA), the apparent K_s is 1.4 ± 0.2 μM oligo(dT), and high concentrations of drug $(n = 0.25–6.0)$ are required to show kinetics of inhibition which resemble parabolic noncompetitivity. Figure 2*b* shows competitive inhibition (cooperative) by carminomycin at lower concentrations $(n = 0.017–0.3)$ with the reverse template, oligo(dA):poly(dT). With this template, however, the apparent K_s is 6.9 ± 0.6 μM, suggesting that the drug may be more effective against the template set having the higher apparent K_s, that is the lower affinity for the enzyme. In a comparative study, using only the 10:50 ratios of oligo (dN):poly (dN'), the IC_{50} (μM) for several anthracyclines are about 20-fold higher with oligo (dT):poly (dA) than with the reverse, higher affinity set: carminomycin (25 and < 1.0, respectively), 4-demethoxydaunomycin (35 and 1.4), and daunomycin (40 and 2).

Figure 2*a* shows that, at 10:50 oligo (dT):poly (dA), the inhibition by carminomycin (and daunomycin, not shown) resembles parabolic noncompetitive kinetics. With an apparent K_s = 1.4 μM relative to the oligo(dT), the 10:50 ratio is close to the V_{max} for the system. In another series of studies, the oligo(dT) concentrations are reduced while maintaining the ratio to poly (dA) constant: 3:30, 5:50, 10:100 and so on. The kinetics of inhibition by carminomycin then undergo a change with these conditions (figure 2*c*). At low drug concentrations ($\leqslant 15$ μM), the inhibition is competitive (like that shown with the reverse template in figure 2*b*), but high drug levels (> 15 μM) again reflect the noncompetitive inhibition like that of the 10:50 set (figure 2*a*). Similar changes in kinetics of inhibition are obtained with daunomycin (not shown). Further reduction of the oligo (dT):poly (dA) concentration (2:50) increases the range of competitive-type inhibition, but these reductions in ratios (approaching K_s) have relatively little effect on the IC_{50} for either carminomycin or daunomycin. In all of the studies described here, the kinetics of cooperative competitive and parabolic noncompetitive, suggest two-site binding in a ternary system; furthermore velocity curves (versus either drug or initiator-template concentrations) are sigmoidal which is at least consistent with substrate (DNA) binding. Data presently available are best interpretable as competitions between enzyme and anthracyclines for binding sites on DNA (or the synthetic substrates) and it seems reasonable to suggest that variations in types of kinetics reported in table 1 may be explained by the nature of the associations of the enzymes to different regions of activated DNA (Fisher *et al.*, 1981). It also seems likely that differences in templates (the nicked, activated DNA used in different laboratories) rather than differences in mechanisms may account for some of the variations in types of inhibition kinetics that have been reported.

DNA DAMAGE

The anthracyclines, as well as most other DNA intercalating agents (see Schwartz, 1979), are now known to generate DNA lesions. There is likely to be a number of different mechanisms for their production and some of these are discussed here as they may contribute to the selective cytotoxicity of the anthracyclines.

For the present purposes, DNA damage can be divided into a phase during which induced lesions may be repaired and into another phase in which the damage is different and no repair is evident. Both types of damage are operationally defined, and in fact the degree to which 'repaired' regions of DNA in mammalian cells are functional as authentic templates for both transcription and translation is still not established. The so-called repairable lesions are usually single-strand breaks (or alkali-labile regions) and, using damage from ionising radiation as an illustration, these lesions are quantitatively repaired within 2 hours to $> 95\%$ completion even at superlethal doses of 100,000 rad (400 or more times the D_{37} for most cultured mammalian cells). The unrepaired lesions remaining at this time are probably double-strand breaks that occur when two hits fall randomly on opposite strands within four or five base-pairs of each other. The initial or primary damage found immediately after exposure of cells to ionising radiation is generally characterised by the random sizes of the DNA fragments, and by rapid apparent repair. The amount of this initial damage per rad is relatively constant for most mammalian cells, depending primarily upon dose, rate, aeration, and intracellular radical-quenching defences.

There is a later manifestation of damage that appears after the initial damage from irradiation is $> 95\%$ repaired. This secondary damage appears as double-stranded units with fragment size distributions that are non-random (Kanter and Schwartz, 1980). These units remain unchanged or they may increase in amount with time (24 hours or more), but there is no evidence that they participate in DNA repair processes. The average molecular weight (M_n) of the secondary fragments (M_{n_2}) is about 4-6×10^5 or about 3 orders of magnitude less than that of repaired or control (M_{n_0}) DNA (3-6×10^8 for HeLa and CCRF-CEM cells). The fragments forming during this secondary phase are related in size to nucleosomal DNA, and appear to be a result of an early irreversible change in lethally intoxicated cells.

Skalka *et al.* (1976) and Yamada *et al.* (1981) demonstrated that DNA of irradiated lymphoid cells breaks down after a few hours to discretely sized fragments with lengths that are integral multiples of nucleosomal DNA (1.1×10^5) where the integers are $1,2,3,\ldots.8$. Correspondingly the M_{n_2} of the secondary fragments is the weighted average for four nucleosomal units (Kanter and Schwartz, 1980). In addition to irradiation, secondary DNA fragments of this size distribution are formed by a number of DNA reactive agents (figure 3). With

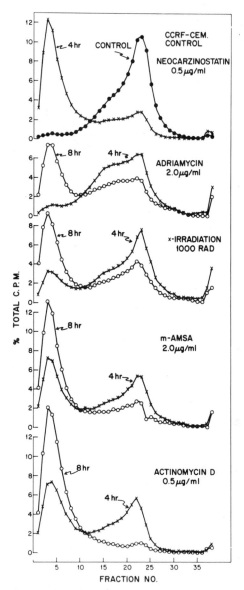

Figure 3 Alkaline sucrose gradient sedimentation profiles of ^{14}C-prelabelled DNA from CCRF-CEM human leukaemia cells at 4 and 8 hours after incubation with different DNA-reactive agents or after 1,000 rad X-irradiation (Kanter and Schwartz, in preparation).

the anthracyclines, non-random secondary damage is obtained in murine P388 and human CCRF-CEM leukaemia an HeLa cells four hours after treatment (Kanter and Schwartz, 1979b; Kanter, 1981; Kanter and Schwartz, in preparation): carminomycin, 4-demethoxydaunomycin, Adriamycin, daunomycin, and

AD32. Secondary damage (Pater and Mak, 1974; Kuo and Hsu, 1978; Kanter, 1980; Wyllie, 1980; Kanter and Schwartz, in preparation) occurs with other DNA reactive agents which include: actinomycin D, neocarzinostatin, bleomycin (Chapter 6), *m*-AMSA (Chapter 1), the nitrosoureas BCNU and CCNU, and even glucocorticoids. In cells treated with the anthracyclines, random-type damage is seen in alkaline sucrose gradients during the first hour or so after drug exposure. Then there are gradual shifts, both to large control-size DNA and to the small secondary fragments, without the intervening period of complete apparent repair, so familiar in irradiated cells. The fractional distribution between amounts of the small fragments (M_{n_2}) and control size DNA is dose-dependent and presumably reflects a probability function of lethal intoxication within a population. Thus, survivor cells regenerate and maintain control-size DNA whereas those fated to die form the secondary fragments of nucleosomal DNA for which there is apparently no repair.

In independent sets of observations by Ross *et al.* (1979) and Zwelling *et al.* (1981), alkaline elution of DNA from L1210 cells treated for an hour with DNA-reactive agents also demonstrates two types of elution patterns. The first type is nearly first order with respect to time, suggesting a random distribution of single-strand breaks. The other pattern indicates extraordinarily rapid elution of a fraction of DNA from filters without further loss. This latter suggests that rapidly eluted DNA consists of small secondary fragments, the retained DNA presumably being like that from control cells. Amounts of the first type of eluted random-sized fragments are dose-dependent and are obtained with relatively low concentrations of Adriamycin, ellipticine, and *m*-AMSA; the second type is described as a function of high drug levels. Thus differential patterns of elution seem likely to distinguish between the initial damage and the secondary DNA fragments of lethally intoxicated cells.

A number of the reports above suggest that the nonrandom sizes of double-strand DNA cuts are internucleosomal chromatin breaks. It is at such relatively exposed junctures that intracellular Mg^{2+}-Ca^{2+} endonucleases may be acting to cleave DNA to produce the discretely-sized double strand fragments (Hewish and Burgoyne, 1973). Suciu (1979) showed that release of chromatin from nuclei of murine liver, spleen, and thymus after X-irradiation is dependent on Mg^{2+} and Ca^{2+} concentrations, and this release correlates with soluble alkaline endonuclease activity for these organs. Protein/DNA ratios and molecular weights of the released DNA indicate that the chromatin is like that of nucleosomal units; whole body irradiation (1,000 rad) increases the release of this chromatin from thymus and spleen but not liver (which correlates with the relative susceptibility of the organs to radiation damage). Radiation-released fragments from thymocytes correspond in size to discrete nucleosomal-sized fragments, as shown by Yamada *et al.* (1981) who used micrococcal nuclease digestion for the comparison. DNA cleavage by bleomycin also occurs at inter-nucleosomal sites, as judged from size distributions (Kuo and Hsu, 1978). Wyllie (1980) reported that integers of nucleosomal units are excised in association with glucocorticoid-induced death of rat thymocytes, apparently after activation of nonlysosomal endonuclease. The increase of this type of endonuclease activity occurs gradually over a period of several hours after either irradiation or steroid treatment.

The DNA fragments formed during the secondary phase of DNA breakdown may condense to form small apoptotic bodies (Kerr *et al.*, 1972; Williams *et al.*, 1974; Don *et al.*, 1977). This process appears to be a programmed sequence for physiological cell death as well as for death from some noxious agents. It is initiated before free lysosomal enzyme activity increases or cell autolysis is apparent (Aoyama *et al.*, 1972). Apoptosis is worth focusing on because, as an orderly process leading to cell death, it supposedly includes early irreversible cellular changes that may be used to distinguish drug sensitive cells within a population from those that are resistant. By using the formation of secondary DNA fragments in tumour cells as a criterion of drug sensitivity, there may be some clinical value for predicting tumour responses to a therapeutic regime. An example of this application is shown in table 2 (Schwartz *et al.*, 1981). In the study, leukaemic cells from patients prior to therapy are incubated for four hours with 1 μg ml^{-1} Adriamycin before DNA damage is measured. The

Table 2   *In vitro* DNA damage by Adriamycin (1 μg/ml; 4 hr) in peripheral blast cells from AML patients prior to therapy (from Schwartz, Preisler, and Kanter, 1981)

Patient	History	*In vitro* DNA lesions*	Patient response
1	New	< 0.15	RD
2	Relapsed	< 0.15	RD
3	Relapsed	< 0.15	RD
4	Relapsed	< 0.15	RD
5	Relapsed	< 0.15	RD
6	New	0.18	CR
7	New	0.31	CR
8	New	0.31	CR
9	Relapsed	0.41	RD
10	New	0.45	CR
11	New	0.46	CR

Peripheral blood leukaemia cells were obtained from 11 patients with acute non-lymphocytic leukaemia. Six patients (new) were studied at the time of their initial diagnosis while the remaining five patients were studied at the time of relapse. All patients except no. 2 received combination remission induction therapy consisting of adriamycin (30 mg/m^2/day on days 1, 2, 3) together with ara-C (100 mg/m^2/day on days 1–10). Patient no. 2 relapsed while on a regimen which included therapy with Adriamycin-ara-C. This patient subsequentially did not enter remission after being treated with daunomycin and 5 azacytidine. CR indicates complete haematologic remission as according to the criteria formulated by Cancer and Acute Leukemia Group B. Patients with resistant disease (RD) survived at least 7 days after their course of chemotherapy was completed so that a determination of the *in vivo* sensitivity of their leukaemic cells to remission induction therapy could be estimated.

*Frank breaks or alkali labile regions in the DNA alkaline unwinding unit by the method of Kanter and Schwartz (1979c).

concentration of drug in the incubation medium is comparable to clinically obtainable plasma levels, and the time is sufficient to produce secondary DNA damage with Adriamycin in human cells. The patients response to therapy and the DNA damage produced *in vitro* show high positive correlations between detectable damage and complete remission (CR). There is also an indication in the data that resistant disease (RD) correlates with prior therapy in relapsed patients and with little or no susceptibility to *in vitro* DNA damage. Extended now to 22 patients, our preliminary analysis indicates that production of *in vitro* detectable DNA damage is the desirable configuration for new patients to obtain CR ($P > 0.97$). In contrast, prognosis of relapsed patients whose cells resist *in vitro* damage is foreboding, having an 80% probability of RD. The amounts of drug taken up and retained by these cells bear no consistent relationship either to clinical response or to the amount of DNA damage produced by Adriamycin (Schwartz and Kanter, 1981). In two murine leukaemia models, Kanter (1974) also obtained evidence of correlations between DNA damage assayed *in vitro* with six anthracyclines and the survival times of the tumour-bearing (EL-4 and P288 cells) syngeneic hosts after treatment with the agents. These correlations are *prima facie* evidence that delayed expressions of DNA damage may be valuable signals of drug selectivity in different cell populations.

BIOTRANSFORMATION

Anthracycline transformation may influence selectivity by several discrete mechanisms: formation of oxygen radicals, activation to radical drug forms, formation of active products, and termination of drug action. Generation of radicals is discussed in detail in Chapter 9 and is mentioned only as it may pertain to selective cytotoxicity and to biotransformation. Briefly, the compounds may react cyclically to generate oxygen radicals under aerobic conditions. In addition, reduction of the quinone triggers biological activation which may account for the anaerobic potency of these and other quinone-containing agents (Schwartz *et al.* 1963; Teicher *et al.*, 1981).

Anthracycline metabolism generates a substantial number of products, only a few of which are detected by fluorescence. In our current studies, we utilise an electrochemical detection (ECD) unit in series with a fluorescence monitor to evaluate HPLC eluates of metabolised anthracyclines. The ECD distinguishes as many as 30 different products of anthracycline biotransformation, most of which have little or no fluorescence (Schwartz *et al.*, 1982). We presume that these low fluorescent products are semiquinones, open-ring structures, or radicals that are relatively stable under our conditions. It is also possible that reductive activation may produce C_4 dimers, analogous to the 7-deoxyalkavinone dimer formed by reductive cleavage of aclacinomycin A (Komiyama *et al.*, 1979a). Because the low fluorescent products have not been identified, the subsequent discussion deals primarily with the fluorescent compounds that have been isolated and identified. It might be kept in mind, however, that the intermediary reactions that precede formation of each of these products are not well understood and that metabolic pathways are still speculative (figure 4).

There are three types of reactions (other than conjugations) that account for the fluorescent aglycones of Adriamycin and daunomycin that have been identified in urine and bile of animals and patients. All of these and others have now been identified from anaerobic drug incubations with liver microsomes as well as with the reconstituted purified component enzymes, cytochrome P-450 reductase. The anaerobic reactions that daunomycin (Dm) is known to undergo may be generalised as:

(1) Reductive cleavage at C4 to demethoxydaunomycin (4-dDm) and at C7 to the 7-deoxy aglycone (7-dO-Dma).
(2) Hydrolytic cleavage at C4 (4-HO; carminomycin) and at C7 (7-HO aglycone; daunomycinone).
(3) Carbonyl reduction in the side chain (C13) to the dihydro form (Dmol; daunomycinol).

These three types of reactions account for all of the metabolites shown in figure 4. As suggested above, the pathways indicated are tentative, and are likely to be both more complex as well as variable in different cell populations. For example, the reduction of hydroxyl groups at C4 (of 7-deO-Cma) and C7 (of daunomycinone) may occur anaerobically with rat liver microsomes, but to a lesser extent with xanthine oxidase. In addition the pathways depend not only upon redox potentials of the drugs and each transformation product but also upon the reducing enzymes and electron donors. Indeed, some of the reduced aglycones themselves act as electron donors in coupled reactions.

The metabolites in broken boxes in figure 4 are those that have been identified (unconjugated) in urine and bile from daunomycin (Takanashi and Bachur, 1975; Chan and Watson, 1978); the same derivative metabolites have also been isolated from Adriamycin (Bachur, 1979). The primary distinction between biotransformation of Adriamycin and daunomycin (insofar as is presently known) is that the rates of transformation are considerably different, usually being the more rapid for the latter.

Reductive cleavage at C7 by microsomes and by cytochrome P450 reductase is the first reaction undergone by daunomycin, producing the 7-deoxy aglycone (7-dO-Dma). The same reaction occurs with carminomycin, 4-demethoxydauno-mycin, and Adriamycin (Schwartz *et al.*, 1982). The anaerobic reduction of daunomycin generates drug radicals and the reaction occurs with NADPH cytochrome P450 reductase (Komiyama *et al.*, 1979*a*) or with other one-electron transferring flavoproteins with appropriate electron donors (such as NADH or NADPH) (Komiyama *et al.*, 1979*b*; Pan and Bachur, 1980; Pan *et al.*, 1981). Under aerobic conditions, Adriamycin and daunomycin undergo no net metabolic change with liver microsomes, and both agents mediate O_2 conversion to O_2^- (Handa and Sato, 1975; Bachur *et al.*, 1977) which may initiate a cascade of oxyradicals (see Chapter 9). Anaerobically the reduction by NADPH and

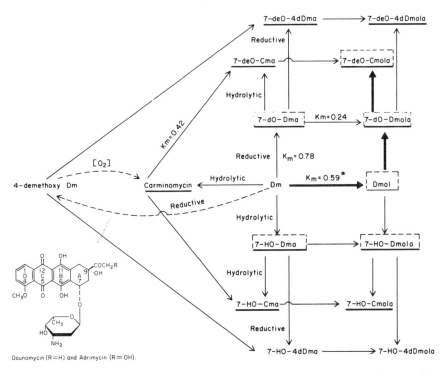

Figure 4 Proposed metabolic pathways of daunomycin (Dm), 4-demethoxydaunomycin and carminomycin (Cm) to aglycones (shown by light arrows). Aglycones form during anaerobic incubations of daunomycin and rat liver microsomes with NADPH and an isocitric dehydrogenase generating system. Underlined compounds have been identified as products of microsomal metabolism; those in broken boxes have been identified in urine and bile of treated patients by Takanashi and Bachur (1975) and Chan and Watson (1978); the heavy arrows indicate their proposed major pathway. Speculative pathways for 4-demethoxy daunomycin are shown as broken arrows. K_m values are those determined *in vitro* by Schwartz and Parker (1981) and (*) by Loveless *et al.* (1978).

cytochrome P450 reductase transforms the agents to semiquinone radicals, permitting spontaneous cleavage to 7-deO-aglycones (Andrews *et al.*, 1982). Either the oxyradicals or the drug radical or both may be responsible in some part for cytotoxicity of the anthracyclines.

The second reaction occurring in the anaerobic liver microsomal system is reduction of the C9 side chain carbonyl (C13) of 7-deO-Dma to form 7-deoxy-daunomycinol aglycone (7-dO-Dmola). The over-all reaction from daunomycin to 7-dO-Dmola follows linear sequential kinetics: daunomycin → 7-deO-Dma → 7-deO-Dmola (Schwartz and Parker, 1981).

In animals, however, the preferred pathway seems to be for daunomycin (and Adriamycin) to be reduced directly to daunomycinol (Dmol) and Adriamycinol (Amol) with subsequent cleavage to the 7-deO aglycones. The side chain reduction to Dmol and Amol is carried out by a soluble aldehyde reductase

(+ NAD(P)H). The enzyme is found in all tissues that have been tested, has a low substrate specificity, and is not inhibited by O_2 (Felsted and Bachur, 1980). Among the properties of Dmol and Amol, both are biologically active and inhibit precursor incorporation into RNA and DNA (Bachur *et al.*, 1976).

A third metabolite formed anaerobically by liver microsomes is 7-HO-Dma. It appears only after the first two aglycones in the sequence reach almost maximal levels during anaerobic incubations, and when Dm is almost completely exhausted. The source of this aglycone is presently uncertain although it is shown as deriving from daunomycin in figure 4 in a sequence of branched kinetics. This pathway is consistent with transformation inhibitions by SKF525A and cyclophosphamide described below. (Dm, daunomycin).

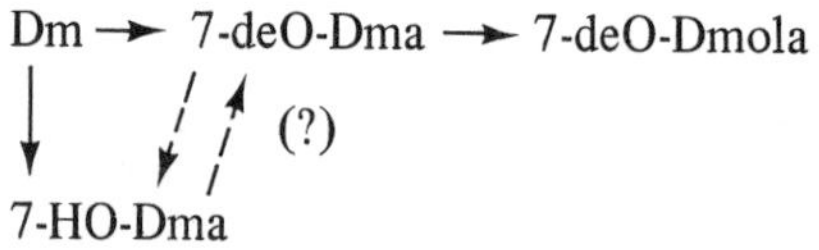

Dm ⟶ 7-deO-Dma ⟶ 7-deO-Dmola

↓ ↗↗ (?)

7-HO-Dma

Our unpublished studies with purified 7-deO-Dma, 7-HO-Dma, and microsomes also yield some evidence that suggest possible formation of 7-deO-Dma from 7-HO Dma, and the reverse but to a lesser extent.

The other aglycones formed from daunomycin (figure 4) have been found wih aged microsomes and purified microsomal enzymes in reconstituted systems (Schwartz *et al.*, 1982). These accumulate later during anaerobic incubations after daunomycin is no longer detected, apparently deriving from the previously formed aglycones. The reactions occur primarily as C_4 reductive and hydrolytic cleavage of the daunomycin-derived aglycones, thus forming the aglycones of 4-demethoxydaunomycin and carminomycin, respectively. Their formation is inhibited $> 90\%$ by SKF525A (β-diethylaminoethyl diphenylpropylacetate 4×10^{-4} M), suggesting these biotransformations require cytochrome P450 rather than the reductase. By contrast, the accumulations of 7-deO-Dma and 7-deO-Dmola are increased 50–300% and those of 7-HO-Dma are essentially unaltered by SKF525A. Similar results are obtained with cyclophosphamide from which the acrolein biotransformation product binds and inactivates cytochrome P450 (Gurtoo *et al.*, 1981). This is a possible site of drug interaction, and suggests one way that cyclophosphamide may alter cytotoxicity of the anthracyclines.

It has now been shown that liver nuclei transform Adriamycin and daunomycin to aglycones through the semiquinone radical mechanism (Bachur *et al.*, 1982). The reaction is also carried out by heart and kidney nuclei. The proximity of these drug radicals to DNA support the possibility of adduction of aglycones to DNA. Synthesis of DNA is associated with nuclear membranes (Infante *et al.*, 1976) and the radicals formed here during transformation of the anthracyclines may contribute to DNA damage (Berlin and Haseltine, 1981).

A number of reports now indicate that xanthine oxidase and other flavoproteins transform the anthracyclines. In the remainder of this discussion on biotransformation, the studies with xanthine oxidase will be reviewed because it and similarly acting enzymes are likely to be active in tumour cells. Furthermore, xanthine oxidase is important in purine metabolic pathways, so that it is possible that the metabolism of anthracycline affects, and is affected by, nucleic acid precursor metabolism (Rajagopalan, 1980). This could be an important consideration in patients pretreated with allopurinol to reduce uric acid levels by blocking this enzyme.

Initial anaerobic reductive cleavage of the anthracyclines by xanthine oxidase appears to be similar to that observed with NADPH-cytochrome P450 reductase, involving drug radicals and then aglycone formation. Komiyama *et al.* (1979*b*) demonstrated that xanthine oxidase reduces both Adriamycin and daunomycin to their respective 7-deoxy-aglycones. For daunomycin the active electron donors are: NADH > purine > xanthine > glycoaldehyde >> NADPA (Pan and Bachur, 1980). With xanthine as the electron donor in an aerated medium, oxygen uptake is stimulated and cleavage occurs only after the system becomes anaerobic. During the aerobic phase with NADH and NADPH when oxygen is consumed, the radical O_2^- is generated. Other flavoproteins catalysing reduction of the anthracyclines are also those having 1-electron transfer systems (Pan *et al.*, 1981): nitrate reductase, NADH-cytochrome *c* reductase, and lipoamide dehydrogenase. Obligate 2-electron transport flavoproteins are inactive (glutathione reductase, L-amino acid oxidase, D-amino acid oxidase, D-glucose oxidase, and lactic dehydrogenase). Oxygen consumption with NADH-xanthine oxidase occurs also with 7-HO-Dma > daunomycin > 7-deO-Dma (Pan *et al.*, 1981), so it is reasonable to expect that the aglycones may contribute to cytotoxicity by foming O_2^- radicals. The highly lipophilic nature of the aglycones suggests the further possibility that they cannot be eliminated from cells (except after conjugation, which may be limited in tumour cells) and they may reside within the lipid domain of membranes for relatively long periods of time.

Assuming that xanthine oxidase is a major anthracycline metabolising enzyme in tumours and some normal tissues, it may be of extreme clinical relevance that allopurinol, a tight binding inhibitor of the enzyme, is routinely administered to many patients to lower uric acids levels prior to and during therapy. The cleavage of daunomycin by the enzyme is inhibited by allopurinol. Figure 5 shows that concentrations (< 0.3 mM) of allopurinol achievable in plasma (Appelbaum *et al.*, 1982) block formation of 7-deO-Dmola with a concurrent increase in 7-deO-Dma. At higher concentrations (> 1.0 mM) formation of 7-deO-Dma is completely inhibited. The clinical consequences of combining allopurinol and anthracyclines may be to increase toxicity of the agents by slowing detoxification in tumour, heart or other tissues. If allopurinol inhibition (at high concentrations) prevents formation of O_2^- and anthracycline semi-quinone radicals (and this is not known at the time of writing), anthracyclines may still act at membrane sites as well as interfering with nucleic acid synthesis by intercalation.

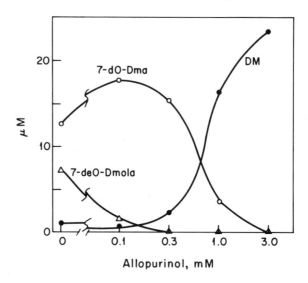

Figure 5   Biotransformation of daunomycin (Dm) to the 7-deoxy-Dm aglycone and 7-deoxy-daunomycin aglycone by xanthine oxidase (milk) with varying concentrations of allopurinol. Incubations are carried out in 1 ml containing 35 μM daunomycin, 0.3 mM hypoxanthine, 50 mM K-phosphate, and 1.0 mM EDTA for 30 min under anaerobic conditions (N_2). Extractions and analysis by HPLC were as described by Schwartz and Parker (1981).

CELL MEMBRANES AND LIPIDS

Most agents pass through plasma membranes, in the absence of pores and carrier-mediated transport, by diffusion through lipid domains of the structure. It is by this mechanism as well as by carrier mediation that anthracyclines appear to enter cells (Skovsgaard, 1978; Dalmark, 1981), so a brief description of membranes and membrane models may have some value here. In the fluid mosaic model, the matrix of a membrane is a bilayer of phospholipids, part of which is fluid under physiological conditions. Both phospholipids and integral proteins are distributed asymmetrically in the two sides of a membrane, with hydrophobic ends (or regions) extending into the maxtrix and hydrophilic regions at or extending from the two surfaces. Most of our present information is based on red cell structure in which the zwitterions phosphatidylcholine and sphingomyelin are concentrated in the external plane of the bilayer. The zwitterion phosphatidylethanolamine is concentrated in the internal or cytoplasmic face of the bilayer along with phosphatidylserine which has a net negative charge at neutral pH.

The asymmetric charge distribution in bilayers of the red cell membrane is the basis for the *coupled bilayer hypothesis* by Sheetz and Singer (1974): each bilayer in a closed membrane may respond differently to various perturbations while still remaining coupled to each other. As a consequence, planar expansion or contraction of one bilayer relative to the other alters the curvature of both bilayers to the extent that slippage is prohibited or at least impeded (as by van der Waals contacts or by transecting proteins between the bilayers). Thus the effect of a perturbation is not confined to its immediate vicinity but instead is broadcast over a far larger segment of the domain. The energy cost of the coupling is distributed by ATP-spectrin interactions at the cytoplasmic face of the plasma membrane (Birchmeier and Singer, 1977), and results in large thermodynamic changes on both sides of a membrane. These may lead to a vast array of consequences, including shape changes of the intact cell. It is beyond our scope here to review the bilayer couple hypothesis in detail, but it is important to describe some of the consequences of the asymmetry of bilayers pertaining to actions of the anthracyclines.

Amphipathic cationic molecules (including the anthracyclines and other DNA intercalating drugs) demonstrate two-phase binding. 'Low' drug concentrations protect erythrocytes from hypotonic lysis and 'high' concentrations promote it (Seeman, 1970). These effects are presently related to the negative potential across the membrane which at pH 7 and low ionic strength decreases the binding of cationic amphipaths. Binding at the cytoplasmic side of the membrane produces the biphasic fragility curves, first decreasing and then increasing fragility as drug concentrations are increased (Schioppacassi *et al.*, 1977). The red cell biphasic fragility curves obtained with anthracyclines occur at concentrations (50 μM) far in excess of those obtainable in therapeutic regimens (Schioppacassi and Schwartz, 1977; Mikkelson *et al.*, 1977) and are described here only as a model to demonstrate membrane interactions. Despite the high drug concentration used to obtain the curves, the changes in fragility of red cells may be important in that they reflect not only species differences but also differences among individuals (Schioppacassi and Schwartz, 1977), and may be related to genetic and/or to environmental factors (such as dietary or pathological effects). It seems also pertinent that chlorpromazine and numerous other sedatives and anaesthetics show the same type of biphasic fragility curve with erythrocytes (Seeman, 1970). And the anthracyclines, in turn, have anaesthetic (Kanter, unpublished) as well as neurotoxic properties (Philips *et al.*, 1975; Cho, 1977; Cho *et al.*, 1979, 1980; Bronson *et al.*, 1982). Neurotoxicity and membrane reactivity are also common to a substantial number of other DNA intercalating agents (see Schwartz, 1979).

The bilayer couple hypothesis also predicts that the shape of red cells and ghosts should change with the acid-base nature of amphipaths. Anionic and impermeable amphipaths bind preferentially to the external side of a membrane, expanding it relative to the inner side and crenating the surface. The cationic amphipaths pass (uncharged) into the membrane, accumulate (protonated) in the cytoplasmic face of the bilayer, and cause it to expand relative to the external plane of a bilayer, thereby inducing erythrocytes to change into cup configurations. The amphipathic tertiary amines chlorpromazine and lidocaine are examples of such agents. Similar changes are observed with Adriamycin and daunomycin at concentrations that may be pertinent to clinically achievable plasma levels (Goldman *et al.*, 1978; Mikkelsen *et al.*, 1977).

These models indicate that the anthracyclines act in cells without nuclei, suggesting that membrane effects independent of DNA binding might contribute to selective cytotoxicity. Using P388 cells Kessel (1979*b*) found that Adriamycin increases the negative potential and enhances glycosylation while membrane hydrophobicity decreases. The changes, prominent in cells sensitive to the agent, are minimal in resistant lines of P388. Tritton and Yee (1982) attempted to distinguish a membranal component of cytotoxicity from those that may be related to DNA actions. In their approach to this critical problem, they immobilised Adriamycin covalently through an *N*-alkylcarbamate linkage to agarose heads. Both growth and clonogenicity of L1210 cells are inhibited by immobilised as well as by free drug but not by the underivatised beads. Release of Adriamycin from the beads was measured fluorometrically in cell extracts after exposure, and it was found that fresh beads, even exhaustively washed after derivatization, leak a small amount of drug (0.005% of total) into the cells. If the same beads are then washed and again exposed repeatedly to cells, however, no intracellular drug is detected and yet the potential for cytotoxicity still remains. The estimated effective concentration of the agarose-bound Adriamycin is about 8×10^{-10} M or 100 to 1,000 times less than required for comparable cytotoxicity by the free drug. This potentiation is critical because it suggests extraordinary drug sensitivity at the plasma membrane. It is not yet known whether the action occurs on the external membranal face or if a portion of the molecule enters the membrane. Even if the drug itself is not released into the cytoplasm, it might be converted by plasma membrane NADH oxidase or other electron-shuttling enzymes (Crane *et al.*, 1980) to a semiquinone with accompanying cleavage and release of aglycone(s) within the membrane. Because of the high lipophilicity of the aglycones and their ability to participate in redox reactions, the aglycones could be trapped within the lipid structure of a membrane, exerting toxic effects for long periods of time.

Drug-related changes in redox functions at membrane sites may have a substantial number of sequelae in a cell. The report by Crane *et al.* (1980) shows marked stimulation of NADH oxidase by Adriamycin. This enzyme participates in regulation of cyclic AMP and cyclic GMP and cyclic phosphodiesterases which are inhibited by NADH. The net effect of the anthracyclines, by altering redox states in membranes, might be to interfere *indirectly* with cell proliferation, hormone regulation, energy-dependent metabolite transport, and cation exchange in heart and other organs as well as in tumour cells (Goldenberg *et al.*, 1978; Dasdia *et al.*, 1979; Harper *et al.*, 1979; and Caroni *et al.*, 1981). It seems doubtful whether the anthracyclines have real *direct* effects on cyclase systems, although cardiac guanylate cyclase is moderately inhibited (50%) by high (1 mM) concentrations of Adriamycin and daunomycin (Levey *et al.*, 1979; Lehotay *et al.*, 1979). Nevertheless, there is a substantial body of experimental evidence that selectivity of anthracyclines may be due, at least to some degree, to actions at membrane sites which may be communicated to a cell's DNA (Williams *et al.*, 1974). Mechanisms of selective action in membranes are still fragmentary, mostly indirect, and cannot yet be directed at specific target cells. Still the different combinations of physicochemical properties within the family of related anthracyclines, together with the array of membranal protein and lipid components that are unique to each population of cells, seem likely to provide the range of distinctions among agents and among cell populations necessary for selective cytotoxicity.

To identify parameters that might contribute to selective membrane activity, it is worth while to review some of the recent work on drug–liposome interactions. Using single wall liposomes prepared from phosphatidylcholine and dicetylphosphate, Goldman *et al.* (1978) demonstrated differential affinity of a set of anthracyclines for the membranes. In general, partitioning of the agents into non-polar lipid regions of the membranes correlates with lipophilicity (at pH 7.2) : Adriamycin $<$ adriamycin-14-acetate $<$ daunomycin $<$ adriamycin-14-octanoate $<$ AD32. Thus Adriamycin which is a weak base (pK_a = 8.2) and the least lipophilic of the five only partially partitions into polar lipid regions at pH 7.2. Increasing the alkyl chain at the 14-position (or neutralising the amino charge as the *N*-trifluoroacetyl in AD32) increases lipid partitioning. In this study Goldman *et al.* (1978) also demonstrated erythrocyte hypotonic lysis and deformations by these anthracyclines. Shape changes to cup-forms are consistent with the bilayer couple hypothesis, and correlate with their data from liposomal partitioning.

Using liposomal preparations, Tritton *et al.* (1977) and Karczmar and Tritton (1979) demonstrated that Adriamycin fluidises pure phosphatidylcholine membranes (that is, the temperature of gel-to-liquid crystal phase transition [T_m] is lowered). In this regard, it is also interesting that Adriamycin at cytotoxic concentrations (0.1–10 μM) increases the membrane fluidity of sarcoma 180 cells within 30 min (Murphree *et al.*, 1981). When small amounts of cardiolipin are incorporated into liposomal membrane, Adriamycin raises the T_m (Karczmar and Tritton, 1979). The T_m is lowered by Adriamycin when phosphatidyl serine replaces cardiolipin. The presence of anionic phospholipids apparently alters the organisation of the bilayer, thereby changing affinity of the drug for the membrane at physiological temperatures (above the T_m). (Below the T_m, the membranes tend to exclude Adriamycin from deep penetration.) In the fluid membrane above the T_m, cardiolipin facilitates two types of penetration, one in which the fluorophore is relatively exposed, the other more deeply embedded in the membrane. Part of the interaction between cardiolipin-containing membranes and Adriamycin might be explained by a complex formed between the drug and cardiolipin which is partially ionic (Duarte-Karim *et al.*, 1976) but also involves pi-electron interactions between unsaturated regions of the lipid chains and the conjugated ring system of the drug chromophore (that is, lipid intercalation) (Schwartz *et al.*, 1978; Schwartz and Kanter, 1979). It is also reasonable to expect that gross changes in the membranes caused by small amounts of cardiolipin reflect an asymmetric localisation of cardiolipin in the inner face of the bilayer (Serpentino, 1977). This condition is similar to that described in relation to the bilayer couple hypothesis (Sheetz and Singer, 1974).

In describing some effects of Adriamycin on liposomes, it was noted above that incorporation of low concentrations of the acid phospholipids cardiolipin and phosphatidylserine may support opposite changes in membrane fluidity (Karczmar and Tritton, 1979). This appears to be another indication that small alterations in membrane composition may cause marked changes in drug action. Assuming that cardiolipin is a privileged target for the anthracyclines, Goormaghtigh *et al.* (1980) correlated binding to cardiolipin, penetration into liposomal membranes, and apparent toxicity to cardiac mitochondria. Myocardial mitochondria are a rich source of this phospholipid and are also organelles prominently manifesting damage during progressive cardiac intoxication by Adriamycin and daunomycin. It may likewise prove to be important that cardiolipin is found, although at lower concentrations, in other organs and some tumours (White, 1973; Bergelson *et al.*, 1974).

The roles that lipids and the radical products of their autoxidation might play as determinants of selectivity are not yet clear. There are a number of correlations that suggest that drug interactions with unsaturated lipids in general, and cardiolipin in particular, may enhance cardiac toxicity. Lipid peroxidation is stimulated by metabolic generation of reactive oxygen species from Adriamycin and daunomycin (Goodman and Hochstein, 1977). In addition, products of lipid peroxidation may add to drug cytotoxicity (Beneditti *et al.*, 1979). Autoxidation of unsaturated lipids (such as cardiolipin, linoleic acid or arachadonic acid) also produce non-enzymatic changes in the anthracyclines. Schwartz and Kanter (1979) postulated that released carbonyls react with the free amino of daunosamine to form Schiff bases that are biologically active. The interaction with arachadonic acid also raises the possibility that anthracyclines may affect prostaglandin levels in some cells (Schwartz, unpublished).

The relative ease of production of oxyradicals by the anthracyclines is a function of the redox potential of the agents, and Lown *et al.* (1982) have shown that anthracyclines of different structures having correspondingly different electrochemical properties may indeed correlate with cardiotoxicity. The depletion of scavengers (Olson *et al.*, 1980) and the low levels of such intrinsic radical-protection systems as glutathione peroxidase, superoxide dismutase, and catalase in heart (Meyers *et al.*, 1977) might therefore permit lipid autoxidation to continue unabated. By inhibiting glutathione reductase, BCNU may enhance radical-dependent cytotoxicity of the anthracyclines (Sagone and Burton, 1979). The cascades of oxyradicals that are generated intracellularly might damage mitochondrial as well as nuclear DNA continuously in myocardial cells (Kanter and Schwartz, 1977), even after drug concentrations fall below effective levels. It is also appropriate to note that α-tocopherol and ubiquinone, both membranal radical-quenching agents, have been proposed as protectants against cardiac toxicity (Bertazzoli *et al.*, 1975; Myers *et al.*, 1976; Doroshow *et al.*, 1979). In hopes that radicals generated within membranes are causally related to cardiac toxicity, the protectants are expected to sop up these radicals without interfering with DNA-related antitumour activity. As indicated above, such a dual mechanism of selective cytotoxicity is not inconsistent with anthracycline actions. However, it should be emphasised that lesions associated with membrane radicals and with DNA-related effects are common to both tumours and myocardia and to other tissues as well, and the effectiveness of such protectants is not established (Mimmough *et al.*, 1979).

DETERMINANTS IN MODELS OF SELECTIVITY

Evaluation of parameters as possible determinants of selectivity is usually approached experimentally by modelling, and some qualitative studies have been described above. Mathematic modelling may be useful to identify and weigh the contribution of various parameters to selective drug actions. Kanter and Schwartz (1979*b*) reported correlations between intracellular drug levels and a functional DNA-dependent effect (inhibition of thymidine incorporation, I) in a human cell line, CCRF-CEM. To develop a model, the structural lesion, DNA damage (n^*), was also evaluated, using 10 anthracyclines (figure 1; table 3). The agents can be separated into three groups and are numbered in the order of increasing lipophilicity. The first group of five are essentially unsubstituted: (1) Adriamycin, (2) 4'-*epi*adriamycin, (3) carminomycin, (4) 4-demethoxydaunomycin, and (5) daunomycin. The next group of three consists of C14-esters of Adriamycin: (6) benzoate, (7) 1-naphthaleneacetate, and (8) octanoate. The last two in the set are the most lipophilic and are *N*-acyl derivatives: (9) *N*-acetyldaunomycin and (10) AD32. In the model, potency (P) of the agents is taken as the reciprocal of the IC_{37} for growth of cells at 50 hours after a 2-hour exposure to the drugs. The potency is assumed to be a function of the specific activity (SA) of the agents, where SA is taken in terms of functional and structural effects on DNA:inhibition of thymidine incorporation (I) and structural damage (n) produced by intracellular amounts of drug (r^*) where $r > 0$:

$$P = f\left(\frac{In}{r}\right)$$

and the parameters are measured 6 hours after allowing a 2-hour period for drug uptake. The expression is evaluated for constants by linearising the function:

$$\ln P = k_a + k_b \ln\left(\frac{In}{r}\right)$$

---

*The use of n here denotes the number of lesions per alkaline unwinding unit in DNA. It is not the same n which is also used in the current literature to denote the frequency of drug-binding sites in DNA. Likewise, r denotes an amount of drug retained intracellularly; it is not the molar ratio of bound drug to DNA nucleotides.

Table 3 Model parameters of anthracyclines (figure 1) with CCRF-CEM human leukaemia cells (from Kanter and Schwartz, 1979*b*)

Agent no.	IC_{37} (P)*	r†	I‡	n§	SA‖
1	0.18 (5.5)	0.16	0.42	0.52	1.4
2	0.13 (7.7)	0.24	0.68	0.85	2.4
3	0.0078 (128)	0.89	>0.99	9.8	>11.0
4	0.014 (71)	0.58	>0.99	6.7	>11.5
5	0.038 (26)	0.43	0.98	2.2	5.0
6	0.14 (7.1)	1.34 (0.067)	0.43	0.27	0.087 (1.7)
7	0.25 (4.0)	1.64 (0.115)	0.14	1.5	0.13 (1.8)
8	0.036 (28)	0.79 1.03	0.79	0.95	0.73 (1.1)
9	10 (0.1)	0.12	0.26	<0.10	<0.22
10	1 (1.0)	1.70	0.77	1.5	0.68

Cells were incubated for two hours with drugs (1 μg ml^{-1}, nos 1–8; 10 μg ml^{-1}, nos 9 and 10) and in drug-free medium for six hours to measure r, I and n. Growth inhibition was estimated from cell no. after 2 hours incubation with agents and 50 hours in drug-free medium. Structures of the agents tested are shown in figure 1 and are numbered in order of increasing lipophilicity.
*Agent concentration (μM) inhibiting growth by 37%. Potency (P) is reciprocal of IC_{37} (in parentheses).
†Agent concentration (nmol/10^6 cells) retained at the end of incubations in drug-free medium. Values in parentheses are corrected to indicate Adriamycin formed from esters 6, 7, 8.
‡Inhibition of 30-min thymidine incorporation at the end of incubation: 1-(treated/control c.p.m.).
§No. of DNA lesions in alkaline unwinding units (Mn_0 =3.1 × 10^8) at the end of incubation.
‖Specific activity: I n/r. Values in parenthesis are corrected for Adriamycin from the esters 6, 7, 8 (r); an estimate of the error in SA may be obtained by comparing these corrected values (1.7, 1.8, 1.1) with that of Adriamycin (no. 1): 1.4.

Taking exponentials on both sides of the equation and inverting to obtain the IC_{37}, the equation is given as:

$$IC_{37} = K_a \left(\frac{r}{In} \right)^{k_b}$$

where

$$K_a = e^{-k_a} = 0.77, \text{ and } k_b = 1.68$$

For seven of the 10 agents, the equation demonstrates that I, n and r are useful parameters to describe potency of the agents under the specific conditions of the experiment (figure 6). Three of the esters (6), (7) and (8) are more potent growth inhibitors than the parameters I, n and r predict. When r for these three is corrected for intracellular Adriamycin concentrations after cleavage of the esters, the specific activity then accurately reflects the potency of two of the esters (6) and (7). (AD32 is so rapidly deesterified to AD41 that no correction is applied.) For nine agents in the set, excluding only the octanoate ester (8),

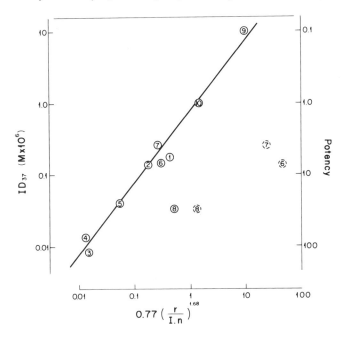

Figure 6 Growth inhibition of CCRF-CEM cells IC_{37} as a function of model parameters of specific activity as described in the text and given in table 3. Potency is $1/ID_{37}$. The anthracyclines are numbered as in figure 1 and table 3. The dashed circles are uncorrected for deesterification.

the equation expresses the IC_{37} for the cells over a range exceeding three orders of magnitude ($R > 0.99$).

In a system having a 1,300-fold span of sensitivities (that is, highest/-lowest IC_{37}) and described by three determinant parameters (I, n and r) each parameter need only vary about 10-fold to account for IC_{37} values over the entire range:

$$(1,300)^{1/3} = 10.9$$

The ranges in table 3 are 14-fold for r, > 7-fold for I, and about 20-fold for n (excluding N-acetyldaunomycin for which $n = 0.1$ is at the lower limits of sensitivity). These relatively small spans indicate that other parameters (such as dissociation constants, redox potentials, pK_a) having similarly narrow ranges may in the future be incorporated into a more comprehensive model.

In using both I and n, the basic assumption is that each one affects the other to a small degree, but that they are largely independent of each other. Inhibition of synthesis, for example, might increase the probabilities that in S-phase cells the exposed DNA strands are more susceptible to attack by free radicals. Alternatively, probabilities that lesions are repaired (during the period that a target cell is able to survive the lesions) might be decreased by I. The probabilities are increased that structural lesions (due to intercalated and oxyradical adducts) inhibit DNA synthesis, blocking proliferating cells during S-phase. The organisational disarray associated with damaged DNA might also prevent cells from completing G2. In either event, the hoard of cells completing mitosis and those entering S-phase in the subsequent mitotic cycle are decreased. Without implying any causal relation between I and n, the model suggests that there is some interdependence or proportion of association between these parameters. The amount of such interdependence (such as the variance of one attributable to the other) may be approximated statistically from the square of the coefficients of correlation. The R^2 for n and I using the entire set of compounds in table 3 is 0.42. This value indicates a considerable degree of association between the variables. Using the same approach, the interdependence of I and r is only 0.33 and that of n and r is even less at 0.12. (If r is not corrected for ester cleavage, $R^2 = 0.001$ for r and I, indicating no statistical interdependence).

The DNA-related parameters, I and n, are not sufficient to predict the potency of Adriamycin-14-octanoate (8). Even after correction for cleavage to the parent compound (70% of the total retained drug), potency is about 20 times higher than expected from the specific activity. If we further assume that specific activity reflects the intracellular de-esterified Adriamycin concentrations, then the uncleaved ester (30% of the total) accounts for 95% of the potency of this agent. Although we have no firm explanation it might be noted that Adriamycin-14-octanoate is the most lipophilic cationic amphipathic agent in the entire set, suggesting that a major contribution to its action may be related to localisation in intracellular membranes or at the cytoplasmic face of plasma membranes. In a study with erythrocytes, Mikkelson *et al.* (1977) demonstrated that Adriamycin, daunomycin and Adriamycin-14-acetate expand the cytoplasmic face of plasma membranes to induce cup formation (10^{-4}–10^{-5}M). At concentrations two orders of magnitude lower (10^{-7}M), however, the octanoate lyses the erythrocytes, a remarkable demonstration of membranal potency for an agent in the Adriamycin class.

The modelling studies described here and in foregoing sections support the idea of dual mechanisms of selectivity. In a complex series of interdependent actions, events at both DNA and membranal sites appear to contribute to selectivity of the anthracyclines. Described above, the structural and functional effects at DNA sites of intracellular drug in proliferating CCRF-CEM cells can be used to predict potency for most of the anthracyclines in the set. The membranal effects are not quantitated, although these should be related in some way to both pK_a and partition coefficients. For the set of 10 agents used by Kanter and Schwartz (1979*b*), the lipophilicity was optimal for carminomycin (3), relative to potency.

SUMMARY

Selective agents, in general, interfere with the physiological functions of cells in one or more specific ways (Schwartz and Mihich, 1973). A drug effect becomes lethal, however, only when critical perturbations persist for longer than the period of time that the cells can survive the lesion. Irreversibility in this context implies that lesions are not or cannot be repaired, replaced by biosynthesis or uptake, or by-passed through alternative metabolic pathways.

In the model of Kanter and Schwartz (1979b), the paired variances for the parameters of specific activity suggested varying degrees of interdependence. To weigh the individual contributions of n and r to the specific activity of the set of agents in table 3, approximations are estimated by comparing variances (R^2) of $I:r = 0.33$ and $n:r$ 0.12. The contribution of I is seen from these values to be 2.75 times greater than n to specific activity. Consequently we expect that I is the more likely to occur with cells that take up this set of agents. But just because it is a more likely biochemical effect, it is also less likely to signal selectivity. That is, the more general a potentially lethal biochemical effect is, the less probable that it distinguishes between similar agents of different efficacies or similar cell populations of different sensitivities. The converse is that the less frequent event (secondary DNA damage) may be more likely to signal selective differences. This conclusion is supported by the predictive value of delayed DNA damage for therapeutic responses of AML patients (Schwartz *et al.*, 1981) and of leukaemia-bearing mice (Kanter, 1974). The proposition here is that this type of damage is a critical event that may discriminate between responsive and non-responsive populations of similar cells (as in acute myeloblastic leukaemia).

For anthracyclines, the components of selectivity include physicochemical characteristics and biochemical reactivity of the agents. But it is the target cell itself that orchestrates selectivity, and its physiology determines sensitivity to cytotoxic insults. Cytotoxicity appears to be primarily dependent on membranal and DNA-related effects. Underlying both, the oxidation-reduction properties of the agents appear to be major (but not the sole) contributing factors. Other factors that contribute to their activities include those that affect DNA intercalation and perhaps lipid partitioning. Selectivity seems to depend more on the DNA-related effects than on those associated with membranal changes, but this may simply reflect the paucity of information concerning the latter. Even if no specific drug effects were to be attributable to the plasma membrane, it is still the first cellular encounter for drugs and so it is the initiating barrier and thus the first determinant of selectivity.

Drug levels and the rates at which active agents are depleted from the surrounding medium are also important determinants of cytotoxicity and may contribute to selective differences between cells. To these may be added the degree of aeration as a factor that may in turn determine the formation and effective life of oxyradicals and drug radicals. Relative anaerobic conditions sustained by some cells in solid tumours may therefore be instruments of selective cytotoxicity. Selectivity is further served by susceptibility to radical-dependent damage, modified in individual cells and in different cell populations by protective quenching and repair mechanisms available to each. If the determinants of selectivity reside in complex sequences (such as membrane barriers, drug activation, DNA and enzyme interactions) staged by a target cell, then it is not surprising that small differences within a population of cells can lead to major differences in therapeutic responses. Selectivity may also be enhanced not only by environmental factors, but by therapeutic combinations as well, so that agents of similar structure may have vastly different activities, with relatively small changes in only a few determinant parameters affecting cell responses by more than 1,000 fold. Thus, minor changes in drug structure, in target cells, or even in pharmacokinetics among different hosts are magnified into Gompertzian drug effects. For these reasons it is understandable that one anthracycline agent may produce a complete therapeutic success in one patient and total failure in another, both with the same *type* of neoplastic disease.

ACKNOWLEDGEMENTS

The author expresses his gratitude to Dr P. M. Kanter and to Dr T. R. Tritton for helpful discussions and for kindly providing access to some of their unpublished studies. Many of the results described here were supported in part by grant CA-24778 awarded by the National Cancer Institute, DHHS, and by a grant awarded by the American Cancer Society.

REFERENCES

Andrews, P. A., Callery, P. S. and Bachur, N. R. (1982). *Fedn Proc.*, **41**, 1564

Albert, A. (1979). *Selective Toxicity*, John Wiley and Sons, New York, pp. 3–662

Aoyama, T., Kawamoto, Y., Furuta, I. and Kondo, T. (1972). *Int. J. Radiat. Biol.*, **21**, 545

Appelbaum, S. J., Meyerson, M., Dorr, R. T. and Perrier, D. (1982). *Cancer Chemother. Pharmac.*, **8**, 93

Arcamone, F. (1981). *Doxorubicin Anticancer Antibiotics*, Academic Press, New York, pp. 1–369

Arlandini, E., Vigevani, A. and Arcamone, F. (1977). *Il Famaco-Ed. Sci.*, **32**, 315

Bachur, N. R. (1979). *Cancer Treatment Rep.*, **63**, 817

Bachur, N. R., Gee, M. V. and Friedman, R. D. (1982). *Cancer Res.*, **42**, 1078

Bachur, N. R., Steele, M., Meriwether, W. D. and Hildebrand, R. C. (1976). *J. med. Chem.*, **19**, 651

Bachur, N. R., Gordon, S. L. and Gee, M. V. (1977). *Molec. Pharmac.*, **13**, 901

Bardos, T. and Kalman, T. (1982). *New Approaches to the Design of Antineoplastic Agents*, Elsevier, New York, in press

Benedetti, A., Casini, A. F., Ferrali, M. and Comporti, M. (1979). *Biochem. J.*, **180**, 303

Benjamin, R. S., Riggs, C. E., Jr and Bachur, N. R. (1977). *Cancer Res.*, **37**, 1416

Bergelson, L. D., Dyatlovitskaya, E. V., Sorokina, I. B. and Gorkora, N. P. (1974). *Biochim. biophys. Acta*, **360**, 361

Berlin, V. and Haseltine, W. A. (1981). *J. biol. Chem.*, **256**, 4747

Bertazzoli, C., Chieli, T., Ferni, G., Ricevuti, G. and Solcia, E. (1972). *Tox. appl. Pharmac.*, **21**, 287

Bertazzoli, C., Chieli, T. and Solcia, E. (1971). *Experientia*, **27**, 1209

Bertazzoli, C., Sala, L., Solcia, E. and Ghione, M. (1975). *IRCS med. Sci.*, **3**, 468

Birchmeier, W. and Singer, S. J. (1977). *J. Cell Biol.*, **73**, 647

Bronson, R. T., Henderson, I. C. and Fixler, H. (1982). *Cancer Treatment Rep.*, **66**, 1349

Caroni, P., Villani, F. and Carafoli, E. (1981). *FEBS Lett.*, **130**, 184
Casazza, A. M., Isetta, A. M., Guiliani, F. and DiMarco, A. (1975). In *Adriamycin Review* (ed. M. Staquet), European Press, Ghent, p. 123
Chan, K. K. and Watson, E. (1978). *J. pharmac. Sci.*, **67**, 1748
Chandra, P. (1975). *Cancer Chemother. Rep.* (3) **6**, 115
Chuang, R. Y. and Chuang, L. F. (1979). *Biochemistry*, **18**, 2069
Cho, E.-S. (1977). *J. Neuropathol. exp. Neurol.*, **36**, 1977
Cho, E.-S., Jortner, B. S., Schaumberg, H. H. and Spencer, P. S. (1979). *Neurotoxicology*, **1**, 583
Cho, E.-S., Spencer, P. S. and Jortner, B. S. (1980). In *Experimental and Clinical Neurotoxicity* (ed. P. S. Spencer and H. H. Schaumberg), Williams and Wilkins, Baltimore, p. 430
Crane, F. L., MacKellar, W. C., Moore, D. J., Ramasarma, T., Goldenberg, H., Grebing, C. and Low, H. (1980). *Biochem. biophys. Res. Commun.*, **93**, 746
Crooke, S. T. and Reich, S. D. (1980). *Anthracyclines: Current Status and New Developments*, Academic Press, New York, pp. 1–444
Dalmark, M. (1981). *Scand. J. clin. Lab. Invest.*, **41**, 633
Danø, K. (1972). *Cancer Chemother. Rep.*, **56**, 321, 701
Danø, K. (1973). *Biochim. biophys. Acta*, **323**, 466
Danø, K., Frederiksen, S. and Hellung-Larsen, P. (1972). *Cancer Res.*, **32**, 1307
Dasdia, T., DiMarco, A., Goffredi, M., Minghetti, A. and Necco, A. (1979). *Pharmac. Res. Commun.*, **11**, 19
DiMarco, A. (1975). *Cancer Chemother. Rep.*, (3) **6**, 91
DiMarco, A., Casazza, A. M., Dasdia, T., Necco, A., Pratesi, G., Rivalta, P., Velcich, A., Zaccara, A. and Zunino, F. (1977). *Chem.-biol. Interact.*, **19**, 291
DiMarco, A., Zunino, F., Silvestrini, R., Gambarucci, C. and Gambetta, R. A. (1971). *Biochem. pharmac.*, **20**, 1323
Don, M. M., Ablett, G. and Kerr, J. F. R. (1977). *Austr. J. exp. Biol. med. Sci.*, **55**, 407
Doroshow, J. H., Locker, G. Y., Baldinger, J. and Myers, C. E. (1979). *Res. Commun. chem. Pathol. Pharmac.*, **26**, 285
Duarte-Karim, J. M., Ruysschaert, J. M. and Hildebrand, J. (1976). *Biochem. biophys. Res. Commun.*, **71**, 658
Felsted, R. L. and Bachur, N. R. (1980). In *Enzymatic Basis of Detoxication Vol. I* (ed. W. B. Jacoby), Academic Press, New York, p. 281
Fisher, P. A., Chen, J. T. and Korn, D. (1981). *J. biol. Chem.*, **256**, 133
Goldenberg, H., Crane, F. L. and Morre, D. J. (1979). *J. biol. Chem.*, **254**, 2491
Goldman, R., Facchinetti, T., Bach, D., Raz, A. and Shinitzky, M. (1978). *Biochim. biophys. Acta*, **512**, 254
Goodman, J. and Hochstein, P. (1977). *Biochem. biophys. Res. Commun.*, **77**, 797
Goodman, M. F., Bessman, M. J. and Bachur, N. R. (1974). *Proc. natn. Acad. Sci. U.S.A.*, **71**, 1974
Goormaghtigh, E., Chatelain, P., Caspers, J. and Ruysschaert, J. M. (1980). *Biochem. Pharmac.*, **29**, 3003
Gurtoo, H. L., Marinello, A. J., Struck, R. F., Paul, B. and Dahms, R. P. (1981). *J. biol. Chem.*, **256**, 11691
Handa, K. and Sato, S. (1975). *Gann*, **66**, 43
Harper, J. R. Jr, Orringer, E. P. and Parker, J. C. (1979). *Res. Commun. chem. Pathol. Pharmac.*, **26**, 277
Henry, D. (1976). In *Cancer Chemotherapy* (ed. A. Sartorelli), American Chemical Society, Washington, p. 15
Hewish, D. R. and Burgoyne, L. A. (1973). *Biochem. biophys. Res. Commun.*, **52**, 504
Infante, A. A., Firshein, W., Hobart, P. and Murray, L. (1976). *Biochemistry*, **15**, 4810
Israel, M., Modest, E. J. and Frei, E. III (1975). *Cancer Res.*, **35**, 1365
Kanter, P. M. (1974). Ph.D. Dissertation, St. Univ. N.Y., Buffalo.
Kanter, P. M. (1981). *Proc. Am. Ass. Cancer Res.*, **22**, 812
Kanter, P. M. and Schwartz, H. S. (1977). *Proc. Am. Assc. Cancer Res.*, **18**, 229
Kanter, P. M. and Schwartz, H. S. (1979a). *Cancer Res.*, **39**, 448
Kanter, P. M. and Schwartz, H. S. (1979b). *Cancer Res.*, **39**, 3661

Kanter, P. M. and Schwartz, H. S. (1979c). *Analyt. Biochem.*, **97**, 77
Kanter, P. M. and Schwartz, H. S. (1980). *Int. J. Radiat. Biol.*, **38**, 483
Karczmar, G. S. and Tritton, T. R. (1979). *Biochim. biophys. Acta.*, **557**, 306
Kerr, J. F. R., Wylie, A. H. and Currie, A. R. (1972). *Br. J. Cancer*, **26**, 239
Kessell, D. (1979a). *Biochem. Pharmac.*, **28**, 3028
Kessell, D. (1979b). *Molec. Pharmac.*, **16**, 306
Komiyama, T., Oki, T., Taiji, T. and Umezawa, H. (1979a). *Gann*, **70**, 403
Komiyama, T., Oki, T. and Inui (1979b). *J. Antibiot.*, **32**, 1219
Kuo, M. T. and Hsu, T. C. (1978). *Nature, Lond.*, **271**, 83
Krishan, A., Dutt, K., Israel, M. and Ganapathi, R. (1981). *Cancer Res.*, **41**, 2745
Lee, Y. C. and Byfield, J. E. (1976). *J. natn. Cancer Inst.*, **57**, 221
Lehotay, D. C., Levey, B. A., Rogerson, B. and Levey, G. S. (1979). *Clin. Res.*, **27**, 616A
Levey, G. S., Lerey, B. A., Ruiz, E. and Lehotay, D. C. (1979). *J. molec. cell. Cardiol.*, **11**, 591
Loveless, H., Arena, E., Felsted, R. L. and Bachur, N. R. (1978). *Cancer Res.*, **38**, 593
Lown, J. W., Chen, H.-H., Plambeck, J. A. and Acton, E. M. (1982). *Biochem. Pharmac.*, **31**, 575
Maniglia, C. A. and Wilson, R. G. (1981). *Chem.-biol. Interact.*, **33**, 319
Maral, R., Bourat, G., Ducrot, J. and Werner, G. H. (1967). *Pathol.-Biol.*, **15**, 903
Marquardt, H., Philips, F. S. and Sternberg, S. S. (1976). *Cancer Res.*, **36**, 2065
Mikkelsen, R. B., Lin, P. S. and Wallach, D. F. H. (1977). *J. molec. Medicine*, **2**, 33
Mimnaugh, E. G., Siddik, Z. H., Draw, R., Sikic, B. I. and Gram, T. E. (1979). *Tox. appl. Pharmac.*, **49**, 119.
Muggia, F. M. (1982). *Proc. Int. Symposium on Anthracycline Antibiotics in Cancer Therapy*, Martius Nijhoff, Hingham, Massachusetts, 1–567
Murphree, S. A., Triton, T. R., Smith, P. L. and Sartorelli, A. C. (1981). *Biochim. biophys. Acta*, **649**, 317
Myers, C. E., McGuire, W. P., Liss, R. H., Ifrim, I., Grotzinger, K. and Young, R. C. (1977). *Science*, **197**, 165
Myers, C. E., McGuire, W. P. and Young, R. (1976). *Cancer Treatment Rep.*, **60**, 961–962
Olson, R. D., MacDonald, J. S., vanBoxtel, C. J., Boerth, R. C., Harbison, R. D., Slonim, A. E. and Oates, J. A. (1980). *J. pharmac. exp. Ther.*, **215**, 450
Pan, S.-S. and Bachur, N. R. (1980). *Molec. Pharmac.*, **17**, 95
Pan, S.-S., Pedersen, L. and Bachur, N. R. (1981). *Molec. Pharmac.*, **19**, 184
Pater, M. M. and Mak, S. (1974). *Nature, Lond.*, **250**, 786
Philips, F. S., Gilladoga, A., Marquardt, H., Sternberg, S. S. and Vidal, P. M. (1975). *Cancer Chemother. Rep.*, **6**, 177
Phillips, D. R., DiMarco, A. and Zunino, F. (1978). *Eur. J. Biochem.*, **85**, 487
Pigram, W. J., Fuller, W. and Hamilton, L. D. (1972). *Nature new Biol.*, **235**, 17
Rajagopalan, K. V. (1980). In *Enzymatic Basis of Detoxication, I.* (ed. W. B. Jakoby), Academic Press, New York, p. 295
Remers, W. A. (1979). *Chemistry of Antitumor Antibiotics*, John Wiley and Sons, New York, p. 63
Ross, W. E., Glaubiger, D. and Kohn, K. W. (1979). *Biochim. biophys. Acta*, **562**, 41
Sagone, A. L. and Burton, G. M. (1979). *Am. J. Hemat.*, 7, 97
Sartiano, G. P., Lynch, W. E. and Bullington, W. D. (1979). *J. Antibiot.*, **32**, 1038
Schioppacassi, G. and Schwartz, H. S. (1977). *Res. Commun. chem. Pathol. Pharmac.*, **18**, 519
Schneider, Y. J., Baurain, R., Zenebergh, A. and Trouet, A. (1979). *Cancer Chemother. Pharmac.*, **2**, 7
Schwartz, H. S. (1974). *Cancer Chemother. Rep.*, **58**, 55
Schwartz, H. S. (1976). *Biomedicine*, **24**, 317
Schwartz, H. S. (1979). *Adv. Cancer Chemother.*, **1**, 1
Schwartz, H. S. and Grindey, G. B. (1973). *Cancer Res.*, **33**, 1837
Schwartz, H. S. and Kanter, P. M. (1979). *Eur. J. Cancer*, **15**, 923
Schwartz, H. S. and Kanter, P. M. (1981). *Cancer Lett.*, **13**, 309
Schwartz, H. S., Kanter, P. M., Marinello, A. J. and Gurtoo, H. L. (1982). In *Proceedings of the International Symposium on Anthracycline Antibiotics in Cancer Therapy* (ed. F. M. Muggia), Hingham, Massachusetts, Martinus Nijhoff, 125

Schwartz, H. S. and Mihich, E. (1973). In *Drug Resistance and Selectivity* (ed. E. Mihich), New York, Academic Press, p. 413

Schwartz, H. S. and Parker, N. B. (1981). *Cancer Res.*, **41**, 2343

Schwartz, H. S., Preisler, H. D. and Kanter, P. M. (1981). *Leukemia Res.*, **5**, 363

Schwartz, H. S., Schioppacassi, G. and Kanter, P. M. (1978). *Antibiotics Chemother.*, **23**, 247

Schwartz, H. S., Sodergren, J. H. and Philips, F. S. (1963). *Science*, **142**, 1181

Seeman, P. (1970). In: *Permeability and Function of Biological Membranes* (ed. L. Bolis), Amsterdam, North-Holland, p. 40

Serpentino, P. M. (1977). PhD Dissertation, Univ. Rochester

Sheetz, M. P. and Singer, S. J. (1974). *Proc. natn Acad. Sci. U.S.A.*, **71**, 4457

Skalka, M., Matyasova, J. and Cejkova, M. (1976). *FEBS Lett.*, **72**, 271

Skovsgaard, T. (1978). *Biochem. Pharmac.*, **27**, 1221

Sternberg, S. S., Philips, F. S. and Cronin, A. P. (1972). *Cancer Res.*, **32**, 1029

Suciu, D. (1979). *Int. J. Radiat. Biol.*, **35**, 119

Supino, R., Necco, A., Dasdia, T., Casazza, A. M. and DiMarco, A. (1977). *Cancer Res.*, **37**, 4523

Takanashi, S. and Bachur, N. R. (1975). *J. Pharmac. exp. Ther.*, **195**, 41

Tanaka, M. and Yoshida, S. (1980). *J. Biochem.*, **87**, 911

Teicher, B. A., Lazo, J. A. and Sortorelli, A. C. (1981). *Cancer Res.*, **41**, 73

Tritton, T. R., Murphree, S. A. and Sartorelli, A. C. (1977). *Biochem. Pharmac.*, **26**, 2319

Tritton, T. R. and Yee, G. (1982). *Science*, **214**, 248

White, D. A. (1973). In *Form and Function of Phospholipids* (ed. G. B. Ansell), New York, Elsevier, p. 441

Williams, J. R., Little, J. B. and Shipley, W. U. (1974). *Nature, Lond.*, **252**, 754

Wyllie, A. H. (1980). *Nature, Lond.*, **284**, 555

Yamada, T., Ohyama, H., Kinjo, Y. and Watanabe, M. (1981). *Radiat. Res.*, **85**, 544

Young, R. C., Ozols, R. F. and Myers, C. E. (1981). *N. Engl. J. Med.*, **305**, 139

Zunino, F., Gambetta, R., DiMarco, A. and Zaccara, A. (1972). *Biochim. biophys. Acta*, **277**, 489

Zunino, F., Gambetta, R., DiMarco, A., Luoni, G. and Zaccara, A. (1976). *Biochem. biophys. Res. Commun.*, **69**, 744

Zunino, F., Gambetta, R., DiMarco, A., Zaccara, A. and Luoni, G. (1975). *Cancer Res.*, **35**, 760

Zwelling, L. A., Michaels, S., Erickson, L. C., Ungerleider, R. S., Nichols, M. and Kohn, K. W. (1981). *Biochemistry*, **20**, 6553

5
Molecular Aspects of the Interaction between Quinoxaline Antibiotics and Nucleic Acids

Michael J. Waring and Keith R. Fox

INTRODUCTION

Quinoxaline antibiotics are of widespread occurrence in nature. They are produced by a variety of species of streptomycetes and are characterised by an octadepsipeptide ring to which are attached two moieties of quinoxaline-2-carboxylic acid, hence their name. They were first isolated in the 1950s (see Katagiri *et al.*, 1975 for a full description) and the best-known member of the group is echinomycin (Corbaz *et al.*, 1957; Keller-Schierlein *et al.*, 1959) whose structure, shown in figure 1, was recently revised.

In all the antibiotics the peptide ring is crossed-bridged by a sulphur-containing covalent linkage which originates from interaction between the side chains of two *N*-methyl cysteine residues. Depending upon the exact constitution of that cross-bridge the antibiotics fall into two families called quinomycins and triostins, the former having a thioacetal cross-bridge and the latter a disulphide (figure 2). It is likely that the triostins serve as biosynthetic precursors of the corresponding quinomycins (Katagiri *et al.*, 1975; Dell *et al.*, 1975; Williamson *et al.*, 1982). Only two amino acids vary between the natural antibiotics, as illustrated in figure 2, and the variations are of a decidedly conservative character (Katagiri *et al.*, 1975). It is noteworthy that four of the six peptide bonds forming the ring bear *N*-methyl substituents and also that the structural formulae of these antibiotics imply substantial elements of twofold rotational symmetry: ostensibly perfect for members of the triostin series having the same

Molecular Aspects of Anti-cancer Drug Action, ed. Neidle & Waring
0333–31556/83/127–156

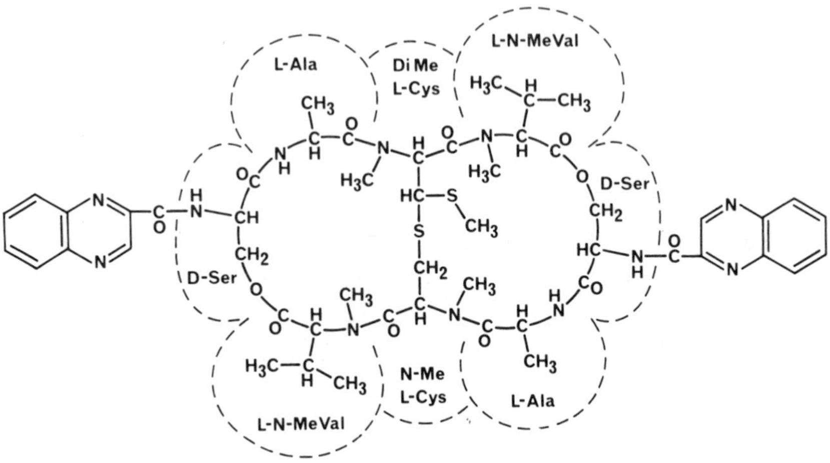

Figure 1   Structural formula of echinomycin (Dell *et al.*, 1975; Martin *et al.*, 1975)

amino acid at positions Y, and only broken for quinomycins like echinomycin (quinomycin A) by the asymmetrically placed *S*-methyl group of the cross-bridge. That quasi-perfect rotational symmetry is highly significant for their mode of action at the molecular level, as we shall see.

The biological properties of the quinoxaline antibiotics have been thoroughly reviewed by Katagiri *et al.* (1975). Briefly, they are characteristically highly active against Gram-positive, anaerobic and acid-fast bacteria, whereas Gram-negative bacteria and fungi are generally resistant (Yoshida *et al.*, 1961; Shoji and Katagiri, 1961; Katagiri *et al.*, 1966), most probably for reasons of relative impermeability, since quinomycin A inhibits the incorporation of labelled uridine into RNA by spheroplasts of *Escherichia coli*, whereas intact cells are insensitive to the drug (Sato *et al.*, 1967a). Moderate activity against protozoa and viruses has also been reported (Ueda *et al.*, 1954; Corbaz *et al.*, 1957;

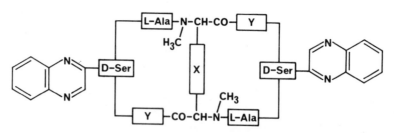

Figure 2   Quinoxaline antibiotics. In quinomycins the cross-bridge (X) between the two *N* methyl cysteine residues is a thioacetal $-CH_2-S-CH(SCH_3)-$ whereas in the corresponding triostins it is a simple disulphide $-CH_2-S-S-CH_2-$. In echinomycin (quinomycin A) and triostin A the amino acid residues at position Y are L-*N*-methyl-valine; in quinomycin C and triostin C they are *N*-γ-dimethyl-*allo*-isoleucine.

Katagiri *et al.*, 1957; Tsunoda, 1962; Sato *et al.*, 1969), but most interest in quinoxaline antibiotics has been focused on their cytotoxicity and promise as antitumour agents (Katagiri and Sugiura, 1961; Matsuura, 1965; Harada *et al.*, 1968; Katagiri *et al.*, 1975).

Their mode of action involves interference with nucleic acid synthesis, attributable to binding to DNA in susceptible cells (Ward *et al.*, 1965; Sato *et al.*, 1967a,b; Waring and Makoff, 1974; Katagiri *et al.*, 1975). Echinomycin is an extremely powerful inhibitor of RNA synthesis in *Bacillus megaterium*, some 4–5 times more potent than actinomycin D, and is at least as selective as actinomycin in inhibiting DNA-directed RNA synthesis (Waring and Makoff, 1974). It is generally agreed that its antitumour and other biological activities result directly, perhaps solely, from its interaction with DNA (Ward *et al.*, 1965; Sato *et al.*, 1967b; Kageyama *et al.*, 1970; Katagiri *et al.*, 1975; Waring, 1977, 1979, 1981; Gale *et al.*, 1981). Echinomycin has found employ as a counterstain in sophisticated cytochemical procedures for studying fluorescent banding patterns of human and mammalian metaphase chromosomes (Schnedl *et al.*, 1980).

Although the strong binding of echinomycin to DNA was originally reported by Ward *et al.* (1965) and Sato *et al.* (1967b) it was almost ten years before the unique character of the interaction was recognised (Waring and Wakelin, 1974), largely as a result of the development of a novel solvent-partition method which enabled accurate measurement of DNA-binding curves for such feebly water-soluble antibiotics as the quinoxalines (Waring *et al.*, 1975; Waring, 1979). Those studies established that echinomycin binds to DNA by a process of bifunctional intercalation, involving simultaneous insertion of both its quinoxaline chromophores into the double helix, and served as a model for later work on the synthesis of bifunctional intercalators such as diacridines, dimeric phenanthridines and dimers of other known intercalative molecules. Some, like echinomycin, have shown promise as potentially powerful chemotherapeutic drugs. Aspects of those developments are described in Chapters 1, 3 and 9.

Our understanding of the molecular nature of the echinomycin–DNA complex and of constraints on *bis*-intercalative binding of quinoxalines in general has been greatly facilitated by the availability of analogues, both natural and synthetic. Apart from the naturally-occurring quinoxaline antibiotics which have been available for some time, another cyclic peptide antibiotic which belongs to a new class of bifunctional intercalating molecules has recently been purified and characterised, luzopeptin (originally code-named BBM 928A and isolated from cultures of *Actinomadura luzonensis*, Fig. 3) (Huang *et al.*, 1980; Konishi *et al.*, 1981; Arnold and Clardy, 1981). In a further series of developments, chromophore-substituted analogues of the natural quinoxaline antibiotics have been obtained by directed biosynthesis involving supplementation of antibiotic-producing cultures with heteroaromatic acids intended to serve as precursors of the chromophoric moieties (Bradley *et al.*, 1980; Gauvreau and Waring, in preparation; A. Cornish, unpublished work). Yoshida *et al.* (1968) first described

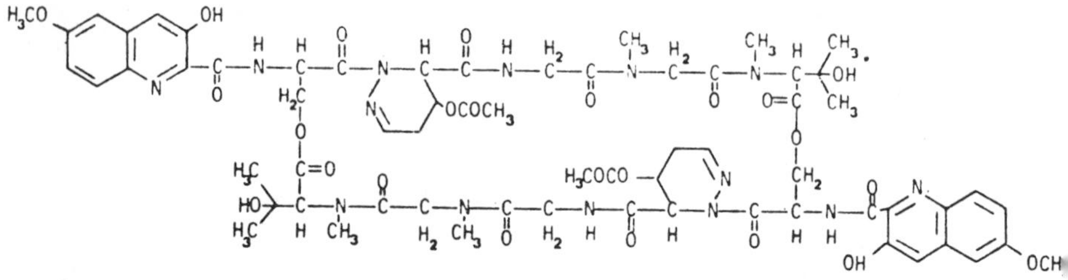

Figure 3   Structural formula of luzopeptin (BBM 928A) according to Konishi *et al*. (1981) and Arnold and Clardy (1981).

the isolation of mono- and *bis*-substituted quinoline analogues of echinomycin, now designated 1QN and 2QN respectively (Fox *et al*., 1980*a*), and their methods have been adapted to enable the production of a wide range of novel analogues having halogenated, methylated, or otherwise modified chromophores (Bojesen *et al*., 1981; Williamson *et al*., 1982; Gauvreau and Waring, in preparation).

A third group of important quinoxaline compounds has originated from chemical synthesis, largely the result of the efforts of Olsen and his collaborators. In the course of experiments which culminated in the total synthesis of triostin A (Chakravarty and Olsen, 1978) these workers synthesised numerous analogues in the triostin series which have proved invaluable in the investigation of structure—activity relationships. Some have replacement amino acids or epimerised centres, for example variants having L-serine in place of the natural D-isomer, but by far the most important analogue is the des-*N*-tetramethyl derivative of triostin A known by the acronym TANDEM (Ciardelli and Olsen, 1977; Ciardelli *et al*., 1978). The interest in TANDEM (figure 4) extends beyond its usefulness as a probe for the probable role of peptide bond methylation in quinoxaline antibiotics; at the time of writing it is the only member of the quinoxaline series for which a definitive structure has been determined by X-ray crystallography, described below. TANDEM is, however, inactive to all intents and purposes in antibacterial tests against *Staphylococcus aureus* (Gauvreau and Waring, in preparation).

BIFUNCTIONAL INTERCALATION

The prime evidence for bifunctional (or *bis*-) intercalative binding of quinoxaline antibiotics to DNA rests upon the results of experiments with closed circular duplex DNA and sonicated rod-like fragments of helical DNA aimed at measuring unwinding and extension of the helix respectively. These two independent approaches are now generally regarded as together providing acceptable *prima facie* evidence of intercalation (Waring, 1981; Gale *et al*., 1981). Lacking any crystallographically-based evidence for intercalative interaction between a quinoxaline antibiotic and a helical oligonucleotide fragment, they represent the

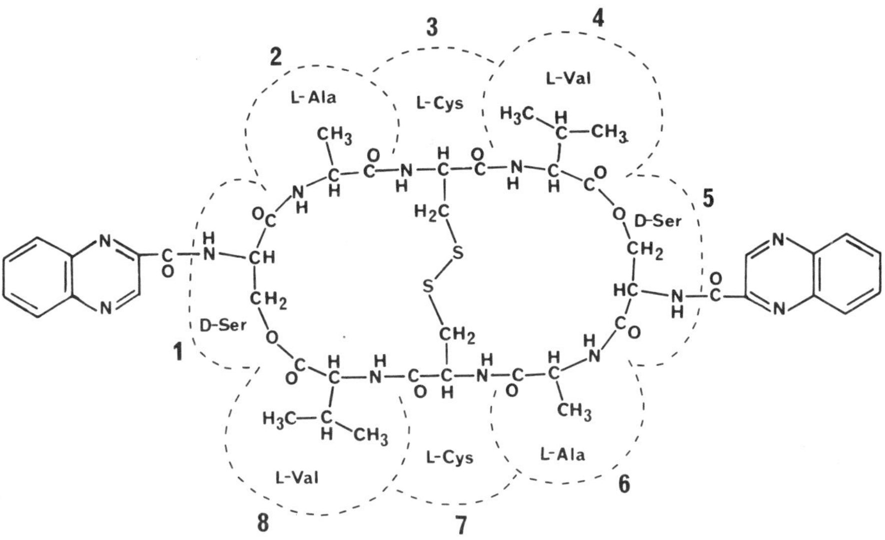

Figure 4 Structural formula of des-*N*-tetramethyltriostin A (TANDEM) with amino acid residues numbered according to the scheme adopted by Olsen and collaborators (see Fox *et al.*, 1980*b*).

most direct tests currently available. In the present context the argument runs that *bis*-intercalation should ideally be associated with twice the helix unwinding and extension produced by a standard (monofunctional) intercalating agent, such as ethidium. In respect of helix extension it is also possible to compare the magnitude of the measured effect with that predicted on a purely theoretical basis, that is, 3.4 Å extension by each intercalated chromophore yielding 6.8 Å as the expectation for an ideal process of bifunctional intercalation. The validity of these ideas has been thoroughly tested in experiments on synthetic *bis*-intercalators (see Chapter 1) and has generally held up well, notwithstanding the disregard of likely additional structural distortions reflected in twisting and tilting of base-pairs (Hogan *et al.*, 1979).

Table 1 presents a summary of findings reported for quinoxalines and the three related quinoline peptides. It can be seen that all except one produce an unwinding effect in the range 43°-50° or more, well up towards the notional value of 52° being twice the angle seen with ethidium. Similarly, all but one cause the helix to extend by 5.7 Å or more per ligand molecule bound, respectably closer to 6.8 Å than to 3.4 Å, and more than double the value of 2.7-2.9 Å recorded for ethidium under comparable conditions (Reinert, 1973; Hogan *et al.*, 1979; Huang *et al.*, 1980). The exceptions occur in respect of the unwinding angle of 2QN, which appears inexplicably low (Fox *et al.*, 1980), and the helix extension (perhaps also the unwinding) determined for luzopeptin. The low values for luzopeptin can plausibly be attributed to the use of a higher salt concentration for those experiments (see legend to table 1); it is known that the

Table 1 Properties of intercalation complexes formed between quinoxaline antibiotics and calf thymus DNA

Compound	Helix-unwinding angle	Helix extension (Å)	Reference
Echinomycin	48°	6.4	Waring and Wakelin (1974)
Quinomycin C	⩾50°	–	Lee and Waring (1978a)
Triostin A	47°	6.1	Lee and Waring (1978a)
Triostin C	⩾47°	⩾6.6	Lee and Waring (1978a)
TANDEM	46°	–	Lee and Waring (1978b)
1QN	⩾44°	5.7	Fox et al. (1980a)
2QN	⩾26°	5.7	Fox et al. (1980a)
Luzopeptin	43°	4.3	Huang et al. (1980)

1QN and 2QN are biosynthetic analogues of echinomycin having one or both of the natural chromophores replaced by quinoline ring systems. All data refer to measurements made in buffer of ionic strength 0.01, except for luzopeptin where the buffer contained 75mM NaCl and 10mM Tris–HCl (pH 7.4). Where necessary, original estimates have been corrected to refer to an assumed unwinding angle for ethidium of 26° (Wang, 1974) and helix extension has been calculated on the assumption that simple intercalation of an aromatic chromophore extends the helix by 3.4 Å.

relevant parameters for echinomycin are diminished by raising the ionic strength (Waring and Wakelin, 1974; Wakelin and Waring, 1976).

Clearly the data summarised in table 1 are uniformly consistent with the conclusion that bifunctional intercalation is a group property of the quinoxaline antibiotics, and the same is broadly true for the tightly-binding analogues investigated so far. Strictly speaking, however, that conclusion is limited to the DNA species investigated (bacteriophage PM2 DNA for the unwinding experiments; sonicated calf thymus DNA for the measurements of extension) and while it may reasonably be extended to include natural DNAs from other sources it might not necessarily apply to synthetic polynucleotides. As explained below, there are substantial differences between the binding properties of different antibiotics when tested with synthetic DNAs and the possibility must be considered that those differences reflect variations in the mechanism of binding. That does not seem to be so, judged by the criterion of helix extension— the only direct test applicable to synthetic polynucleotides with present-day techniques (Fox et al., 1981a). For poly (dA-dT) the results are gratifyingly clear-cut (figure 5): echinomycin extends the helix by 7.3 Å per antibiotic molecule bound, assuming the B-form of DNA and other factors as in the legend to table 1. Other quinoxalines and their analogues also yield values close to the theoretical 6.8 Å: triostin A, 6.5 Å; 2QN, 6.3 Å (Fox et al., 1981a); TANDEM, 6.8 Å (Fox et al., 1982); and [Ala³, Ala⁷] TANDEM, an analogue in which both L-Cys residues of the octapeptide ring are replaced by L-Ala so that it lacks a cross-bridge altogether, 6.4 Å (Fox et al., 1980b). For poly (dG-dC), echinomycin again produces a near-theoretical extension of 6.7 Å (figure 5) whereas with

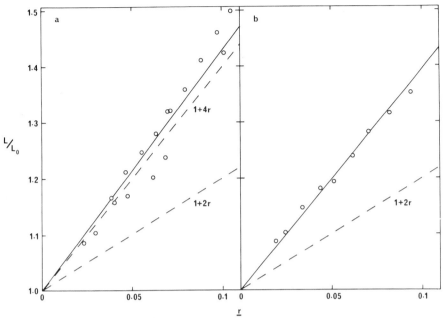

Figure 5   Effect of echinomycin on the relative contour length of (a) poly (dA-dT) and (b) poly (dG-dC). The ordinate represents the fractional increase in helix length; the abscissa shows the level of binding in terms of antibiotic molecules bound per nucleotide (r). The lines drawn through the points were fitted by the method of least squares; their slopes are 4.29 ± 0.33 for poly (dA-dT) and 3.95 ± 0.16 for poly (dG-dC). Also shown are theoretical lines (broken) representing slopes of 2 and 4 corresponding to idealised monofunctional and bifunctional intercalation respectively. From Fox *et al.* (1981*a*).

triostin A the calculated extension is only 4.3 Å; however, controls reveal that the behaviour of fragments of this polynucleotide seems to depart from the ideal, witness the anomalously low extension provoked by binding of ethidium (2.0 Å) against which the effect of triostin A is more than double (Fox *et al.*, 1981*a*). Clearly it would be premature to conclude anything definite about monofunctional or bifunctional intercalative binding of triostin to poly (dG-dC), though its effect on natural DNAs and poly (dA-dT) speaks unambiguously for bifunctional reaction.

EQUILIBRIUM BINDING TO DNA AND RECOGNITION OF NUCLEOTIDE SEQUENCES

The solvent-partition method (Waring *et al.*, 1975) has proved readily adaptable to measuring the interaction between DNA and most quinoxaline antibiotics or their analogues thus far examined. Scatchard plots of the binding data are typically curved, as expected for a reaction in which the ligand molecule occludes more than one potential binding site in a polymeric molecule which can be considered a linear lattice of sites (Crothers, 1968; McGhee and von Hippel, 1974). Examples of curves describing the binding of echinomycin to a variety of natural DNAs are illustrated in figure 6. They reveal that the affinity of the antibiotic for DNAs from different sources varies over a wide range and that it is not simply related to the gross base-composition, though there is an undoubted tendency for G+C-rich DNAs to bind the antibiotic more tightly (Wakelin and Waring, 1976).

Binding parameters derived from these curves are included in table 2, together

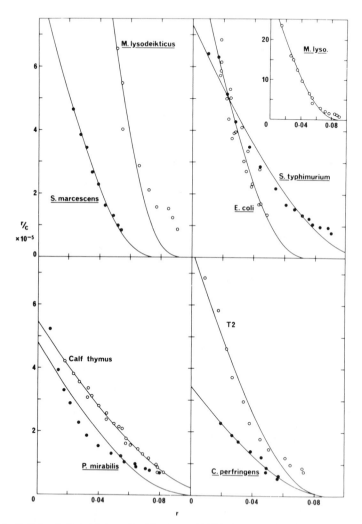

Figure 6   Scatchard plots representing the interaction between echinomycin and DNA from different sources. The binding ratio, r, is as defined in the legend to figure 5; c represents the free antibiotic concentration in solution. The curves are theoretical, computed to provide a best fit to equation (10) of McGhee and von Hippel (1974). Each panel is drawn at the same scale; inset shows the data for *Micrococcus lysodeikticus* DNA plotted on a different scale in order to display points determined at low values of r. From Wakelin and Waring (1976).

with data for the other quinoxaline antibiotics, four analogues, and results for the interaction with two alternating polynucleotides composed exclusively of GC or AT nucleotide-pairs. It is immediately evident that each ligand displays a characteristic pattern of preferences which can only be attributed to some form of sequence-specificity and that that selectivity is remarkably sensitive to small changes in the molecular structure of the molecule. It is difficult to make sense of the variations in n, the parameter of apparent site-size, but the values of $K(0)$ indicate large changes in intrinsic affinity for the different DNAs. The differences seem to be exaggerated with the two synthetic DNAs, such that certain ligands display an overwhelming preference for one over the other: the outstanding example is TANDEM, which binds at least 10,000 times more tightly to poly (dA-dT) than to poly (dG-dC) – indeed interaction with the latter polymer is barely detectable at all (Lee and Waring, 1978b). Other noteworthy comparisons which emerge from table 2 are the reversal of preferences for the two synthetic DNAs resulting from mere alteration of the cross-bridge (echinomycin versus trio-stin A), the marked preference of TANDEM for (A+T)-rich DNAs as opposed to the preference for (G+C)-rich DNAs displayed by all other antibiotics, and the remarkable changes in relative binding constants for poly (dG-dC) and poly (dA-dT) caused by substituting one or two quinoline moieties for the natural quinoxaline ring systems in echinomycin. Additionally, 2QN binds more tightly to both alternating polydeoxynucleotides than to any natural DNA. By contrast, substitution of the quinoxaline rings in triostin A with 6-chloro groups produces very little change in DNA-binding parameters.

Although in general the McGhee-von Hippel (1974) model for non-interactive, excluded-site binding to a linear lattice of identical sites provides a reasonable fit to experimental data (figure 6) the agreement is not always satisfactory, and the evidence of sequence preferences indicates one likely inadequacy of the model. Moreover, various lines of evidence point to the possibility of monofunctional intercalation occurring at the same time as bifunctional reaction. With this in mind, Shafer and Waring (1980) have reinterpreted the echinomycin–DNA binding data according to a statistical mechanical model (Shafer, 1980) which allows for simultaneous binding by both modes: the analysis provides very good agreement with the experimental data and is consistent with the favoured model of *bis*-intercalation in which the two chromophores of a bound ligand molecule are required to be separated by two DNA base pairs (contrast with figure 5, Chapter 1). Extension of this approach to the other bifunctional quinoxalines has led to better estimates of the occluded site-size for a *bis*-intercalated ligand, which commonly appears larger than that predicted by the simpler McGhee-von Hippel model, in the region of 6 or 7 base-pairs (Shafer and Waring, 1982).

Another deviation from agreement with the theory of binding to non-interacting sites has been consistently observed in Scatchard plots for binding to poly (dA-dT). They nearly always display the "humped" shape characteristic of cooperative binding at low ligand/nucleotide ratios, and the effect is most marked with TANDEM, which has an extraordinary affinity for this polynucleotide

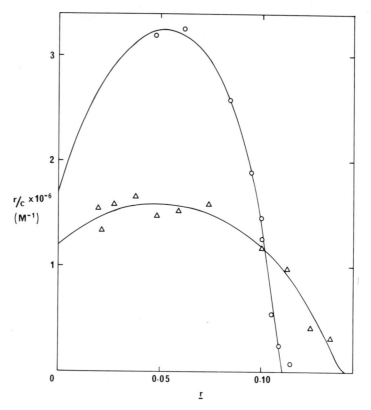

Figure 7   Binding of TANDEM to poly (dA-dT) in buffer of ionic strength 0.2 (○) and 1.0 (△) at 20°. The curves drawn were computed to fit the experimental points so as to satisfy equation (15) of McGhee and von Hippel (1974) describing a cooperative binding process. From Fox *et al*. (1982).

(figure 7). The structural basis for this effect is not known, but it may well depend upon interaction between adjacent bound ligand molecules. It is not seen in a Scatchard plot for the binding of TANDEM to a synthetic self-complementary octanucleotide duplex which seems to present a single isolated binding site with a non-cooperative association constant about 6 μM^{-1} (figure 8).

Because virtually all the equilibrium binding data for quinoxaline antibiotic-DNA interaction have been collected using the solvent-partition method it is important to verify that concordant results can be obtained by other means where possible. In two instances this goal has been achieved. 2QN, the *bis*-quinoline analogue of echinomycin, is weakly fluorescent but sufficiently so to enable spectrofluorometric determination of binding curves based on the quenching of its fluorescent emission when bound to DNA (Fox *et al.*, 1980a). Curves thus determined for binding to all the DNAs represented in table 2 are in good agreement with those measured by solvent partition. [Ala³, Ala⁷] TANDEM, though only very weakly bound by natural DNAs, interacts

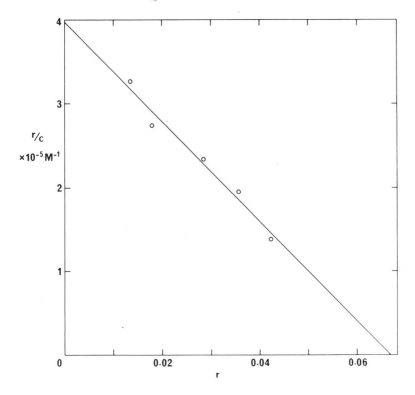

Figure 8   Scatchard plot for the interaction of TANDEM with d(GpGpTpApTpApCpC) at 5°C. From Fox *et al.* (1982).

moderately strongly with poly (dA-dT) and is sufficiently water-soluble to permit determination of binding data by ultraviolet spectrophotometry (Fox *et al.*, 1980*b*). Again, the results agree well with those obtained by solvent partition. They too exhibit the humped shape indicative of cooperative interaction.

STRUCTURE–ACTIVITY RELATIONS AND DETERMINANTS OF RECOGNITION-SPECIFICITY

Early efforts to establish structure–activity relationships for DNA binding by quinoxaline antibiotics revealed that any drastic modification of the peptide ring, such as hydrolysis of the lactone links or destruction of the cross-bridge, led to loss of activity (Wakelin and Waring, 1976; Lee and Waring, 1978*b*). Subsequent work with analogues of TANDEM showed that the D-configuration at the serine residues to which the chromophores are attached is also essential: even if one is epimerised, and the other remains D-, activity is lost (Lee and Waring, 1978*b*; Fox *et al.*, 1980*b*). However, as noted above the *bis*-alanyl analogue of TANDEM which lacks a cross-bridge does retain a reasonable affinity for poly (dA-dT) (Fox *et al.*, 1980*b*). We can briefly summarise the known variations in the peptide structure which can be tolerated as follows: Obviously conservative amino acid replacements at the *N*-methyl-valine positions (Y in figure 2) are acceptable; the evidence indicates that they affect solubility and the absolute magnitude of affinity constants but not the broad pattern of sequence-selectivity (Lee and Waring, 1978*a*; Waring, 1979). The cross-bridge (X in figure 2) can vary as between quinomycins and triostins but changes in binding specificity occur (table 2). The four methyl substituents on peptide bonds can be dispensed with, but only at the expense of a considerable loss of binding energy; again there results a substantial alteration in sequence preferences (table 2; Lee and Waring, 1978*b*; Fox *et al.*, 1982).

It is in many ways easier to focus attention on the role of the chromophores. In the first place, they are not "merely inert plugs that serve only to help locate the peptide portion of the molecule within one of the grooves of the DNA helix after the fashion of a staple" (Fox *et al.*, 1980*a*). If, for example, they are removed from TANDEM altogether (leaving charged α-amino groups on the D-serine residues) the resulting cyclic peptide fails to manifest any but the most meagre evidence of interaction with DNA, and it has no perceptible effect on the supercoiling of circular DNA (figure 9). Substitution of the serines with benzyloxycarbonyl substituents does not restore activity (Lee and Waring, 1978*b*). If the quinoxalines of echinomycin are replaced by quinoline rings the pattern of specificity changes, as we have seen (table 2), and some sort of preference for alternating purine-pyrimidine nucleotide sequences seems to intervene (Fox *et al.*, 1980*a*).

There is evidently a good deal of subtlety as regards the acceptability of replacement chromophores in antibiotic analogues, and in the DNA recognition properties they impart. Three halogenated derivatives of triostin A have been prepared and their DNA-binding characteristics are outlined in table 3. Because of their insolubility in aqueous systems it was not feasible to determine true binding parameters for the 6-bromo and 7-chloro derivatives, but reliable estimates of their relative binding constants for different DNAs were calculated as described in the legend. On the whole there are no systematic differences from the parent antibiotic as regards their interaction with natural DNAs, but some large changes occur in the relative binding constants for the synthetic polydeoxynucleotides. *Bis*-7-chloro-triostin A seems to have an extraordinarily high affinity for poly (dA-dT). The most remarkable thing about this analogue is that it seems to be quite inactive in tests for antibacterial activity, whereas the 6-chloro isomer and the 6-bromo analogue are respectively slightly less active and slightly more active than triostin A itself (A. Cornish, personal communication). It may not be coincidence that TANDEM, the other agent with a notably high affinity for poly (dA-dT) is essentially inactive as an antibacterial drug.

Table 2 Parameters of interaction between quinoxaline antibiotics and nucleic acids.

DNA	(G+C) (%)	Echinomycin		Quinomycin C		Triostin A		TANDEM		1QN		2QN		Bis-6-chloro-triostin A	
		$K(0)$	n	$K(0)$	n	$K(0)$	n	$K(0)$	n	$K(0)$	n	$K(0)$	n	$K(0)$	n
Micrococcus lysodeikticus	72	3.10	9.85	8.26	8.34	1.13	7.73	0.02	12.13	3.93	8.25	1.15	7.50	0.79	7.00
Escherichia coli	50	0.98	11.88	—	—	0.92	7.54	0.05	10.55	3.36	9.76	1.66	8.40	0.45	8.02
Calf thymus	42	0.55	7.17	2.60	8.30	0.70	7.29	0.08	8.89	1.87	7.59	0.88	6.70	0.67	7.38
Clostridium perfringens	30	0.34	9.46	0.91	9.56	0.49	8.52	0.18	10.02	0.56	6.82	0.36	5.48	0.24	6.15
poly(dG-dC)	100	0.55	5.48	3.36	6.12	0.43	7.41	<0.002	—	3.87	7.42	3.27	5.93	—	—
poly (dA-dT)	0	0.31	5.65	0.57	7.02	0.97	4.76	>15	7.40	0.34	4.32	5.46	4.65	—	—

$K(0)$, in units of μM^{-1}, and n represent respectively the association binding constant and calculated site-size (no. of DNA nucleotides occluded by a bound antibiotic molecule) derived from analysis of Scatchard plots according to equation (10) of McGhee and von Hippel (1974), except for TANDEM-poly (dA-dT) interaction where their cooperative equation (15) was used. Data from references as follows: echinomycin (Wakelin and Waring, 1976); quinomycin C, triostin A (Lee and Waring, 1978a); TANDEM (Lee and Waring, 1978b); 1QN, 2QN (Fox et al., 1980a); bis-6-chloro-triostin A (unpublished measurements of K. R. Fox, A. Cornish and M. J. Waring).

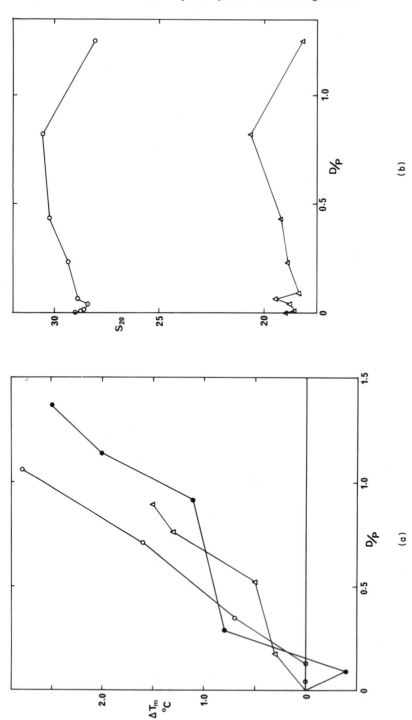

Figure 9 (a) Effect of the octapeptide nucleus of TANDEM on the thermal denaturation of DNA. The ordinate shows the elevation of the transition midpoint ("melting temperature") for poly (dA-dT) (○), calf thymus (●), and *Clostridium perfringens* DNA (△). The abscissa represents the molar ratio of TANDEM peptide to DNA nucleotides. (b) Failure of the peptide moiety of TANDEM to affect the supercoiling of closed circular duplex PM2 DNA (○) or the sedimentation of nicked PM2 DNA (△). Unpublished experiments of K. R. Fox and M. J. Waring.

Table 3 Binding of halogenated triostins to DNA

DNA	Triostin A	Bis-6-chloro-triostin A	Bis-6-bromo-triostin A		Bis-7-chloro-triostin A	
	K_{rel}	K_{rel}	r	K_{rel}	r	K_{rel}
M. lysodeikticus	1.61	1.18	0.054±0.001	1.32	0.037±0.005	1.86
E. coli	1.31	0.67	0.051±0.002	1.10	0.029±0.007	1.40
Calf thymus	1	1	0.049±0.001	1	0.021±0.005	1
C. perfringens	0.70	0.36	0.037±0.001	0.52	0.027±0.005	1.33
poly (dG-dC)	0.61	–	0.061±0.006	1.97	0.016±0.006	0.67
poly (dA-dT)	1.39	–	0.042±0.005	0.65	0.077±0.008	29.4

K_{rel} represents the binding constant of each ligand for a given type of DNA expressed relative to that for calf thymus DNA set equal to 1.0. Data for triostin A and the 6-chloro derivative from table 2. Values for the 6-bromo and 7-chloro derivatives were determined by performing simple solvent-partition experiments in which samples of the various DNAs were equilibrated with a stock antibiotic solution in *iso*-amyl acetate/*n*-heptane (7:3, v/v); r (as defined in table 1) was then determined and used to calculate a relative binding constant assuming a site-size of 8 nucleotides. Unpublished experiment of K. R. Fox, A. Cornish and M. J. Waring.

DYNAMICS OF INTERACTION DETERMINED BY KINETIC ANALYSIS

While the equilibrium studies described above have yielded valuable information about the mechanism of interaction between quinoxaline antibiotics and DNA, and have hinted at the origins of their sequence selectivity, they cannot distinguish between differences in the association or dissociation reactions independently. Without the use of DNAs of defined sequence and/or radioactively labelled antibiotics for low-level binding experiments it seems unlikely that equilibrium studies will be able to provide much more information on the precise nature of preferred binding sites. Accordingly, recent efforts to investigate further the sequence preferences of these ligands have been directed towards investigating the kinetics of their dissociation from DNA (Fox *et al.*, 1981*b*; Fox and Waring, 1981; Fox *et al.*, 1982). The technique employed for this work was based on detergent-induced dissociation, involving the addition of a detergent (sodium dodecyl sulphate; SDS) to a drug–DNA complex. In this procedure it is assumed that the drug dissociates from the DNA in a normal fashion, unaffected by the presence of the detergent, except that the free drug is sequestered by the detergent, preventing its reassociation with the DNA helix (Müller and Crothers, 1968; Behr *et al.*, 1969; Gabbay *et al.*, 1976).

Decay curves which illustrate the dissociation of echinomycin from the synthetic DNAs poly (dG-dC) and poly (dA-dT) are shown in figure 10. They are adequately described by a single exponential, in contrast to the dissociation from a natural DNA (calf thymus) shown in figure 11 which requires a minimum of three exponentials to fit the experimental data. Whereas the dissociation from the synthetic DNAs appears to be a single-step process, the time constant for which depends on the base composition, the complexity of the dissociation profile seen with natural DNAs is interpreted as arising from the presence of several different classes of intercalative binding sites from which the drug dissociates simultaneously. The longest component in the decay then represents the dissociation of the drug from its tightest, that is, most preferred, binding site(s). A consequence of this hypothesis is that, in a natural DNA containing binding sites of varying affinity for the drug, the fraction of molecules bound to each class of sites should vary according to the binding ratio. At low binding levels the drug will bind selectively to its preferred binding site(s); however, as the binding ratio is increased a greater proportion of the ligands will be located at the weaker binding sites. If the dissociation from each class of binding sites conforms to a single exponential characterised by a distinct time constant τ, the relative contribution of each exponential to the total observed decay should vary as a function of the binding ratio (r). This is indeed found to be the case. Figure 11 displays the results of such an experiment. The amplitude of the slowest component (A_3) decreases as the binding ratio increases, although there is no systematic variation in A_1 or A_2. Fox *et al.* (1981*b*) showed that under these circumstances a plot of A_3 against $1/r$ should yield a straight line with slope approximately equal to the r value at which the high affinity sites become fully

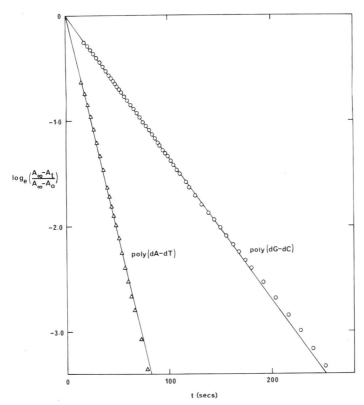

Figure 10   Kinetics of detergent-promoted dissociation of echinomycin from synthetic polynucleotides. The ordinate represents the natural logarithm of the fractional absorbance change. The lines fitted to the experimental points are single exponentials characterised by time constants of 24.3 sec for poly (dA-dT) and 74.1 sec for poly (dG-dC). From Fox *et al.* (1981*b*).

saturated. For echinomycin and calf thymus DNA this graph has a slope of 0.0031±0.0005, indicating that the tightest binding sites saturate at an *r* value around 0.003. If we take the most conservative estimate of average site-size as predicted by McGhee-von Hippel analysis (4 base pairs, corresponding to saturation of the DNA at an *r* value of 0.125) then 2.4% of the total available sites must be of this high affinity form. This very simple analysis suggests that of the 136 distinguishable permutations of four base pairs, three or four bind the drug more tightly than the rest.

If it is correct that the complex dissociation profile seen with natural DNAs is a result of non-equivalence in potential binding sites due to the sequence-selectivity of the drug, the relative proportions of the various time constants should also vary as a function of the composition of the DNA. Table 4 records the parameters for dissociation of echinomycin from a variety of natural DNAs having different base-composition. It can be seen that the observed time constants do indeed fall into three distinct categories whose relative amplitudes

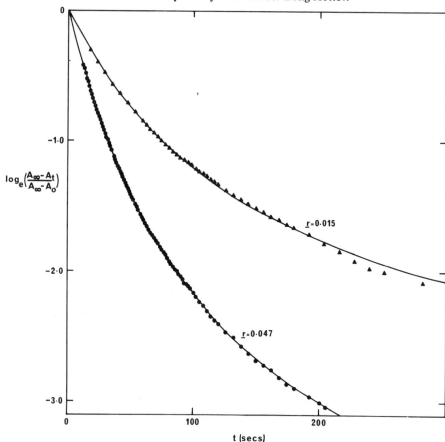

Figure 11 Kinetics of dissociation of echinomycin from calf thymus DNA at different values of the starting antibiotic/nucleotide ratio r. The lines represent three-exponential fits characterised by the following values of τ (time constants, sec.) and A (relative amplitude, %): for $r = 0.014$, $\tau_1 = 32.2$, $\tau_2 = 79.1$, $\tau_3 = 585$ sec., and $A_1 = 40\%$, $A_2 = 41\%$, $A_3 = 19\%$; for $r = 0.047$, $\tau_1 = 18.4$, $\tau_2 = 53.1$, $\tau_3 = 363$ sec., and $A_1 = 55\%$, $A_2 = 38\%$, $A_3 = 7\%$. From Fox *et al.* (1981*b*).

vary considerably. For *Clostridium perfringens* and *Micrococcus lysodeikticus* DNAs the dissociation curve was completely described by two exponentials alone. It appears that the fast-dissociating class of sites is not present in significant amounts in *M. lysodeikticus* DNA whereas the slow-dissociating sites are below the limit of detection in *C. perfringens* DNA. There appears to be quite a good correlation between the contribution of τ_3 to the total decay and the measured equilibrium binding constant for the various DNAs. This is especially noteworthy for bacteriophage T2 DNA, the equilibrium binding constant for which appears larger than would have been predicted on the basis of its gross G+C content; it has 35% G+C residues yet binds echinomycin better than does calf thymus DNA (42% G+C) (Wakelin and Waring, 1976). The measured value of A_3 for T2 DNA is consistent with this unexpectedly large

Table 4 Dissociation of echinomycin from various DNAs

DNA	%G+C	τ_1 (sec)	τ_2 (sec)	τ_3 (sec)	A_1	A_2	A_3	$K(0)$ (μM^{-1})
C. perfringens	30	17.1±2.5	58.3±3.4		56±5	44±5		0.34
Bacteriophage T2	34	22.0±0.7	77.0±2.0	296±21	35±1.5	42±1	22±1	0.78
Calf thymus	42	16.5±0.6	51.0±2.0	297±35	49±5	42±4.5	9±1	0.55
E. coli	50	12.7±3.5	51.0±5.5	216±7	25±3.5	43±3	32±2	0.98
M. lysodeikticus	72		42.5±2.0	216±7		27±2	63±2	3.10
poly (dG-dC)	100	80.6±1.9						0.55
poly (dA-dT)	0	25.0±2.8						0.31

The time constants τ_1, τ_2, and τ_3 and their respective relative amplitudes A_1, A_2, and A_3 (expressed as per cent of total absorbance change) were determined from observed decays as illustrated in figure 11. Data from Fox *et al.*, (1981*b*). Values for $K(0)$, the equilibrium binding constants, are taken from Wakelin and Waring (1976).

binding constant. The values of A_3 are generally higher for the more (G+C)-rich DNAs, suggesting that this slow component represents dissociation from sites rich in G+C residues. However, it is unlikely to be due to dissociation from GCGC (or CGCG) sequences since τ_3 is about 5 times larger than the time constant observed with poly (dG-dC) itself. The observation that the value of A_3 is greatest for the DNA containing 72% (G+C) is fully consistent with the suggestion that the preferred sequence(s) may contain three GC base-pairs and one AT pair (Lee and Waring 1978a).

For a truly monomolecular reaction, such as would be expected for the dissociation of these ligands from DNA, the measured rate constant should be independent of the starting concentration of reactant. However, contrary to expectation the dissociation rate constants were found to vary with DNA concentration, the effect being most pronounced for τ_3 (Fox et al., 1981b). This has been explained by invoking an alternative bimolecular pathway for dissociation, termed facilitated dissociation, in which the drug is "helped" off the DNA by interaction with another nucleic acid molecule. At low DNA concentration the monomolecular pathway should predominate, but as the DNA concentration is raised the bimolecular pathway will become more important, speeding up the overall dissociation rate. The model proposed to account for this effect involves some interaction, albeit weak, between the "back" of the peptide part of an intercalated antibiotic molecule and another DNA molecule. The phenomenon appears not to be restricted to DNA but is equally effective with RNA, for which the quinoxaline antibiotics normally display little or no affinity (Fox et al., 1981b). A weak, nonspecific, interaction of this type involving the peptide ring may be implicated in the weak binding measured with synthetic depsipeptides, such as ELSERTA and ACIMBO (Lee and Waring, 1978b) for which an intercalative mode of binding can be virtually ruled out, and could explain the weak interaction observed between quinoxalines and polyrA·polyrU (Lee and Waring, 1978b; Fox et al., 1980b) since this probably does not involve intercalation either.

Kinetic studies have also been performed, albeit in less detail, with quinomycin C, triostin A, triostin C, and TANDEM (Fox and Waring, 1981; Fox et al., 1982). Relevant parameters of decay are recorded in table 5. In every case the dissociation from the synthetic DNAs is described by a single exponential whereas with natural DNAs a three-exponential function is required for a complete description. The dissociation time constants recorded for quinomycin C are, as anticipated on the basis of its higher affinity constant, larger than those of echinomycin. The dissociation profiles of triostins A and C also required three exponentials but were distinctly faster than those seen with the quinomycins. Moreover, the resolution between the various time constants is generally less marked for the triostins than for the quinomycins. As the different time constants contributing to the total decay become better resolved we can conclude that the selectivity of the ligand must be more marked. The ratio τ_3/τ_1 is greater for both echinomycin and quinomycin C than for triostin C; for the latter the ratio is only 7.3 whereas for echinomycin and quinomycin C it reaches 18 and 25 respectively. This is in

Table 5 Parameters of dissociation of quinoxaline antibiotics from nucleic acids

	Calf thymus DNA							Poly (dA-dT)		Poly (dG-dC)	
	τ_1 (sec)	τ_2 (sec)	τ_3 (sec)	A_1	A_2	A_3	$K(0)$ (μM^{-1})	τ (sec)	$K(0)$ (μM^{-1})	τ (sec)	$K(0)$ (μM^{-1})
Echinomycin	16.5±0.6	51.0±2.0	297±35	49±5	42±4.5	9±1	0.55	25.0±2.8	0.27	80.6±1.9	0.55
Quinomycin C	27.7±3.4	148±15	690±73	18±4	52±1.0	29±1	2.60	36.0	0.57	—	3.36
Triostin A	*	5.6±0.7	31.6±2.6	*	82±2.5	18±2.5	0.70	*	0.97	—	0.43
Triostin C	9.8±0.7	27.5±3.5	72±1.0	45±3	41±4	14±1	3.15	6.6±0.5	—	9.76	—

*Too fast to measure. Kinetic data from Fox and Waring (1981). Equilibrium binding constants from Lee and Waring (1978a).

accord with the view already expressed that the triostins are less sequence-specific than the quinomycins (Lee and Waring, 1978a).

All the antibiotics, but not TANDEM, dissociate faster from the synthetic DNAs than the slowest component of their dissociation from natural DNAs. This provides further evidence that the preferred binding site(s) probably contain a mixture of all four bases. The situation with TANDEM is slightly different. Its dissociation from poly (dA-dT) is well described by a single exponential, the decay constant for which depends on the binding ratio (figure 12). The dissociation of TANDEM from natural DNA is relatively fast, so that only about 14% of the total process could be accounted for (Fox et al., 1982). This observable part of the decay from calf thymus DNA yielded a time constant not very different from that predicted for poly (dA-dT) at very low binding levels. It is therefore tempting to speculate that the slowest dissociation of TANDEM from natural DNA corresponds to release of the ligand from (A+T)-rich sequences.

Quite a good correlation can be drawn between the dissociation behaviour of these antibiotics and their biological activity (Fox and Waring, 1981), as was noted for the actinomycins by Müller and Crothers (1968) and Crothers (1971) and for anthracycline antibiotics by Gabbay et al. (1976) and Wilson et al. (1976). The relative potencies against *Staphylococcus aureus* of echinomycin: triostin C:triostin A: TANDEM are 1:0.25:0.08:<0.005 (Lee, 1977) while the ratios of their longest dissociation time constants (τ_3) are 1:0.24:0.11:0.06. Data for quinomycin C are less precise: Katagiri et al. (1975) record that it is generally 2–5 times more potent than echinomycin against Gram-positive bacteria. The kinetic experiments yield a value for τ_3 which is 2.3 times as large as that of echinomycin, consistent with the trend. These observations substantiate the hypothesis that the antimicrobial/cytotoxic properties of quinoxaline antibiotics can be attributed to their binding to DNA so as to provide a strong but reversible impediment to the progress of RNA polymerase along its template (Waring and Makoff, 1974; Gale et al., 1981).

By constructing Arrhenius plots from measurements of the variation with temperature of decay constants it has proved possible to investigate thermodynamic aspects of the dissociation of triostin C and echinomycin from calf thymus DNA (Fox et al., 1981b; Fox and Waring, 1981). The results suggest that the faster dissociation which distinguishes triostins from quinomycins can be accounted for by their larger entropy of activation, resulting in greater freedom of movement in the activated complex and rendering its formation correspondingly more probable. Fox and Waring (1981) suggested that this might be a direct consequence of the nature of the cross-bridge affecting the flexibility of the molecule, so that the shorter, more rigid thioacetal cross-bridge of the quinomycins might constrain the flexibility of the peptide backbone more severely than the disulphide bridge of the triostins.

As well as providing valuable insights into the nature and abundance of preferred DNA-binding sites, the kinetic studies have also yielded some clues as to the effect of drug binding on the DNA structure. These have arisen from

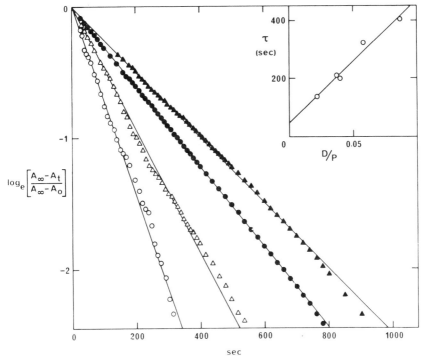

$$\log_e \left[\frac{A_\infty - A_t}{A_\infty - A_o} \right]$$

Figure 12 Dissociation of TANDEM from poly (dA-dT). The ordinate represents the natural logrithm of the fractional absorbance change. The different sets of data correspond to complexes having values for the starting TANDEM/nucleotide ratio as follows: ▲, $r = 0.087$ ($\tau = 403$ sec); ●, $r = 0.058$ ($\tau = 327$ sec); △, $r = 0.036$ ($\tau = 212$ sec); ○, $r = 0.022$ ($\tau = 140$ sec). The lines fitted to the experimental points represent single exponentials characterised by the indicated time constants. Inset is a plot showing the effect of input (molar ratio of drug molecules to DNA nucleotides) on the time constant for the dissociation reaction. From Fox *et al.* (1982).

observations on the variation of time constants with binding ratio for echinomycin dissociating from both calf thymus and *E. coli* DNAs (see figure 11) (Fox *et al.*, 1981*b*), and the variation of the decay time when TANDEM dissociates from poly (dA-dT) (figure 12; Fox *et al.*, 1982). With echinomycin and natural DNAs the dissociation time constants themselves become longer at low binding ratios, the effect becoming quite pronounced at *r* values below 0.025. It has been postulated that this variation is due to the cumulative effect of progressive perturbations in the DNA structure caused by drug binding. Relaxation back from this perturbed structure must occur very slowly, since the very long decay observed from low input ratio complexes is not present in the later stages of dissociation from high input ratio complexes. This implies that the DNA possesses some "long-term memory", so that the initial binding ratio determines the decay constants for virtually the whole dissociation profile. The effect is even more pronounced for TANDEM and poly (dA-dT), the decay constant for which becomes longer at higher input ratios (figure 12). Similar behaviour has been described for actinomycin D and poly (dG-dC) (Krugh *et al.*, 1979); however, no such variation was detected with echinomycin and poly (dA-dT). These observations cannot be explained by any simple models invoking cooperativity in the dissociation process (Fox, 1980) since the profiles of decay followed single exponential kinetics under all the conditions investigated. Once again the behaviour of a complex appears to depend on its history, suggesting that the DNA is able to memorise its starting characteristics.

CONFORMATIONAL STUDIES

The unique features of the interaction between quinoxaline antibiotics and nucleic acids revealed by the experiments summarised above have stimulated a good deal of interest in molecular models. In a real sense, the constraints arising from the need to satisfy the steric requirements of bifunctional intercalation as well as nucleotide sequence recognition render this sytem unusually propitious for an attempt to relate three-dimensional structure to function, especially as regards specificity in DNA-binding. To date efforts have largely been concentrated on precisely defining the structure of the ligands using methods of crystallography, nuclear magnetic resonance (NMR), energy calculations and interactive molecular model-building.

For echinomycin, the results of detailed NMR experiments (Cheung *et al.*, 1978; Williamson and Williams, 1981) have been reinforced by potential energy calculations and molecular model-building studies (Ughetto and Waring, 1977) to establish that the octapeptide ring behaves as a more or less rigid disk. To it are attached, at opposite ends and about 10 Å apart, the quinoxaline chromophores. They extend outwards approximately perpendicular to the plane of the peptide disk, with their own planes nearly parallel (figure 13), so that the space between them is almost exactly correct to accommodate two sandwiched DNA base-pairs. This is, of course, the minimum number of base-pairs needed to render the model compatible with the concept of neighbour-exclusion (see pp. 21 and 24). On the same side of the peptide disk as the chromophores lie the valine *N*-methyl and alanine carbonyl groups. On the other side, and presumably directed away from the DNA and outwards towards the surrounding solvent in the *bis*-intercalated complex, lie the *S*-methyl of the thioacetal cross-bridge together with the *N*-methyl and carbonyl substituents of the cysteine residues (figure 13).

Curiously, the situation in regard to triostin A is different and somewhat more complicated. NMR spectra have revealed that this antibiotic can adopt two distinct, slowly interconvertible conformations in solution (Blake *et al.*, 1977). One, the so-called polar isomer (designated 'p') is favoured in solvents such as pyridine and dimethyl sulphoxide; the other, designated 'n' is found to predominate in nonpolar solvents such as chloroform, CCl_4, and benzene–chloroform mixtures. Each conformer is internally symmetrical, as judged by the equivalence of resonances from pairs of symmetry-related protons in the

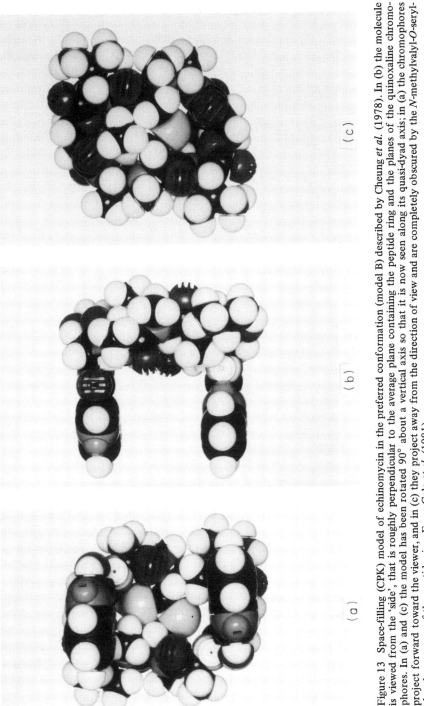

(a)

(b)

(c)

Figure 13   Space-filling (CPK) model of echinomycin in the preferred conformation (model B) described by Cheung *et al.* (1978). In (b) the molecule is viewed from the 'side', that is roughly perpendicular to the average plane containing the peptide ring and the planes of the quinoxaline chromophores. In (a) and (c) the model has been rotated 90° about a vertical axis so that it is now seen along its quasi-dyad axis; in (a) the chromophores project forward toward the viewer, and in (c) they project away from the direction of view and are completely obscured by the *N*-methylvalyl-*O*-seryl-alanyl sequences of the peptide ring. From Gale *et al.* (1981).

molecular formula (compare figures 1 and 2). Doubtless the p conformer is the one most likely to be relevant to the recognition of DNA in aqueous solution; the conformation of its peptide ring appears similar to that of echinomycin (Kalman *et al.*, 1979). These authors have speculated that the two conformations arise from reversing the chirality of the disulphide bond, whereas Kawano *et al.* (1977) attributed the difference to *cis–trans* isomerisation of *N*-methylated peptide bonds.

Besides these NMR studies on the conformations of the antibiotics themselves there have been a few investigations using this technique into the nature of the complexes formed between triostin A and DNA bases (Kyogoku *et al.*, 1978, 1981). The resonances of triostin p in chloroform solution were observed to shift on adding equimolar quantities of adenine and guanine derivatives (but not uracil or cytosine derivatives) whereas the resonances of triostin n (nonpolar form) remained unchanged. Kyogoku *et al.* (1978, 1981) postulated hydrogen bonding between the alanine NHs and the amine and carbonyl groups of guanine, or the amino and N7 groups of adenine. Extrapolation of these studies to the interaction with DNA must be questioned since not only were the experiments performed with base analogues, lacking both sugar and phosphate groups, but in one case the hydrogen bonding scheme proposed would necessitate disrupting the Watson–Crick base pairing (Kyogoku *et al.*, 1978). In the other case the model would involve intercalation of the polar conformer of triostin A from the wide groove (Kyogoku *et al.*, 1981) whereas it is generally believed that these peptide antibiotics intercalate from the narrow groove (Wakelin and Waring, 1976). These studies have, however, served to highlight the different binding properties of the two conformers of triostin A.

More recently our understanding of possible structures adopted by the quinoxaline antibiotics has been strengthened by the determination of a crystal structure for TANDEM (Viswamitra *et al.*, 1981; Hossain *et al.*, 1982). The salient features of this crystal structure (figure 14) are as follows. The octapeptide ring is held in a relatively rigid conformation by means of symmetry-related internal hydrogen bonds between the carbonyl groups of the alanines and the amide N atoms of the valines, yielding a segment of β-pleated sheet structure. The chromophores and the disulphide cross-bridge are disposed on opposite sides of the peptide ring, with the carbonyl oxygens of cysteine and valine on the same side as the cross-bridge. By rotating the chromophores through 15–20° about the N–C_α bonds which connect them to the peptide ring (asterisked in figure 14) it is possible to bring them parallel so as to accommodate two base-pairs between them. The molecule has a chirality in a right-handed helical sense, and an approximate axis of twofold rotational symmetry.

A crystal structure has also been reported for luzopeptin A (Arnold and Clardy, 1981). Again the structure contains two symmetry-related internal hydrogen bonds bridging the decadepsipeptide ring, between the glycine amide nitrogens and the sarcosine carbonyl oxygens (see figure 3).

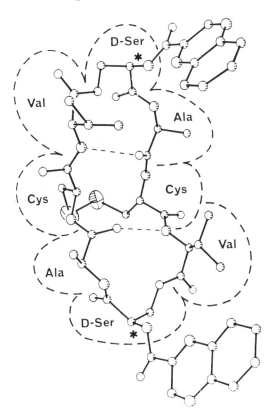

Figure 14   Molecular structure of TANDEM determined by X-ray crystallography at room temperature to a resolution of 0.85 Å. Dotted lines represent intramolecular hydrogen bonds connecting the NH groups of L-Val to the carbonyl O atoms of L-Ala. These bonds hold the peptide ring in a relatively stable conformation. Asterisks indicate the $N-C_\alpha$ (D-Ser) bonds through which the chromophores are attached to the octapeptide. From Viswamitra *et al.* (1981).

THE MOLECULAR BASIS OF SEQUENCE RECOGNITION

While great caution must be exercised when extrapolating from molecular model studies it is interesting to speculate how the different nucleotide sequence preferences of the various members of the quinoxaline series might originate. Early ideas that the position of the cross-bridge might sterically affect interactions with substituents such as the 2-amino group of guanine (Lee and Waring, 1978*b*; Waring, 1979) now seem less likely in view of the crystal structure of TANDEM which shows the disulphide bridge positioned away from the chromophores (figure 14). There is also the observation that [Ala³, Ala⁷] TANDEM, lacking the offending cross-bridge, still displays a significant preference for binding to poly (dA-dT) (Fox *et al.*, 1980*b*). We are best advised to examine the probable nature of interactions influencing the binding reaction in the light of the TANDEM crystal structure, with a judicious eye to likely differences from the natural antibiotics.

To account for the impressive specificity of TANDEM for poly (dA-dT) Viswamitra *et al.* (1980) searched for possible hydrogen bond contacts between the functional groups of TANDEM, facing the helix, and the base pairs. For binding in the minor groove the NHs of the L-alanine residues seem likely candidates as hydrogen bond donors if only because planes drawn through them, parallel to the chromophores, are separated by 3.5–4 Å. The most likely hydrogen bond acceptors in the minor groove are the 2-keto groups of thymine. Involvement of these groups in hydrogen bonding would suggest that the base-pair sequence sandwiched between the two chromophores is ApT. It is interesting that according to the alternating-B DNA model for poly (dA-dT) proposed by Klug *et al.* (1979) the sequence into which the chromophores would then be inserted, TpA, should be easier to unstack than its isomeric sequence ApT. On the basis of these arguments the ability of TANDEM to intercalate specifically into AT sequences appears to result from a complex and subtle interplay of forces between the DNA as receptor and the specific steric structure of the chromopeptide, held rigidly in a suitable conformation for binding.

Extrapolating from this model for the interaction of TANDEM with poly (dA-dT) to the natural quinoxaline antibiotics we first note that the molecular structures of quinomycins and triostins must differ from that of TANDEM because the methylated peptide substituents prevent formation of the internal hydrogen bonds. The result of this could well be to open out the peptide backbone, exposing the carbonyl oxygen atoms of the alanines, making them available as potential hydrogen bond acceptors to interact with the 2-amino groups of guanine nucleotides in the minor groove. These antibiotics would then be able to recognise both AT (or conceivably TA) as well as GC (or CG) sequences at the sandwiched base pairs.

Despite the insights gained from the crystal structure of TANDEM and the various lines of evidence suggesting the rigidity of the peptide backbone, there is no evidence to prove that its conformation does not alter on interaction with the DNA. Indeed circular dichroism experiments have suggested that the steric relationship between the chromophores of echinomycin and the peptide ring might alter on interacting with DNA (Costantino *et al.*, 1978). It might also be foolhardy to attempt a precise description of quinoxaline antibiotic–DNA complexes while there exists the current climate of uncertainty as to the exact local conformation of DNA related to its nucleotide sequence (Wang *et al.*, 1979; Klug *et al.*, 1979; Leslie *et al.*, 1980). We may be sure that imminent developments in our knowledge of polynucleotide structure (and its variability) will add a fresh dimension to problems of sequence recognition by antibiotic ligands.

ACKNOWLEDGEMENTS

Original work reported here was supported by grants from the Medical Research Council, the Science and Engineering Research Council, the Royal Society and the Cancer Research Campaign. We are grateful to our many colleagues and to the authorities of CIBA-Geigy Ltd and Imperial Chemical Industries for their willing cooperation and assistance in setting up experiments.

REFERENCES

Arnold, E. and Clardy, J. (1981). *J. Am. Chem. Soc.*, **103**, 1243
Behr, W., Honikel, K. and Hartmann, G. (1969). *Eur. J. Biochem.*, **9**, 82
Blake, T. J., Kalman, J. R. and Williams, D. H. (1977). *Tetrahedron Lett.*., **30**, 2621
Bojesen, G., Gauvreau, D., Williams, D. H. and Waring, M. J. (1981). *J.C.S. Chem. Commun.*, **1981**, 46
Bradley, S. G., Gauvreau, D. and Waring, M. J. (1980). *Devl ind. Microbiol.*, **21**, 245
Chakravarty, P. K. and Olsen, R. K. (1978). *Tetrahedron Lett.*, **1978**, 1613
Cheung, H. T., Feeney, J., Roberts, G. C. K., Williams, D. H., Ughetto, G. and Waring, M. J. (1978). *J. Am. chem. Soc.*, **100**, 46
Ciardelli, T. L., Chakravarty, P. K. and Olsen, R. K. (1978). *J. Am. chem. Soc.*, **100**, 7684
Ciardelli, T. L. and Olsen, R. K. (1977). *J. Am. chem. Soc.*, **99**, 2806
Corbaz, R., Ettlinger, L., Gäumann, E., Keller-Schierlein, W., Kradolfer, F., Neipp, L., Prelog, V., Reusser, P. and Zähner, H. (1957). *Helv. chim. Acta*, **40**, 199
Costantino, P., De Santis, P., Ughetto, G. and Waring, M. J. (1978). *FEBS Lett.*, **88**, 349
Crothers, D. M. (1968). *Biopolymers*, **6**, 575
Crothers, D. M. (1971). *Prog. molec. subcell. Biol.*, **2**, 10
Dell, A., Williams, D. H., Morris, H. R., Smith, G. A., Feeney, J. and Roberts, G. C. K. (1975). *J. Am. chem. Soc.*, **97**, 2497
Fox, K. R. (1980). Thesis, Univ. Cambridge
Fox, K. R., Gauvreau, D., Goodwin, D. C. and Waring, M. J. (1980*a*). *Biochem. J.*, **191**, 729
Fox, K. R., Harrison, N. L. and Waring, M. J. (1981*a*). *FEBS Lett.*, **133**, 305
Fox, K. R., Olsen, R. K. and Waring, M. J. (1980*b*). *Br. J. Pharmac.*, **70**, 25
Fox, K. R., Olsen, R. K. and Waring, M. J. (1982). *Biochim. biophys. Acta.*, **696**, 315
Fox, K. R., Wakelin, L. P. G. and Waring, M. J. (1981*b*). *Biochemistry*, **20**, 5768
Fox, K. R. and Waring, M. J. (1981). *Biochim. biophys. Acta*, **654**, 279
Gabbay, E. J., Grier, D., Fingerle, R. E., Reimer, R., Levy, R., Pearce, S. W. and Wilson, W. D. (1976). *Biochemistry*, **15**, 2062
Gale, E. F., Cundliffe, E., Reynolds, P. E., Richmond, M. H. and Waring, M. J. (1981). *The Molecular Basis of Antibiotic Action*, 2nd edn. Wiley, London
Harada, Y., Sunagawa, N. and Katagiri, K. (1968). *Gann*, **59**, 513
Hogan, M., Dattagupta, N. and Crothers, D. M. (1979). *Biochemistry*, **18**, 280
Hossain, M. B., Van Der Helm, D., Olsen, R. K., Jones, P. G., Sheldrick, G. M., Egert, E., Kennard, O., Waring, M. J. and Viswamitra, M. A. (1982). *J. Am. chem. Soc.*, in press
Huang, C. H., Mong, S. and Crooke, S. T. (1980). *Biochemistry*, **19**, 5537
Kageyama, M., Hasegawa, M., Inagaki, A. and Egami, F. (1970). *J. Biochem.*, **67**, 549
Kalman, J. R., Blake, T. J., Williams, D. H., Feeney, J. and Roberts, G. C. K. (1979). *J. chem. Soc. Perkin* **1**, 1313
Katagiri, K., Endo, H., Tada, M. and Nikaido, H. (1966). *A. Rep. Shionogi Res. Lab.*, **16**, 58
Katagiri, K., Okamoto, S., Sato, K., Shamoaka, N., Tawara, K. and Sasaki, S. (1957). *A. Rep. Shionogi Res. Lab.*, **7**, 191
Katagiri, K. and Sugiura, K. (1961). *Antimicrob. Agents Chemother.*, **1961**, 162
Katagiri, K., Yoshida, T. and Sato, K. (1975). *Antibiotics III. Mechanism of Action of Antimicrobial and Antitumour Agents*, (ed. J. W. Corcoran and F. E. Hahn). pp. 234–251, Springer Verlag, Berlin

Kawano, K., Higuchi, N. and Kyogoku, Y. (1977). In *Peptide Chemistry* (ed. T. Nakajima) (English edn), Peptide Research Foundation, Osaka, Japan, 93

Keller-Schierlein, W., Mihailovic, M. L. and Prelog, V. (1959). *Helv. chim. Acta*, **42**, 305

Klug, A., Jack, A., Viswamitra, M. A., Kennard, O., Shakked, Z. and Steitz, T. A. (1979). *J. molec. Biol.*, **131**, 669

Konishi, M., Ohkuma, H., Sakai, F., Tsuno, T., Koshiyama, H., Naito, T. and Kawaguchi, H. (1981). *J. Am. chem. Soc.*, **103**, 1241

Krugh, T. R., Hook, J. W., Lin, S. and Chen, F. M. (1979). In *Stereodynamics of Molecular Systems*, (ed. R. H. Sarma), Pergamon, New York, pp. 423–435

Kyogoku, Y., Higuchi, N., Watanabe, M. and Kawano, K. (1981). *Biopolymers*, **20**, 1959

Kyogoku, Y., Yu, B. S., Akutsu, H., Watanabe, M. and Kawano, K. (1978). *Biochem. biophys. Res. Commun.*, **83**, 172

Lee, J. S. (1977). Thesis, Cambridge Univ.

Lee, J. S. and Waring, M. J. (1978*a*). *Biochem. J.*, **173**, 115

Lee, J. S. and Waring, M. J. (1978*b*). *Biochem. J.*, **173**, 129

Leslie, A. G. W., Arnott, S., Chandrasekaran, R. and Ratliff, R. L. (1980). *J. molec. Biol.*, **143**, 49

McGhee, J. D. and von Hippel, P. H. (1974). *J. molec. Biol.*, **86**, 469

Martin, D. G., Mizsak, S. A., Biles, C., Stewart, J. C., Baczynskyj, L. and Meulman, P. A. (1975). *J. Antibiot.*, **28**, 332

Matsuura, S. (1965). *J. Antibiot.*, A. **18**, 43

Müller, W. and Crothers, D. M. (1968). *J. molec. Biol.*, **35**, 251

Reinert, K. E. (1973). *Biochim. biophys. Acta*, **319**, 135

Sato, K., Niinomi, Y., Katagiri, K., Matsukage, A. and Minagawa, T. (1969) *Biochim. biophys. Acta*, **174**, 230

Sato, K., Shiratori, O. and Katagiri, K. (1967*b*). *J. Antibiot.*, A. **20**, 270

Sato, K., Yoshida, T. and Katagiri, K. (1967*a*). *J. Antibiot.*, A. **20**, 188

Schnedl, W., Dann, O. and Schweizer, D. (1980). *Eur. J. Cell Biol.*, **20**, 290

Shafer, R. H. (1980). *Biopolymers.*, **19**, 419

Shafer, R. H. and Waring, M. J. (1980). *Biopolymers*, **19**, 431

Shafer, R. H. and Waring, M. J. (1982). *Biopolymers*, submitted

Shoji, J. and Katagiri, K. (1961). *J. Antibiot.*, A. **14**, 335

Tsunoda, A. (1962). *J. Antibiot.*, A. **15**, 60

Ueda, M., Tanigawa, Y., Okami, Y. and Umezawa, H. (1954). *J. Antibiot.*, A. **7**, 125

Ughetto, G. and Waring, M. J. (1977). *Molec. Pharmac.*, **13**, 579

Viswamitra, M. A., Kennard, O., Cruse, W. B. T., Egert, E., Sheldrick, G. M., Jones, P. G., Waring, M. J., Wakelin, L. P. G. and Olsen, R. K. (1981). *Nature, Lond.*, **289**, 817

Wakelin, L. P. G. and Waring, M. J. (1976). *Biochem. J.*, **157**, 721

Wang, A. H. J., Quigley, G. J., Kolpak, F. J., Crawford, J. L., van Boom, J. H., van der Marel, G. and Rich, A. (1979). *Nature, Lond.*, **282**, 680

Wang, J. C. (1974). *J. molec. Biol.*, **89**, 783

Ward, D., Reich, E. and Goldberg, I. H. (1965). *Science*, **149**, 1259

Waring, M. J. (1977). In *Drug Action at the Molecular Level*, (ed. G. C. K. Roberts) Macmillan, London, pp. 167–189

Waring, M. J. (1979). In *Antibiotics Vol. 5/Part 2 Mechanism of Action of Antieukaryotic and Antiviral Compounds*, (ed. F. E. Hahn), Springer-Verlag, Berlin, pp. 173–194.

Waring, M. J. (1981). *A. Rev. Biochem.*, **50**, 159

Waring, M. J. and Makoff, A. (1974). *Pharmac.*, **10**, 214

Waring, M. J. and Wakelin, L. P. G. (1974). *Nature, Lond.*, **252**, 653

Waring, M. J., Wakelin, L. P. G. and Lee, J. S. (1975). *Biochim. biophys. Acta*, **407**, 200

Williamson, M. P., Gauvreau, D., Williams, D. H. and Waring, M. J. (1982). *J. Antibiot.*, **35**, 17

Williamson, M. P. and Williams, D. H. (1981). *J.C.S., Chem. Commun.*, 165

Wilson, D. W., Grier, D., Reimer, R., Bauman, J. D., Preston, J. F. and Gabbay, E. J. (1976). *J. med. Chem.*, **19**, 381

Yoshida, T., Katagiri, K. and Yokozawa, S. (1961). *J. Antibiot.*, A. **14**, 330

Yoshida, T., Kimura, Y. and Katagiri, K. (1968). *J. Antibiot.*, **21**, 465

6

Bleomycin

Lawrence F. Povirk

INTRODUCTION

Phleomycin, a historical precursor of bleomycin, was isolated from *Streptomyces verticillus* by Umezawa and colleagues in 1956 (Maeda *et al.*, 1956). A screening of other *S. verticillus* strains for similar compounds yielded the bleomycin(s), a family of basic glycopeptides, which had similar antitumour activity but less renal toxicity than phleomycin (Umezawa *et al.*, 1966; Ishizuka *et al.*, 1967).

The toxic side effects of bleomycin do not overlap those of the more frequently-used antitumour drugs; bleomycin is not myelosuppressive, but causes pulmonary fibrosis at high doses (Blum *et al.*, 1973). For this reason it is under intensive clinical investigation in combination therapy with nearly all the major antitumour drugs. Currently, bleomycin is used mainly in treatment of squamous cell carcinomas, testicular cancers and cancers of the head and neck. It is also used in treatment of Hodgkins disease and other lymphomas (see review by Friedman, 1978).

Bleomycin is believed to act by degrading cellular DNA in a reaction requiring $Fe(II)$ and O_2 as cofactors (Onishi *et al.*, 1975; Sausville *et al.*, 1978a, b). Recent evidence strongly suggests that the ultimate agent of DNA damage is some form of activated reduced oxygen, produced as a consequence of oxidation of bleomycin-chelated $Fe(II)$ to $Fe(III)$ in a quaternary $DNA \cdot bleomycin \cdot iron \cdot oxygen$ complex.

Molecular Aspects of Anti-cancer Drug Action, ed. Neidle & Waring
0333-315561/83/157–181

STRUCTURE

Primary structure of bleomycin and its analogues

Bleomycin (figure 1) is a glycopeptide containing several unusual amino acids and sugars, a pyrimidine, and a planar bithiazole ring system (Takita *et al.*, 1978*a*; Koyama *et al.*, 1968). The bleomycin molecule may be formally divided

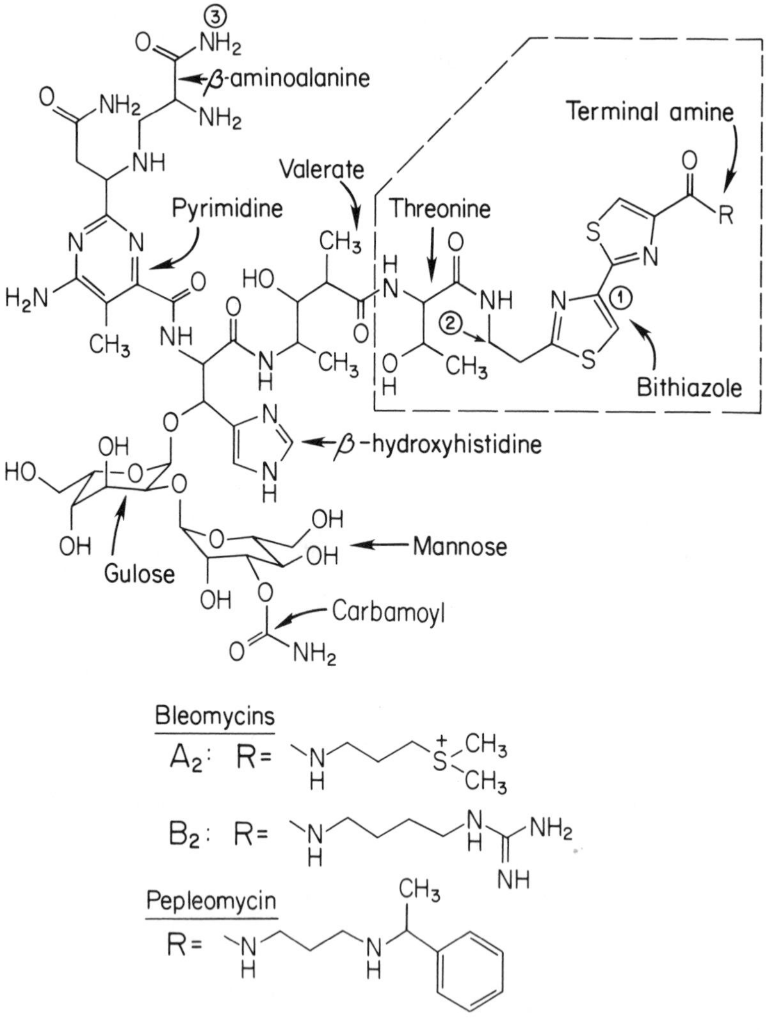

Figure 1 Primary structure of bleomycin as determined by Takita *et al.* (1972, 1978*a*), showing (1) double-bond which is replaced by a single-bond in phleomycin; (2) site of attachment of an additional talose sugar derivative in tallysomycin, and (3) amide which is deaminated in deamidobleomycin. Enclosed area is the DNA-binding region, designated tri-peptide S when it contains the terminal amine of bleomycin A_2.

between its DNA-binding domain (boxed area in figure 1) and its metal-chelating domain (the remainder of the molecule). In the case of bleomycin A_2, the DNA-binding region, designated "tripeptide S", can be conveniently split off from the parent molecule by mild acid hydrolysis (Umezawa, 1973).

The various bleomycins differ only in their terminal amines. Clinical bleomycin consists mainly of bleomycins A_2 and B_2, plus smaller amounts of other components. Bleomycins with unusual terminal amines can be produced by adding appropriate precursors to the fermentation broth. Alternatively, the terminal amine can be replaced chemically after isolation, as in the case of pepleomycin (Tanaka, 1978). In bleomycinic acid (Umezawa *et al.*, 1973), the terminal amine is replaced by a hydroxyl; this gives the molecule a negatively charged carboxylic acid group, rendering it inactive. Most chemical studies on bleomycin have been performed with bleomycin A_2. Bleomycin has an absorbance maximum at 287 nm ($\epsilon = 14{,}000$ M^{-1}; Dabrowiak *et al.*, 1978*a*).

Phleomycin is identical to bleomycin, except that it lacks one double bond in the bithiazole moiety (Takita *et al.*, 1972). This structural difference makes it less stable than bleomycin (Umezawa *et al.*, 1966). Tallysomycin (structure not shown) is an antibiotic which is similar in structure to bleomycin, but which contains two slightly modified amino acids and an additional talose sugar derivative attached to the peptide chain between the threonine and bithiazole moieties (Kawaguchi *et al.*, 1977). Although their relative activities vary, all bleomycin analogues are thought to function in a similar manner, that is, by causing strand breaks in cellular DNA.

Deamidobleomycin is an inactive metabolite formed in mammalian cells by enzymatic deamination of the β-aminoalanine moiety (Umezawa *et al.*, 1972, 1974).

Roy *et al.* (1981) have recently described a method for producing a stable radiolabelled bleomycin, by chemical exchange of one methyl group in the dimethylsulphonium moiety of the bleomycin A_2 terminal amine (see figure 1). Attempts at total chemical synthesis of bleomycin are in progress (Hecht *et al.*, 1979).

Metal–bleomycin complexes

Bleomycin forms one-to-one chelate complexes with several metals, including copper, nickel, zinc, iron, indium, gallium and cobalt (Nunn, 1976; Dabrowiak, 1980). Various metal–bleomycin complexes have been examined, primarily as models for the biologically-important iron complexes. Despite this intensive effort, none of their structures are known with certainty (see review by Dabrowiak, 1980).

Cu(II)·bleomycin is an extremely stable complex, and bleomycin is recovered from culture filtrates in this form (Umezawa *et al.*, 1966). While neither free bleomycin nor any metal–bleomycin complex has yet been crystallised, Iitaka *et al.* (1978) have succeeded in crystallising the Cu(II) complex of P-3A, an intermediate in bleomycin biosynthesis. The three-dimensional structure, as determined by X-ray crystallography, indicates a square-pyramidal chelation site with five nitrogen ligands (figure 2). Although it has not been proved that Cu(II)·P-3A

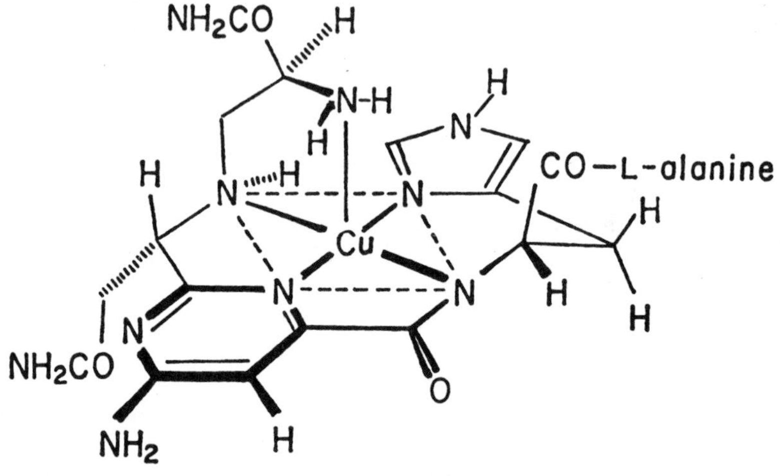

Figure 2   Structure of the Cu(II) complex of P-3A, an intermediate in bleomycin biosynthesis, as determined by X-ray crystallography. Reprinted from Iitaka *et al.* (1978) with permission.

and Cu(II)·bleomycin have the same chelation geometry, several lines of evidence suggest that the two are similar. In Cu(II)·bleomycin, participation of the primary amine of β-aminoalanine (the axial ligand of figure 2) and of the imidazole of β-hydroxyhistidine as ligands is confirmed by their disappearance as titratable groups when the Cu(II) complex is formed (Muraoka *et al.*, 1976). Ultraviolet spectroscopy, specifically the red shift of the Π–Π* electronic transition at 250 nm, suggests participation of the pyrimidine (Dabrowiak *et al.*, 1978*a*). The identity of specific nitrogen ligands of the pyrimidine and imidazole rings as those shown in figure 2, may be inferred from steric requirements (Takita *et al.*, 1978*b*). Furthermore, in the ^{13}C spectrum of Cu(II)·bleomycin, nearly all the carbon resonances of the β-aminoalanine, pyrimidine and β-hydroxyhistidine moieties (that is, the groups which contain the metal ligands of the P-3A complex) are missing, due to severe line broadening induced by their proximity to the paramagnetic Cu(II) ion (Dabrowiak *et al.*, 1978*b*). The resonances of the threonine, bithiazole and terminal amine, although slightly broadened, remain resolvable, suggesting that these groups, which are not contained in P-3A, are not involved in metal chelation. Resonances of the sugar moieties are also resolvable with the exception of the mannose carbamoyl. Involvement of the carbamoyl in metal chelation is also suggested by the distinct differences in electrochemistry and in circular dichroism spectra between Cu(II)·bleomycin and Cu(II)·isobleomycin; in isobleomycin the carbamoyl is displaced from the

3-O to the 2-O position of mannose (Ishizu *et al.*, 1981). Thus, Takita *et al.* (1978) conclude that the structure of Cu(II)·bleomycin is probably the same as that of Cu(II)·P-3A, except that the bleomycin complex has a loosely-bound mannose carbamoyl oxygen at the sixth coordination site which is vacant in the P-3A complex.

The Zn(II)·bleomycin complex has the advantage that all ^{13}C and ^{1}H NMR resonances remain unbroadened, since Zn(II) is diamagnetic. The bleomycin moieties whose resonances show significant chemical shift changes upon Zn(II) chelation are essentially the same moieties whose ^{13}C resonances disappear upon Cu(II) chelation (Dabrowiak *et al.*, 1978*b*). Thus, many of the same moieties and possibly the same individual ligands are involved in Cu(II) and Zn(II) chelation. Oppenheimer *et al.* (1979) have shown by analysis of ^{1}H NMR coupling constants that Zn(II) chelation confers conformational rigidity on the β-aminoalanine and β-hydroxyhistidine moieties, constraining some carbon–carbon bonds in sterically unfavourable rotamers which are not present in metal-free bleomycin. In the same study, rather large chemical shifts were seen for the valerate protons, suggesting that this moiety, which is not present in the P-3A fragment, may be involved in Zn(II) chelation.

Co(II)·bleomycin has been studied because it undergoes a two-step oxidation by O_2 to Co(III)·bleomycin (Nunn, 1977*a*), a reaction which may be analogous to the biologically important oxidation of Fe(II)·bleomycin. The anaerobic ESR spectrum of Co(II)·bleomycin suggests axial symmetry, and the presence of superhyperfine splitting suggests that only one axial ligand is a nitrogen (Sugiura, 1978). The EPR spectrum of Co(II)·P-3A is similar to that of Co(II)·bleomycin although it lacks superhyperfine splitting. Addition of O_2 produces drastic changes in the Co(II)·bleomycin EPR spectrum, changes which closely mimic changes seen for other Co(II) complexes (including myoglobin) upon formation of their respective O_2 adducts (Sugiura, 1978). These results strongly suggest formation of a ternary bleomycin·Co(II)·O_2 complex, in which O_2 is one of the metal ligands. This complex slowly decays to produce two forms of Co(III)·bleomycin (DeRiemer *et al.*, 1979). One of these is extremely stable and has been used for radiodiagnostic tumour imaging.

The demonstration of an oxygen-sensitive one-to-one Fe(II)·bleomycin complex was central to the implication of Fe(II) as a cofactor in bleomycin--induced DNA degradation (Sausville *et al.*, 1978*b*). Under anaerobic conditions, Fe(II)·bleomycin is a stable complex with an absorption maximum at 480 nm ($\epsilon = 380$ M^{-1}). It has no EPR signal (Sugiura and Kikuchi, 1978). The ^{13}C NMR spectra of both Fe(II)·bleomycin and Fe(III)·bleomycin are similar to that of Cu(II)·bleomycin in that all resonances except those of the threonine, bithiazole and terminal amine groups are missing due to line broadening (Dabrowiak *et al.*, 1979). These results suggest that most of the same moieties are involved in Fe(II) and Fe(III) chelation as are involved in Cu(II) and Zn(II) chelation.

Fe(II)·bleomycin forms stable ternary complexes with the oxygen analogues NO, CO, and C_2H_2NC, all having enhanced visible absorbance (Burger *et al.*,

1979a). Bleomycin•Fe(II)•NO is EPR active. The NMR spectrum of bleo-mycin•Fe(II)•CO, unlike that of Fe(II)•bleomycin, has sharp ¹H resonances, indicating that CO coordination renders Fe(II) in the complex diamagnetic. From this NMR spectrum, Oppenheimer *et al.* (1979) tentatively assigned the six ligands of the complex (figure 3). The imidazole of β-hydroxyhistidine and

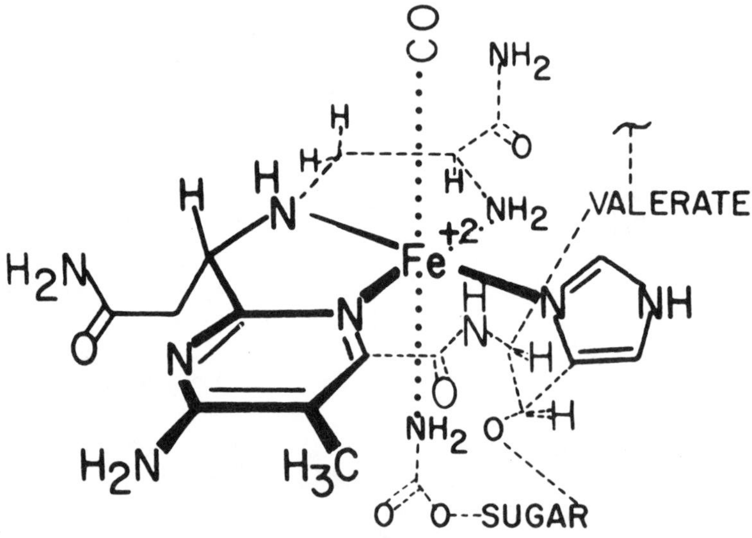

Figure 3   Tentative structure of bleomycin•Fe(II)•CO, based primarily on ¹H NMR of the complex. The portions of the structure drawn in dashed lines indicate ligands whose position in the complex is not certain. There are some recent data which are not entirely consistent with this model structure (see text). Reprinted from Oppenheimer *et al.* (1979) with permission.

the primary and secondary amines of β-aminoalanine are implicated as ligands by the *p*H-independence of their chemical shifts, the pyrimidine by the similarity of its methyl chemical shift to that of Zn(II)•bleomycin, and the carbamoyl by the strong downfield shift of the C-3′ proton of the mannose moiety. CO is the sixth ligand. If bleomycin•Fe(II)•CO and its analogue bleomycin•Fe(II)•O₂ have the same chelation geometry, participation of the carbamoyl as a ligand is contrary to the suggestion by Takita *et al.* (1978b) (based on analogy to the Cu(II)•P-3A complex), that the carbamoyl is displaced by O₂ in the bleomycin•Fe(II)•O₂ complex. As a further complication, a later study by Oppenheimer *et al.* (1980) comparing bleomycin•Cu(I)•CO and bleomycin•Fe(II)•CO suggested participation of the amide nitrogen between the pyrimidine and imidazole, as a ligand in bleomycin•Fe(II)•CO (contrary to figure 3). Thus there is now an unresolved paradox: seven ligands for which there is experimental evidence, but only six available ligand sites.

Fe(III)·bleomycin, formed by addition of ferric salts to bleomycin at acid pH, exists in a pH-dependent equilibrium ($pK = 4.3$) between low spin ($S = \frac{1}{2}$) and high spin ($S = \frac{5}{2}$) forms (Dabrowiak *et al.*, 1979; Burger *et al.*, 1979*a*). At neutral pH, the low spin form predominates, but low pH or presence of $H_2PO_4^-$ anion favours the high spin form. At very low pH ($pK = 2.8$) the complex dissociates. The low spin Fe(III)·bleomycin EPR spectrum is very similar to that of heme oxygenases. Although this similarity is surprising considering the much greater aromaticity of the heme complex, it tends to confirm that the bleomycin complex consists largely of nitrogenous ligands.

Fe(III)·bleomycin can also be formed by O_2 oxidation of Fe(II)·bleomycin, with formation of unstable intermediates which have been implicated in bleomycin-induced DNA damage (Sugiura and Kikuchi, 1978; Dabrowiak *et al.*, 1979; Kuramochi *et al.*, 1981; Burger *et al.*, 1981). These reactions are discussed in more detail below.

CHEMISTRY OF DNA DEGRADATION

Binding to DNA

Binding of bleomycin to DNA has been shown by a variety of techniques including nuclear magnetic resonance (NMR), quenching of the bithiazole fluorescence (Chien *et al.*, 1977), equilibrium dialysis (Povirk *et al.*, 1979; Roy *et al.*, 1981) and circular dichroism (Krueger *et al.*, 1973). Clearly the bithiazole and terminal amine are the moieties most strongly associated with the DNA. These moieties show the largest changes in ^1H NMR chemical shifts when bleomycin or Zn(II)·bleomycin binds to poly(dA-dT)·poly(dA-dT) (Chen *et al.*, 1980; Glickson *et al.*, 1981). Furthermore the tripeptide S fragment has nearly the same affinity for DNA as the whole bleomycin molecule (Chien *et al.*, 1977; Povirk *et al.*, 1979). Whether the bithiazole intercalates between DNA bases, or simply binds in the minor groove, is disputed. Bleomycin and tripeptide S (but not phleomycin) induce the DNA lengthening and unwinding characteristic of classical intercalators, and the negative electric dichroism of oriented DNA·bleomycin complexes indicates that the bithiazole rings are parallel to the aromatic rings of DNA bases (Povirk *et al.*, 1979, 1981*a*). However, in NMR studies of the poly(dA-dT)·poly(dA-dT)·bleomycin complex, the nonexchangeable bithiazole protons do not show the large upfield shifts seen for protons of other intercalated ring systems (Chen *et al.*, 1980; Glickson *et al.*, 1981).

Studies with synthetic polynucleotides (Asakura *et al.*, 1978; Kasai *et al.*, 1978) indicate that (1) bleomycin interacts most strongly with guanine-containing nucleic acids, (2) alternating purine-pyrimidine sequences are preferred and (3) the positive charge on the terminal amine contributes to binding, presumably by electrostatic attraction to DNA phosphates.

There is also some degree of interaction between DNA and the chelating portion of bleomycin. In the absence of metals, the potential chelating ligands are free to rotate in solution even while the bithiazole and terminal amine are tightly bound to DNA (Chien *et al.*, 1977). However, in the Zn(II)·bleomycin·[(poly (dA-dT)·poly(dA-dT)] complex the valerate and threonine residues show interaction with both Zn(II) and poly(dA-dT)·poly(dA-dT), as indicated by shifts in ^{1}H NMR resonances (Glickson *et al.*, 1981). EPR studies of Cu(II)·bleomycin bound to oriented DNA fibres indicate that the chelate complex is bound stereospecifically, that is, with the plane of symmetry (figure 2) parallel to the plane of DNA bases (Hutchinson *et al.*, 1981). The EPR spectrum of bleomycin· Fe(II)·NO shows perturbations when bound to DNA (Antholine and Petering, 1979*b*; Sugiura and Ishizu, 1979). Interestingly these perturbations are not seen with RNA (Sugiura *et al.*, 1981), which binds bleomycin, but is not degraded by the drug and does not competitively inhibit bleomycin-induced DNA degradation (Asakura *et al.*, 1975; Kasai *et al.*, 1978; Müller *et al.*, 1972). Perturbations of the bleomycin·Co(II)·O$_2$ EPR spectrum are also seen in the presence of DNA (Sugiura, 1978).

Association constants of 10^5–10^6 M^{-1} have been reported by various workers for the bleomycin·DNA complex in solutions with ionic strengths of several millimolar (Chien *et al.*, 1977; Kasai *et al.*, 1978; Povirk *et al.*, 1979; Roy *et al.*, 1981). Little quantitative data on bleomycin·DNA binding at physiological ionic strengths has been obtained, but it is clear that high salt strongly inhibits binding; even at only 50 mM NaCl, half-maximal quenching of Cu(II)·bleomycin fluorescence requires a DNA concentration of 2 mM nucleotides (Povirk *et al.*, 1981*a*).

Nature of DNA damage

Release of bases and base-propenals

Free bases are major products of the bleomycin-DNA reaction (Müller *et al.*, 1972; Haidle *et al.*, 1972). Some early reports indicated release of thymine only, but subsequent quantitative studies with specifically labelled *Escherichia coli* DNA (Povirk *et al.*, 1978) showed release of all four bases, even at very low levels of DNA degradation. The yield varied in the order thymine > cytosine > adenine > guanine (relative yield 1:0.63:0.15:0.08); this same order was seen in extensively degraded HeLa and calf thymus DNA (Sausville *et al.*, 1978*b*; Takeshita *et al.*, 1978). Povirk *et al.* (1978) also reported the release of low-molecular-weight "minor products" containing thymine, cytosine or purine label. Using DNA containing uniformly labelled [^{14}C]thymidine, Burger *et al.* (1980) isolated a thymine-containing product (almost certainly the same as the "minor product" of Povirk *et al.*) and showed that it consisted of thymine linked to a three-carbon deoxyribose fragment, and that it contained the aldehyde-like thiobarbituric acid-reactive group which had been previously reported by Kuo and Haidle (1973) but not identified. Furthermore, when Fe(II) was used as a cofactor instead of mercaptoethanol, the yield of the thymine-sugar fragment was comparable to that of free thymine. Giloni *et al.* (1981) showed by chemical synthesis, NMR, and mass spectrometry that the adenine- and thymine-sugar fragment species were 3(adenin-9'-yl)-2-propenal and 3(thymin-1'-yl)-2-propenal, respectively. These authors isolated similar cytosine- and guanine-containing compounds and concluded that all four products have a common structure: base-CH=CH—CHO. The base specificity of base-propenal production was similar to that previously found for release of free bases. 2-Mercaptoethanol converted the base-propenals to free base plus a mercaptan-propenal adduct; this reaction explains the absence or low yield of these products (Povirk *et al.*, 1978) when 2-mercaptoethanol is used as a cofactor. However, the release of free bases in the absence of mercaptan appears not to result from further hydrolysis of base-propenals, but represents a separate independent reaction (Giloni *et al.*, 1981).

In the presence of Fe(II) and 2-mercaptoethanol, the total number of bases (plus "minor products") released can exceed the number of bleomycin molecules by a factor of 3 (Povirk, 1979; Ekimoto *et al.*, 1980). Thus bleomycin action is catalytic and bleomycin is usually not inactivated in the reaction with DNA.

Single-strand breaks

Because of their ease of detection, strand breaks were the first specific lesions reported in bleomycin-treated DNA (Suzuki *et al.*, 1969). Exposure of bleomycin-treated DNA to alkali increases the apparent number of strand breaks by a factor of approximately 2 (Povirk *et al.*, 1977; Lloyd *et al.*, 1978a); thus, bleomycin produces roughly equal numbers of true breaks and alkali-labile bonds. The yield of true breaks plus alkali-labile bonds is equal to the yield of released bases plus "minor products" (Povirk, 1979). Bleomycin-induced alkali-labile bonds may correspond to apurinic and apyrimidinic sites formed as a consequence of base release, as they have the same characteristic alkaline hydrolysis rate as apurinic sites induced by heating DNA and are cleaved by certain apurinic/apyrimidinic endonucleases (Schyns *et al.*, 1978; Lloyd *et al.*, 1978a). Furthermore, Giloni *et al.* (1981) found that acid hydrolysis of bleomycin-treated DNA containing uniformly labelled [^{14}C]thymidine produced [^{14}C]levulinic acid, a possible hydrolysis product of deoxyribose at an apyrimidinic site.

Conversely, true strand breaks are probably coincident with release of base-propenals. The bleomycin-induced strand break has a 5'-phosphate end group (Kuo and Haidle, 1973). At the 3'-end, however, a deoxyribose fragment, which Giloni *et al.* (1981) have identified as glycolic acid, remains attached to the terminal 3'-phosphate ($-PO_4-CH_2-COOH$). Thus, the products of bleomycin-induced strand breakage indicate an oxidative cleavage of the $(C-3')-(C-4')$ bond in deoxyribose, leaving a two-carbon sugar fragment (glycolic acid) at the 3'-end, a base with an attached three-carbon sugar fragment (base-propenal) and a 5'-phosphate end group (see figure 6).

Presence of the glycolic acid sugar fragment at the 3'-end explains (1) the anomalous mobility of bleomycin-cut 5'-end-labelled DNA on sequencing gels (D'Andrea and Haseltine, 1978; Takeshita *et al.*, 1978), (2) The failure of bleomycin to release low-molecular weight labelled products from [thymidine-5'-^{3}H]

DNA (Haidle *et al.*, 1972) and (3) the failure of bleomycin-cut DNA to serve as a primer for *E. coli* DNA polymerase I, either with or without phosphatase treatment (Kappen and Goldberg, 1978). The sugar fragment also renders the 3'-end partially resistant to snake venom exonuclease (Giloni *et al.*, 1981), but it can be removed by *E. coli* exonuclease III (Niwa and Moses, 1981).

Bleomycin-induced DNA strand breakage is sequence-specific (D'Andrea and Haseltine, 1978; Takeshita *et al.*, 1978). d(GpC) and d(GpT) sequences are most susceptible to cleavage, with the pyrimidine sugar being attacked. Preference for a guanine residue on the 5'-side of the break reflects the preferential binding of bleomycine to guanine-containing nucleic acids (Kasai *et al.*, 1978). Because the sequence specificity is largely determined by the two bases adjacent to the breaks, and because intercalators are known to prefer alternating purine-pyrimidine sequences (Krugh, 1972), both Takeshita *et al.* (1978) and Povirk *et al.*, (1979) have suggested that the specificity may be due to the preferential intercalation of the bithiazole at d(GpC) and d(GpT) sites. However, Povirk *et al.* (1981*a*) later reported that Cu(II)·phleomycin, unlike Cu(II)·bleomycin, showed no physical chemical evidence (DNA lengthening) for intercalation, while Takeshita *et al.* (1981) reported that phleomycin showed nearly the same sequence specificity for strand breakage as bleomycin. Phleomycin did show enhanced breakage of supercoiled DNA (Povirk *et al.*, 1981*b*); this enhancement is also seen with bleomycin (Povirk *et al.*, 1979) and is expected from an intercalator, but could have other explanations as well. Intercalation by phleomycin is not expected on structural grounds, since its bithiazole moiety lacks a double bond and is non-planar. Thus, the possible role of intercalation in DNA strand breakage by both drugs is still unclear.

Double-strand breaks

By subjecting bleomycin-treated closed circular DNAs to gel electrophoresis, the yield of single- and double-strand breaks can be accurately determined even at very low levels of DNA degradation. Such experiments (Povirk *et al.*, 1977; Lloyd *et al.*, 1978*a, b*) indicate that the number of double-strand breaks is far greater than can be accounted for by random coincident single-strand breaks (figure 4). To explain double-strand breakage D'Andrea and Haseltine (1978) have proposed two selective cleavages at opposite base-paired d(GpC) sites, while Lloyd *et al.* (1978*b*) have postulated simultaneous breakage of both DNA strands by a bleomycin dimer. However, these models are not supported by the kinetic data, which show that the ratio of single- to double-strand breaks (approximately 9:1) is constant throughout the course of the reaction and over a wide range of drug concentrations and DNA breakage levels (Povirk *et al.*, 1977; Lloyd *et al.*, 1978*a, b*; Haidle *et al.*, 1979). These results suggest, though they do not prove, that the double-strand break is the result of a single bleomycin-DNA interaction. One plausible mechanism for simultaneous breakage of both DNA strands is a bleomycin-initiated free radical cascade, such as occurs in lipid peroxidation (Mead, 1976), but there is no empirical evidence for such a process.

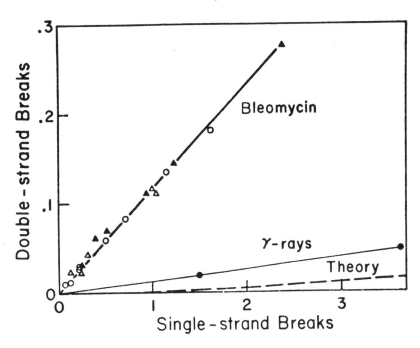

Figure 4 Relative numbers of single- and double-strand breaks induced by bleomycin and by γ-rays in ColE₁ DNA compared to the numbers expected from a random distribution of single-strand breaks ("theory"). ColE₁ DNA was treated with various concentrations of bleomycin in the presence of 2-mercaptoethanol, and single- and double-strand breakage was determined from the conversion of form I (supercoiled) DNA to form II (nicked circular) and form III (linear) DNA. From Povirk *et al.* (1977).

Thus, the mechanism of double-strand breakage by bleomycin remains uncertain.

The ratio of single- to double-strand breaks is different for different bleomycin analogues. Phleomycin, which appears not to intercalate, induces almost no double-strand breakage (Huang *et al.*, 1981). Thus, alteration of DNA secondary structure or stabilisation of the DNA·bleomycin complex by bithiazole intercalation may be important for double-strand breakage. Similarly, addition of high salt (0.3 M), which would be expected to reduce the stability of the DNA·bleomycin complex, reduces the fraction of double-strand breaks (Lloyd *et al.*, 1978*b*), and tallysomycin, which binds more tightly to DNA than bleomycin (Strong and Crooke, 1978), produces a higher fraction of double-strand breaks (Mirabelli *et al.*, 1980).

Bleomycin-treated PM2 DNA shows discrete bands on agarose gel electrophoresis (Lloyd *et al.*, 1978*b*). Thus, certain as yet undetermined DNA sequences appear to be particularly susceptible to bleomycin-induced double-strand breakage.

Oxidation-reduction of iron-bleomycin

Although stimulation of bleomycin-induced DNA degradation by Fe(II) was first reported by Ishida and Takahashi (1975) and a requirement for O_2 was shown by Onishi *et al.* (1975), not until the detailed studies of Sausville *et al.* (1978*a, b*) was it appreciated that Fe(II) was a necessary cofactor in the reaction and that oxidation of Fe(II)·bleomycin by dissolved O_2 was a critical reaction step. Following this discovery, the redox properties of iron·bleomycin complexes have received considerable attention. Fe(III)·bleomycin, produced either by mixing Fe(III) and bleomycin at slightly acid *p*H or by air oxidation of Fe(II)·bleomycin, is readily reduced to Fe(II)·bleomycin by the sulphydryl reagents dithiothreitol, 2-mercaptoethanol, cysteine and glutathione, a likely candidate for the *in vivo* cofactor of bleomycin (Povirk, 1979; Antholine and Petering, 1979*a*). This reaction is slow (although somewhat faster than reduction of unchelated Fe(III)) and is normally the rate-limiting step in DNA degradation (Ekimoto *et al.*, 1980). Even in the absence of added iron, DNA degradation by bleomycin plus sulphydryls is strongly inhibited by the iron-specific chelator deferoxamine (Sausville *et al.*, 1978*a*). Thus, stimulation of bleomycin by sulphydryls alone (Suzuki *et al.*, 1969; Onishi *et al.*, 1975) appears to be due to their ability to reduce traces of Fe(III) (present as a contaminant in bleomycin (Lown and Sim, 1977) and in most DNA preparations) to Fe(II) which is then chelated by bleomycin.

In the presence of DNA, oxidation of Fe(II)·bleomycin is accompanied by release of DNA bases and formation of thiobarbituric acid-reactive base-propenals (see above). Under optimal conditions, the total molar yield of these products is approximately one-third of the initial Fe(II)·bleomycin concentration (Povirk, 1979). Addition of sulphydryls recycles the Fe(III)·bleomycin back to Fe(II)·bleomycin and increases the yield of free bases (plus "minor products") by a factor of approximately 8 (Povirk, 1979; Ekimoto *et al.*, 1980). Thus, in the presence of DNA each bleomycin molecule goes through an average of 8 oxidation-reduction cycles before being inactivated. In the absence of DNA, however, a single Fe(II)·bleomycin oxidation inactivates most of the bleomycin, presumably because the reactive group attacks bleomycin itself for lack of DNA substrate (Burger *et al.*, 1979*a*).

Fe(II)·bleomycin oxidation occurs in three kinetically resolvable steps (figure 5). First, a ternary bleomycin·Fe(II)·O_2 complex, having no EPR signal, is formed very rapidly, at a rate which is proportional to the O_2 concentration. This complex has, like its NO, CO and C_2H_5NC analogues, much greater visible absorbance than Fe(II)·bleomycin (Burger *et al.*, 1979*a, b*). Furthermore, these oxygen analogues, two of which are known to be Fe(II) ligands in ternary complexes with bleomycin, reversibly inhibit DNA degradation by Fe(II)·bleomycin plus O_2 (Burger *et al.*, 1979*a*). Thus, it is very likely that O_2 is bound as a metal ligand in the bleomycin·Fe(II)·O_2 complex, as it appears to be in bleomycin·Co(II)·O_2 (Sugiura, 1978).

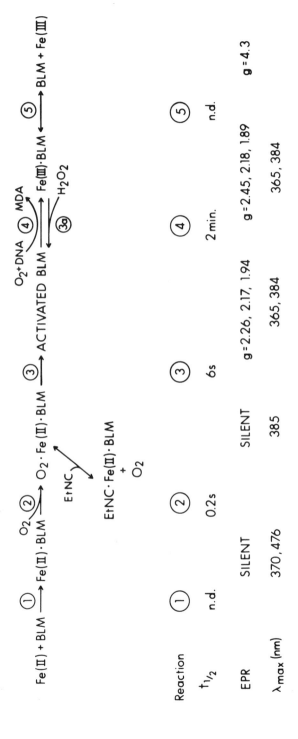

Figure 5 Kinetic steps in the activation of Fe(II)·bleomycin by O_2, showing the optical and EPR spectral properties of each product, and the reaction half-lives ($t_{1/2}$) at 6°C (see text, Section II C). BLM, bleomycin; EtNC, ethyl isocyanide, C_2H_5NC; MDA, malondialdehyde-like thiobarbituric acid-reactive material produced from DNA; n.d., not determined. Reprinted from Burger *et al.* (1981) with permission.

Bleomycin·Fe(II)·O$_2$ is then converted with first-order kinetics to a roughly one-to-one mixture of Fe(III)·bleomycin and "activated bleomycin", a transient ferric species originally noted by Sugiura and Kikuchi (1978) and by Dabrowiak *et al.* (1979), which is believed to be the ultimate, or at least the proximate, agent of DNA damage (Kuramochi *et al.*, 1981; Burger *et al.*, 1981). Activated bleomycin and Fe(III)·bleomycin have identical optical spectra, but can be distinguished quantitatively by their EPR spectra (figure 5). Activated bleomycin slowly decays to Fe(III)·bleomycin, with kinetics which closely parallel the appearance of DNA damage. Furthermore, if the mixture of activated bleomycin plus Fe(III)·bleomycin is preincubated in the absence of DNA, its potential to cause DNA damage (thiobarbituric acid-reactive material and base release) decays in parallel with decay of activated bleomycin (Kuramochi *et al.*, 1981; Burger *et al.*, 1981). It is these kinetics which implicate activated bleomycin as the agent of bleomycin-induced DNA damage. The Fe(III)·bleomycin formed from decay of activated bleomycin has the same EPR spectrum as that formed immediately from bleomycin·Fe(II)·O$_2$ or that formed by mixing Fe(III) and bleomycin.

The structure of activated bleomycin is not known with certainty but it is clearly an Fe(III) complex, as indicated by its ^{57}Fe EPR spectrum. It contains bound oxygen normally derived from O$_2$ as one of its metal ligands, as indicated by broadening of certain features of its EPR spectrum when ^{17}O$_2$ is used for activation (Burger *et al.*, 1981). However, activated bleomycin can also be formed anaerobically from Fe(III)·bleomycin plus H$_2$O$_2$. Kuramochi *et al.* (1981) have proposed bleomycin ferric peroxide, bleomycin·Fe(III)$^-$OOH, as the most likely structure. However, Burger *et al.* (1981) conclude that either bleomycin·Fe(III)$^-$OOH or bleomycin·Fe(III)$\overset{\cdot\cdot}{\underset{\cdot\cdot}{O}}$ would be consistent with all the spectroscopic data. Interaction of activated bleomycin with DNA is indicated by a perturbation of its EPR spectrum (Burger *et al.*, 1981). If activated bleomycin is bleomycin·Fe(III)$^-$OOH, its production from bleomycin·Fe(II)·O$_2$ would require a one-electron reduction. The source of this reducing equivalent is unknown, but it could come either from a second Fe(II)·bleomycin molecule, or from sulphydryl reducing agents.

There are many apparent similarities between bleomycin and cytochrome P-450 in their mechanisms of oxygen activation. Like Fe(II)·bleomycin, ferrous P-450 is thought to bind O$_2$ as an Fe(II) ligand. A one-electron reduction converts this complex to a P-450 ferric peroxide complex (White and Coon, 1980) which is similar to activated bleomycin in electronic structure and oxidation state (Burger *et al.*, 1981). Splitting of the oxygen–oxygen bond in P-450 ferric peroxide generates a species which is believed to be capable of abstracting a hydrogen from a noncovalently bound substrate molecule, resulting in its oxidation via a free radical intermediate. It should be noted, however, that oxygen–oxygen bond cleavage (the critical free radical-generating step in the cytochrome P-450 reaction) has not been demonstrated in activated bleomycin, and that bleomycin oxidizes deoxyribose by four electron equivalents (figure 6), whereas the P-450 system oxidises its substrate by only two electron equivalents (White and Coon, 1980).

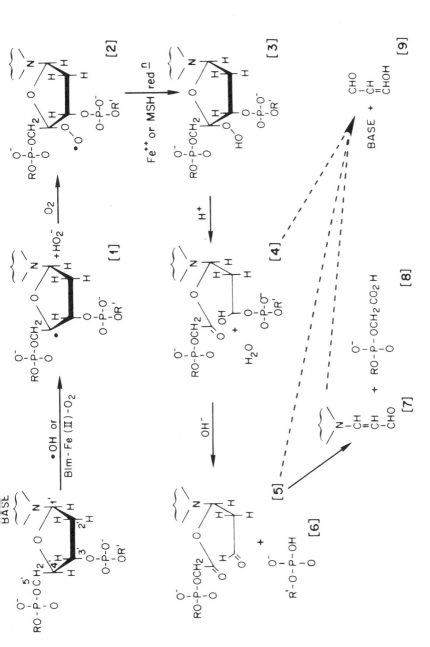

Figure 6 Proposed mechanism of deoxyribose cleavage by Fe(II)•bleomycin. Reprinted from Grollman and Takeshita (1980) with permission.

The yield of thiobarbituric acid-reactive aldehyde-like functions from DNA is equal to one-half the number of activated bleomycin molecules formed (Burger *et al.*, 1981). Since an approximately equal number of free bases were presumably also released independently and without production of aldehyde-like functions (Kuramochi *et al.*, 1981; Burger *et al.*, 1980) the efficiency of production of total DNA lesions by activated bleomycin may approach 100%.

Mechanism of deoxyribose cleavage

While great progress has been made in elucidating the chemistry of bleomycin action, events occurring between formation of activated bleomycin and the final appearance of DNA products (see above), remain open to speculation. The most detailed model of deoxyribose cleavage is that proposed by Johnson *et al.* (figure 6). The first step in their model reaction is abstraction of a hydrogen from the C-4' of deoxyribose by some free radical. Although activated bleomycin appears not to be a free radical, it may generate a short-lived free radical which reacts with DNA so rapidly that it cannot be detected kinetically. Hydrogen abstraction would result in a deoxyribose radical which could react with molecular oxygen to form a peroxyl radical on C-4' (Grollman and Takeshita, 1980). This proposed reaction step is consistent with the finding that, in addition to its role in formation of activated bleomycin, O_2 is also required as a cofactor in the generation of thiobarbituric acid-reactive material from activated bleomycin plus DNA (Burger *et al.*, 1981). Reduction and hydrolysis of the peroxyl radical would lead to cleavage of the (C-3')-(C-4') bond, breakage of the sugar-phosphate backbone and release of a base-propenal. Except for the final products, none of the intermediates in this scheme have been isolated and identified (Giloni *et al.*, 1981). However, Burger *et al.* (1980) have found that, at $-5°C$, the thiobarbituric acid-reactive aldehyde-like function remains attached to macromolecular DNA; this finding is consistent with formation of a species such as intermediate [5] in the reaction scheme (figure 6).

Much effort has been directed towards implicating free radicals in bleomycin action. Models involving free radicals are attractive because they provide a plausible mechanism of carbon–carbon bond cleavage via hydrogen abstraction, and because radicals are expected from a reaction between Fe(II) and O_2. Direct generation of free radical species from bleomycin·Fe(III)⁻OOH is also at least formally possible. Superoxide and hydroxyl free radicals have been detected by spin-trapping experiments in reaction mixtures containing Fe(II), bleomycin and O_2 (Sugiura and Kikuchi, 1978), but their presence alone does not imply a role in DNA degradation. Studies of bleomycin-induced DNA degradation employing scavengers of free radical intermediates as inhibitors have yielded contradictory results (Lown and Sim, 1977; Sausville *et al.*, 1978b). In view of the sequence specificity of bleomycin-induced DNA strand breakage, the specific cleavage of the (C-3')-(C-4') bond and the high efficiency of production of DNA products by activated bleomycin, it is unlikely that bleomycin functions simply by generating hydroxyl or other diffusible free radicals from Fe(II) plus O_2 in the vicinity of DNA. Rather, these results strongly suggest that whatever its chemical nature, and whether or not it is a free radical, the reactive group which ultimately attacks deoxyribose is formed at a specific site in a DNA·bleomycin·iron·oxygen complex of defined stereochemical structure. The functional similarity of bleomycin to cytochrome P-450 (see above) suggests that generation of a specific deoxyribose free radical, by DNA-bound activated bleomycin or an immediate product thereof, is at least plausible as an initial step in deoxyribose cleavage.

CELLULAR RESPONSE

Cell killing

With some important exceptions, the response of cultured mammalian cells to bleomycin is similar to their response to ionising radiation, which is also thought to kill cells by damaging their DNA. If a population of exponentially growing cells is treated with bleomycin, they tend to accumulate in the G2 phase of the cell cycle (Tobey, 1972); a similar mitotic delay is seen in irradiated cells (Okada, 1970). Furthermore, synchronised cultured cells show two periods of increased sensitivity to killing by bleomycin, one in mitosis, and one in late G1/early S phase; and two periods of relative resistance, one in early G1 and one in late S (Barranco and Humphrey, 1971). This is the same pattern of sensitivity as is seen for ionising radiation (Terasima and Tolmach, 1963). Radiation-sensitive cells derived from individuals with the genetic disease ataxia telangiectasia are also bleomycin-sensitive (Taylor *et al.*, 1979; Cohen *et al.*, 1981), suggesting that the sensitivity to both agents may be related to DNA damage.

However, the dose–response curves for cell survival (as measured by colony forming ability) are distinctly different for the two agents. For irradiated cells, a plot of log(survival) versus dose is essentially linear, with an initial shoulder at very low doses (Okada, 1970). However, for bleomycin-treated cells, the same curve is consistently concave upwards (Barranco and Humphrey, 1971; Terasima *et al.*, 1972). For example, bleomycin at $10\ \mu\mathrm{g\ ml}^{-1}$ results in 50% survival of CHO cells, but $90\ \mu\mathrm{g\ ml}^{-1}$ is required to reduce survival to 25%. The two simplest explanations for such results are (1) an inherently resistant fraction of cells and (2) a nonlinear relation between drug dose and damage to the target molecule, such as would result from saturation of transport processes or depletion of cofactors. The phenomenon may be partly explained by resistant cells in a certain phase of the cell cycle (Barranco and Humphrey, 1971). However, Terasima *et al.* (1972) found that synchronised L5 cells in G1 phase showed the same concave upward survival curve as exponentially growing cells. They also found that killing by two bleomycin treatments was increased if the treatments were spaced at least 2 hours apart. These results suggest that cells rapidly acquire resistance during bleomycin treatment, and lose this resistance within a few hours. The molecular basis of the resistance is unknown.

Surprisingly, *E. coli* mutants which are genetically deficient in DNA repair processes, show little increased sensitivity to bleomycin. In exponentially growing cells, *recA, lexA, uvrA* and *polA* mutations have no effect on sensitivity (Endo, 1970; Onishi *et al.*, 1973). Treatments such as amino acid starvation, which allow cells to complete chromosome replication but prevent initiation of a new round of replication, render the cells an order of magnitude more resistant to bleomycin (Yamamoto and Hutchinson, 1979). Starved *recA* cells are four times more sensitive than starved wild type, but the other repair mutants are not. These results provide some evidence for involvement of DNA damage in cytotoxicity, since the sensitivity is strongly dependent on the state of the chromosome, but the lack of sensitivity in repair mutants is confusing.

Inhibition of DNA synthesis

In mammalian cells bleomycin selectively inhibits DNA synthesis, compared to RNA and protein synthesis (Suzuki *et al.*, 1968). The dose–response curves for DNA synthesis inhibition are, like the survival curves, concave upward (Watanabe *et al.*, 1973). Ataxia talangiectasia cells, which are sensitive to killing by bleomycin and by ionising radiation, are more resistant than normal cells to inhibition of DNA synthesis by both these agents (Cramer and Painter, 1981; Edwards *et al.*, 1981). These results suggest that the ataxia cells are deficient in some cellular mechanism which recognises the presence of DNA damage and delays DNA synthesis until the damage can be repaired. Thus, bleomycin-induced inhibition of DNA synthesis is probably caused by DNA damage, but the molecular events which mediate this process are unknown.

Bacterial semiconservative DNA synthesis is almost totally refractory to bleomycin, a result which is difficult to explain (Cohen and I, 1976; Yamamoto and Hutchinson, 1982).

DNA strand breaks

DNA strand breaks seen in bleomycin-treated mammalian cells (Terasima *et al.*, 1970) are presumably the result of direct action of the drug on cellular DNA, although this has not been proved. Some breaks could also result from cleavage of bleomycin-induced apurinic/apyrimidinic sites by cellular endonucleases. Iqbal *et al.* (1976) have examined bleomycin-induced DNA strand breakage in CHO cells by the technique of alkaline elution, which permits measurement of DNA damage at doses for which there is still significant cell survival. These authors found a surprising heterogeneity in DNA damage. Some portions of DNA were heavily degraded while other portions remained virtually intact. The fraction of intact DNA roughly correlated with the fraction of surviving cells, suggesting that the heterogeneity might be due to a resistant population of cells. Using synchronized cells and alkaline sucrose gradients, Clarkson and Humphrey

(1976) found only a slight variation in the extent of strand breakage by bleomycin in various phases of the cell cycle. Studies by either method suggest that most bleomycin-induced single-strand breaks are repaired within 30 min, but that a fraction of nonrepairable breaks, apparently associated with the heavily degraded portion of the DNA, may also be present.

Since double-strand DNA breaks are generally thought to be much more toxic than single-strand breaks, and since bleomycin induces direct double-strand breakage *in vitro*, the examination of double-strand breakage in mammalian cells is an important problem, but one that has been difficult to approach experimentally. Studies utilising neutral sucrose sedimentation provide some evidence for induction of double-strand breaks by bleomycin, and for their subsequent repair (Saito and Andoh, 1973). However, because of the supralethal bleomycin doses used and the lack of quantitative data on the number of breaks produced, few conclusions can be drawn regarding the relation between double-strand breaks and cytotoxicity.

Rather more definitive results were obtained by Yamamoto and Hutchinson (1981) in *E. coli* cells. They found that, at a bleomycin dose sufficient to kill 90% of the cells, DNA sedimented in neutral sucrose at the position expected for a full length linear *E. coli* chromosome. However, DNA from untreated cells sedimented even faster, as would be expected for a circular DNA molecule. These authors concluded that their data were consistent with the hypothesis that bleomycin kills cells by selectively introducing one or a few double-strand breaks in a confined region of the *E. coli* chromosome. This hypothesis could also explain the lack of sensitivity of repair mutants (see above), since such concentrated DNA damage might be essentially nonrepairable. Evidence for selective breakage of membrane-associated DNA has been reported by Miyaki *et al.* (1974). The extensive, apparently random DNA degradation previously reported in bacterial cells (Onishi *et al.*, 1973) may have been a result, rather than a cause, of cell death.

Thus, while the generally held contention that bleomycin cytotoxicity results from DNA strand breakage has not been refuted, there are certain selectivities in bleomycin-induced damage to both mammalian and bacterial DNA which have yet to be explained.

Chromosome aberrations

Like other DNA-damaging agents, bleomycin induces measurable chromosome aberrations even at sublethal doses ($< 1\ \mu g\ ml^{-1}$) (reviewed by Vig and Lewis, 1978). Chromosome aberrations found in peripheral lymphocytes of patients undergoing bleomycin treatment provide the most direct evidence for DNA damage by the drug in a clinical situation (Bornstein *et al.*, 1971).

In cultured human lymphocytes (Tamura *et al.*, 1974), bleomycin produces aberrations even if treatment occurs in G2 and the DNA is not replicated between bleomycin treatment and harvesting of mitotic cells. This "S-independence" of its cytogenetic effect may be related to the direct production of strand breaks by the drug (Kelly and Bender, 1978). Primarily chromatid-type aberrations (chromatid breaks and gaps) are induced if cells are treated in G2. If cells are treated in G1, chromosome-type aberrations (primarily dicentric chromosomes and acentric fragments) also appear. Conspicuous by their absence are detectable numbers of sister chromatid exchanges, except at cytotoxic doses (Tamura *et al.*, 1974; Dresp *et al.*, 1978; Perry and Evans, 1975).

Perhaps related to the absence of chromatid exchanges is the surprising lack of mutagenesis by bleomycin. Although there has been one report of muta-genesis in yeast (Moore, 1978) the drug does not produce detectable mutation in the highly sensitive Ames *Salmonella*/microsome mutagenesis test system (Benedict *et al.*, 1977). Hutchinson *et al.* (1981) have suggested that bleomycin-induced mutagenesis in bacteria may be masked by lethality, due to the apparent confinement of severe damage to a small fraction of the DNA.

In summary, bleomycin induces in cells many of the effects expected from a DNA-damaging agent. The drug is more intriguing, however, for the effects which it fails to induce, especially mutagenesis and sister chromatid exchange. Since these phenomena form the basis for two of the most sensitive and widely used screening procedures for environmental DNA-damaging agents, bleomycin presents an interesting problem in genetic toxicology.

MOLECULAR ASPECTS OF CLINICAL THERAPY

Deamination and metal chelation *in vivo*

Although bleomycin is administered in metal-free form, there is evidence that some or all of the drug is converted in the serum to a stable Cu(II)·bleomycin complex (Kanao *et al.*, 1973). Since Cu(II)·bleomycin is as active as bleomycin in inhibiting cellular DNA synthesis, yet does not degrade DNA *in vitro* (Suzuki *et al.*, 1968; Shirakawa *et al.*, 1971) intracellular removal of Cu(II) from the complex and replacement by Fe(II) must be postulated. Takahashi *et al.* (1977) have found that rat ascites hepatoma cells contain both low-molecular weight sulphydryl compounds capable of reducing Cu(II)·bleomycin to Cu(I)·bleomy-cin, and high-molecular weight proteinaceous ligands capable of removing Cu(I) from Cu(I)·bleomycin.

Most chelated metals can be displaced from their bleomycin complexes by Cu(II), but Co(III) cannot (Nunn, 1977*b*). Co(III)·bleomycin does not kill cultured cells, and so is apparently not dissociated intracellularly (Nunn and Lunec, 1979). Because of this extreme stability, [57]Co(III)·bleomycin has been used experimentally as a radiodiagnostic agent, and has been found to accumulate preferentially in tumours (Rasker *et al.*, 1975). Although its molecular basis is unknown, selective accumulation in tumours may explain the fact that clinically effective doses of bleomycin are much smaller than the equivalent doses required to kill cultured cells (Nunn and Lunec, 1979).

A second major factor in bleomycin selectivity is the intracellular inactivation of the drug. When [^{3}H] bleomycin was administered to mice, a comparison of the radioactivity and the biological activity (as determined in an antibacterial assay) recovered from various organs indicated that most organs contained predominantly inactive bleomycin. Only in the skin and the lung were significant amounts of active bleomycin found (Umezawa *et al.*, 1972, 1974). Furthermore, when bleomycin was incubated with extracts from various tissues, lower amounts of bleomycin-inactivating activity were found in lung and skin than in liver, spleen and kidney. Thus, failure of the lung to inactivate bleomycin may be related to pulmonary toxicity, and failure of skin tissue to inactivate the drug may be related to the therapeutic effect of bleomycin on squamous cell tumours. At least in some cases, the therapeutic effect against specific tumours could be correlated with their inability to inactivate the drug. A putative bleomycin-inactivating enzyme has been partially purified from mouse liver (Umezawa *et al.*, 1974) which converts bleomycin to its inactive analogue deamidobleomycin by deaminating the amide of β-aminoalanine (see figure 1). Thus, differential inactivation of bleomycin may be due to differences between various tissues in their content of this or a similar enzyme. Bleomycin-resistant cultured cells, containing increased deaminating activity, have recently been isolated (Akiyama *et al.*, 1981).

Pulmonary toxicity

The usual limiting toxicity in bleomycin treatment is an unpredictable and sometimes lethal pulmonary fibrosis, dependent on cumulative bleomycin dosage (Blum *et al.*, 1973). The molecular mechanism of this effect is essentially unknown and it may or may not be related to DNA damage. Raisfeld (1979) has found that intratracheal administration of certain bleomycin terminal amines alone caused significant bleomycin-like pulmonary fibrosis in mice. Pepleomycin, a bleomycin with a synthetic terminal amine (see figure 1) has been reported to cause less pulmonary toxicity than other bleomycins (Tanaka, 1977).

On the other hand a similar pulmonary fibrosis is seen in poisoning by paraquat, which like bleomycin catalyses oxidation–reduction reactions which result in superoxide and other oxygen free radicals. Although the exact mechanism of paraquat toxicity is also disputed (see review by Smith *et al.*, 1978), it is reasonable to propose that the toxicity of both agents may be mediated by their oxidation–reduction activities, either directly through production of oxygen radicals, or indirectly through saturation of cellular detoxification systems.

Preliminary studies by Newman *et al.* (1980) in rats suggest that release of angiotensin-converting enzyme into pulmonary lavage may be a useful early indicator of bleomycin-induced lung damage.

Can bleomycin therapy be improved?

In spite of great advances in our understanding of bleomycin action, clinical use of the drug is still based largely on previous clinical experience. Some attempts at cell-cycle based therapy have been made; vincristine is administered to arrest cycling tumour cells in mitosis, followed by bleomycin which is selectively toxic towards mitotic cells (Livingston *et al.*, 1973).

Changes in the bleomycin molecule itself have not led to dramatic improvements in its effectiveness. Not surprisingly, even small alterations in the chelating moieties inactivate the drug (Umezawa *et al.*, 1972; Asakura *et al.*, 1975). With natural bleomycin analogues, Huang *et al.* (1981) and Mirabelli *et al.* (1980) have found that changes in the threonine/bithiazole region can have dramatic effects on the relative production of single- and double-strand DNA breaks. Thus, these authors suggest development of analogues which would produce double-strand breaks only in certain cell types; it remains to be shown, however, that double-strand breaks are, in fact, the predominant lethal lesion.

Although a coherent picture of bleomycin action at the chemical level is now emerging, our knowledge of the interaction of the drug with whole cells is still fragmentary. A clearer understanding of the process of bleomycin uptake into cells, of the reasons for accumulation in certain cell types, and of the intracellular activation and inactivation mechanisms could facilitate rational design of more effective forms of the drug.

ACKNOWLEDGEMENTS

I thank Drs Richard M. Burger, Irving H. Goldberg, Arthur P. Grollman and Norman J. Oppenheimer for helpful comments. I am especially indebted to Dr Franklin Hutchinson who initiated and encouraged my interest in bleomycin. The preparation of this review was supported in part by grant GM 12573 and Fellowship CA 06475 from the US Public Health Service.

REFERENCES

Akiyama, S., Ikezaki, K., Kuramochi, H., Takahashi, K. and Kumano, M. (1981). *Biochem. biophys. Res. Commun.*, **101**, 55
Antholine, W. E. and Petering, D. H. (1979*a*). *Biochem. biophys. Res. Commun.*, **90**, 384
Antholine, W. E. and Petering, D. H. (1979*b*). *Biochem. biophys. Res. Commun.*, **91**, 528
Asakura, H., Hori, M. and Umezawa, H. (1975). *J. Antibiot.*, **28**, 537
Asakura, H., Umezawa, H. and Hori, M. (1978). *J. Antibiot.*, **31**, 156
Barranco, S. C. and Humphrey, R. M. (1971). *Cancer Res.*, **31**, 1218
Benedict, W., Baker, M. S., Haroun, L., Choi, E. and Ames, B. N. (1977). *Cancer Res.*, **37**, 2209
Blum, R. H., Carter, S. K. and Agre, K. (1973). *Cancer*, **31**, 903
Bornstein, R. S., Hungerford, D. A., Hallei, G., Engstrom, P. F. and Yarbro, J. W. (1971). *Cancer Res.*, **31**, 2204
Burger, R. M., Peisach, J., Blumberg, W. E. and Horwitz, S. B. (1979*a*). *J. biol. Chem.*, **254**, 10906
Burger, R. M., Horwitz, S. B., Peisach, J. and Wittenberg, J. B. (1979*b*). *J. biol. Chem.*, **254**, 12299
Burger, R. M., Berkowitz, A. R., Peisach, J. and Horwitz, S. B. (1980). *J. biol. Chem.*, **255**, 11832
Burger, R. M., Peisach, J. and Horwitz, S. B. (1981). *J. biol. Chem.*, **256**, 11636
Chen, D. M., Sakai, T. T., Glickson, J. D. and Patel, D. J. (1980). *Biochem. biophys. Res. Commun.*, **92**, 197

Chien, M., Grollman, A. P. and Horwitz, S. B. (1977). *Biochemistry*, **16**, 3641
Clarkson, J. M. and Humphrey, R. M. (1976). *Cancer Res.*, **36**, 2345
Cohen, M. M., Simpson, S. J. and Pazos, L. (1981). *Cancer Res.*, **41**, 1817
Cohen, S. S. and I, J. (1976). *Cancer Res.*, **36**, 2768
Cramer, P. and Painter, R. B. (1981). *Nature, Lond.*, **291**, 671
Dabrowiak, J. C., Greenaway, F. T., Longo, W. E., Van Husen, M. and Crooke, S. T. (1978*a*). *Biochim. biophys. Acta*, **517**, 517
Dabrowiak, J. C., Greenaway, F. T. and Grulich, R. (1978*b*). *Biochemistry*, **17**, 4090
Dabrowiak, J. C., Greenaway, F. T., Santillo, F. S. and Crooke, S. T. (1979). *Biochem. biophys. Res. Commun.*, **91**, 721
Dabrowiak, J. C. (1980). *J. inorg. Biochem.*, **13**, 317
D'Andrea, A. D. and Haseltine, W. A. (1978). *Proc. natn. Acad. Sci. U.S.A.*, **75**, 3608
DeRiemer, L. H., Meares, C. F., Goodwin, D. A. and Diamanti, C. I. (1979). *J. med. Chem.*, **22**, 1019
Dresp, J., Schmid, E. and Bauchinger, M. (1978). *Mutat. Res.*, **56**, 341
Edwards, M. J., Taylor, A. M. R. and Flude, E. J. (1981). *Biochem. biophys. Res. Commun.*, **102**, 610
Ekimoto, H., Kuramochi, H., Takahashi, K., Matsuda, A. and Umezawa, H. (1980). *J. Antibiotics*, **33**, 426
Endo, H. (1970). *J. Antibiotics*, **23**, 508
Friedman, M. A. (1978). *Recent Results Cancer Res.*, **63**, 152
Giloni, L., Takeshita, M., Johnson, F., Iden, C. and Grollman, A. P. (1981). *J. biol. chem.*, **256**, 8608
Glickson, J. D., Pillai, R. P. and Sakai, T. T. (1981). *Proc. natn. Acad. Sci. U.S.A.*, **78**, 2967
Grollman, A. P. and Takeshita, M. (1980). *Adv. Enzyme Reg.*, **18**, 67
Haidle, C. W., Weiss, K. K. and Kuo, M. T. (1972). *Molec. Pharmac.*, **8**, 531
Haidle, C. W., Lloyd, R. S. and Robberson, D. L. (1979). In *Bleomycin: Chemical, Biochemical and Biological Aspects*, (ed. S. Hecht), Springer-Verlag, New York
Hecht, S., Burlett, D. J., Mushika, Y., Kuroda, Y. and Levin, M. D. (1979). In *Bleomycin: Chemical, Biochemical and Biological Aspects*, (ed. S. Hecht), Springer-Verlag, New York,
Huang, C.-H., Mirabelli, C. K., Jan, Y. and Crooke, S. T. (1981). *Biochemistry*, **20**, 233
Hutchinson, F., Povirk, L. F. and Yamamoto, K. (1981). In *Molecular Actions and Targets for Cancer Chemotherapeutic Agents*, (eds. A. C. Sartorelli, J. S. Lazo and J. R. Bertino), Academic Press, New York
Iitaka, Y., Nakamura, H., Nakatani, T., Muraoka, Y., Fujii, A., Takita, T. and Umezawa, H. (1978). *J. Antibiotics*, **31**, 1070
Iqbal, Z. M., Kohn, K. W., Ewig, R. A. G. and Fornace, A. J. (1976). *Cancer Res.*, **36**, 3834
Ishida, R. and Takahashi, T. (1975). *Biochem. biophys. Res. Commun.*, **66**, 1432
Ishizu, K., Murata, S., Myoshi, K., Sugiura, Y., Takita, T. and Umezawa, H. (1981). *J. Antibiotics*, **34**, 994
Ishizuka, M., Takayama, H., Takeuchi, T. and Umezawa, H. (1967). *J. Antibiotics, Ser. A*, **20**, 15
Kanao, N., Tomita, S., Ishihara, S., Murakami, A. and Okada, H. (1973). *Chemotherapy*, **21**, 1305
Kappen, L. S. and Goldberg, I. H. (1978). *Biochemistry*, **17**, 729
Kasai, H., Naganawa, H., Takita, T. and Umezawa, H. (1978). *J. Antibiotics*, **31**, 1316
Kawaguchi, H., Sukuira, H. T., Tomita, K., Konishi, M., Saito, K., Kobaru, S., Numata, K., Fukisawa, K., Miyaki, T., Hatori, M. and Koshiyama, H. (1977). *J. Antibiotics*, **30**, 779
Kelley, J. E. T. and Bender, M. A. (1978). *J. supramolec. Struct.*, **S2**, 44, abstr.
Koyama, G., Nakamura, H., Muraoka, Y., Takita, T., Maeda, K. and Umezawa, H. (1968). *Tetrahedron Lett.*, **44**, 4635
Krueger, W. C., Pschigoda, L. M. and Reusser, F. (1973). *J. Antibiotics*, **26**, 424
Krugh, T. R. (1972). *Proc. natn. Acad. Sci. U.S.A.*, **69**, 1911
Kuo, M. T. and Haidle, C. W. (1973). *Biochim. biophys. Acta*, **335**, 109
Kuramochi, H., Katsuoshi, T., Takita, T. and Umezawa, H. (1981). *J. Antibiotics*, **34**, 576
Livingston, R. B., Bodey, G. P., Gottlieb, J. A. and Frei, E. III (1973). *Cancer Chemother. Rep.*, **57**, 219

Lloyd, R. S., Haidle, C. W. and Hewitt, R. R. (1978a). *Cancer Res.*, 38, 3191
Lloyd, R. S., Haidle, C. W. and Robberson, D. L. (1978b). *Biochemistry*, 17, 1890
Lown, J. W. and Sim, S. (1977). *Biochem. biophys. Res. Commun.*, 77, 1150
Maeda, K., Kosaka, H., Yagishita, K. and Umezawa, H. (1956). *J. Antibiotics, Ser. A.*, 9, 82
Maeda, K., Kosaka, H., Yagishita, K. and Umezawa, H. (1963). *J. Antibiotics, Ser. A.*, 16, 86
Mead, J. F. (1976). In *Free Radicals in Biology*, Vol. 1, (ed. W. A. Pryor) Academic Press, New York,
Mirabelli, C. K., Huang, C. H. and Crooke, S. T. (1980). *Cancer Res.*, 40, 4173
Miyaki, M., Kitayama, T. and Ono, T. (1974). *J. Antibiotics*, 27, 647
Moore, C. W. (1978). *Mutat. Res.*, 58, 41
Müller, W. E. G., Yamazaki, Z., Breter, H.-J. and Zahn, R. K. (1972). *Eur. J. Biochem.*, 31, 518
Muraoka, Y., Kobayashi, H., Fujii, A., Kunishima, M., Fujii, T., Nakayama, Y., Takita, T. and Umezawa, H. (1976). *J. Antibiotics*, 29, 853
Newman, R. A., Kimberly, P. J., Stewart, J. A. and Kelly, J. (1980). *Cancer Res.*, 40, 3621
Niwa, O. and Moses, R. E. (1981). *Biochemistry*, 20, 238
Nouel, J. P. (1976). *Gann Monogr. Cancer Res.*, 19, 301
Nunn, A. D. (1976). *J. Antibiotics*, 29, 1102
Nunn, A. D. (1977a). *Int. J. nuclear Med., Biol.*, 4, 204
Nunn, A. D. (1977b). *Eur. J. nuclear Med.*, 2, 53
Nunn, A. D. and Lunec, J. (1979). In *Bleomycin: Chemical, Biochemical and Biological Aspects*, (ed. S. Hecht) Springer-Verlag, New York.
Okada, S. (1970). In *Radiation Biochemistry*, (eds. K. K. Altman, G. B. Gerber and S. Okada) Vol. 1, Academic Press, New York,
Onishi, T., Shimada, K. and Takagai, Y. (1973). *Biochim. biophys. Acta*, 312, 248
Onishi, T., Iwata, H. and Takagi, Y. (1975). *J. Biochem.*, 77, 745
Oppenheimer, N. J., Rodriguez, L. O. and Hecht, S. M. (1979). *Proc. natn. Acad. Sci. U.S.A.*, 76, 5616
Oppenheimer, N. J., Chang, C., Rodriquez, L. O. and Hecht, S. M. (1981). *J. biol. Chem.*, 256, 1514
Perry, P. and Evans, H. J. (1975). *Nature, Lond.*, 258, 121
Povirk, L. F. (1979). *Biochemistry*, 18, 3989
Povirk, L. F., Wübker, W., Köhnlein, W. and Hutchinson, F. (1977). *Nucleic Acids Res.*, 4, 3573
Povirk, L. F., Köhnlein, W. and Hutchinson, F. (1978). *Biochim. biophys. Acta*, 521, 126
Povirk, L. F., Hogan, M. and Dattagupta, N. (1979). *Biochemistry*, 18, 96
Povirk, L. F., Hogan, M., Dattagupta, N. and Buechner, M. (1981a). *Biochemistry*, 20, 667
Povirk, L. F., Dattagupta, N., Warf, B. C. and Goldberg, I. H. (1981b). *Biochemistry*, 20, 4007
Raisfeld, I. (1979). In *Bleomycin: Chemical, Biochemical and Biological Aspects*, (ed. S. Hecht) Springer-Verlag, New York,
Rasker, J. J., van de Poll, M. A. P. C., Beekhuis, H., Woldring, M. G. and Nieweg, H. O. (1975). *J. nuclear Med.*, 16, 1058
Roy, S. N., Orr, G. A., Brewer, C. F. and Horwitz, S. B. (1981). *Cancer Res.*, 41, 4471
Saito, M. and Andoh, T. (1973). *Cancer Res.*, 33, 1696
Sausville, E. A., Peisach, J. and Horwitz, S. B. (1978a). *Biochemistry*, 17, 2740
Sausville, E. A., Stein, R. W., Peisach, J. and Horwitz, S. B. (1978b). *Biochemistry*, 17, 2747
Schyns, R., Mulquet, M. and Verly, W. G. (1978). *FEBS Lett.*, 93, 47
Shirakawa, I., Azegami, M., Ishii, and Umezawa, H. (1971). *J. Antibiotics*, 24, 761
Smith, L. L., Rose, M. S. and Wyatt, I. (1978). *Ciba Fdn Symp.*, 65, 321
Strong, J. E. and Crooke, S. T. (1978). *Cancer Res.*, 38, 3322
Sugiura, Y. (1978). *J. Antibiotics*, 31, 1206
Sugiura, Y. (1979). *Biochem. biophys. Res. Commun.*, 87, 643
Sugiura, Y. and Kikuchi, (1978). *J. Antibiotics*, 31, 1310
Sugiura, Y. and Ishizu, K. (1979). *J. Inorg. Biochem.*, 11, 171
Sugiura, Y., Takita, T. and Umezawa, H. (1981). *J. Antibiotics*, 34, 249

Suzuki, H., Nagai, K., Yamaki, H., Tanaka, N. and Umezawa, H. (1968). *J. Antibiotics*, 21, 379

Suzuki, H., Nagai, K., Yamaki, H., Tanaka, N. and Umezawa, H. (1969). *J. Antibiotics*, 22, 446

Takahashi, K., Yoshioka, O., Matsuda, A. and Umezawa, H. (1977). *J. Antibiotics*, 30, 861

Takeshita, M., Grollman, A. P., Ohtsubo, E. and Ohtsubo, H. (1978). *Proc. natn. Acad. Sci. U.S.A.*, 75, 5983

Takeshita, M., Kappen, L. S., Grollman, A. P., Eisenberg, M. and Goldberg, I. H. (1981). *Biochemistry*, 20, 7599

Takita, T., Muraoka, Y., Yoshioka, T., Fujii, A., Maeda, K. and Umezawa, H. (1972). *J. Antibiotics*, 25, 755

Takita, T., Muraoka, Y., Nakatani, T., Fujii, A., Umezawa, Y., Naganawa, H. and Umezawa, H. (1978a). *J. Antibiotics*, 31, 801

Takita, T., Muraoka, Y., Nakatani, T., Fujii, A., Naganawa, H. and Umezawa, H. (1978b). *J. Antibiotics*, 31, 1073

Tamura, H., Sugiyama, Y. and Sugahara, T. (1974). *Gann*, 65, 103

Tanaka, W. (1977). *Japanese J. Antibiotics*, 30, S-41

Taylor, A. M. R., Rosney, C. M. and Campbell, J. B. (1979). *Cancer Res.*, 39, 1046

Terasima, T. and Tolmach, L. J. (1963). *Biophys. J.*, 3, 11

Terasima, T., Yashukawa, M. and Umezawa, H. (1970). *Gann*, 61, 513

Terasima, T., Takabe, Y., Katsumata, T., Watanabe, M. and Umezawa, H. (1972). *J. natn. Cancer Inst.*, 49, 1093

Tobey, R. A. (1972). *J. cell. Physiol.*, 79, 259

Umezawa, H. (1973). *Biomedicine*, 18, 459

Umezawa, H., Maeda, K., Takeuchi, T. and Okami, Y. (1966). *J. Antibiotics, Ser. A*, 19, 200

Umezawa, H., Takeuchi, T., Hori, S., Sama, T., Ishizuka, M., Ichikawa, T. and Komai, T. (1972). *J. Antibiotics*, 25, 409

Umezawa, H., Takahashi, Y., Fujii, A., Saino, T., Shirai, T. and Takita, T. (1973). *J. Antibiotics*, 26, 117

Umezawa, H., Hori, S., Sawa, T., Yoshioka, T. and Takeuchi, T. (1974). *J. Antibiotics*, 27, 419

Vig, B. K. and Lewis, R. (1978). *Mutat. Res.*, 55, 121

Watanabe, M., Takabe, Y., Katsumata, T., Terasima, T. and Umezawa, H. (1973). *J. Antibiotics*, 26, 417

White, R. E. and Coon, M. J. (1980). *Ann. Rev. Biochem.*, 49, 315

Yamamoto, K. and Hutchinson, F. (1979). *J. Antibiotics*, 32, 1181

Yamamoto, K. and Hutchinson, F. (1982). To be published.

7

Action of Platinum Antitumour Drugs

J. J. Roberts and M. P. Pera, Jr.

INTRODUCTION

Studies leading to the present understanding of the biological activities of
platinum coordination complexes, and to the use of such compounds in cancer
chemotherapy, began in the 1960s, when Rosenberg and co-workers discovered
that certain electrolysis products generated from platinum electrodes passing
current through growth medium selectively inhibited cell division in bacteria
(Rosenberg *et al.*, 1965, 1967). Several years later, the same group reported that
certain platinum coordination complexes displayed significant antitumour activity
in vivo (Rosenberg *et al.*, 1969). The discovery of the antitumour activity of
platinum coordination complexes eventually led to the introduction of one
compound, *cis*-dichlorodiamine platinum (II) (*cis*-DDP, or cisplatin), into
therapeutic trials in the treatment of human malignancy. While the initial trials
demonstrated the activity of the drug against human tumours, renal toxicity
appeared to be a serious limitation to the use of the compound (Gottlieb and
Drewinko, 1975), since it occurred at therapeutic doses and was only partially
reversible. However, clinical oncologists continued to explore ways of
administering the drug, and it was found that fluid administration or induction
of diuresis allowed oncologists to avoid prohibitive nephrotoxicity during cisplatin
therapy (Einhorn and Donohue, 1977; Hayes *et al.*, 1977). The agent was found
to enhance long-term survival significantly in combination therapy of testicular
cancer (Einhorn and Williams, 1980) (table 1), and to be active in the treatment
of ovarian carcinoma and squamous cell carcinoma of the head and neck (for a
review of the clinical activity of *cis*-DDP, see Prestayko *et al.*, 1979). Several
analogues of *cis*-DDP that showed good antitumour activity and reduced toxicity
in animal studies have now entered clinical trials (figure 1).

Molecular Aspects of Anti-cancer Drug Action, ed. Neidle & Waring
0333–315561/83/183–231

Table 1   Therapeutic results of *cis*-DDP (cisplatin) plus vinblastine plus bleomycin chemotherapy in disseminated testicular cancer.

No. of patients	78
Complete remission	53 (68%)
Partial remission	23 (30%)
NED with surgery	11 (14%)
Relapses	10 (13%)
No. continuously NED	53 (68%)

Of 78 patients, 53 (68%) were continuously disease free (NED). Given the natural history of the disease, it is likely that these patients are cured. (Einhorn and Williams, 1981)

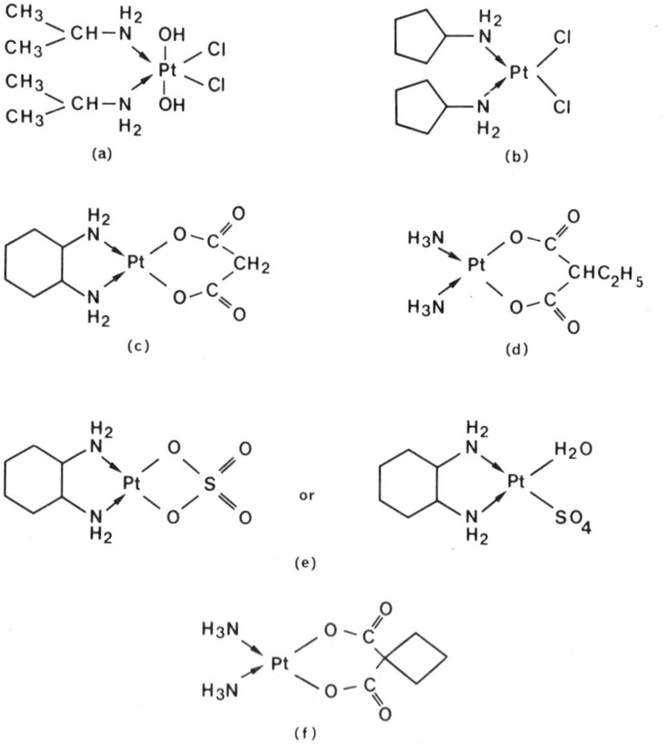

Figure 1   Some second-generation platinum complexes with promising antitumour activity. (a) *cis*-dichloro-bis(isopropylamine)*trans*-dihydroxyplatinum (IV); (b) *cis*-dichloro-bis(cyclopentylamine)platinum (II); (c) Malanato 1,2-diamino-cyclohexane platinum (II); (d) Ethylmalonato diamine platinum (II); (e) sulphato 1,2-diaminocyclohexane platinum (II); (f) cyclobutanedicarboxylic acid diamine platinum (II).

Since the first description of the biological properties of platinum complexes, there has been a considerable effort to understand the molecular basis of their actions. The clear demonstration of the usefulness of *cis*-DDP in treatment of human tumours has added impetus to this work, because ultimately an understanding of the mechanism of drug action should provide a more rational basis for improvement in therapy. Much evidence is available now to support the view that platinum drugs exert their cytotoxic effects through an interaction with DNA. The main purpose of this chapter is to assess critically the evidence supporting this hypothesis. We briefly discuss the chemistry and structure-activity relationships of this class of drugs, and then summarise some of their biological properties other than antitumour activity. We will consider other hypotheses that have been devised to account for the biological properties of these drugs. Then we will consider in detail the evidence that indicates DNA as a major target for drug action.

This chapter is intended not only to document the considerable progress made in understanding the action of the platinum drugs, but also to reveal some of the limits of our present understanding.

CHEMISTRY AND STRUCTURE–ACTIVITY RELATIONSHIPS OF PLATINUM ANTITUMOUR DRUGS

Chemical features of platinum drugs

The only well-characterised oxidation states of platinum in aqueous solution are 2+ and 4+, and both states dictate a strict stereochemical disposition of the groups of ligands that surround the metal ion. In the case of Pt (II), only a square planar geometry is found, whereas in Pt (IV) only octahedral coordination of six ligands is possible. Interconversion between the two oxidation states is facile, reduction from 4+ to 2+ usually occurring at a potential of about zero volts on the hydrogen scale, although the exact redox potential clearly depends upon the nature of the ligands about the platinum ion.

The chemistry of both oxidation states is dominated by the ability of the compounds to undergo ligand substitution or exchange reactions. Generally, both oxidation states of platinum form their most stable complexes with the more polarisable atoms. The following order of thermodynamic stability usually holds:

$$S > I > Br > Cl > N \sim O > F$$

If the incoming ligand can bind to the metal ion at more than one site, the thermodynamic stability is greatly enhanced. This is commonly referred to as the "chelate effect".

The kinetic stability of ligands bound to platinum (II) varies over a very wide range. Furthermore, the rate of substitution of a given ligand is strongly influenced by the nature of the ligand opposite it in the square planar array. This is called the *trans* effect. Generally, the more strongly bound the ligand, the more effective it will be at stabilizing ligands *trans* to it.

A consequence of these facts is that *cis*-DDP is a bifunctional reagent with both chloride ligands open to substitution by incoming groups that are either in large excess or form thermodynamically more stable links to platinum. The amine groups are both kinetically and thermodynamically inert to substitution. Thus the remaining two positions in the square plane are blocked to substitution by incoming groups. However, if the chloride ions were replaced by ligands with a very strong *trans* directing influence, it is possible that the amines would become labilised and replaced. Similarly, in the *trans* isomer, the chloride ions are readily substituted, whereas the amine groups are difficult to replace. Hence both compounds are bifunctional reagents, but with very different requirements for the stereochemical disposition of two incoming groups (Thomson *et al.*, 1972; Thomson, 1974). It is a most important feature of the chemistry of square planar platinum (II) compounds that substitution reactions proceed with retention of configuration.

In aqueous solution, both chloride ions are slowly lost from the coordination sphere of the Pt(II) ion, and water or hydroxide ion becomes bound. Thus a distribution of species is set up involving the presence of unhydrolysed and of partially and fully hydrolysed species. However, this equilibrium is labile so that, if the chloride ion concentration is raised to that of, say, isotonic saline, the majority species in solution will be unhydrolysed *cis*-DDP (Thomson, 1977). Recently it has been shown that, under certain conditions of pH and metal ion concentration, hydroxy-bridged polymeric species can be formed. Both dimeric and trimeric species have been isolated from solution, and their structures have been determined by X-ray crystallography (Lock *et al.*, 1977). It is not clear whether such species could be formed in the high chloride levels of plasma, although it seems unlikely at this stage.

Structure–activity relationships

A number of platinum compounds have been synthesised and screened for anti-tumour activity in various rodent tumour systems (Connors *et al.*, 1972; Braddock *et al.*, 1975; Bradner *et al.*, 1980; Harrap *et al.*, 1980). The influence of configuration and charge, and the nature of ligands surrounding the complex, may be assessed from this body of work.

Configuration and net charge on the complex play a primary role in determining drug activity. *Trans* compounds are more reactive than *cis* compounds, the *trans* $Pt(NH_3)_2Cl_2$ aquating four times faster and ammonating some thirty times faster than the *cis* isomer. This means that *trans* compounds are likely to react faster and with a wider variety of body constituents than the *cis* isomers. Hence the unique activity of the *cis* isomers must reside in the particular reactions with target molecules which their configuration dictates. The results of Connors *et al.* (1972) and Braddock *et al.* (1975) clearly indicated that activity is confined to neutral complexes with *cis* leaving ligands. A comparison of *cis* and *trans* compounds of the type (PtA_2X_2), where A is the stable ligand, and X the leaving group, showed activity either (1) only for *cis* and not *trans* or (2) not in either *cis* or *trans* compounds. No example of activity in the *trans* but not in *cis* isomer has been found.

In general charged complexes showed no activity even when they have *cis* leaving groups (particularly chlorides). Presumably, such complexes are unable to cross the cell membrane.

The nature of the leaving group X in PtA_2X_2 determines the rate of substitution reactions, and enormous variations in reactivity may be observed. The introduction of very highly labile groups, such as H_2O or NO_3, gives rise to very toxic compounds with little antitumour activity. The toxicity of such complexes is possibly due to the formation of dimeric and trimeric bridged compounds. On the other hand, complexes containing very strongly bonded groups, such as thiocyanate, are inactive in biological systems (Cleare and Hoeschele, 1973). Active compounds may contain halides (Cl or Br), oxyanions (SO_4), carboxylates, or dicarboxylates as leaving groups. The carboxylate- or dicarboxylate-containing compounds undergo substitution *in vitro* only at a very slow rate (Clear *et al.*, 1980). In contrast, the oxyanion compounds would be expected to undergo facile conversion to the corresponding chloride species in plasma. Thus there is no simple relationship between reactivity and antitumour effect, except that highly inert or highly reactive complexes are not useful drugs.

Most studies of the role of the non-labile group A in determining antitumour activity of platinum complexes have centred on amine systems. The nature of the amine group coordinated to platinum can indeed influence antitumour activity, but the relationships observed in a given tumour model do not apply to all systems. Active compounds have been synthesised that consist of straight and branched chain alkyl amines, alicyclic amines, heterocyclic compounds and chelating amines. The low solubility of some of the compounds has complicated testing and may account for some of the variability of the results obtained in different tumour systems. Nevertheless, the results of Burchenal *et al.* (1977), showing that a subline of L1210 mouse leukaemia resistant to cisplatin retained its sensitivity *in vivo* to certain 1,2-diaminocyclohexane complexes, revealed that differences in the pharmacological or molecular properties of complexes containing different A ligands could be significant in therapy (figure 2).

BIOLOGICAL EFFECTS OF PLATINUM ANTITUMOUR DRUGS INDICATIVE OF INTERACTION WITH DNA

Before turning to a detailed consideration of the mechanism of action of platinum drugs, we discuss some properties of these agents, apart from antitumour activity, that are shared with a variety of physical and chemical agents that interact with DNA.

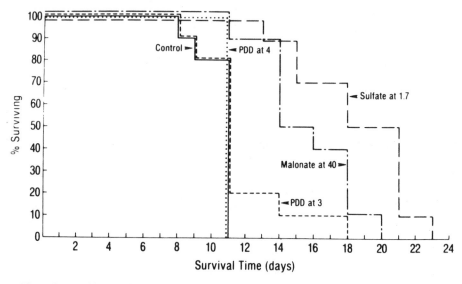

Figure 2   Sensitivity of a *cis*-DDP-resistant L1210 mouse leukaemia strain to two 1,2-diaminocyclohexane platinum compounds *in vivo*. The survival times of tumour-bearing mice treated with *cis*-DDP or the malonate or sulphate 1,2-diaminocyclohexane platinum compounds are shown. A similar lack of cross resistance was seen in these cells *in vitro* (Burchenal *et al.*, 1977).

Filament formation in bacteria

Probably the first observation of an effect of a platinum coordination complex in a biological system and one which gave a clear indication of its biochemical mode of action, came during experiments to investigate the effect of an electric current on growing bacteria (Rosenberg *et al.*, 1965). It was noticed that when a low alternating current was passed through platinum electrodes to growing Gram-negative bacteria in nutrient media, cell division was inhibited and the bacteria grew into long filaments. Subsequently, it was discovered that some of the platinum dissolves under these conditions to give first of all the ionic species ammonium hexachloroplatinate. This compound can itself, at high concentration, inhibit cell division, but aged solutions were found to be far more efficient in producing filaments but only if exposed to visible light. The photochemical change which occurred in a solution of hexachloroplatinate, which gave rise to a more active agent, was shown to involve the replacement of the chloride ligands by NH_3 with the loss of one negative charge per replacement to give finally a stable neutral species. The new species was shown convincingly to inhibit cell division but not growth, in contrast to the parent ionic species which was a bacteriocide and not a bacteriostat.

Filamentous growth in bacteria may be indicative of the ability of an agent to react with DNA leading to a selective inhibition of DNA synthesis, with no accompanying effect on other biosynthetic pathways such as RNA or protein synthesis. A variety of agents, such as UV- and X-irradiation and cytotoxic alkylating agents, can also elicit this response as a result of their common ability to damage DNA.

Induction of lysogeny

Further important evidence for direct attack on DNA came from the ability of platinum compounds to induce the growth of phage from lysogenic strains of *E. coli* (Reslova, 1971). The release of the phage DNA to direct synthesis of new phage is normally a rare event. However, agents which can react with DNA can cause the phage DNA to be released and phage particles to be synthesised with consequent cell lysis. Reslova (1971) was able to show that there exists an excellent correlation between the antitumour activity of platinum compounds and their ability to induce lysogenic *E. coli* to enter the lytic cycle.

Inactivation of viruses and transforming DNA

The interactions of platinum compounds with viruses have further indicated the relatively greater importance of reactions with DNA as against those with protein in producing biological effects. Kutinova *et al.*, (1972) demonstrated the inactivation of the infectious activity of extracellular papovavirus SV40 by *cis*-DDP. An indication of the mechanism of inactivation was derived by following the capacity of the inactivated virus to induce either tumour or viral antigens. The capacity of the virus to induce the tumour antigen was found to be less sensitive to *cis*-DDP than the viral antigen formation or the infectivity of the virus. This finding indicated that the viral DNA and not the protein coat is the primary target for the platinum compound. The inactivation of *Bacillus subtilis* transforming DNA by platinum compounds, likewise indicated the effect of these agents on the biological function of DNA (Reslova *et al.*, 1972).

Mutagenic properties of neutral platinum complexes

The genotoxic nature of platinum compounds and the importance of the geometrical arrangement of ligands for biological activity, also emerges from studies on the mutagenic properties of these agents in a number of prokaryotic and eukaryotic systems (Trosko, 1974; Beck and Brubaker, 1975; Monti-Bragadin *et al.*, 1975; Lecointe *et al.*, 1977; Zwelling *et al.*, 1979a; Johnson *et al.*, 1980). The *cis* derivatives were in all cases appreciably more mutagenic than the corresponding *trans* isomers. A comparison of cytotoxicity and mutagenicity, either per DNA interstrand crosslinked lesion (Zwelling *et al.*, 1979a) or per total number of lesions (Johnson *et al.*, 1980), for *cis*- and *trans*-DDP showed that these two biological effects differ in their sensitivity to DNA binding. Thus, in a study of mutation at the HGPRT locus in Chinese hamster V79 cells, equitoxic doses of *trans*-DDP were much less mutagenic than *cis*-DDP, even though interstrand crosslinking for the two isomers was comparable (Zwelling *et al.*, 1979a). The study by Johnson *et al.* (1980) in Chinese hamster ovary (CHO) cells revealed that the mutagenicity per molecule bound to DNA immediately after 16–24 hr treatment was at least 750 times larger for *cis*-DDP than for the *trans* isomer, while cytotoxicity per DNA lesion was a factor of 9 greater for *cis* than for *trans*-DDP. However, *cis*-DDP was a less efficient mutagen than ultraviolet light or *N*-methyl-*N*-nitrosourea, in that the number of mutants per DNA lesion produced by the platinum drug was several orders of magnitude lower than that produced by these other mutagens (table 2).

Table 2   Relative mutagenicities of *cis* and *trans* DDP induced DNA lesions in the Chinese hamster ovary cell-HGPRT system.

Compound	Lesions/nucleotide/μM concentration	Mutants/10^8 cells/μM concentration	Mutants/lesions
cis-Pt(NH$_3$)$_2$Cl$_2$	2.8×10^{-5}	31.5	9.4×10^{-4}
trans-Pt(NH$_3$)$_2$Cl$_2$	0.87×10^{-5}	0.013	1.3×10^{-6}
N-methyl-*N*-nitrosourea	5.7×10^{-8}	2.8	4.1×10^{-2}
N-ethyl-*N*-nitrosourea	0.38×10^{-8}	0.84	1.8×10^{-1}
UV	1.43×10^{-5}	400 mutants /10^6 cells	2.3×10^{-2}

Note the marked difference in the mutagenicity of the two isomers, and their low mutagenicity relative to other agents. Mutants/lesion in the HGPRT locus = [(mutants/cell/μM) $\div$ (lesions/nucleotide/μM)] $\div$ 1,200 nucleotides/HGPRT locus/cell. UV treatment with 5 J m^{-2} (Johnson *et al.*, 1980)

Induction of cellular transformation *in vitro* and carcinogenic activity *in vivo*

Another property platinum antitumour drugs share with agents that interact with DNA is induction of cancer. *In vitro* studies of morphological transformation, a property often correlated with tumour induction *in vivo*, showed *cis*-DDP was capable of transforming secondary Syrian hamster embryo cells (Turnbull *et al.*, 1979). The finding that *trans*-DDP transformed 10T½ mouse cells and 3T3 cells was of interest in light of the low mutagenicity of this compound (Fornace and Little, 1980). In carcinogenesis studies *in vivo*, *cis*-DDP produced lung adenomas in A/Jax mice, skin papillomas in CD1 mice when administered in combination with croton oil, and sarcomas at the site of injection in F344 rats (Leopold *et al.*, 1979). When administered at maximally tolerated doses, *trans*-DDP did not induce tumours in the lung adenoma or skin papilloma systems, as expected from its low mutagenicity (Leopold *et al.*, 1981).

ALTERNATIVE HYPOTHESES FOR THE MECHANISM OF ACTION OF PLATINUM ANTITUMOUR DRUGS

The above survey indicates that many biological properties of platinum compounds might arise from interaction with the genetic material of the cell, to produce toxicity through interference with replication or heritable changes through alteration in readout of the genetic code. However, the evidence for DNA as a target of drug action, discussed below, is not yet conclusive. Before we consider such evidence, it is important to evaluate other hypotheses concerning drug action.

Modification of the host immune response to tumour

Some years ago, Rosenberg (1975) suggested that it was unlikely that selectivity of platinum drugs could result solely from a cytotoxic effect. It was argued that the regression of advanced animal tumours such as sarcoma 180 and Walker 256 carcinoma following single, sublethal, doses of platinum drugs could possibly be the result of a drug-induced enhancement of immune defences against the tumour. The elimination of large tumour burdens by sublethal drug doses certainly indicates considerable selectivity, and enhanced immunity against tumour cells might have a role, particularly in allograft tumour systems such as those mentioned above. However, much evidence indicates that, *in vivo*, platinum antitumour drugs suppress both cellular and humoral immune responses (Berenbaum, 1971; Khan and Hill, 1972; Khan *et al.*, 1972; Khan and Hill, 1973; Brambilla *et al.*, 1974; Bowen *et al.*, 1974; Mally *et al.*, 1979; Wierda and Pazdernik, 1979). The interesting suggestion that cell surface DNA was present solely on tumour cells, and was rendered highly antigenic following reaction with platinum drugs (Aggarwal *et al.*, 1975), was not upheld by subsequent observations (McAllister *et al.*, 1977).

It must be noted that in no syngeneic animal tumour system has enhanced antitumour immunity been demonstrated following treatment with platinum drugs. Indeed, in the L5178Y/DBA2 system, tumour inhibition *in vivo* was shown to result solely from the expected direct cytotoxicity of the drug (Szumiel *et al.*, 1981) as measured by *in vitro* colony formation assay.

Recently Kleinerman *et al.* (1980*a*) have shown that treatment of human monocytes with *cis*-DDP *in vitro* results in an enhanced spontaneous cytotoxicity towards target red cells. The same phenomenon was seen in cells taken from patients undergoing *cis*-DDP chemotherapy *in vivo* (Kleinerman *et al.*, 1980*b*). While the relationship between such enhanced spontaneous monocyte killing *in vitro* and macrophage control of the growth of syngeneic tumours *in vivo* remains to be established, this unusual response might play a role in the elimination of residual tumour cell populations.

Alterations in nuclear protein phosphorylation

Harrap *et al.* (1980) have proposed that changes in the level of phosphorylation of nuclear proteins are related to the cytotoxic actions of alkylating agents and the platinum drugs. Increased phosphorylation of nuclear proteins following treatment with a variety of platinum drugs was correlated with tumour sensitivity and resistance, and with drug toxicity in host tissues *in vivo*, even in tissues made up primarily of non-dividing cell populations. The alterations in chromosome protein phosphorylation generally ran in parallel to cytotoxic effects. The changes in the level of phosphorylation occurred late, however, from 24 to 48 hours after treatment, and may well be a consequence of cell death and chromatin dissolution rather than a cause thereof. Further understanding of the specific nature of the proteins affected, and their function within the cell, is required before the relevance of the observed alterations is understood.

Alterations in cyclic AMP metabolism

One of the arguments for the importance of DNA as a target for alkylating agents is that in the absence of any specificity of reaction, the level of reaction with proteins is too low, on a per molecule basis, to inactivate sufficient proteins to cause cell death. However, Tisdale and Phillips (1975*a*) demonstrated that at low doses of *cis*-DDP or other difunctional agents, the low K_m form of cyclic adenosine (3′,5′-monophosphate) (cAMP) phosphodiesterase was inhibited, resulting in a dose-dependent cAMP accumulation in treated Walker 256 carcinoma cells. Because of the importance of cAMP in cell cycle regulation, and because of the apparent unusual sensitivity of the low K_m form of cAMP phosphodiesterase to certain difunctional agents, the findings of Tisdale and Phillips (1975*a,b,c*) that correlated tumour cell sensitivity to altered cyclic nucleotide metabolism merit further study. In particular, it is necessary to consider exactly what biochemical effects link the transient increase in cAMP levels to inhibition of cell proliferation.

Mutation as a cause of cell death

Certain evidence from studies of the binding of platinum drugs to nucleic acid bases has been interpreted to be indicative of a bifunctional chelation of guanine to platinum via the N7 – O6 position (see below). Binding of platinum at the O6 position of guanine would be expected to result in a disruption of normal base pairing and substitution mutation upon DNA replication as in the case of methylation at O6 of guanine (Loveless, 1969). Such binding would not occur with *trans*-DDP due to steric factors, and this would explain the observed differences in mutagenic efficiency with *cis* and *trans*-DDP (Rosenberg, 1977). Recent findings have indicated that cellular homogenates from tumours resistant to cisplatin contain activity which releases O6 methyl guanine from DNA *in vitro*, whereas sensitive lines contain less of this activity (Rosenberg, 1981).

However, these finding are difficult to reconcile with those of Laurent *et al.* (1981) who found that the sensitivity of human cells to *cis*-DDP could not be related to the known capacity of cells to repair O6 methylated guanine. There are other difficulties with hypotheses relating mutation to cytotoxicity. Although *cis*-DDP is mutagenic, the frequency of mutation induction with this drug is quite. low compared to other mutagenic agents (see above). Furthermore, although each *cis*-DDP reaction with DNA is at least 750 times more efficient in causing mutation in CHO cells than those produced by its *trans* isomer, lesions caused by the *cis* compound are only 9 times more effective in causing toxicity (Johnson *et al.*, 1980). Thus, while it is clear that *cis*-DDP is a far more effective mutagen than the *trans* isomer, it is difficult to argue that for this pair of compounds, the same lesions are responsible for mutation and cytotoxicity.

REACTION OF PLATINUM DRUGS WITH DNA

Although the above hypotheses have raised several interesting questions, the bulk of the evidence at present favours the theory that DNA damage and synthesis of DNA on a damaged template is directly responsible for the cytotoxic activity of platinum drugs. Before presenting the experimental evidence, it is useful first to outline the general argument.

Platinum antitumour drugs are bifunctional electrophilic species which react in a relatively nonselective fashion with nucleophilic sites in cellular macromolecules. On a per molecule basis, DNA is the only target that is large enough to undergo one or more reactions per molecule at pharmacologically relevant doses. Reaction with DNA results in disturbances in the rate of DNA synthesis, or the quality of nascent DNA synthesised, which leads to chromosome damage and cell death. Cells possess to varying degrees the capacity to excise damage from template DNA, and moreover, cells vary in their capacity to synthesise DNA on a damaged template. Thus cytotoxicity and cellular sensitivity to platinum drugs will be a function of the extent of reaction with DNA, the capacity to remove damage from the template, and the capacity to synthesise DNA on a template containing damage.

Thus, evidence for this mechanism comes from studies of the interaction of the drugs with nucleic acids and nucleic acid components, from observations on the effects of the drugs on DNA synthesis, and from observations relating to repair of DNA lesions and cytotoxicity. First we will consider the reactions of the drug with nucleic acids.

Reactions of neutral platinum complexes with nucleic acid components *in vitro*

Reactions with bases
Changes in the ultraviolet absorption spectrum of salmon sperm DNA after reaction with either *cis*- or *trans*-DDP provided conclusive evidence that both platinum compounds bind to the organic bases of DNA (Horacek and Drobnik,

1971; Mansy *et al.*, 1973). Spectrophotometric studies further confirmed that the three bases, guanine, adenine and cytosine would all react with both isomers, the rate of reaction with guanine being faster than with the other two bases (Mansy *et al.*, 1973). The ^{14}C-labelled *cis*-dichloro(ethylenediamine) platinum (II) similarly reacted preferentially with guanosine as compared with adenosine and cytosine (Robbins, 1973). A very slow reaction occurs with thymine (Mansy *et al.*, 1973) while substituted thymines react to give blue and purple coloured complexes of uncertain structure (Thomson *et al.*, 1977). The above indication that guanine is the base in DNA with which reaction is most likely to occur preferentially was supported by studies on the extent of reaction of *cis*-DDP to DNAs of varying (G+C)/(A+T) ratios (Stone *et al.*, 1974), and by the ratio of the binding of a radioactively labelled platinum compound to the purine bases in DNA (Munchausen and Rahn, 1975).

Sites of reaction on bases

By blocking the various possible binding sites in the purine bases by either methylation or protonation, Mansy *et al.*, (1973) defined the sites most likely to be involved in reaction with either *cis*- or *trans*-DDP. They concluded that the *cis* isomer forms a bidentate chelate with either the 6-NH_2 and N7, or the 6-NH_2 and N1 of adenine, and the 4-NH_2 and N3 of cytosine. The *trans* isomer, on the other hand, interacts monofunctionally at the N7 and N1 of adenine and the N3 of cytosine. Both isomers react monofunctionally with the N7 of guanine and hypoxanthine. Robbins (1973) used a similar approach and blocked the N7, N9 or both positions of guanine or its derivatives, and concluded that the N7 position of guanine is a primary point of attachment for *cis*-DDP, but that reaction probably occurs also at a second site. More recently, X-ray diffraction studies of the complexes formed between *cis* $Pt(NH_3)_2X_2$ and various bases have confirmed some of the conclusions obtained from the early spectrophotometric studies. The product of the reaction between inosine and $[Pt(NH_3)_2I_2]$ consists of two hypoxanthine rings bound to the platinum ion via the N7 position, as in figure 3 (Goodgame *et al.*, 1975). A similar structure results from the interaction of $PtCl_2$ (en) with guanosine (Gellert and Bau, 1975; Cramer and Dahlstrom, 1977); again the N7 position becomes occupied by the metal.

Other atoms which have been shown to have an affinity for platinum include the deprotonated N1 and N3 positions of guanine and thymine respectively as well as the deprotonated NH_2 group of cytosine (Bau, 1981).

There is no evidence yet from crystallographic studies that the O6 position of guanine, the 6-NH_2 group of adenine or the 4-NH_2 group of cytosine can be occupied by platinum (II) ions. A bidentate binding reaction between the N7 and the O6 positions of guanine for the *cis* but not the *trans* platinum (II) compounds was an attractive possibility to account for the difference between the biological effectiveness of the two isomers. However, such bidentate binding to a single base has not yet been identified unambiguously and the evidence for it, based originally on analyses of electronic binding energies (Millard *et al.*, 1975;

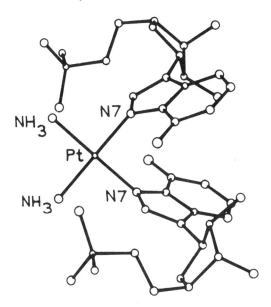

Figure 3   The structure of the reaction product of inosine and cis[Pt(NH$_3$)$_2$I$_2$] (Goodgame *et al.*, 1975).

Macquet and Theophanides, 1975), has been questioned by others (for example, Kelman *et al.*, 1977). On the basis of ^{195}Pt-^{1}H spin-spin coupling by NMR spectroscopy, *mono* and *bis*- (NH$_3$)$_2$Pt-inosine complexes were identified as being bound through the N 7 position (Kong and Theophanides, 1974). Raman difference spectroscopy of the products of reaction of the diaquo forms of *cis*-DPP and dichloroethylenediamine platinum with inosine or 1-methylinosine and 5'-inosine phosphate also favoured coordination of platinum at N7 and excluded reactions of the C6-oxygen site (Chu and Tobias, 1976; Tobias *et al.*, 1975). Roos *et al.* (1974), on the basis of CD spectral changes, showed that *cis*-DDP would bind to two adenine bases in a stacked configuration. They postulated that either the C-(6)NH$_2$ and N1 or N7 and C-(6)NH$_2$ groups were involved.

DNA crosslinking reactions

Cis platinum compounds can induce the formation of interstrand crosslinks in isolated DNA (Horacek and Drobnik, 1971; Pascoe and Roberts, 1974*a,b*; Harder, 1975; Ganguli and Theophanides, 1979) or in the DNA of whole cells (Roberts and Pascoe, 1972; Zwelling *et al.*, 1978). There is as yet no direct evidence to indicate which of the many possible binding sites discussed above, are involved in such a reaction. A possibility, suggested from examination of a DNA model, is that crosslinking could occur between the 6-amino groups in adenines in opposing strands of DNA in a dA-dT sequence (Thomson, 1974). These groups would be 3.5 Å apart, which is close to the distance between the *cis* leaving groups in *cis*-DDP, 3 Å. There is some evidence to indicate that *cis*-DDP can link two NH_2 groups in this way in a simple nucleotide (Roos *et al.*, 1974: Kleinwächter and Zaludova, 1977). Alternatively the amino groups of guanines or of cytosines in opposing strands of DNA in a dC-dG DNA sequence are theoretically amenable to crosslinking. The preferential interstrand crosslinking of DNAs rich in guanine and cytosine was noted from studies of the renaturability of crosslinked DNAs of different G+C content (Ganguli and Theophanides, 1979). Another method of investigating DNA crosslinking by platinum drugs is illustrated in figure 4. The introduction of either inter or intrastrand crosslinks but not monofunctional adducts into DNA will prevent the intercalation of 9-aminoacridine into DNA. A study of inhibition of intercalation as a function of (G+C) content of the DNA or of copolymers indicated that crosslinking increases with the (G+C) content of the DNA. The A-T, but not A-A sequence was also deduced to be a crosslinking site, a finding consistent with the earlier chemical evidence favouring crosslinking between amino groups of adenine (Mansy *et al.*, 1973; Roos *et al.*, 1974). The extent of inhibition of

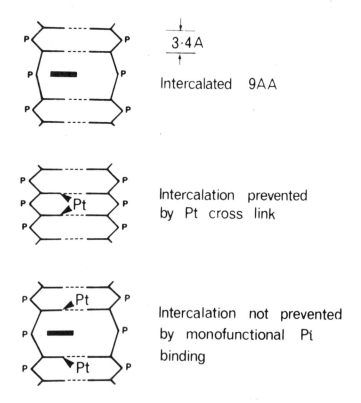

Figure 4   Schematic depiction of the results of studies on the inhibition of 9-aminoacridine intercalative binding to DNA following cisplatin treatment. Results are consistent with intrastrand crosslink formation (Roos, 1977).

intercalation was, however, too great to be accounted for by the relatively few interstrand crosslinks likely to be present in the platinum-modified DNAs or copolymers (see below) and it was therefore thought that the effect was most likely to be the result of intrastrand crosslinks in DNA.

The frequency of interstrand crosslinks was originally estimated by Pascoe and Roberts (1974*a,b*), from a combination of measurements of the amount of platinum bound to HeLa cell crosslinked DNA molecules of an estimated size: it was shown to be a relatively rare event, accounting for less than one per cent of the total number of reactions with cellular DNA. Recent quantitative studies of crosslinking of Chinese hamster cell DNA, using a variety of techniques, confirmed the rarity of crosslinks in whole cells at the time of their maximum development (see below).

Despite the clear evidence for the formation of interstrand DNA crosslinks both in isolated DNA and in the DNA of whole cells, many recent observations indicate the likely formation of intrastrand crosslinking between adjacent bases in DNA by platinum compounds. Moreover the importance of this reaction as an inactivating event in bacteriophage was noted, first, by Shooter *et al.* (1972) for T7 bacteriophage and, more recently, by Filipski *et al.*, (1980) for phage λ (see below). For such a crosslinking reaction to occur, local perturbation of the double helix is required, and indeed, some observations of X-ray photoelectron spectral changes (Millard *et al.*, 1975) or CD spectral changes (Macquet and Butour, 1978; Tamburio *et al.*, 1977) support such a modification. Marked enhancement of the CD spectrum of DNA was observed even at low levels of reaction and this effect increases with increasing (G+C) content of several different DNAs (Macquet and Butour, 1978). The binding of platinum to the N7 position of guanine could weaken the G–C hydrogen bonding which, it is suggested, makes the N1 position of guanine available for further reaction (Kelman *et al.*, 1977). Goodgame *et al.* (1975) also argued that crosslinking of guanine moieties via the N7 position would require a considerable distortion of the DNA structure and therefore be unlikely to occur directly *in vivo*. It was therefore proposed that a firm binding to N7 of guanine would initially be formed, with possibly a weaker second bond with the O6 position in the same guanine molecule. However, during DNA replication, it could be envisaged that the weak link to O6 could be broken and a new one established with the N7 position of another adjacent guanine moiety. Covalent binding of both *cis-* and *trans*-DDP to closed circular PM2 DNA alters the degree of supercoiling and shortens the DNA, as revealed by electron microscopy, presumably by disrupting and unwinding the complex (Macquet and Butour, 1978*b*; Cohen *et al.*, 1980).

An investigation of the effect of cisplatin on the *Bam*HI restriction enzyme digest pattern of λ DNA was interpreted as evidence for binding to adjacent guanine bases in DNA (Kelman and Buchbinder, 1978). Consistent with this concept was the ability of cisplatin to react with poly dG·poly dC more avidly than with poly (dGdC) (Stone *et al.*, 1974; Stone *et al.*, 1976). Some elegant experiments by Cohen *et al.* (1980) add further strong support to the ability of *cis*- but not *trans*-DDP to form intrastrand crosslinks between adjacent guanines in a $dG_4 \cdot dC_4$ sequence in circular pSM1 DNA at low Pt/P ratios. The restriction endonuclease *Pst*I cleaves the circular DNA at four sequence specific loci to produce fragments designated A–D in order of decreasing size (figure 5a). Cleavage of DNA that has been incubated with *cis*-DDP is prevented and produces instead partially-fragmented pieces that can be identified in agarose gels (figure 5b). These partials include the four dimers BA, AC, DB and CD and the four trimers DBA, BAC, ACD and CDB, all of which eventually appear over the time course of the experiment. Since *cis*-DDP unwinds the double helix, not surprisingly, the *Pst*I enzyme is prevented from cutting the DNA. Significantly,

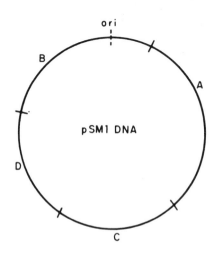

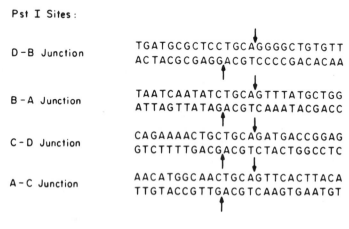

Pst I Sites:

D – B Junction

TGATGCGCTCCTGCAGGGGCTGTGTT
ACTACGCGAGGACGTCCCCGACACAA

B – A Junction

TAATCAATATCTGCAGTTTATGCTGG
ATTAGTTATAGACGTCAAATACGACC

C – D Junction

CAGAAAACTGCTGCAGATGACCGGAG
GTCTTTTGACGACGTCTACTGGCCTC

A – C Junction

AACATGGCAACTGCAGTTCACTTACA
TTGTACCGTTGACGTCAAGTGAATGT

(a)

Figure 5a

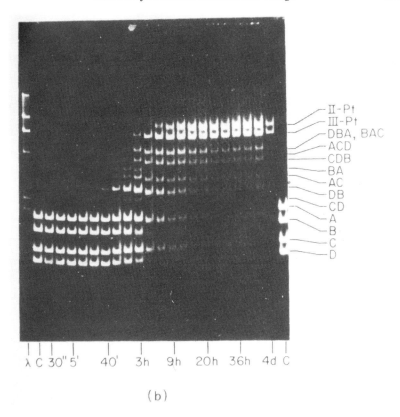

(b)

Figure 5 Inhibition of *Pst*I restriction enzyme cleavage at a specific site in pSM1 DNA by *cis*-DDP. pSM1 DNA was incubated with *cis*DDP for varying periods. The DNA was then digested with *Pst*I restriction endonuclease and run on an agarose gel. *Pst*I cleavage of untreated control pSMI DNA produced four restriction fragments, designated A, B, C and D (Channel 2, 'C'). Extensive incubation of the DNA with *cis*-DDP inhibits restriction cleavage at all sites due to drug induced helix unwinding. Such inhibition results in the appearance of dimers and trimers of the restriction fragments in the gel (CD, DB, AC, BA, ACD, CDB, DBA, BAC fragments in Channel 14, 9 hr incubation). However, at low levels of binding (short incubation periods), cleavage is selectively inhibited at the D–B junction, as evidenced by the appearance of the D–B dimer only (1 hr time point, Channel 9). All cleavage sites have the same recognition sequence. The unique feature of the D–B site is a $(dG)_4$-$(dC)_4$ sequence adjacent to the recognition sequence (a). Selective inhibition of cleavage at this site is thus consistent with intrastrand crosslinking of adjacent guanine or cytosine residues. Trans-DDP did not selectively inhibit cleavage at this site (Cohen *et al.*, 1980).

however, the fragments first to disappear during drug incubation are the B and D fragments with the concomitant appearance of the platinated BD dimer. This indicated that at low levels of Pt–DNA binding, the restriction enzyme could cleave the DNA at three of the four normal cutting sites but not the B–D junction. At comparable binding levels the *trans* isomer did not produce this effect. Since all four cutting sites have the same six base-pair sequence, it is argued that the selective inhibition by *cis*-DDP of cleavage at the B–D junction must involve base pairs adjacent to the restriction sequence, namely the occurrence of the unique $(dG)_4 \cdot (dC)_4$ cluster. The authors further point out that examination of a model of this sequence reveals that *cis*-DDP can bind two adjacent guanine bases at N7 or two adjacent cytosine bases at N3 but only if base pairing is disrupted such as to produce a two-fold shortening of the DNA as observed by electron microscopy (Macquet and Butour, 1978; Cohen *et al*. 1979).

Reaction with DNA of cells in culture

Studies with cultured cells have indicated the relevance of platinum–DNA binding to cytotoxicity. Pascoe and Roberts (1974*a,b*) studied the interaction of several platinum compounds with macromolecules at measured levels of cell kill.

To assess the possible importance of DNA, RNA and protein as primary targets for platinum (II) compounds, these binding data (expressed as moles per gram of macromolecule) were used to construct curves of log survival against the amount of drug bound to each type of macromolecule. The resulting graphs were then characterised in a manner similar to those relating log cell survival to dose of drug given to the cells. The shoulder width of the binding curve was given the value B_q and the slope of the straight line portion B_0. For both *cis*- and *trans*-DDP the binding coefficients were higher for RNA than DNA. However, the true significance of these binding coefficients can only be appreciated if account is taken of the molecular weights of the molecules concerned. If one assumes no selectivity in the binding to any particular RNA or protein molecule, then it is possible to calculate the number of platinum molecules bound to each macromolecule at a given toxic dose. The results of such a calculaton, performed at the concentration of *cis*-DDP which reduced the surviving fraction of HeLa cells from f to $0.37f$ (this is theoretically the concentration at which one inactivating event occurs, on the average, in each cell) show the number of molecules bound to DNA is strikingly more than that to either RNA or protein, clearly indicating that DNA is the most susceptible cellular target for *cis*-DDP. The binding data further indicate that at this concentration of *cis*-DDP approximately only one molecule of protein out of 1,500 molecules will have received one platination reaction. Unless there is considerable specificity in the reaction of platinum drugs with a particular protein enzyme molecule, then this level of reaction would be too low to inactivate enzyme activity. Moreover, the level of reaction with rRNA, tRNA or mRNA would not be expected, again, in the absence of any selectivity of reaction, to inactivate all such molecules and lead to interference with protein synthesis.

Similar DNA binding and cell survival studies have been carried out in Chinese hamster cells in culture with other platinum compounds that have shown encouraging activity against experimental animal tumours (Roberts and Fraval, 1978). Differences of up to tenfold were found in the molar concentrations of these agents that were required to produce equitoxic effects on cells in culture after incubation for 1 hour.

The levels of reaction with DNA at equitoxic doses (B_0 values), on the other hand, were, for most compounds, of the same order and differed by only a few fold (table 3).

Table 3 Comparison of the toxicity of various platinum complexes toward Chinese hamster V79 379A cells in culture and plasmacytoma ADJ/PC6 tumour cells *in vivo* in relation to DNA binding (Roberts, 1981)

Compound	ADJ/PC6 A mouse plasmacytoma		Chinese hamster V79 379 A cells	
	LD_{50}	ID_{90}	D_0	B_0
	(mg/kg)	(mg/kg)	(μM/1 h)	(nmol/g)
cis-Pt(II)Cl$_2$(NH$_3$)$_2$	13	1.6	15	8.5
cis-Pt(IV)Cl$_2$(*iso*-C$_3$H$_7$NH$_2$)$_2$(OH)$_2$	54	4.2	48	2.5
cis-Pt(II)(1,1-CBDCA)(NH$_3$)$_2$	180	14.5	120	3.0
cis-Pt(II) (mal) (1, 2-dac)	ND	ND	23	2.5
cis-Pt(II) (SO$_4$)H$_2$O(1, 2-dac)	14	0.4	65	17.5
cis-Pt(II)Cl$_2$(C$_5$H$_9$NH$_2$)$_2$	480	2.4	120	ND

ND, Not determined; CBDCA, cyclobutanedicarboxylic acid; dac, 1, 2-diaminocyclohexane.

Reaction with the DNA of cells *in vivo*

It is clearly essential, for an understanding of the mechanism of the tumour-inhibitory action of platinum compounds, to establish that the sensitivity of tumour cells *in vivo* is related to the extent of reaction of platinum with their DNA in a manner similar to that for cells treated *in vitro* with these agents. In a preliminary study, mice bearing the transplanted ADJ/PC6 plasmacytoma were treated with *cis*-DDP and two other active platinum congeners, CHIP and *cis*-diammine (1:1-cyclobutanedicarboxylato) platinum (II) at doses that had an equal inhibitory effect on the tumour (ID_{90}) (Roberts, 1981). Despite the difference in the actual amounts of the materials administered to the mice, the doses did not differ by more than a factor of two when expressed in terms of their molar concentrations. Interestingly, the amounts of the platinum drugs bound to the tumour DNA at these equitoxic concentrations were all remarkably similar.

Pera *et al.* (1982) studied the reaction of *cis*-DDP and hydroxymalonato diamine platinum (II) with DNA of B16 melanoma and bone marrow in C57BL mice. Inhibition of tumour growth, and colony formation assays for melanoma cells as well as bone marrow stem cells, were used to quantitate toxicity and to

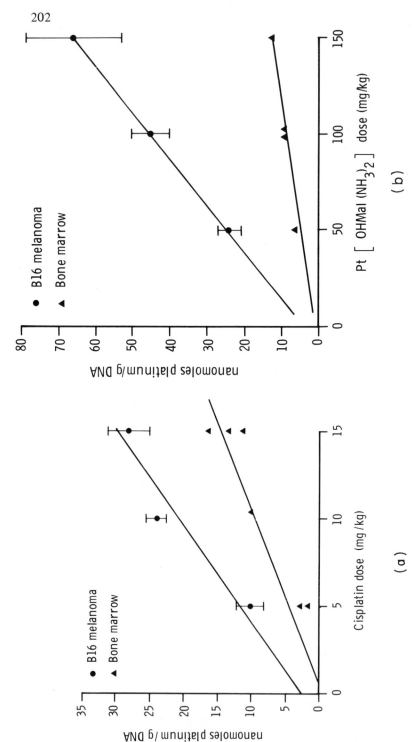

Figure 6 Binding of platinum to B16 melanoma or bone marrow DNA following treatment of C57BL mice with cisplatin (a) or hydroxymalonato diammine platinum (II) (b). The latter drug produced more selective killing of melanoma cells, and the improved selectivity was associated with a greater reaction with tumour DNA, possibly due to enhanced delivery of active drug to the tumour (Pera et al., 1982).

indicate antitumour selectivity. Pt[OHmal(NH$_3$)$_2$] produced greater selective inhibition of tumour growth and more selective tumour cell killing compared with *cis*-DDP. In the case of *cis*-DDP, binding of platinum to DNA at measured levels of survival *in vivo* was similar to values previously observed in cultured cells, a finding that again strengthens arguments concerning the mechanism of action of the drugs based on *in vitro* work. The greater selective toxicity of Pt[OHMal(NH$_3$)$_2$] towards the B16 melanoma was associated with an increased binding of platinum to tumour DNA, relative to *cis*-DDP. The enhanced DNA binding in the tumour seen with Pt[OHMal(NH$_3$)$_2$] was not seen in the marrow (figure 6, table 4). Thus the increased antitumour specificity of the newer congener probably results from pharmacological factors that enhance delivery of active drug to tumour cells.

Table 4 Amount of platinum bound to marrow and tumour DNA of C57BL mice treated with *cis*-DDP or Pt[OHMal(NH$_3$)$_2$]

	CFU-S		B16 LCFC	
	D_{37}	Amount bound	D_{37}	Amount bound
cis-DDP	5 mg/kg	4nmol/g	11mg/kg	22nmol/g
Pt[OHMal(NH$_3$)$_2$]	40 mg/kg	5nmol/g	20mg/kg	12.5nmol/g

Each drug was given to the mice at doses producing 37% survival of bone marrow stem cells (CFU-S) or B16 lung colony forming cells (LCFC) (Pera *et al.* 1982).

Role of crosslinking reactions

The structural requirement for bifunctionality and the principal biochemical effects of the platinum compounds, as discussed below, suggested a parallel between the platinum drugs and the classical bi-functional alkylating agents such as the nitrogen mustards. The latter compounds have been thought for some time to produce an inhibition of DNA synthesis by their ability to introduce crosslinks into the DNA of mammalian cells. It has, however, been a matter of contention as to whether the principal lesion is a crosslink between strands of the DNA helix or crosslinks between bases on one strand of DNA, or possibly, even between DNA and protein. In order to estimate crosslinks in the DNA, one strand of DNA was given a density and radioactivity label by growing cells in the presence of [^{3}H] BUdR. Crosslinking between a "labelled heavy" strand and a "light unlabelled" strand produced a "labelled hybrid" species, and this species could be separated in an alkaline caesium chloride gradient (Roberts and Pascoe, 1972).

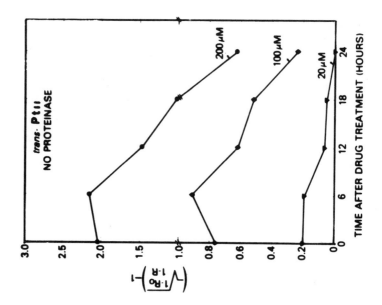

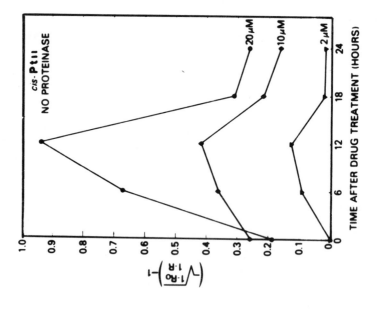

Figure 7

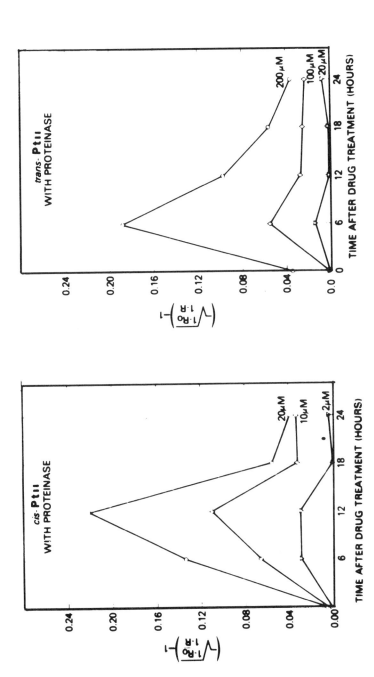

Figure 7 Alkaline elution measurement of DNA–protein and DNA–DNA interstrand crosslinking in L1210 mouse leukaemia cells following treatment with *cis*- and *trans*-DDP. In the absence of proteinase, total crosslinking is measured; the inclusion of proteinase in the procedure eliminates DNA–protein crosslinks, allowing estimation of DNA-interstrand crosslinks. Note the different time course of total crosslinking with the two isomers and the lower proportion of DNA interstrand crosslinks produced by *trans*-DDP (Zwelling *et al.*, 1979*b*).

The possible role of the crosslinking reaction in determining the cytotoxic action of platinum drugs, was ascertained from a study aimed at answering the question, does the extent of interstrand crosslinking of DNA correlate with the cytotoxic activity of a range of platinum compounds? (Pascoe and Roberts, 1974*a*,*b*). The relative toxicities of the *cis* and *trans* isomers of the platinum (II) neutral complexes, can be defined by the slopes of the survival curves (D_0) obtained by treating HeLa cells in culture. Comparison of these two sets of values indicated that the relative abilities of *cis*- and *trans*-DDP compounds to crosslink DNA *in vivo* (but not *in vitro*) were related to their cytotoxic action. Thus the relative ability of *cis*- and *trans*-DDP to kill cells, measured either on the basis of dose ($D_0^{trans}/D_0^{cis} = 18$) or DNA binding ($B_0^{trans}/B_0^{cis} = 5.6$) is of the same order as the 12-fold difference in the doses required to produce equal amounts of crosslinking with the two compounds. These studies therefore suggest that interstrand crosslinking with both the platinum (II) compounds may be important in inducing their cytotoxic effects and that the *cis* isomer is most effective in inducing the reaction.

A reinvestigation of crosslinking of DNA by platinum (II) compounds using a different method from that described above, namely alkaline elution (Zwelling *et al.*, 1978), has confirmed the greater ability of *cis*-DDP as compared with *trans*-DDP to crosslink cellular DNA. These investigators made the further interesting observation that incubation of treated mouse leukaemic L1210 cells in a drug-free medium resulted in an increase in the number of DNA crosslinks. Crosslinking effects developed, following treatment with concentrations as low as 1 μm for *cis*- and 5 μm for *trans*-DDP which permitted over 80% survival of colony-forming ability. The maximum crosslinking effect by *cis*-DDP required about 12 hr post-treatment incubation before it was fully developed, whereas the crosslinking effect of *trans*-DDP was fully developed by 6 hr after exposure to the drug. The crosslinking effects of both agents were reversed upon further incubation of the cells, presumably due to the operation of a DNA excision repair process.

A further study employing alkaline elution showed that the crosslinking effect produced by *cis*- and *trans*-DDP could be separated into two components, one proteinase-sensitive and due to DNA-protein crosslinking, another proteinase-resistant and due to DNA interstrand crosslinking (Zwelling *et al.*, 1979*b*). DNA-protein crosslinks were at maximum levels immediately after drug removal, while DNA–DNA interstrand crosslinks reached maximum levels 6–12 hrs after drug removal (figure 7). Toxicity of the two agents in L1210 leukaemia cells, and V79 Chinese hamster cells, correlated well with DNA interstrand crosslinking, but not with DNA–protein crosslinking (Zwelling *et al.*, 1979*a*,*b*).

Because the biophysical basis of the elution technique is poorly understood, quantitation of interstrand crosslinking with this technique is based upon largely unproven assumptions. Therefore studies were undertaken to compare elution results with the alkaline caesium chloride technique described above, alkaline sucrose sedimentation, and estimation of renaturable DNA in cell lysates. Direct quantitation of crosslink frequency following cisplatin treatment was thus obtained over a wide dose range using methods based upon known biophysical properties of DNA. The results showed clearly that alkaline elution following proteinase digestion gave an accurate measure of interstrand crosslinking, and showed also that crosslinks were only a small fraction of the total drug–DNA reaction products (Roberts and Friedlos, 1981*a*; Pera *et al.*, 1981*b*) (figure 8).

It has been shown that cells can be protected from the toxic effect of *cis*-DDP by preventing the formation of DNA crosslinks by incubating cells in the presence of thiourea immediately after treatment (Zwelling *et al.*, 1979*c*) a finding which further supports a cytotoxic role for DNA interstrand crosslinks. Further investigations using mouse leukaemia cells and human fibroblasts having varying sensitivity to *cis*-DDP have shown that cellular sensitivity often correlates with

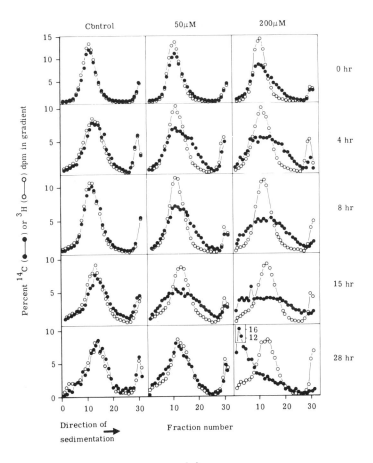

(a)

Figure 8a

interstrand crosslink formation (Zwelling *et al.*, 1981; Laurent *et al.*, 1981; Erickson *et al.*, 1981). However, DNA–protein crosslinking also correlated with cytotoxicity in most of these studies. It is possible then that sensitivity was simply a function of intracellular concentration of active drug metabolites, rather than the extent of formation of a specific DNA lesion. In one study, DNA–DNA crosslinking, but not DNA–protein crosslinking, was shown to correlate with *cis*-DDP toxicity towards normal and transformed human fibroblasts (Erickson *et al.*, 1981) (figure 9). The authors argued that DNA–protein crosslinking reflected the overall level of active drug metabolites in the cell. Sensitivity, then, was not related to drug uptake, but specifically to interstrand crosslinking. The assumption that DNA–protein crosslinking invariably gives an accurate reflection of uptake of active forms of the drug is open to question. Additional data on the overall reaction of drug with DNA would strengthen these arguments.

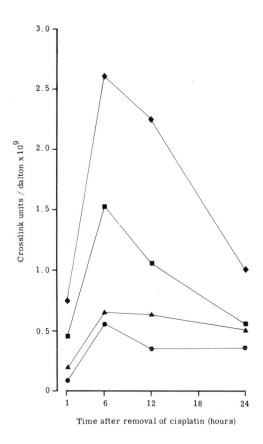

(b)

Figure 8b

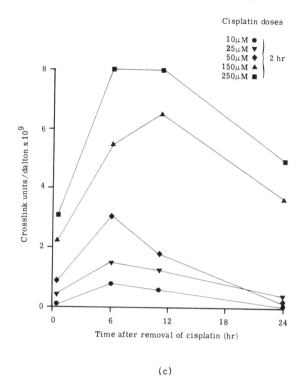

(c)

Figure 8   Time course of formation and disappearance of DNA interstrand crosslinks in Chinese hamster V79 cells measured by three independent and sensitive techniques. In alkaline sucrose gradients (a), ^{14}C-labelled DNA from treated cells shows a shift in sedimentation profile towards the high molecular weight end of the gradient; the effect was maximum at 6 hr and thereafter was reversed, with some evidence of DNA degradation at later time points. Estimation of crosslinking by measurement of renaturing DNA in cell lysates (b) showed a similar time course (cis-DDP in doses: •, 10 μm; ▲, 20 μm; ■, 50 μm; ♦, 100 μm). Measurement of crosslinking by alkaline elution (c) showed the same pattern; quantitation of the number of crosslinks was similar to that seen in the renaturing DNA assay (Pera *et al.*, 1981*a*).

Studies of certain mouse leukaemia L1210 lines resistant to *cis*-DDP (Zwelling *et al.*, 1981; Strandberg, 1981) as well as studies of Walker carcinoma cells (C. J. Rawlings and J. J. Roberts, unpublished) have indicated that there is not always a simple correlation between crosslink formation and cell kill. Indeed, a study by Shooter *et al.* (1972) on the inactivation of bacteriophage indicated that DNA interstrand crosslinking reactions are unlikely to be important cytotoxic events in this system. These authors compared *cis*- and *trans*-DDP and *cis*-dichloro(ethylenediamine)platinum(II) for their ability to inactivate T7 bacteriophage. They also carried out parallel studies on the extent of interstrand crosslinking of phage DNA with all agents and the extent of overall reaction of the ^{14}C-labelled ethylenediamine derivative with double stranded T7 phage DNA. It was thus possible to calculate that at the dose of labelled drug that reduced

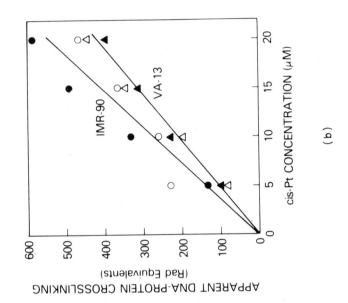

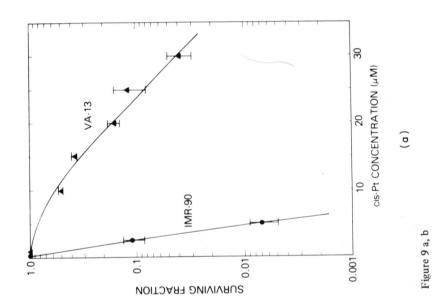

Figure 9 a, b

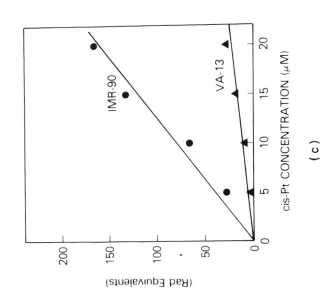

Figure 9   Correlation of DNA–DNA interstrand crosslinking, but not DNA–protein cross-linking with cisplatin toxicity to human fibroblasts. (a) Shows cell survival (colony forming ability) for two human fibroblast lines against cisplatin dose. The IMR-90 line is clearly more sensitive. (b) Shows that for a given dose of cisplatin, DNA–protein crosslinking was similar in the two cell lines. However, in c, it is apparent that *cis*-DDP treatment produced more DNA interstrand crosslinks in the more sensitive IMR-90 cell strain (Erickson *et al.*, 1981).

survival to 37% (that is, the mean lethal dose, D_0) there are five molecules bound to each T7 phage particle. However, it could be shown that at a dose which induces on average one crosslink into the phage DNA, there were 35 molecules of platinum drug bound to the phage. Crosslinking is, therefore, a relatively rare event under these conditions of treatment. Moreover, since there were only five platination reactions with the phage particle at the mean lethal dose, four of which were with the nucleic acid, it follows that the crosslinking reaction could not be an inactivating event. In other words, when one has a level of reaction which would inactivate all phages, only a small proportion of them would contain a crosslink in their DNA.

A further indication of the lack of importance of a DNA interstrand crosslink as an inactivating event for bacteriophage came from the finding that while the two *cis* platinum compounds were appreciably more effective than the *trans* isomer in activating the bacteriophage, nevertheless, all three compounds crosslink DNA with approximately equal efficiency. Filipski *et al.* (1980) confirmed and extended the findings of Shooter *et al.* (1972). Again it was shown that only one crosslink was produced by *cis*-DDP at doses yielding 5 inactivating events in each phage. To determine the nature of the inactivating events, bacteriophage DNA was treated with *cis*- and *trans*-DDP, and the crosslinked and uncrosslinked fractions were separated. It was found that following *trans*-DDP treatment, uncrosslinked DNA was active in transfection, whereas the crosslinked fraction was inactivated. However, in the case of phage DNA treated with *cis*-DDP, both the crosslinked and uncrosslinked DNA were found to be inactivated. Thus, *cis*-DDP produces a DNA lesion within one strand of the double helix, possibly an intrastrand crosslink, that results in phage inactivation (figure 10).

To summarise, interstrand crosslinking often correlates with cytotoxicity, but certain exceptions exist. It is entirely possible that intrastrand crosslinking might provide an even better correlation, but it is not yet possible to measure this lesion in mammalian cells.

BIOCHEMICAL EFFECTS OF DRUG-DNA INTERACTION

Inhibition of DNA synthesis

The significance of the interaction of platinum compounds with cellular DNA is apparent from studies of the effects of drugs on macromolecular synthesis. Simultaneously and independently it was found that *cis*-DDP selectively and persistently inhibited the rate of DNA synthesis as compared with effects on RNA and protein synthesis in human AV3 cells in culture (Harder and Rosenberg, 1970) and in Ehrlich ascites cells *in vivo* (Howle and Gale, 1970). These observations were confirmed in HeLa cells in culture and extended to show that such selective inhibition of DNA synthesis occurs with low doses of drugs which produced only minimal cytotoxicity as measured by effects on colony-forming ability (Pascoe and Roberts, 1974*a,b*). Moreover, those compounds which had

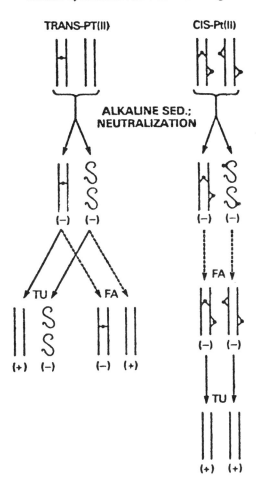

Figure 10 A model to account for phage inactivation by *cis* and *trans*-DDP, showing the postulated role of intrastrand crosslinks in the action of the *cis* isomer. Phage DNA was treated with *cis* or *trans* DDP. The DNA was then separated into crosslinked and uncrosslinked fractions in alkaline sucrose gradients on the basis of size. Thereafter the DNA was neutralised. Crosslinked DNA produced by *cis* or *trans* DDP treatment renatured, but was inactive on transfection, as was single-stranded DNA. (− indicates no activity in the transfection assay; + indicates activity). When the single-stranded DNA fractions were reannealed in formamide (FA), uncrosslinked *trans*-DDP-treated DNA regained activity. The uncrosslinked, *cis*-DDP-treated DNA, however, did not regain activity following reannealing. Thiourea (TU) treatment, which removes platinum from DNA, could restore transfection ability to DNA containing *trans*-DDP induced interstrand crosslinks, or DNA containing *cis*-DDP single strand adducts. The results are consistent with the interpretation that *cis*-, but not *trans*-DDP produces a lesion in one strand of the double helix, possibly an intrastrand crosslink, that results in phage inactivation (Filipski *et al.*, 1980).

been found to be effective against sarcoma 180 and which also caused filament formation, displayed similar inhibiting effects on DNA synthesis whereas the inactive compounds showed no effects until very high dose levels were employed (Harder and Rosenberg, 1970). Selective inhibition of DNA synthesis has also been demonstrated in phytohaemaglutinin-stimulated human peripheral lymphocytes (Howle *et al.*, 1971), in the folate-stimulated kidney, and intestinal mucosa of normal- and tumour-bearing rats (Taylor *et al.*, 1976).

The likely basis for this selective biochemical effect on DNA synthesis came from the observation that the inhibition of DNA synthesis was persistent and progressive with time after removal of the drug. It is now clear, particularly by comparison with analogous effects produced by direct reacting agents such as mustard gas (Roberts *et al.*, 1971), that both effects are consistent with the view that the primary chemical lesion is in the DNA of the cell, which is then inhibited as a template for DNA replicaton. Thus modifications to the DNA template will block DNA replication but not affect transcription or translation. Under conditions of low cell killing the selective inhibition by platinum compounds of DNA synthesis but not of RNA or protein synthesis, leads to the formation of giant cells, a feature observed in cells treated with a variety of agents also known to block DNA replication selectively.

It should be noted that a depression in the rate of incorporaton of thymidine into DNA need not necessarily result directly from inhibition of DNA synthesis. Heinen and Bassleer (1976) have questioned the interpretation of the results of Howle and Gale (1970) discussed above. The former investigators used quantitative cytochemical methods to show that mouse Ehrlich cells treated with *cis*-DDP *in vivo* ceased division with a G2 content of DNA; RNA and protein synthesis had continued and the cells were grossly enlarged. The depression in DNA synthesis observed earlier (Howle and Gale, 1970) was attributed to a G2 block. However, the studies of Heinen and Bassleer (1976) were performed six days after drug treatment. In contrast, studies of synchronised V79 Chinese hamster cells treated in G1 with *cis*-DDP showed that depression of DNA synthesis in these cells was the result of a decrease in DNA synthetic rate, rather than a decreased rate of entry of cells into S (Fraval and Roberts, 1978*a*). As a result, S phase was prolonged in these cells (figure 11). Following G1 treatment with *cis*-DDP, synchronised HeLa cells also showed a decrease in the amount of thymidine incorporation into DNA during S phase, but the effect was not immediately manifested in these cells (Fraval and Roberts, 1978*b*) (figure 12). Thus cells differ in the way in which their replication machinery responds to *cis*-DDP-induced damage. Such differences might account for variations in cellular sensitivity (see below) and further studies in this area are warranted.

Johnson *et al*. (1978) used an *in vitro* T7 DNA replication system which copies exogenous T7 DNA by a mechanism that closely mimics *in vivo* DNA replication to demonstrate the relative inhibitory effect of either pyrimidine dimers or bound ^{195m}Pt-labelled *cis*- or *trans*-DDP. It could be shown that *cis*-DDP and pyrimidine dimers inhibited DNA replication to the same extent per

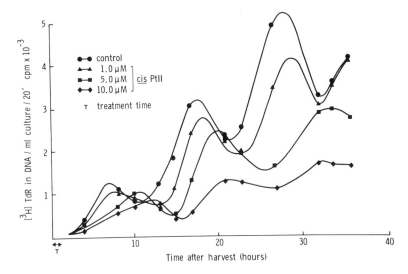

Figure 11 Effects of *cis*-DDP treatment in G1 phase on DNA synthesis in synchronised Chinese hamster V79 cells (Fraval and Roberts, 1978*a*).

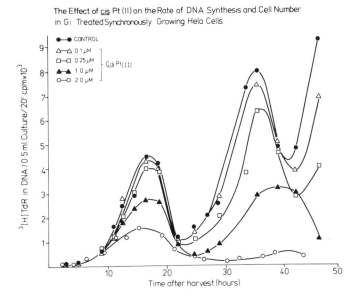

Figure 12 Effects of *cis*-DDP treatment in G1 phase on DNA synthesis in synchronised HeLa cells (Fraval and Roberts, 1978*b*).

lesion (63% inhibition per 3×10^{-4} lesions per nucleotide phosphate) and the *trans*-DDP was fivefold less inhibitory. RNA polymerase is similarly inhibited by the binding of both *cis*- and *trans*-DDP to DNA but the *cis* isomer inhibits much more strongly (Srivasta *et al*., 1978). The reaction of either platinum complex with the enzyme, on the other hand, had no effect on transcription. The chain length of the RNA produced through the RNA polymerase reaction decreases as the amount of platinum available for DNA binding increases. The effect was presumed to be due to platinum atoms on the DNA template acting to stop chain elongation.

The alternative possibility that DNA synthesis is inhibited because of the inactivation of enzymes involved in DNA replication seems unlikely, not only because of the failure of *cis*-DDP to block protein synthesis, but also because of its failure to inactivate DNA polymerase *in vitro* except with very high concentrations (Harder *et al*., 1976). Other studies on the reversible interaction of *cis* and *trans* platinum coordination complexes with various enzymes also make it unlikely that inhibition of enzymes is involved in the mechanism of action of these agents (Teggins and Friedman, 1974; Friedman and Teggins, 1974; Srivasta *et al*., 1978). The extent of binding to the enzyme by either *cis* or *trans* isomer was more dependent on the enzyme than on the configuration of the complex.

When platinum dichloroethylenediamine-modified calf thymus DNA was used as a substrate for DNA or RNA polymerase, results at variance with those found *in vivo* were obtained, since both RNA and DNA synthesis were reduced similarly (Giraldi and Taylor, 1974). On the other hand, an analogous study using salmon sperm DNA treated with *cis*- and *trans*-DDP clearly suggested that the basis for the selectivity of the *cis* isomer may reside in its ability to react in a specific defined molecular configuration with DNA (Harder *et al*., 1976).

Relationship between DNA replication on a damaged template and cytotoxicity

Further evidence relating drug induced alterations in DNA synthesis to cytotoxicity in bacteria and mammalian cells comes from studies of how cells cope with DNA template damage. Studies of excision repair of platinum induced damage, discussed below, indicate that the majority of DNA–platinum products, involving one strand of a double helix, are chemically stable and are only slowly removed from DNA by an enzymatic process. Thus it could be supposed that persistent lesions in DNA are circumvented during DNA replication by a process analogous to that which facilitates the survival of excision defective bacteria after UV-irradiation. Evidence has been adduced that in bacteria, gaps are initially left in the daughter DNA opposite thymine dimers in the template strand of DNA; these gaps are subsequently filled by a process involving recombination, which is controlled by the *rec* genes, and/or *de novo* DNA synthesis, which is probably controlled by the *exr*, (now called *lex*) gene. It was found by Drobnik *et al*. (1975) that mutation of Exr[+] to Exr[−] resulted in an increase of 3–6 times in the sensitivity of *E.coli* towards UV and *N*-methyl-*N'*-nitroso-*N*-nitroguanidine

(MNNG) and less than a twofold increase in the sensitivity to the X-irradiation. On the other hand, the increase in sensitivity of colony forming ability to *cis*-DDP due to this mutation is 13–23 times. Beck and Brubaker (1973) reached a similar conclusion with regard to the important role of this particular repair pathway from a comparison of the effect of treatment with UV light and *cis*-DDP on cell survival and filamentation in recombination repair deficient mutants of *E. coli* K-12. The recombination repair deficient mutant *rec* A13 and the double mutant *uvr*A6 lexl, which is known to undergo extensive autodegradation of DNA after treatment with UV-irradiation, were particularly sensitive to *cis*-DDP.

It now seems certain that mammalian cells also possess varying capacities to replicate their DNA on a template containing unexcised damage. It is also apparent, however, that the mechanism of any such repair process differs from that which is thought to occur in bacteria. Newly-synthesised DNA in some UV-irradiated and chemically-treated mammalian cells is initially smaller than that in control cells and it has been proposed that gaps are left in new DNA opposite lesions.

On the other hand, other evidence suggests that gaps are not formed in newly synthesised DNA but that there is simply a delay in the rate of synthesis at the site of each lesion. It has further been proposed that the lesion can be circumvented during DNA replication by a mechanism involving strand displacement. However, irrespective of the mechanism involved in synthesising past radiation or chemically-induced lesions in DNA, it has been found that in some cells the process is amenable to inhibition by the trimethylxanthine, caffeine. Thus it has been shown that the rate of elongation of newly synthesised DNA in UV-irradiated or chemically treated cells, was dramatically impaired in the presence of caffeine. As a consequence of this inhibition, many cell lines, competent in this replicative by-pass, are rendered extremely sensitive to the lethal effects of these agents by post treatment incubation in the presence of non-toxic concentrations of caffeine (Roberts, 1978).

There is now ample evidence to indicate that UV- or X-irradiation or chemically-induced cell death is a function of the amount of chromosome damage which can be observed at the first or second mitosis after treatment. Post treatment incubation in the presence of caffeine enhances dramatically the chromosome damaging effects of UV-irradiation and chemicals in both plant and animal cells.

The various cellular effects of *cis*-DDP and their modifications by caffeine, suggest that lesions are introduced onto DNA by platinum compounds which are also circumvented by a caffeine sensitive process (van den Berg and Roberts, 1975*a*).

Post-treatment incubation of cells in medium containing 0.75 mM caffeine dramatically potentiated the toxicity of *cis*-DDP and increased the number of cells containing chromosome damage. Caffeine not only increased the number of *cis*-DDP treated cells containing chromosomal aberrations but it also enhanced the severity of the damage observed. The delayed appearance of chromosome abnormalities after *cis*-DDP treatment also suggests that DNA replication is necessary for their formation, and in this respect *cis*-DDP resembles UV-irradiation and alkylating agents rather than X-irradiation (van den Berg and Roberts, 1975*a*).

The proposal has therefore been made that inadequate replication of DNA on a DNA damaged template is responsible for both cell death and chromosome damage and that post-treatment incubation of cells in medium containing caffeine, enhances these two effects of DNA damage by inhibiting a process which would permit replication to proceed past lesions. Support for this notion has come from studies on both the rate of DNA synthesis and the size of DNA synthesised in both asynchronous and synchronised populations of *cis*-DDP treated cells in the presence and absence of caffeine.

The dose-dependent depression in the rate of DNA synthesis in *cis*-DDP treated, asynchronously growing, Chinese hamster cells can be seen as a dose-dependent delay in the time of appearance of the peak rate of DNA synthesis (mid-S) in G1-treated, synchronously growing, Chinese hamster cells (Fraval and Roberts, 1978*a*). As a consequence of this dose-dependent extension in the time for passage through the S-phase, cells were correspondingly delayed in the time at which they underwent cell division.

It was found that post-treatment incubation in medium containing a non-toxic concentration of caffeine, reversed the *cis*-DDP induced inhibition of DNA synthesis in asynchronous populations of cells (van den Berg and Roberts, 1976), while post treatment incubation in the presence of caffeine of G1-treated synchronous cells leads to a reversal of the *cis*-DDP induced delay in the peak rate of DNA synthesis. Under these conditions of *cis*-DDP and caffeine treatment, the peak rate of DNA synthesis now approximates in time of appearance to that in the control cells (Fraval and Roberts, 1978*a*) (figure 13).

The immediate, dose-dependent, selective and persistent inhibition of DNA synthesis induced in *cis*-DDP treated Chinese hamster cells as measured by the decreased uptake of (^{3}H)TdR into DNA, as discussed above, can also be visualised as a dose-dependent decrease in the size of pulse-labelled, newly-synthesised DNA in treated cells (van den Berg and Roberts, 1976). If, however, compensation is made for the reduction in the rate of DNA synthesis by increasing the labelling period in *cis*-DDP treated cells, then the alkaline sucrose gradient sedimentation profile of labelled DNA in treated cells is very similar to that of DNA in control, untreated cells. It was concluded from such studies that the replicating machinery is delayed at the site of platinum-induced lesions in the template strand, but, given sufficient time, it can circumvent the lesions without forming discontinuities (gaps) in the newly-synthesised DNA. Alternatively, if gaps are first formed opposite platinum reaction sites in DNA, then they must be rapidly filled and are too transitory for detection.

The size of newly synthesised DNA in *cis*-DDP treated cells may be contrasted with the size of such DNA in cells treated similarly with *cis*-DDP and labelled with (^{3}H)TdR in the presence of non-toxic concentrations of caffeine. Under

Effect of cis Pt II with or without caffeine on DNA synthesis in chinese hamster cells.

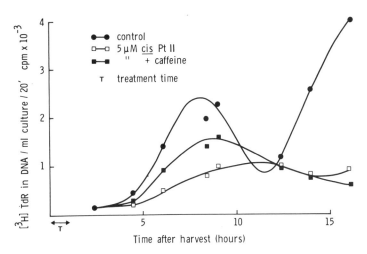

Figure 13 Effects of *cis*-DDP treatment in G1 phase on DNA synthesis in synchronised Chinese hamster V79 cells, with or without post-treatment incubation in 0.75 mM caffeine (Fraval and Roberts, 1978*a*).

these conditions the size of nascent DNA was markedly reduced as compared with that in untreated control or *cis*-DDP only treated cells (figure 14). The decrease in size of DNA was not the result of a decrease in overall rate of DNA synthesis since, as indicated above, the rate of DNA synthesis in *cis*-DDP treated cells is faster in the presence of caffeine than in its absence. The size of the DNA synthesised during 4 hr in the presence of caffeine in *cis*-DDP treated cells was dependent on the initial dose of DDP. It thus appears that caffeine interferes with the mechanism by which the cell replicates its DNA past lesions on the DNA template. Some support for this suggestion was obtained from a comparison of the distance between platinum-induced lesions on the template strand of DNA and the size of the newly-synthesised DNA in cells treated with various doses of *cis*-DDP and postincubated in the presence of caffeine. The distance between platinum atoms on one strand of DNA was calculated from atomic absorption measurements of the platinum bound to DNA isolated from *cis*-DDP treated cells and this was found to correspond closely to the size of the newly-synthesised DNA (table 5). It was concluded, therefore, that in Chinese hamster cells all unexcised platination reactions are normally circumvented during DNA replication by a caffeine-sensitive process. A model depicting the effects of platinum adducts on DNA synthesis in the absence and presence of caffeine is shown in figure 15.

It is significant that in the absence of caffeine, the process of DNA elongation in V79 cells treated with minimally toxic doses of *cis*-DDP eventually produced DNA molecules of molecular weight greater than 5×10^8. Nonetheless, a signifi-

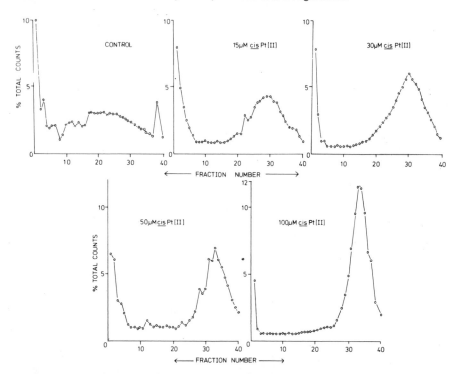

Figure 14 Dose-dependent reduction in the size of nascent DNA synthesised in *cis*-DDP treated Chinese hamster V79 cells in the presence of caffeine. Alkaline sucrose gradient sedimentation profiles of DNA synthesised during a 2-hr period in the presence of 0.75 mM caffeine in cells treated for 2 hr with *cis*-DDP are shown. Sedimentation analysis was performed after a 30 min incubation period in the absence of labelled precursor, but in the presence of 0.75 mM caffeine. Sedimentation (7,000 r.p.m./16 hr/20°C) is from right to left and gradient conditions are such that template-sized DNA sedimented predominantly to the bottom of the tube (van den Berg and Roberts, 1978).

Table 5 Relationship between dose of cisplatin, level of binding to DNA, and molecular weight of DNA synthesised in the presence of 0.75 mM caffeine and variable molecular weights for DNA.

Dose of *cis*-DDP (μM)	Binding to DNA (μmol/g)	Calculated spacing between DNA platinations (daltons x 10^{-7}*)	MW of DNA synthesised (daltons) x 10^{-7}
30	0.040	5	3.5
50	0.076	2.6	1.85
100	0.111	1.8	0.95

Cis-DDP was given at the doses indicated in Chinese hamster V79 cells (van den Berg and Roberts, 1976).

*Calculation based on DNA of infinite length.

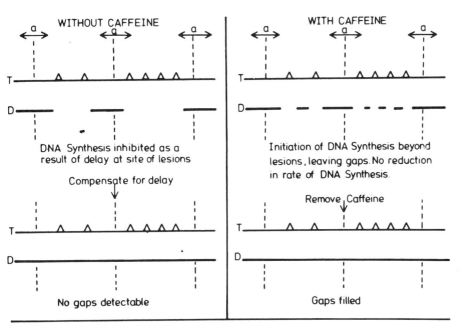

Figure 15 A model depicting a possible mechanism for caffeine potentiation of *cis*-DDP toxicity in Chinese hamster V79 cells. T, template DNA; D, daughter strand. The central feature of the model is that in the absence of caffeine, DNA synthesis results in production of full molecular weight DNA. In the presence of caffeine, synthesis is not delayed at the lesions, but gaps are left. (van den Berg and Roberts, 1976).

cant fraction of cells died. Probably a simple delay in DNA synthesis is insufficient to account for cell killing. It may be that this newly synthesised DNA is not properly integrated into structures of larger size, for chromosomal breakage is clearly manifested under these conditions. This is an area which requires further exploration.

EXCISION REPAIR OF PLATINUM-INDUCED DNA DAMAGE

Bacterial studies

The main photoproduct in the DNA of UV-irradiated *E. coli* or mammalian cells is a pyrimidine dimer which can be recognised either by an endonuclease which inserts a nick in the DNA adjacent to the dimer, or by a combination of a glycosylase that removes one of the thymine residues and an apurinic acid endonuclease, which subsequently makes an incision in the DNA backbone at the apyrimidinic site. Subsequent steps in this excision repair process involve removal of the dimer attached to an oligonucleotide and resynthesis of the removed section of the DNA. The contribution which this excision repair process makes to the ultimate survival of bacterial cells treated with various agents can be assessed by determining the sensitivities of strains of *E. coli* carrying mutations in genes known to code for steps in this repair pathway. It was concluded from such studies that damage introduced into DNA by certain bifunctional agents, such as nitrogen mustard, mitomycin C and psoralen-plus-visible light, as well as the damage introduced by certain bulky monofunctional carcinogenic agents such as 4-nitroquinoline-1-oxide and 7-bromomethylbenz(*a*)anthracene, was also eliminated by enzymes encoded by genes known to code for the enzymes required for the repair of UV-induced thymine dimers. From analogous studies on the sensitivities of such DNA repair deficient *E. coli* mutants to *cis*-DDP, it was similarly concluded that the excision repair process(es) contribute to only a small extent to the recovery of strains of *E. coli* from the DNA damaging effects of this agent (Drobnik *et al.*, 1973; Beck and Brubaker, 1973). Some kinds of damage to the DNA of T-odd bacteriophage can be repaired by the enzymatic excision repair system of the host bacterium (host-cell repair). The role played by the *hcr* locus is important for the survival of cells after UV-irradiation but not after treatment with MNNG and only minimally so after treatment with *cis*-DDP. The relatively minor importance of the *hcr* locus for the inactivation of *E. coli* by *cis*-DDP was confirmed by the observation that the bacteriophages T3 or T4 BOL after treatment with *cis*-DDP give the same inactivation curves in both hcr^+ and hcr^- strains of the indicator bacteria (Drobnik *et al.*, 1975). Similarly it was found that platinum-treated transforming DNA does not appear to be more sensitive when assayed in a strain of *Haemophilus influenzae* which carried the *uvrl* mutation and hence lacks the activity of the specific UV endonuclease which incises UV-irradiated DNA (Munchausen, 1974). Thus, in bacteria, mutations in known excision-repair pathways do not affect cellular response to the platinum drug.

Mammalian cell studies

In an early experiment, alkaline sucrose gradient sedimentation of prelabelled cellular DNA following treatment of cells with *cis*-DDP did not reveal any accumulation of low molecular weight DNA (van den Berg and Roberts, 1975*b*). From a knowledge of the extent of reaction of the platinum compound with DNA at the concentration employed, it could be concluded, either that lesions were not generally recognised by an endonuclease which inserted 'nicks' into DNA, or, alternatively, if the lesions were recognised by an endonuclease, then completion of the later stages of the excision repair process(es) led to the rapid restoration of high molecular weight DNA. Moreover, since any apurinic sites in DNA would be converted into DNA single strand breaks under these alkaline conditions, there was no obvious evidence from these studies for the removal of substituted purines by means of an N-glycosylase although such sites might be rapidly repaired.

More recent investigations using alkaline elution (Zwelling *et al.*, 1979*b*) or a combination of alkaline elution, alkaline sucrose sedimentation and DNA renaturation, (Pera *et al.*, 1981*a*) have clearly demonstrated repair of DNA–protein and DNA–DNA interstrand crosslinks in cultured mammalian cells following *cis*-DDP treatment (figure 8). Interstrand crosslinking in Chinese hamster V79 cells was demonstrated by a decrease in the rate of filter elution of DNA from X-irradiated, treated cells, a shift in alkaline sucrose gradients of DNA molecules towards the high molecular weight end of the gradient, or an increase in the rapidly renaturing fraction of DNA in cell lysates. All of these drug-induced phenomena reached a maximum at 6–12 hr after drug treatment, then declined. The half-life of DNA interstrand crosslinks usually appeared to be between 12 and 24 hr. Some DNA degradation occurred, but it was insufficient to account for crosslink reversal. Because crosslinks induced by platinum are stable in isolated DNA under physiological conditions, this reversal may be attributed to repair, though the mechanism remains unknown.

Fraval and Roberts (1979) demonstrated removal of platinum adducts from DNA of exponentially growing Chinese hamster V79 cells. The half life of total drug–DNA reaction products was approximately 28 hrs. Again, such products are stable chemically under physiological conditions. Thus removal of the DNA adducts could be attributed to repair.

Relationship between DNA excision repair and cytotoxicity

Although correlations have been established in some cases between the extent of drug–DNA interaction and cytotoxicity, the relationship between excision repair of DNA damage and toxicity is not always clear.

The rare skin condition, xeroderma pigmentosum (XP), is characterised by extreme sensitivity to sunlight and a predisposition to skin cancer. Cells taken from persons suffering from this condition are more sensitive to UV-irradiation than normal cells and are deficient in excision repair of UV-induced damage. These same cells are also sensitive to other DNA damaging agents such as hydrocarbon epoxides, 4-nitroquinoline-1-oxide and 7-bromomethylbenz(*a*)anthracene and sensitivity is again associated with decreased levels of various manifestations of DNA excision repair. (For review see Roberts, 1978). It has now been found (Fraval *et al.*, 1978) that these repair deficient XP cells are also more sensitive than normal fetal lung cells to *cis*-DDP, when the lethal effects of the drug are expressed as a function of reaction with DNA rather than as a function of dose of reagent. It could therefore be reasoned that this increased sensitivity of XP cells is similarly due to their decreased ability to excise *cis*-DDP induced DNA damage.

Cells from patients with the genetic disease Fanconi's anaemia, show unusual sensitivity to the cytotoxic and clastogenic effects of difunctional alkylating agents. These cells are also unusually sensitive to *cis*-DDP (M. F. Pera and J. J. Roberts, unpublished). Such sensitivity is not the result of decreased binding of platinum to DNA, but it remains to be determined if repair of various lesions in these cells is abnormal.

If DNA synthesis on a damaged template is responsible for toxicity, cells that are allowed to repair DNA prior to S-phase and cell division should show less toxicity than cells entering the proliferative cycle immediately following treatment. In an initial study, Fraval and Roberts (1979) treated stationary Chinese hamster V79 cells with *cis*-DDP and measured toxicity and platinum-DNA interaction after various holding periods in the nondividing state. The cells slowly excised platinum, and plating efficiency increased. There was a similar relationship between the amount of platinum bound to DNA and cell survival whether one compared the two immediately following treatment with several drug doses, or after varying periods of recovery.

However, the cells used in these initial studies, Chinese hamster V79 cells, do not tolerate the nondividing state well, and it was later found that they do not recover from *cis*-DDP toxicity under some conditions. Therefore, the experiment was repeated using human fibroblasts (Pera *et al.*, 1981*b*) under conditions where minimal DNA synthesis occurred. These cells were healthy in the non-dividing condition. The fibroblasts recovered from *cis*-DDP toxicity if held in the nondividing state, and they excised platinum lesions from DNA by a first order process with a half life of 2.5 days (figure 16). DNA–DNA interstrand crosslinks, measured by alkaline elution and by estimation of renaturable DNA in cell lysates, were found to be repaired with a half life of about 36 hr (figure 17). DNA protein crosslinks were removed at a similar rate and there was no evidence for any accompanying degradation. Thus the reversal of crosslinking was attributed to repair rather than introduction of DNA strand breaks. The relationship between cell survival and the amounts of platinum remaining bound to DNA at the time cells were plated out for estimation of cell survival, was similar to that observed in cells treated with several drug doses and plated immediately. The results strongly supported the hypothesis that damage present on the DNA template at the time of entry into the proliferative cycles was responsible for cellular toxicity. Further, the results showed that the repair process was actually effective in achieving biological recovery.

When, however, attempts have been made to relate inherent drug sensitivity to repair capacity, clearcut results have not always emerged. Walker 256 carcinoma cells display unusual sensitivity to difunctional alkylating agents and to *cis*-DDP. There is a subline of this tumour that shows a 30-fold increase in resistance to *cis*-DDP. Nevertheless, the two sublines bind the same amount of platinum on DNA following treatment with a given dosage, they excise platinum lesions from their DNA at a similar rate, and they remove DNA-protein crosslinks and DNA interstrand crosslinks from DNA at a similar rate (C. J. Rawlings and J. J. Roberts, unpublished). Following sulphur mustard treatment, nascent DNA synthesis on a damaged template appears to be similar in both lines (J. J. Roberts and F. Friedlos, unpublished). It is of course possible that these lines might vary in their capacity to remove a specific lesion that was not measured –

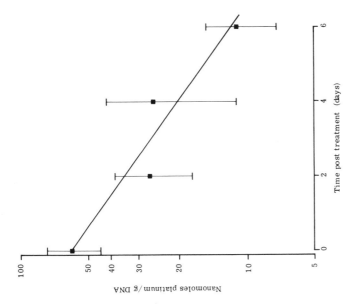

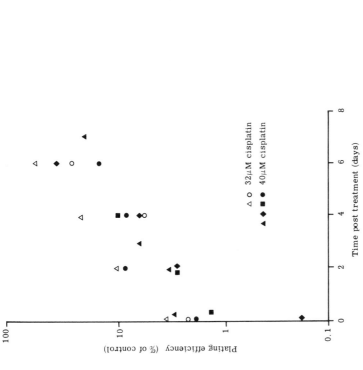

Figure 16 Recovery of non dividing human fibroblasts from *cis*-DDP toxicity and loss of DNA bound platinum adducts. Stationary phase human fibroblasts were treated with *cis*-DDP and were either plated immediately or held in the non dividing state for various time periods. During holding in the non dividing state, the cells recovered from drug toxicity (a). Measurement of platinum bound to DNA after various holding periods in the non dividing condition showed first order loss of the platinum DNA reaction products (b). (Pera *et al.*, 1981*b*).

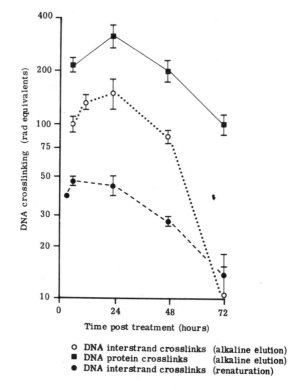

Figure 17   Repair of *cis*-DDP-induced DNA protein crosslinks and DNA interstrand cross-links in human fibroblasts during holding in the non-dividing state. Stationary cultures were treated with 40 μm *cis*-DDP, and crosslinking was measured by alkaline elution or estimation of renaturable DNA in cell lysates after various periods of holding in the non dividing state (Pera *et al.*, 1981*b*).

an intrastrand crosslink, for example – but if this is not the case, then some target other than simply DNA exists in certain cells that confers unusual sensitivity to bifunctional agents. This could conceivably be at the level of chromatin or a DNA-nuclear matrix complex.

CONCLUSIONS

This chapter has reviewed current evidence concerning the mechanism of action of platinum antitumour drugs. The chemistry of the active agents indicates that they are bifunctional electrophiles that will interact with nucleophilic centres in the cell. Many of their biological properties could result from interaction with the cellular genetic material. Alternative hypotheses have been advanced to account for the antitumour activities of these agents, but as yet, these hypotheses have generally not been examined as rigorously as those involving DNA–drug interaction. It is reasonable, therefore, to consider how strongly the experimental evidence to date supports the view that DNA is the target of drug action.

In vitro studies have revealed a number of interactions of platinum drugs with DNA, and recent work has stressed the importance of intrastrand crosslinking between adjacent guanine residues, a reaction that occurs with *cis*, but not *trans*, platinum complexes. Thus progress has been made at the molecular level in accounting for the structural requirements for bifunctionality and *cis* geometry in terms of reaction with DNA. Actual isolation and chemical identification of the reaction products of platinum with DNA is required to clarify the interpretations of some of the more indirect experimental approaches used thus far.

In cells treated *in vitro*, the extent of reaction of platinum per molecule of RNA or protein would appear to be too low to account for cell killing. In the case of DNA, there is sufficient reaction at relevant doses to account for cytotoxicity. Of course, this argument assumes no specificity at all in the binding of drug to protein, and this assumption is difficult to prove. It is now clear that drug reaction with tumour or host cell DNA *in vivo* is similar to that observed in cultured cells at comparable levels of survival. Increased reaction of platinum with tumour DNA has been shown to be associated with improved antitumour selectivity of a *cis*-DDP analogue in one model. Thus, differences in the pharmacological properties of platinum congeners may be expected to result in improved therapy. However, at present, there is little indication that the mechanism of action of platinum analogues is markedly different from that of *cis*-DDP. In cultured cells, it is often the case that interstrand crosslinking reactions correlate with cytotoxicity. There are, however, enough exceptions to this generalisation to question the primacy of interstrand crosslinking as a cytotoxic event produced by *cis*-DDP. Clearly, measurement of interstrand crosslinking alone does not enable one to predict cytotoxic effects.

The biochemical effects of platinum drugs certainly are consistent with the DNA hypothesis. Selective inhibition of DNA synthesis generally occurs rapidly and persists in most cells studied thus far, though further investigation of this phenomenon is certainly warranted, particularly in cells of varying sensitivity to the drug. Further interference by caffeine with the production of full molecular weight DNA results in a dramatic potentiation of toxicity. What remains unclear is how drug interference with DNA elongation leads to cell death in the absence of caffeine, since DNA eventually attains a size greater than 5×10^8 daltons in treated cells at modest dosages, and such cells do complete a round of DNA synthesis. It would seem likely that the experimental protocols employed thus far are inadequately sensitive to follow the coordination and organisation of DNA synthesis, or the integration of nascent DNA fragments 5×10^8 daltons in size into larger structures.

Correlation of DNA excision repair phenomena with toxicity has been observed in some cases. The demonstrated increased sensitivity of cells from patients with xeroderma pigmentosum or Fanconi's anaemia should now be examined in regard to possible repair defects. Recovery studies, wherein treated cells held in a nondividing state are shown to repair DNA and recover from drug toxicity, argue strongly for the importance of drug–DNA interactions. Nonetheless clear examples relating tumour cell sensitivity and resistance to DNA repair phenomena have not yet been forthcoming.

Thus evidence pinpointing DNA as the target for the action of platinum antitumour drugs can still be considered inconclusive. If the hypothesis is correct, many factors will determine the outcome of drug treatment; uptake of active forms of the drug; the production of critical DNA lesions, the nature of which remain to be defined; the ability of the replication machinery to cope with the damage and synthesise new DNA, a process which our current techniques follow only at a limited level of sophistication; and, the capacity of the cell to excise the damage. Given the complexity of the problem, it is perhaps impressive that much evidence has been obtained to support the argument. We still have not explained why chemotherapy based on an agent like *cis*-DDP, which reacts in a relatively nondiscriminating fashion with macromolecules and presumably throughout all cells, can achieve cure of certain advanced, disseminated malignances.

REFERENCES

Aggarwal, S. K., Wagner, R. W., McAllister, P. K. and Rosenberg, B. (1975). *Proc. natn. Acad. Sci. U.S.A.*, **72**, 928

Bau, R. (1981). In *International Conference on the Chemistry of Platinum Group Metals.*, Royal Society of Chemistry, Bristol, p.5

Beck, D. J. and Brubaker, R. R. (1973). *J. Bact.*, **116**, 1247

Beck, D. J. and Brubaker, R. R. (1975). *Mutat. Res.*, **27**, 181

Berenbaum, M. C. (1971). *Br. J. Cancer.*, **25**, 208

Bowen, J. R., Gale, G. R., Gardner, W. A. and Bonner, M. Jr (1974). *Agents Actions.*, **4**, 108

Braddock, P. D., Connors, T. A., Jones, M., Khokhar, A. R., Melzack, D. H., and Tobe, M. L. (1975). *Chem.-biol. Interact.*, **11**, 145

Bradner, W. T., Rose, W. C., and Huftalen, J. B. (1980). In *Cisplatin: Current Status and New Developments* (ed. A. W. Prestayko, S. T. Crooke and S. K. Carter), Academic Press New York, p.171

Brambilla, G., Cavanna, M., and Maura, A. (1974). *Cancer Chemother. Rep. I.*, **58**, 633

Burchenal, J. H., Kalaher, K., O'Toole, T. and Chisholm, J. (1977). *Cancer Res.*, **37**, 3455

Chu, G. Y. H. and Tobias, R. J. (1976). *J. Am. chem. Soc.*, **98**, 2641

Cleare, M. J. and Hoeschele, J. D. (1973). *Platinum Metals Rev.*, **17**, 2

Cleare, M. J., Hydes, P. C., Hepburn, D. R. and Malerbi, B. W. (1980). In *Cisplatin: Current Status and New Developments.*, (ed. A. W. Prestayko, S. T. Crooke and S. K. Carter), Academic Press, New York, p.149

Cohen, G. L., Bauer, W. R., Barton, J. K. and Lippard, S. J. (1979). *Science.*, **203**, 1014

Cohen, G. L., Ledner, J. A., Bauer, W. R., Ushay, H. M., Caravana, C., and Lippard, S. J. (1980). *J. Am. chem. Soc.*, **102**, 2487

Connors, T. A., Jones, M., Ross, W. C. J., Braddock, P. D., Khokhar, A. R. and Tobe, M. L. (1972). *Chem.-biol. Interact.*, **5**, 415

Cramer, R. E. and Dahlstrom, P. L. (1977). *J. clin. Haemat. Oncol.*, **7**, 330

Drobnik, J., Urbankova, M. and Krekulova, A. (1973). *Mutat. Res.*, **17**, 13

Drobnik, J., Blahuskova, A., Vasilukova, S. and Krekulova, A. (1975). *Chem.-biol. Interact.*, **11**, 365

Einhorn, L. H. and Donohue, J. (1977). *Annls Intern. Med.*, **87**, 293

Einhorn, L. H. and Williams, S. D. (1980). *Cancer (Phila.).*, **46**, 1339

Erickson, L. C., Zwelling, L. A., DuCore, J. M., Sharkey, N. A. and Kohn, K. W. (1981). *Cancer Res.*, **41**, 2791

Filipski, J., Kohn, K. W. and Bonner, W. M. (1980). *Chem.-biol. Interact.*, **32**, 321

Fornace, A. J. Jr. and Little, J. B. (1980). *Carcinogenesis.*, **1**, 989

Fraval, H. N. A. and Roberts, J. J. (1978*a*). *Chem.-biol. Interact.*, **23**, 99

Fraval, H. N. A. and Roberts, J. J. (1978*b*). *Chem.-biol. Interact.*, **23**, 111

Fraval, H. N. A., Rawlings, C. J. and Roberts, J. J. (1978). *Mutat. Res.*, **51**, 121

Fraval, H. N. A. and Roberts, J. J. (1979). *Cancer Res.*, **39**, 1793

Friedman, M. E., and Teggins, J. E. (1974). *Biochim. biophys, Acta.*, **350**, 263

Ganguli, P. K. and Theophanides, T. (1979). *Eur. J. Biochem.*, **101**, 377

Gellert, R. W. and Bau, R. (1975). *J. Am. chem. Soc.*, **97**, 7379

Giraldi, T. and Taylor, D. M. (1974). *Biochem. Pharmac.*, **23**, 1659

Goodgame, D. M. L., Jeeves, I., Phillips, F. L. and Skapski, A. C. (1975). *Biochem. biophys. Acta.*, **378**, 153

Gottlieb, J. A. and Drewinko, B. (1975). *Cancer Chemother. Rep.*, **59**, 621

Harder, H. C. and Rosenberg, B. (1970). *Int. J. Cancer.*, **6**, 207

Harder, H. C. (1975). *Chem.-biol. Interact.*, **10**, 27

Harder, H. C., Smith, R. G. and Leroy, A. F. (1976). *Cancer Res.*, **36**, 3821

Harrap, K. R., Jones, M., Wilkinson, C. R. McD.Clink, H., Sparrow, S., Mitchley, B. C. V., Clarke, S. and Veasey, A. (1980). In *Cisplatin: Current Status and New Developments* (ed. A. W. Prestayko, S. J. Crooke and S. K. Carter), New York, Academic Press, p.193

Hayes, D. M., Cvitkovic, E., Golbey, R. B., Scheiner, E., Helson, L. and Krakoff, I. H. (1977). *Cancer (Phila.).*, **39**, 1372

Heinen, E. and Bassleer, R. (1976). *Biochem. Pharmac.*, **25**, 1871

Horacek, P. and Drobnik, J. (1971). *Biochem. biophys. Acta.*, **254**, 341

Howle, J. A. and Gale, G. R. (1970). *Biochem. Pharmac.*, **19**, 2757

Howle, J. A., Thompson, H. S., Stone, A. E. and Gale, G. R. (1971). *Proc. Soc. exp. Biol. Med.*, **137**, 820

Johnson, N. P., Hoeschele, J. D., Kuemmerle, N. B., Masker, W. E. and Rahn, R. O. (1978). *Chem.-biol. Interact.*, **23**, 267

Johnson, N. P., Hoeschele, J. D., Rahn, R. O., O'Neill, J. P. and Hsie, A. W. (1980). *Cancer Res.*, **40**, 1463

Kelman, A. D., Peresie, H. J. and Stone, P. J. (1977). *J. clin. Hemat. Oncol.*, **7**, 440

Kelman, A. D., and Buchbinder, M. (1978) *Biochemie.*, **60**, 893

Khan, A. and Hill, J. M. (1972). *Transplantation*, **13**, 55

Khan, A., Albayrak, A. and Hill, J. M. (1972). *Proc. Soc. exp. Biol., Med.*, **141**, 7

Khan, A. and Hill, J. M. (1973). *Proc. Soc. exp. Biol. Med.*, **142**, 324

Kleinerman, E. S., Zwelling, L. A. and Muchmore, A. V. (1980*a*) *Cancer Res.*, **40**, 3099

Kleinerman, E. S., Zwelling, L. A., Howser, D., Barlock, A., Young, R. C., Decker, J. M., Bull, J. and Muchmore, A. V. (1980*b*). *Lancet.*, **ii**, 1102

Kleinwachter, V. and Zaludova, R. (1977). *Chem-biol. Interact.*, **16**, 207

Kong, P-C. and Theophanides, T. (1974). *Inorg. Chem.*, **13**, 1167

Kutinova, L., Vonka, V. and Drobnik, J. (1972). *Neoplasma.*, **19**, 453

Laurent, G., Erickson, L. C., Sharkey, N. A. and Kohn, K. W. (1981). *Cancer Res.*, **41**, 3347

Lecointe, P., Macquet, J-P., Butour, J-L. and Paoletti, C. (1977). *Mutat. Res.*, **48**, 139

Leopold, W. R., Miller, E. C. and Miller, J. A. (1979). *Cancer Res.*, **39**, 913

Leopold, W. R., Batzinger, R. P., Miller, J. A., Miller, E. C. and Earhart, R. H. (1981). *Proc. Am. Ass. Cancer Res.*, **22**, 108

Lock, C. J. L., Bradford, J., Faggiani, R., Speranzini, R. A., Turner, G. and Zvagulis, M. (1977). *J. clin. Hemat. Oncol.*, **7**, 63

Loveless, A. (1969). *Nature, Lond.*, **223**, 206

Macquet, J-P. and Theophanides, T. (1975). *Bioinorganic Chem.*, **5**, 59

Macquet, J-P. and Butour, J-L. (1978). *Eur. J. Biochem.*, **83**, 375

Mally, M. B., Taylor, R. C. and Calewert, D. M. (1979). *Chemotherapy*, **25**, 117

Mansy, S., Rosenberg, R. and Thomson, A. J. (1973). *J. Am. chem. Soc.*, **95**, 1633

McAllister, P. K., Rosenberg, B., Aggarwal, S. K. and Wagner, R. W. (1977). *J. clin. Hemat. Oncol.*, **7**, 717

Millard, M. M., Macquet, J-P. and Theophanides, T. (1975). *Biochim. biophys. Acta.*, **402**, 166

Monti-Bragadin, C., Tamaro, M. and Banfi, E. (1975). *Chem.-biol. Interact.*, **11**, 469

Munchausen, L. L. (1974). *Proc. natn. Acad. Sci. (U.S.A.)*, 71, 4519

Munchausen, L. L. and Rahn, R. O. (1975). *Biochem. biophys. Acta.*, **414**, 242

Pascoe, J. M. and Roberts, J. J. (1974*a*). *Biochem. Pharmac.*, **23**, 1345

Pascoe, J. M. and Roberts, J. J. (1974*b*). *Biochem. Pharmac.*, **23**, 1359

Pera, M. F. Jr., Rawlings, C. J., Shackleton, J. and Roberts, J. J. (1981*a*). *Biochim. biophys. Acta.*, **655**, 152

Pera, M. F., Rawlings, C. J. and Roberts, J. J. (1981*b*). *Chem.-biol. Interact.*, 37, 245

Pera, M. F., Sessford, D. and Roberts, J. J. (1982). *Biochem. Pharmac.*, 31, 2273

Prestayko, A. W., D'Aoust, J. C., Issell, B. F. and Crooke, S. T. (1979). *Cancer Treat. Rev.*, 6, 17

Resolva, S. (1971). *Chem.-biol. Interact.*, 4, 66

Resolova, S., Srogl, M. and Drobnik, J. (1972). *Advances in Antimicrobial and Antineoplastic Chemotherapy.*, 2, 209

Robbins, A. B. (1973). *Chem.-biol. Interact.*, 6, 35

Roberts, J. J., Brent, T. P. and Crathorn, A. R. (1971). *Eur. J. Cancer*, 7, 515

Roberts, J. J., and Pascoe, J. M. (1972). *Nature, Lond.*, **235**, 282

Roberts, J. J. (1978). *Adv. Radiat. Biol.*, 7, 211

Roberts, J. J. and Fraval, H. N. A. (1978). *Biochimie*, 60, 869

Roberts, J. J. (1981). In *Molecular Actions and Targets for Cancer Chemotherapeutic Agents.*, (ed. A. C. Sartorelli, J. S. Lazo, and J. R. Bertino), Academic Press, New York, p.17

Roberts, J. J. and Friedlos, F. (1981*a*). *Biochim. biophys. Acta.*, **665**, 146

Roberts, J. J. and Friedlos, F. (1982). *Chem. biol. Interact.*, 39, 181

Roos, I. A. G., Thomson, A. J. and Mansy, S. (1974). *J. Am. Chem. Soc.*, **96**, 6484

Roos, I. A. G. (1977). *Chem.-biol. Interact.*, 16, 39

Rosenberg, B., Van Camp, L. and Krigas, T. (1965). *Nature, Lond.*, **205**, 698

Rosenberg, B., Van Camp, L., Grimley, E. G. and Thomson, A. J. (1967). *J. biol. Chem.*, **242**, 1347

Rosenberg, B., Van Camp, L., Trosko, J. E. and Mansour, V. H. (1969). *Nature, Lond.*, **222**, 385

Rosenberg, B. (1975). *Cancer Chemother.Rep.*, **59**, 589

Rosenberg, B. (1977). *J. clin. Hemat. Oncol.*, 7, 817

Rosenberg, B. (1981). In *International Conference on the Chemistry of Platinum Group Metals.*, Royal Society of Chemistry, Bristol, p.A1.

Shooter, K. V., Howse, R., Merrifield, R. K. and Robbins, A. B. (1972). *Chem.-biol. Interact.*, 5, 289

Srivasta, R. C., Froehlich, J. and Eichorn, G. L. (1978). *Biochimie*, 60, 879

Stone, P. J., Kelman, A. D. and Sinex, F. M. (1974). *Nature, Lond.*, **251**, 736

Stone, P. J., Kelman, A. D., Sinex, F. M., Bhargava, M. M. and Halvorson, H. O. (1976). *J. molec. Biol.*, **104**, 793

Strandberg, M. C. (1981). *Proc. Am. Ass. Cancer Res.*, **22**, 202

Szumiel, I., Niepokojczycka, E. and Godlewska, E. (1981). *Br. J. Cancer*, **43**, 116

Tamburio, A. M., Celotti, L., Furcan, D. and Guantieri, V. (1977). *Chem.-biol. Interact.*, 16, 1

Taylor, D. M., Tew, K. D. and Jones, J. D. (1976). *Eur. J. Cancer*, 12, 249

Teggins, J. E. and Friedman, M. E. (1974). *Biochim. biophys. Acta.*, **350**, 273

Thomson, A. J., Williams, R. J. P. and Reslova, S. (1972). *Struct. Bonding (Berlin).*, **11**, 1746

Thomson, A. J. (1974). In *Platinum Coordination Complexes in Cancer Chemotherapy.*, **48**, (ed. T. A. Connors and J. J. Roberts), Springer-Verlag, Berlin, p.38

Thomson, A. J., Roos, I. A. G. and Graham, R. D. (1977). *J. clin. Hemat. Oncol.*, 7, 242

Tisdale, M. J. and Phillips, B. J. (1975*a*). *Biochem. Pharmac.*, **24**, 1271

Tisdale, M. J. and Phillips, B. J. (1975*b*). *Biochem. Pharmac.*, **24**, 205

Tisdale, M. J. and Phillips, B. J. (1975*c*). *Biochem. Pharmac.*, **24**, 211

Tobias, R. S., Chu, G. Y. H. and Peresie, H. J. (1975). *J. Am. chem. Soc.*, 97, 5305

Trosko, J. E. (1974). In *Platinum Coordination Complexes in Cancer Chemotherapy.*, 48, (ed. T. A. Connors and J. J. Roberts), Springer-Verlag, Berlin, p. 108

Turnbull, D., Popescu, N. C., DiPaolo, J. A. and Myhr, B. C. (1979). *Mutat. Res.*, 66, 267

van den Berg, H. W. and Roberts, J. J. (1975*a*). *Mutat. Res.*, 33, 279

van den Berg, H. W. and Roberts, J. J. (1975*b*). *Chem.-biol. Interact.*, 11, 493

van den Berg, H. W. and Roberts, J. J. (1976). *Chem.-biol. Interact.*, 12, 375

Wierda, D. and Pazdernik, T. L. (1979). *Eur. J. Cancer.*, 15, 1013

Zwelling, L. A. Kohn, K. W., Ross, W. C., Ewig, R. A. G. and Anderson, T. (1978). *Cancer Res.*, 38, 1762

Zwelling, L. A., Bradley, M. O., Sharkey, N. A., Anderson, T. and Kohn, K. W. (1979*a*). *Mutat. Res.*, 67, 271

Zwelling, L. A., Anderson, T. and Kohn, K. W. (1979*b*). *Cancer Res.*, 39, 365

Zwelling, L. A., Filipski, J. and Kohn, K. W. (1979*c*). *Cancer Res.*, 39, 4989

Zwelling, L. A., Michaels, S., Schwartz, H., Dobson, P. O. and Kohn, K. W. (1981). *Cancer Res.*, 41, 640

8
Molecular Structure and Antitumour Activity of Alkylating Agents

Derry E. V. Wilman and Thomas A. Connors

CHEMISTRY OF THE ALKYLATING AGENTS

An alkylating agent is a compound capable of replacing a proton in another molecule by an alkyl radical. When the alkyl radical has the form RCH_2-, the overall reaction may be written as follows:

$$RCH_2-X + H-R' \longrightarrow RCH_2-R' + HX \tag{1}$$
$$\text{Alkylating agent} \qquad\qquad \text{Alkylated molecule}$$

Although R may be a complex, multifunctional species, and may even contain one or more aromatic rings, the final attachment of R to the substrate R'H must be through a saturated carbon atom. The alkylation reaction can occur by two extreme mechanisms.

One of these involves the initial formation of a carbenium ion, which then reacts rapidly with the substrate to give the observed alkylated product:

$$RCH_2-X \xrightarrow[\text{slow}]{} \overset{+}{R}CH_2 + X^- \xrightarrow[\text{fast}]{HR'} RCH_2-R' + H^+ \tag{2}$$

The reaction rate is dependent only on the concentration of the alkylating agent and therefore obeys first-order kinetics. This type of process is called a first-order nucleophilic substitution, or S_N1, reaction.

At the other extreme is the bimolecular nucleophilic substitution, or S_N2, reaction. This process obeys second-order kinetics and involves a transition state complex containing both reactants:

$$R^- - + \overset{\overset{R'}{|}}{C}H_2-X \longrightarrow \left[\overset{\overset{R'}{|}}{R^{\delta-}\ldots\ldots CH \ldots\ldots X^{\delta-}} \right] \longrightarrow R^-\overset{\overset{R'}{|}}{C}H_2 + X^- \tag{3}$$

Molecular Aspects of Anti-cancer Drug Action, ed. Neidle & Waring
0333-315561/83/233-282

In practice, it is not possible to separate alkylating agents into two classes on the basis of whether they react by an S_N1 or S_N2 mechanism. This is because the type of reaction that actually occurs with any one molecule depends on a wide variety of external factors. For instance, the nature of the groups R, R' and X affects the electron distribution of the reactants, and hence the ease or difficulty of formation of a transition complex and the ability of the leaving group X to be detached. Normally, the nature of the medium readily affects the course of an alkylation reaction; however, since in biological systems we are concerned solely with aqueous media, variations due to solvent are negligible. The *p*H of the medium also plays a part. With nitrogen mustard derivatives, for example, the half-life is much shorter in alkaline than in acid solution.

Those compounds which react by an S_N1 type mechanism, such as *bis*(2-chloroethyl)sulphide (I), the 'mustard gas' of the First World War, are very unstable and tend to react rapidly with tissue. Compounds for which the S_N2 mechanism predominates, such as 1,4-*bis*(methanesulphonyloxy)butane (II, busulphan, myleran) and triethylenethiophosphoramide (III, thiotepa), tend to react more slowly. On the other hand, although *p*-[N,N-*bis*(2-chloroethyl amino)] phenylbutyric acid (IV, chlorambucil) reacts by an S_N1 mechanism it does so relatively slowly due to the electron-withdrawing character of the aromatic ring.

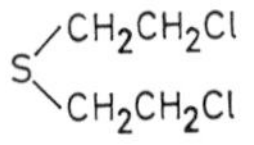

The alkylating agents fall into a number of categories depending on the nature of the reactive leaving group. The chemistry of each category will now be considered in some detail as a prelude to the discussion of the role of alkylating agents in cancer chemotherapy. A number of reviews of this subject have appeared over the years (Golumbic *et al.*, 1946; Price, 1958; Price *et al.*, 1969; Ross, 1953, 1962; Sartorelli and Johns, 1975; Warwick, 1963).

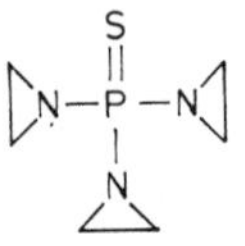

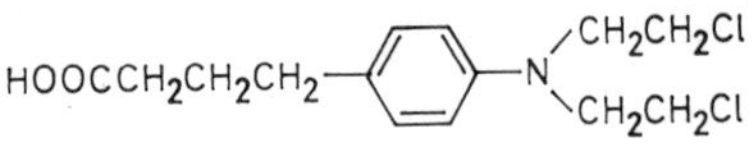

Nitrogen mustards

All mustards derivatives contain a 2-chloroethyl group. Sulphur mustard (I) contains two of these groups, as do the nitrogen mustards (V), but the latter have the added advantage that the third substituent can be varied in order to introduce some degree of selectivity.

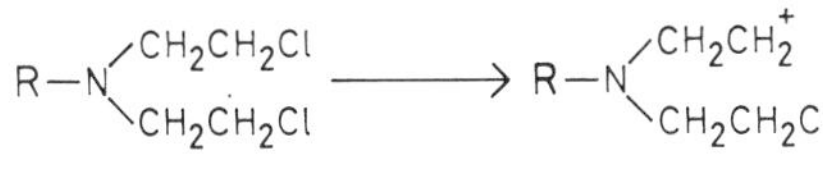

$$R-N\overset{CH_2CH_2Cl}{\underset{CH_2CH_2Cl}{\big<}} \longrightarrow R-N\overset{CH_2CH_2^+}{\underset{CH_2CH_2Cl}{\big<}}$$

V VI

Figure 1 Schematic formation of a carbenium ion from a nitrogen mustard.

As indicated in figure 1, nitrogen mustards yield a carbenium ion (VI) by loss of chloride ion. Since two chloroethyl groups are present, both of which are potential alkylating agents, such compounds are difunctional. It is also possible to make a 'one-arm mustard' (VII) which is monofunctional. Compounds of this type do not exhibit the same sort of biological properties as difunctional ones.

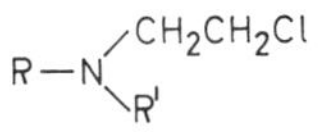

$$R-N\overset{CH_2CH_2Cl}{\underset{R'}{\big<}}$$

VII

Aliphatic 2-chloroethylamines (VIII) rapidly yield a cyclic intermediate in aqueous solution, an aziridinium ion (IX) which can react with anions via the bimolecular mechanism shown in figure 2 (Golumbic *et al.*, 1946). The reaction is assumed to proceed according to this mechanism for the following reasons:

(1) In solution one equivalent of chloride ion is eliminated, but only a trace of hydrogen ion.
(2) There is an instantaneous reaction with thiosulphate, which may be used to determine the concentration of the aziridinium ion.
(3) The rearrangement of X to XII is readily accounted for if the cyclic structure XI is considered as an intermediate (Ross, S. D., 1947; Ross, W. C. J., 1953).

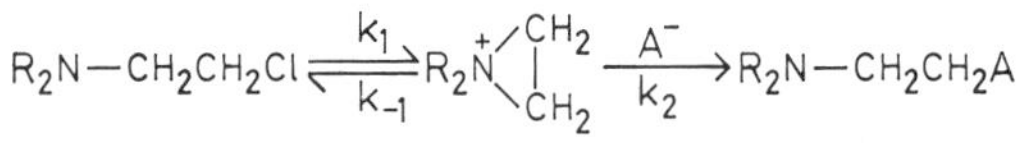

$$R_2N-CH_2CH_2Cl \underset{k_{-1}}{\overset{k_1}{\rightleftarrows}} R_2\overset{+}{N}\overset{CH_2}{\underset{CH_2}{\big<}} | \overset{A^-}{\underset{k_2}{\longrightarrow}} R_2N-CH_2CH_2A$$

VIII IX

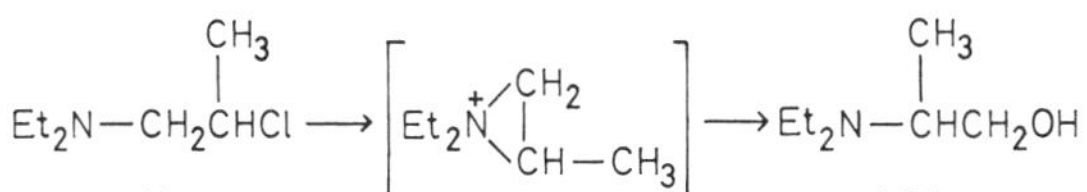

$$Et_2N-CH_2CHCl \longrightarrow \left[Et_2\overset{+}{N}\overset{CH_2}{\underset{CH-CH_3}{\big<}} | \right] \longrightarrow Et_2N-CHCH_2OH$$
$$\overset{|}{CH_3} \qquad\qquad\qquad\qquad\qquad\qquad\qquad\qquad \overset{|}{CH_3}$$

X XI XII

Figure 2 The formation of cyclic intermediates from aliphatic 2-chloroethylamines.

Direct evidence for a cyclic aziridinium ion intermediate is available from nuclear magnetic resonance spectroscopy in the case of 2-chloroethyldiethylamine (VIII, R = C_2H_5) (Levins and Papanastassiou, 1965; Rutman *et al.*, 1969). The subsequent reaction with anions is relatively slow in comparison with the initial cyclisation.

Thus, we have the situation that, in solution, aliphatic nitrogen mustards form a cyclic aziridinium ion by a rapid unimolecular reaction (k_1). The aziridinium ion can then alkylate functional groups in a slow step (k_2).

In the case of the N-(2-chloroethyl)arylamines the nitrogen atom is not sufficiently basic to allow formation of a stable aziridinium ion because of the electron-withdrawing properties of the aromatic ring. Everett and Ross (1949) concluded that the following evidence does not favour the existence of an aziridinium ion:

(1) Halide and hydrogen ions are produced at equal rates.
(2) There is negligible instantaneous reaction with thiosulphate.
(3) The basicity of the nitrogen atom correlates approximately with the ease of hydrolysis.
(4) Compounds such as XIII and XIV have been shown to undergo hydrolysis without structural rearrangement.

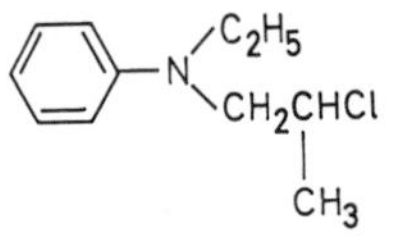

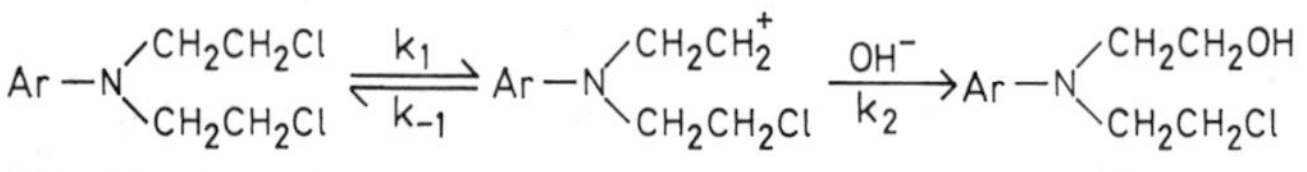

Figure 3   Unimolecular hydrolysis of aryl nitrogen mustards.

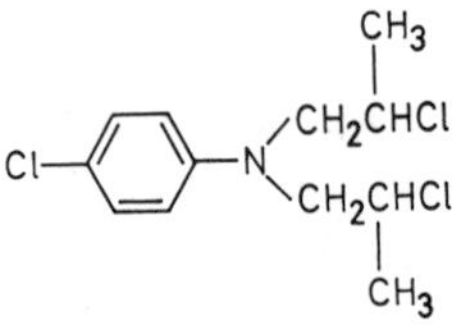

This led to the assumption that the reaction scheme in figure 3 was operative. Although this scheme does not indicate the formation of an aziridinium ion, Everett and Ross (1949) stated that the possibility of the transitory existence of an aziridinium ion is not excluded by the available evidence. This unimolecular mechanism was criticised by Chapman and Jones (1954) who suggested an internal bimolecular mechanism, wherein the unshared electron pair on the β-nitrogen atom cooperates in the decomposition as shown in figure 4. This 'internal S_N2' mechanism of course corresponds to first-order kinetics, as does the true S_N1 process.

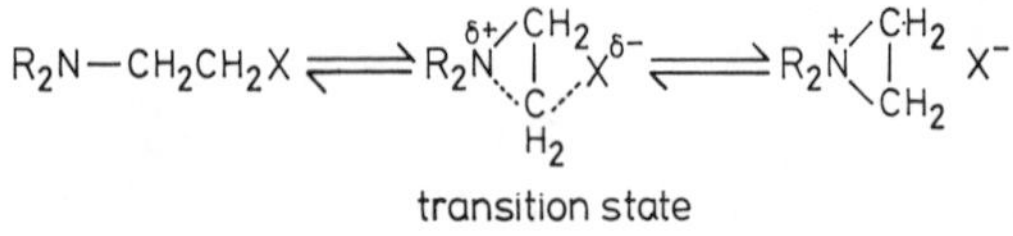

transition state

Figure 4   Internal bimolecular hydrolysis of aryl nitrogen mustards.

Although he did not discount the possibility of a transient aziridinium ion, Ross (1962) continued to support his earlier view. Triggle (1964) explained the differences in solvolysis rates between *N,N-bis*-2-chloroethylaniline (XV, aniline mustard) and 3-phenyl-*n*-propylchloride (XVI) on the basis that the nitrogen atom takes part in the nucleophilic displacement reaction by forming an aziridinium ion. However, Triggle also stated that it was not possible to distinguish between the two reaction pathways merely by observing first-order kinetics in the liberation of chloride ion. The equivalent simultaneous liberation of chloride and hydrogen ions can be explained if $k_2 > k_1$.

Bardos and co-workers (1965) further modified these views and proposed that S_N1 and S_N2 alkylations of a typical heterocyclic nucleus such as 4-(*p*-nitrobenzyl)pyridine (NBP) can take place as shown in figure 5. According to their proposed mechanism, the high-energy transition state of the hydrolytic reaction is a solvated carbonium ion-nitrogen dipole, XVIII, whose collapse gives the final observed product. The transition state of the alkylation reaction is an 'S_N2 complex', XIX, consisting of an aziridinium ion partly opened by the nucleophilic reagent. Both reactions pass through the same reactive intermediate, XVII.

Although they found a correlation between antitumour activity and the extent of hydrolysis by the S_N1 mechanism which substantiated the general concepts originally put forward by Ross (1949, 1962), the correlation between antitumour activity and alkylating activity as determined by the NBP method was less satisfactory. The use of 4-(*p*-nitrobenzyl)pyridine is extremely difficult because small variations in experimental procedure can lead to large differences in results, but when used properly this reagent does provide a reliable measure of alkylating ability within a given series of mustards. For instance, Workman and co-workers (1976) have shown a very good correlation between half-lives derived on the basis of decreased extent of alkylation of NBP and those derived by titrimetric determination in a series of conjugates of *p*-hydroxyaniline mustard.

Williamson and Witten (1967) published evidence for a cyclic intermediate in the case of chlorambucil (IV) and attributed their findings to an initial delay in the first-order reaction plot in buffered aqueous solution. Chlorambucil also liberated chloride ion (mercuric nitrate titration) at a rate greater than expected on the basis of its alkylating activity toward NBP.

Price and co-workers (1969) stated that if the rate constant k_1 for the cyclisation (figure 2) was very low in comparison with the rate constant k_2 for ring opening, there would be no accumulation of cyclic intermediate. The rate-determining step then becomes the first-order cyclisation (k_1) and the reaction

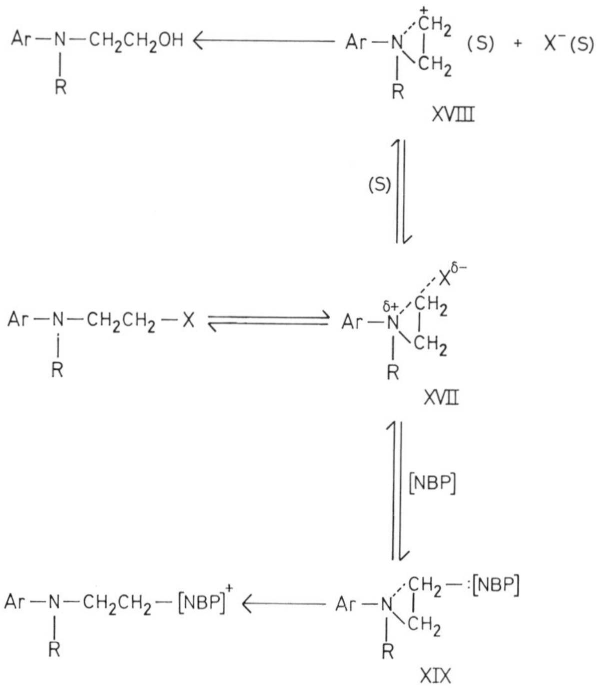

Figure 5 Proposed alkylation mechanism for a heterocyclic nucleus.

is superficially an S_N1 type process. Thus, none of the various types of 2-chloroethyl compounds show any variation in mechanism for the alkylating step, which remains the attack of a nucleophile (A^-) on the cyclic intermediate.

Benn and co-workers (1970) considered the simplest and most reasonable interpretation of the available data to be one involving a conventional aziridinium ion as the reactive intermediate, in accord with the views of Williamson and Witten (1967) and of Price and co-workers (1969). They found that chlorination of XX with phosphorus pentachloride in chloroform gave the corresponding chloride, XXI, which was also the major product of chlorination of XXII together with smaller amounts of the unrearranged chloride XXIII. Acetolysis of the chloride mixture from XXII (that is, XXI and XXIII) gave an even greater proportion of rearranged isomer. These isomerisations accorded with their expectation that the aziridinium ion XXIV would open to give XXI in preference to XXIII.

In an effort to confirm this finding they used nitrogen mustard derivatives containing deuterium at the 2′-position of each chloroethyl group, as in XXV. The aniline mustard analogue XXV (R = H), although it gave retention of configuration on conversion into a dithioether by reaction with potassium p-thiocresolate in *tert*-butyl alcohol, gave the diol and diacetate with deuterium atoms distributed essentially equally between the 1′- and 2′-positions on hydrolysis or acetolysis respectively. The more reactive p-methoxy derivative XXV

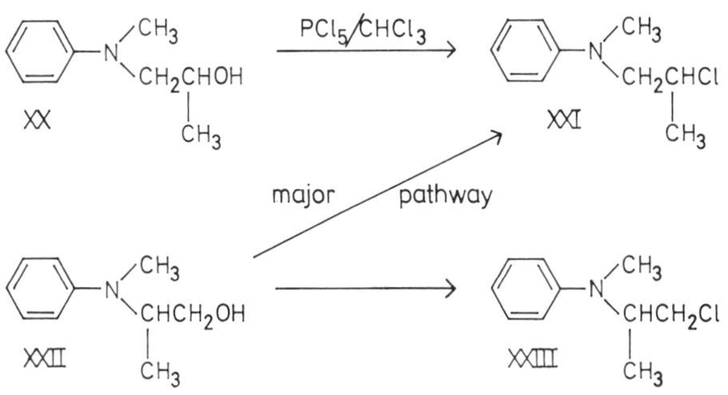

XX XXI XXII XXIII

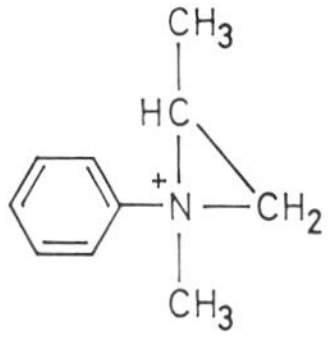

XXIV

($R = p-CH_3O$) yielded a dithioether with extensive scrambling of the label. Acetolysis of XXV ($R = 2,4$-di-NO_2) gave the diacetate with complete scrambling.

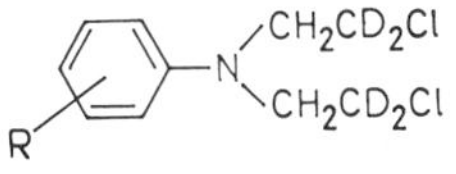

XXV

From these experiments and the other available data, Benn and his colleagues (1970) suggested that the nucleophilic displacement of the side-chain halogen from aryl nitrogen mustards may proceed by one of the two competitive pathways, as shown in figure 6. One pathway involves direct replacement and the other proceeds by way of a reactive aziridinium ion intermediate. The aziridinium ion pathway is preferred except in the case of very powerful nucleophiles (Nu).

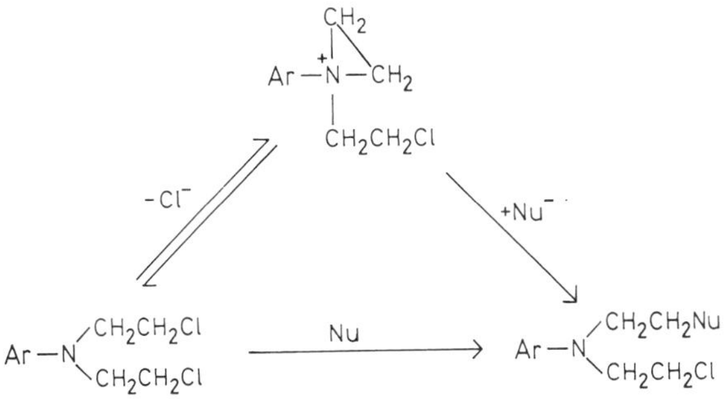

Figure 6 Proposed mechanism of nucleophilic displacement from aryl nitrogen mustards (Benn *et al.*, 1970).

Aziridines

The similarity between the aziridinium ion formed in the reaction of 2-chloro-ethylamines and the aziridine ring led to the investigation of compounds of general structure XXVI for antineoplastic activity. The aziridine ring system is very sensitive to acid. In acidic solution a protonated structure is formed which can react with anions by either a unimolecular (i) or bimolecular (ii) mechanism (figure 7).

Electron-withdrawing substituents on the carbon atoms of the aziridine ring and a polar medium favour the S_N1 mechanism (i). If R in XXVI is an electron-withdrawing group such as aryl or acyl, the compound tends to react via the S_N2 mechanism because of the increased susceptibility of the methylene carbon to nucleophilic attack. Electron-releasing substituents on the aziridine ring generally cause loss of chemical and biological activity (Kahn and Ross, 1969/70). However, Bardos and co-workers (1969) have shown that analogues of XXVII with methyl and ethyl substituents on the aziridine rings generally possess reduced toxicity and antitumour activity. The biological activity of these types of aziridine derivatives appears to vary considerably depending on the nature and position of the ring substituents, but tends not to correlate well with chemical reactivity.

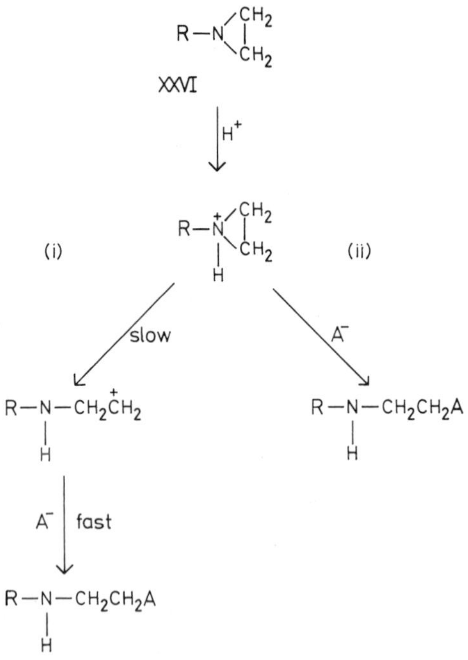

Figure 7   Reactions of the aziridine ring system with anions.

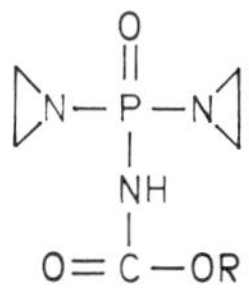

$$\underset{\substack{| \\ \text{NH} \\ | \\ \text{O}=\text{C}-\text{OR}}}{\underset{\substack{\\ \\}}{\triangleright\text{N}-\overset{\overset{\text{O}}{\|}}{\text{P}}-\text{N}\triangleleft}}$$

XXVII

Oxiranes

The oxiranes, or epoxides, contain a three membered oxygen ring similar to the nitrogen containing aziridine ring. However, linkage to the remainder of the molecule is through a carbon atom rather than the heteroatom. Biological activity is observed only in those compounds in which epoxide rings are at opposite ends of a carbon chain, as in 1,2:3,4-diepoxybutane (XXVIII) (Ross, 1950).

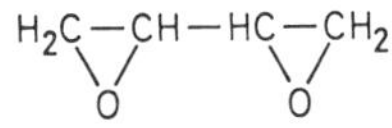

$$\underset{\text{O}}{\text{H}_2\text{C}-\text{CH}}\overset{}{-}\underset{\text{O}}{\text{HC}-\text{CH}_2}$$

XXVIII

Oxiranes react with nucleophiles via an S_N2 reaction mechanism similar to that of aziridines (figure 8). Substitution of the oxirane ring generally leads to the loss of chemical and biological activity.

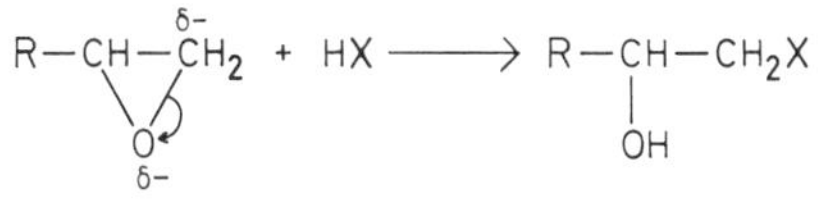

$$\text{R}-\underset{\underset{\delta-}{\text{O}}}{\overset{\delta-}{\text{CH}-\text{CH}_2}} + \text{HX} \longrightarrow \text{R}-\underset{\text{OH}}{\text{CH}-\text{CH}_2\text{X}}$$

Figure 8 Reaction of oxiranes with nucleophiles.

The reactivity of oxiranes may be increased by the presence of neighbouring hydroxyl groups. This anchimeric effect results from hydrogen bonding which facilitates the opening of the epoxide ring. Triptolide is a good antileukaemic agent because it has an hydroxyl group *cis* to an epoxide. If the hydroxyl group is *trans* however all antitumour activity is lost (Kupchan and Schubert, 1974).

Methanesulphonyloxyalkanes

Alkyl esters of methanesulphonic acid are alkylating agents because of the readiness with which the alkyl-oxygen bonds undergo fission. They alkylate by an S_N1 or S_N2 mechanism without formation of an intermediate ring system, depending on the nature of the substituents on the α-carbon. For the purposes of this review, only the esters of di- and polyhydric alcohols are of interest, that is, compounds of general structure XXIX.

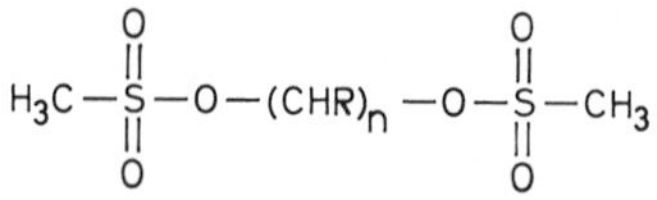

$$H_3C-\overset{\overset{O}{\|}}{\underset{\underset{O}{\|}}{S}}-O-(CHR)_n-O-\overset{\overset{O}{\|}}{\underset{\underset{O}{\|}}{S}}-CH_3$$

XXIX

Butyl ethers

Although β-nitrogen and β-sulphur atoms considerably enhance the hydrolysis rate of alkyl halides relative to the parent unsubstituted compounds, the same is not true of β-oxygen, which causes a large reduction in reaction rate. However, oxygen has a marked tendency to form a relatively stable five-membered oxonium ring that can bring about alkylation of strong external nucleophiles. Tisdale and co-workers (1972) have shown activity against the Walker 256 carcinoma with *bis* (4-bromo-butyl)ether (XXX, R = Br) and *bis* (4-methanesulphonyloxy-butyl)ether (XXX, R = OSO$_2$CH$_3$) and have proposed the mechanism of hydrolysis shown in figure 9.

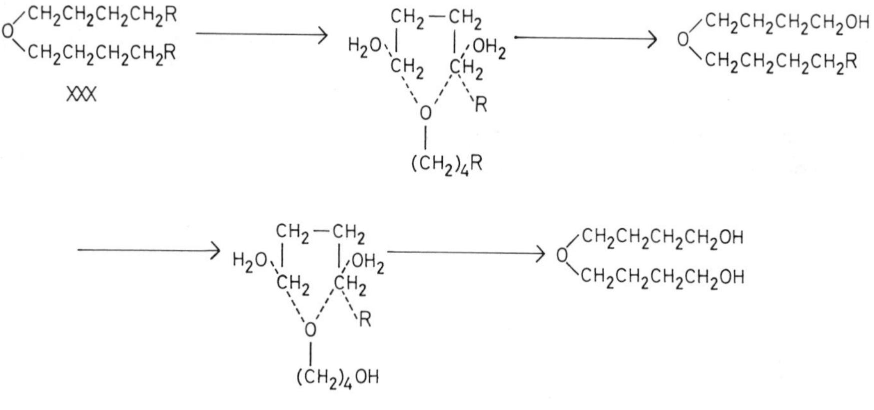

Figure 9 Proposed mechanism for the hydrolysis of butyl ethers (Tisdale *et al.*, 1972).

Apart from the mustards, aziridines, oxiranes, methanesulfonates and butyl ethers, a number of other types of compounds probably act ultimately via an alkylation mechanism; these will be considered individually later.

SITES OF BIOLOGICAL ACTIVITY

Having discussed the mechanisms of chemical reaction of the principal types of alkylating agents useful in cancer chemotherapy, it is now necessary to consider their sites of biological action. The substrates which may be alkylated in biological systems are water, chloride ions, carboxylate and hydroxyl groups, thiols, amines, nitrogenous bases and phosphate ions.

Water is obviously the major substrate for alkylating agents because of its high concentration relative to other nucleophiles. Thus, although the alkylation of water has no direct biological significance it represents a detoxification pathway since it leads to inactivation of the alkylating agent. Chloride ions in serum also react, but this is of no significance in itself. In the case of the mustards, however, it means that the drug will remain effectively unchanged in the circulating blood for a considerably longer period than if it were in a chloride-free medium.

Reaction at the terminal and side-chain carboxyl groups of proteins is possible, but is not readily demonstrated because of the lability of the resultant esters. There is no evidence that protein alkylation is significantly involved in the cytotoxic activity of alkylating drugs. Taking into account the low dose levels of these drugs required to produce clinical and biological effects, it is difficult to see how esterification of a small proportion of the available carboxyl groups could explain these effects.

The hydroxyl groups present in sugar molecules only ionise above pH 12, which is far higher than physiological pH; hence they are not likely to be alkylated. Neither are all thiol groups significantly ionised at physiological pH. The free thiol-containing amino acid cysteine, as well as small peptides like glutathione which constitute a metabolic pool in some cell types, are likely to be alkylated. As is the case with carboxyl groups, biological thiol groups capable of being alkylated are present in excess and can therefore 'detoxify' the drug. Primary amino groups are not available for reaction at physiological pH because under these conditions they are protonated and therefore electrophilic rather than nucleophilic.

Probably the most important moieties to undergo alkylation with a consequent significant biological effect are the heterocyclic bases in polynucleotides, which are highly susceptible to alkylation at physiological pH. Nucleotide phosphate groups would also be expected to be susceptible to alkylation, but this effect has not always been demonstrated because the resultant esters are acid-labile and were often hydrolysed by the acid used to break down the nucleotide chain during the assay. It is also possible that alkylation of heterocyclic bases in nucleotides takes place by initial phosphate esterification followed by alkyl group transfer from oxygen to nitrogen.

So, at the cellular pH of about 7.4, possible sites for alkylation are the ionized carboxyl groups of amino acids and proteins, the ionised thiol groups in free cysteine and in cysteine residues of proteins, and the ring nitrogens of histidine. In nucleic acids and nucleotide coenzymes, the ionized primary and secondary hydroxyl groups of phosphoric acid, the nitrogen atoms of the purine bases, the O6 atom of guanine and the O4 of thymine are capable of being alkylated.

The *in vivo* effects of alkylating agents fall into three main categories:

(1) The cytostatic effect, in which there is delay or prevention of mitosis.

(2) The mutagenic effect, in which division occurs but the daughter cells have altered properties.

(3) The cytotoxic effect, in which the cells are so severely damaged that they cannot survive.

These are further discussed in Chapter 10.

Following the demonstration by Haddow and co-workers (1948) that difunctionality is an essential requisite for high cytotoxicity in alkylating agents, Goldacre and co-workers (1949) suggested that the biological activity of these compounds depends on their ability to react concomitantly at two sites on the same biological macromolecule. Reaction could occur intramolecularly or intermolecularly, and the latter process could lead to interstrand linkage if it involved a double-stranded molecule such as DNA. Such a mechanism seemed especially useful in explaining chromosome fragmentation, bridge formation at anaphase, and chromosome 'stickiness', the most prominent cellular effects of cytotoxic alkylating agents. Diepoxides and polyaziridines were known to cross-link wool and cellulose respectively, and were therefore considered logical candidates for evaluation as antineoplastic agents.

The interactions of 'mustards' with nucleic acids were investigated first by Elmore and co-workers (1948) who had the idea along with many others that attack was probably on the phosphate groups of RNA and DNA. Stacey and co-workers (1958) found that the *in vivo* reaction of melphalan with DNA yielded no acid and concluded that the major alkylation product was a phosphate ester. Brookes and Lawley (1960) discounted this phosphate alkylation theory on the grounds that 7-(2-hydroxyethylthioethyl)guanine was the main product from the reaction of di-2-chloroethylsulphide with guanine. A mechanism involving attack on guanine accounts satisfactorily for the observed difference between RNA and DNA alkylation. Alkylated RNA appears stable in neutral aqueous solution at $37°C$, whereas alkylated DNA decomposes to give 7-alkylguanine because the deoxyribose linkage in alkylated deoxyguanylic acid is labile under these conditions (Lawley, 1957). Conversely, esterification of phosphate groups in nucleic acids to phosphotriesters would be expected to destabilise RNA as opposed to DNA, because dimethyl and dibenzyluridine-3'-phosphate are unstable at $37°C$ (Brown *et al.*, 1955).

By the use of a highly radioactive label, Brookes and Lawley have shown in a number of studies that *bis*(2-chloroethyl)-^{35}S-sulphide and 2-chloroethyl-2-hydroxyethyl-^{35}S-sulphide react very similarly with DNA both *in vivo* and *in vitro*. The only detectable difference, following hydrolysis of the nucleotide chain, is the formation of the *bis*(7-guanyl) derivative XXXI from *bis*(2-chloroethyl)-^{35}S-sulphide. This cross-link accounts for about 25% of the alkylated guanine in DNA.

The steric situation of the N7 atoms of guanine in double helical DNA, as conceived by Crick and Watson (1954), makes them readily available for reaction. The necessary separation for two N7 atoms is about 8 Å. To achieve this the

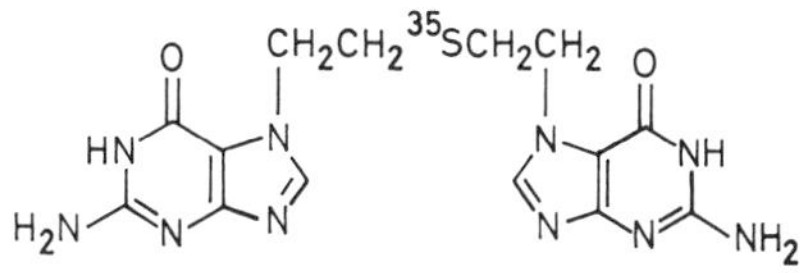

XXXI

DNA must have a GC sequence at the site of reaction. This allows readily for guanine cross-linking, but is not possible if the sequence is reversed. The cross-link means that separation of the individual strands during cell division becomes impossible.

Another observed effect is deletion of an N7-alkylated guanine from the DNA molecule. Following N7-alkylation of guanine, the initially formed quaternary compound XXXII is in equilibrium with the tautomer XXIII where the charge is situated at N9. This tautomeric form is unstable and breaks down by fission of the bond to yield the remainder of the DNA molecule (Ross, 1962).

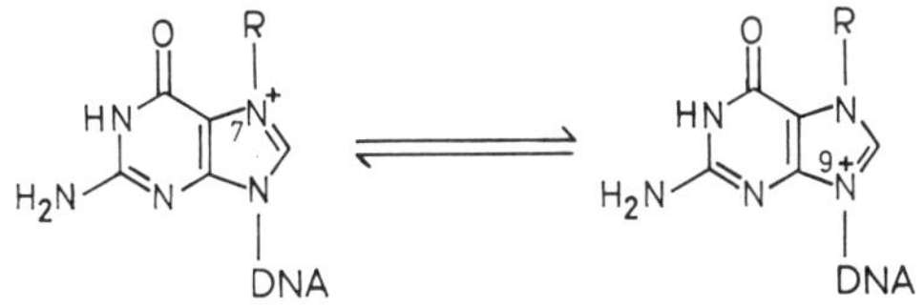

XXXII   XXXIII

The first observation of O6-alkylation in guanine was made by Loveless (1969) who showed that compound XXXIV was formed when the carcinogens and mutagens ethyl methanesulphonate and ethylnitrosourea reacted with deoxyguanosine, whereas the non-carcinogenic and non-mutagenic esters dimethyl sulphate and methyl methanesulphonate gave no similar methylated products. O6-alkylation also displaced the proton from N7, removing the possibility of base pairing via hydrogen bonding at this site. Many further examples of this type of alkylation have since been reported.

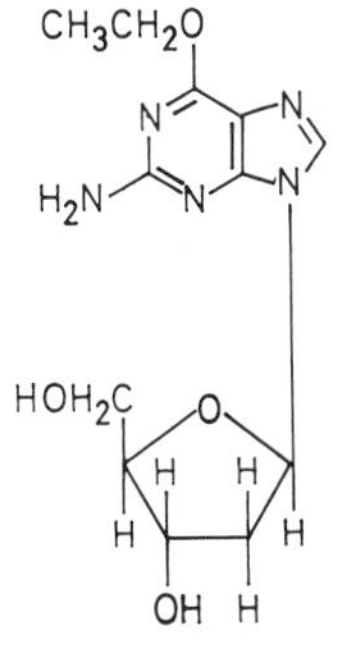

XXXIV

Although on the basis of present evidence it is not possible to state unequivocally that DNA is the only important target for alkylating agents, DNA cross-linking ability surely plays a major role in their cytotoxic action (see Chapter 10). Many effects can be related to interference with normal DNA function. However, it is noteworthy that a correlation has not been shown to exist in the whole animal between DNA alkylation and tumour sensitivity. Certain tumours are known to be 100 times less sensitive to alkylating agents than others, and yet undergo DNA alkylation to the same extent. This may be due to the fact that less sensitive tumours are more efficient than sensitive tumours in repairing alkylated DNA.

APPLICATIONS OF ALKYLATING AGENTS IN CANCER CHEMOTHERAPY

Despite the fact that the alkylating agents are generally cytotoxic and tend to show very little tumour specificity, a number of them have had considerable success in cancer chemotherapy. All normal proliferating tissues are sensitive to cytotoxic agents, the most prominent side effects being bone marrow depression, anaemia, susceptibility to infection, gastrointestinal toxicity, nausea, vomiting, and diarrhoea. Alkylating agents were used initially in the treatment of haematologic tumours, but newer types are finding application in the chemotherapy of solid tumours. Many of these agents are used in combination with other drugs that act via different mechanisms and have different side effects.

Nitrogen mustards

Sulphur mustard (I), although a useful research tool, has found no clinical application due to its highly vesicant nature. The nitrogen mustards are somewhat less vesicant, and a number of them have been used clinically. Nitrogen mustard (XXXV) was the first alkylating agent to reach clinical status (Gilman, 1963). Its greatest success was against lymphomas, and it is still used today to treat patients with lymphosarcoma, Hodgkin's disease, and lung cancer. Due to its very high reactivity and hence rapid detoxification, it is useful for intra-arterial and intrapleural injections. The simple alkyl nitrogen mustards offer little opportunity for significant structural modification with a view to improved selectivity. However, substitution by aryl groups expands this opportunity considerably, as will be illustrated below.

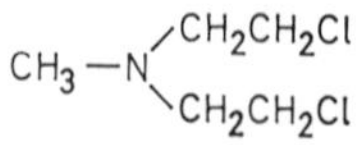

$$CH_3-N\begin{array}{c} {}^{\nearrow CH_2CH_2Cl} \\ {}_{\searrow CH_2CH_2Cl} \end{array}$$

XXXV

Aryl nitrogen mustards were investigated initially by Ross (1949) and coworkers at the Chester Beatty Research Institute and subsequently by many others. The first of these compounds to show clinical usefulness was *N, N-bis* (2-

chloroethyl)-β-naphthylamine (XXXVI) which underwent extensive clinical trial in the early 1950s against Hodgkin's disease, chronic myeloid and lymphatic leukaemias, and polycythaemia. Satisfactory remissions were obtained in about 50% of the cases treated (Ross, 1962). More widespread clinical acceptance of this drug was forestalled by the fear that it might contain traces of, or might decompose to, the potent bladder carcinogen β-naphthylamine.

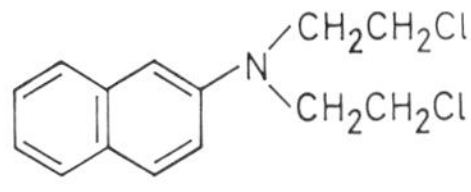

XXXVI

As part of an investigation of the effect of increased anionic, cationic, lipophilic, or hydrophilic character on the activity of derivatives of *N,N-bis*(2-chloroethyl)aniline (aniline mustard, XV), a series of carboxylic acid derivatives (XXXVII, $n = 0$ to 4) was prepared. Of these the most effective tumour inhibitor was *p*-[*N,N-bis*(2-chloroethyl)amino]phenylbutyric acid (chlorambucil, XXXVII, $n = 3$) (Everett *et al.*, 1953). In addition to being water-soluble as a sodium salt, this compound is also surface active because of its high content of fat-soluble groups. Chlorambucil is of particular value in the treatment of chronic lymphoid leukaemia, Hodgkin's disease and other lymphomas, choriocarcinoma, and cancer of the testis, ovary and breast.

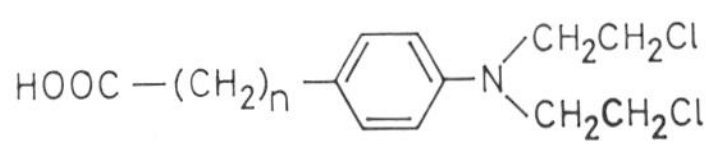

XXXVII

Little is known about the metabolism and mechanism of action of chlorambucil, although many suggestions have been made. Recent investigations have shown that carbon dioxide is a major metabolite, indicating that the most significant metabolic route may be a 'β-elimination' step to form *p*-[*N,N-bis*(2-chloroethyl)amino]phenylacetic acid (XXXVII, $n = 1$) (Godeneche *et al.*, 1975). This metabolic has actually been isolated from the serum of treated rats (McLean, 1974/75). It is interesting that chlorambucil and the metabolite are equally active against the Walker tumour.

The antitumour activity of chlorambucil against an alkylating agent-sensitive strain of the Yoshida sarcoma is enhanced by the administration of prednisolone four hours after the cytotoxic drug. The prednisolone ester of chlorambucil, prednimustine (XXXVIII), has the same effect as this combination, and both the chlorambucil/prednisolone combination and prednimustine are significantly toxic against a Yoshida tumour line with acquired resistance to chlorambucil alone (Harrap *et al.*, 1977). Additionally it has proved possible to potentiate the antitumour effect of prednimustine by giving a subsequent dose of prednisolone though no enhancement is achieved when chlorambucil is given in place of prednisolone. Prednimustine also exhibits steroidal activity in cells which possess receptor-binding capacity. Whether the alkylating and hormonal properties of prednimustine depend on an initial hydrolysis to its components is under investigation (Harrap *et al.*, 1977).

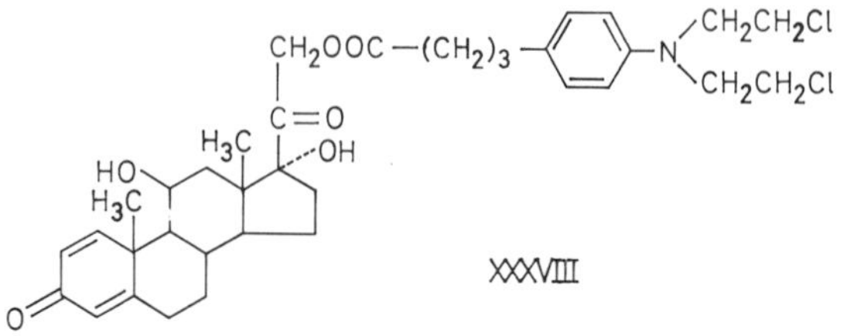

XXXVIII

The alanine derivative of aniline mustard, *p*-[*N,N-bis*(2-chloroethyl)amino] phenylalanine (XXXIX), can be prepared in the optically active *D*- and *L*- forms or as the racemate. It was synthesised originally in support of the concept that attachment of a mustard group to a naturally occurring 'carrier' (in this case phenylalanine) might increase its effectiveness, because of an increased affinity for certain biological sites or a tendency to become involved in various metabolic pathways of the tumour and the host. In Great Britain the *D*- and *L*- forms were synthesised separately, and the *L*- form (melphalan) was found to be considerably more potent than the *D*- form (medphalan) (Bergel and Stock, 1954). In the Soviet Union the racemic *DL*- form (merphalan or sarcolysin) was prepared (Larionov *et al.*, 1955) and shown to have half the potency of melphalan. Extension of this work showed that the *DL*- form of the *meta* isomer is about ten times more active than melphalan against Cloudman S91 melanoma and four times more active against adenocarcinoma 755 (Gram *et al.*, 1959). The *DL*- form of the orthoisomer (merophan) is at least twice as active against the Walker 256 carcinoma as melphalan (Connors and Ross, 1960; Ross, 1962). Melphalan finds clinical use against myeloma, macroglobulinaemia, and tumours of the ovary and breast. Early results of an adjuvant chemotherapy trial of melphalan following surgery for breast cancer have been remarkably good, particularly in premenopausal women (Fisher *et al.*, 1975). Merophan produces excellent results in Burkitt's lymphoma (Clifford, 1966).

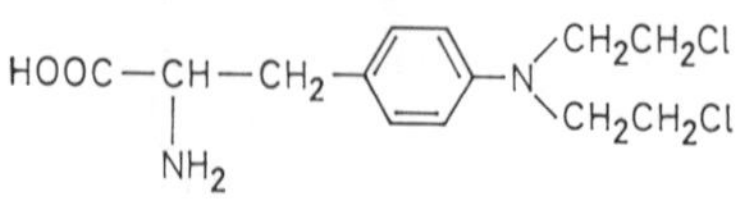

XXXIX

Mitoclomine (XL, R = CH_2CH_2Cl) is an alkylating analogue of vitamin K_5 (XL, R = H). Vitamin K_5 has been shown to concentrate in some tumour systems and lymphocytes (Fox *et al.*, 1971).

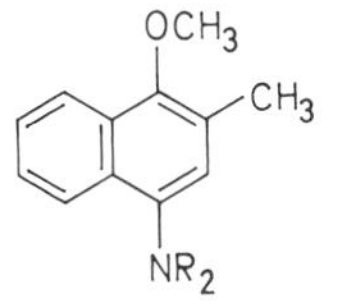

XL

Two piperazine derivatives with activity against chronic myeloid leukaemia and polycythaemia are believed to act via an alkylation mechanism. They are 1,4-*bis*(3-bromopropionyl)piperazine (pipobroman, XLI, X = Br) and 1,4-*bis*(3-methanesulphonyloxypropionyl)piperazine (piposulphan, XLI, X = OSO_2CH_3).

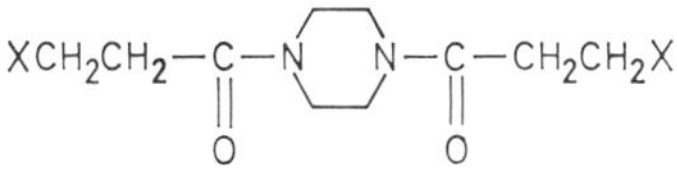

XLI

The *bis*(2-chloroethyl)amino group has been attached to many other naturally occurring structures or analogues thereof, such as amino acids and peptides, pyrimidines, terpenes and steroids. To date, however, none of these has found any extensive clinical application.

Probably the most useful drug of the nitrogen mustard type in clinical use today is cyclophosphamide, a cyclic phosphoric acid amide containing a nitrogen mustard substituent. In view of the suggested mode of action of this drug, it will be more appropriate to discuss it later in the section dealing with enzyme activation and detoxification.

Aziridines

Phosphoric acid derivatives of aziridines (or ethyleneimines as these three-membered ring compounds are sometimes referred to in the older literature) are also important in cancer chemotherapy. The most clinically useful aziridine derivative is triethylenethiophosphoramide (ThioTEPA, III). Its main use today is in the local treatment of non-invasive bladder tumours. Triethylenemelamine (TEM, XLII), was investigated initially on the basis of the cross-linking hypothesis of antitumour activity (Goldacre *et al.*, 1949) because of its ability to cross-link wool fibres under mild conditions. Due to problems of unpredictable absorption and unpleasant side effects it is no longer used clinically.

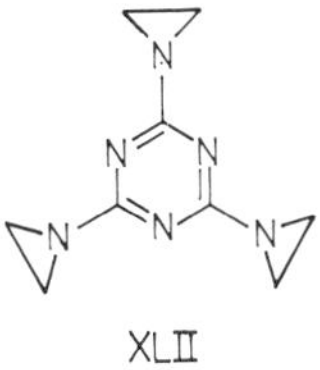

XLII

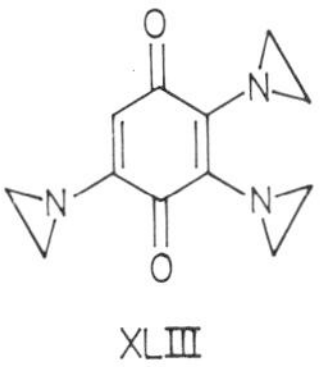

XLIII

Many aziridine derivatives of benzoquinone have been investigated, the most useful of which has been 2,3,4-triethyleneiminobenzoquinone (trenimon, XLIII).

Of particular interest in relation to the aziridines is the unexpected antitumour activity of agents that appear at first glance to be monofunctional. Tetramin (4-aziridino-3-hydroxybut-1-ene, XLIV) has shown activity against a range of tumours (White, 1959). It is thought that this compound may have a difunctional nature because of the possibility of *in vivo* epoxidation of the double bond following alkylation of a receptor by the aziridine group.

The presence of a 2,4-dinitrobenzene moiety in 1-aziridinyl-2,4-dinitrobenzene (CB 1837, XLV) (Hendry *et al.*, 1951, Ross and Mitchley, 1964) is known to confer protein-associating properties on this compound. CB 1837 has a therapeutic index of 10 when assayed against the Walker 256 carcinosarcoma, which compares favourably with the other clinically useful agents shown in table 1.

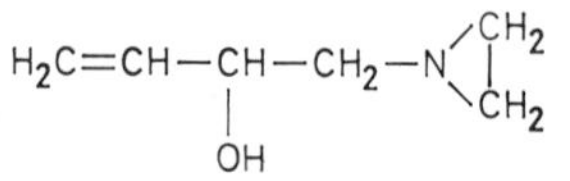

$$H_2C=CH-CH-CH_2-N\begin{smallmatrix}CH_2\\|\\CH_2\end{smallmatrix}$$
$$|$$
$$OH$$

XLIV

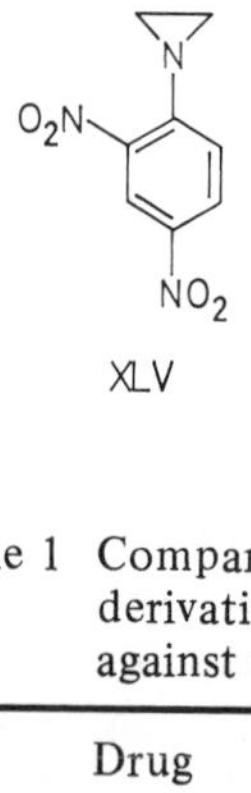

XLV 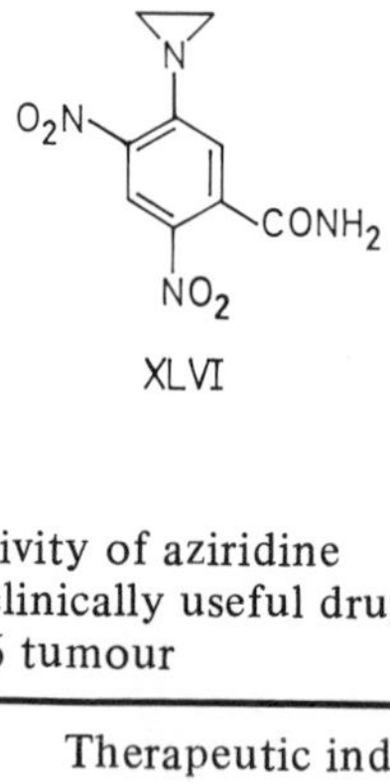   XLVI

Table 1 Comparison of the activity of aziridine
derivatives and other clinically useful drugs
against the Walker 256 tumour

Drug	Therapeutic index
Chlorambucil	7
Melphalan	9
Merophan	14
Triethylenemelamine	9.5
CB 1837	10
CB 1954	70

The therapeutic index is calculated by dividing the dose which
kills 50% of the animals in a group by the dose which produces
90% inhibition of tumour growth.

Khan and Ross (1969/70, 1971/72) examined a large number of analogues of CB 1837, many of which had a low or negligible therapeutic index. One major exception was 5-aziridinyl-2,4-dinitrobenzamide (CB 1954, XLVI), which had a therapeutic index of 70 against the Walker tumour. This compound shows extreme selectivity in its antitumour action, being almost totally inactive in all other experimental animal systems (Cobb *et al.*, 1969). Studies on the mechanism of action of CB 1954 have been pursued with considerable vigour because it is so unusual for a compound to show such a high degree of selectivity.

The mechanism of action of CB 1954 appears to be something other than simple alkylation of DNA, since it does not affect other alkylating agent-sensitive tumours. However, a Walker tumour with acquired resistance to CB 1954 is also cross-resistant to melphalan, implying some similarity in their modes of action. A possible explanation of this similarity is that CB 1954 may undergo *in vivo* metabolism to a difunctional agent (Connors and Melzack, 1971).

The selective toxicity of CB 1954 (XLVI) for the Walker tumour *in vivo* was confirmed *in vitro* (Connors *et al.*, 1972c). In each case its antitumour activity was reversed by 4-aminoimidazole-5-carboxamide (AIC) when the latter is given before or soon after CB 1954. The ribotide of AIC is a key intermediate in purine biosynthesis. Other purines also produced reversal, whereas pyrimidines did not. This suggests that the drug, in addition to its alkylating properties, may be a purine antimetabolite. However, anthranilamide was also effective in protecting against CB 1954, albeit at a considerably higher dose, suggesting that reversal can be achieved by compounds that are structurally similar to the drug and compete for the same receptor site. That this is in fact unlikely, however, is indicated by the structure of partial reduction products isolated from the urine of rats treated with CB 1954 (Jarman *et al.*, 1976). Of the two possible isomeric products XLVII and XLVIII, the latter is the more active, with a therapeutic index of 16 compared with less than one for XLVII. The best protective agent to date is 4-amino-2-phenylimidazole-5-carboxamide (Hickman and Melzack, 1975). The *in vivo* antitumour activity of CB 1954 and a series of its analogues has been shown to correlate with their *in vitro* protein binding (Hickman, 1974).

CB 1954 (XLVI) was active against dog lymphosarcomas, but a preliminary clinical trial was halted when 19 patients failed to show unequivocal regression. More recently CB 1954 and its methyl substituted amide analogues have been shown to be potent radiosensitisers (Stratford *et al.*, 1982).

An assortment of naturally occurring antibiotics containing an aziridine ring have been isolated from *Streptomyces caespitonis*. Clinically the most interesting is mitomycin C (XLIX, R = H). In addition to the aziridine structure the mitomycins contain two other potentially carcinostatic groups, the quinone and urethane moieties. Unlike the synthetic aziridine derivatives the mitomycins are chemically stable even under acidic conditions.

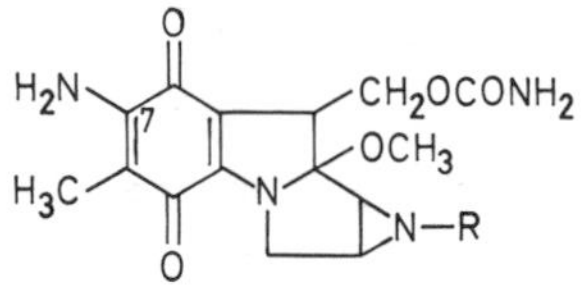

XLIX

The postulated mechanism of action of mitomycin C (Szybalski and Iyer, 1967) involves enzymatic reduction of the quinone moiety followed by spontaneous elimination of the tertiary methoxy group (Chapter 9). The resulting indole system contains two active centres, the aziridine ring and the methylene group of the methylurethane side chain. In addition, the 7-amino group may also be involved in DNA interaction. Unlike other alkylating agents, however, the mitomycins do not alkylate the N7 of guanine in DNA.

Mitomycin C (XLIX, R = H) is used clinically in the treatment of embryonal carcinoma of the testis and tumour-associated hypercalcaemia. Porfiromycin (XLIX, R = CH_3), the *N*-methyl derivative of mitomycin C, is also undergoing clinical investigation (Izbicki *et al.*, 1972).

Investigation of the structural requirements necessary for activity in this series has resulted in the synthesis of a wide range of potential bioreductive alkylating agents based on a quinonoid structure (Lin *et al.*, 1976).

Oxiranes and methanesulphonyloxyalkanes

Only oxiranes with two terminal epoxide groups show *in vivo* antitumour activity. Of those, $1,2:3,4$-diepoxybutane (XXVIII) and $1,1'$-*bis*-$(2,3$-epoxypropyl)-$4,4'$-bipiperidine (eponate) have undergone clinical investigation, but with disappointing results.

$1,2:5,6$-Dianhydro-*D*-mannitol (LI) is believed to be responsible for the antitumour activity of two drugs, $1,6$-dibromo-$1,6$-dideoxy-*D*-mannitol (L) and $1,6$-dimethanesulphonyl-$1,6$-dideoxy-*D*-mannitol (mannitol myleran, LII) (Elson *et al.*, 1968), a suggestion first made by Davis and Ross (1963). Dibromomannitol is used in the treatment of chronic myeloid leukaemia and an analogue, dibromodulcitol (LIII), is under clinical investigation.

```
      CH2Br              CH2              CH2OSO2CH3
       |             O    |                   |
  HO—C—H             \   C—H            HO—C—H
       |                  |                   |
  HO—C—H           HO—C—H             HO—C—H
       |                  |                   |
   H—C—OH            H—C—OH              H—C—OH
       |                  |                   |
   H—C—OH            H—C                  H—C—OH
       |                  |  \ O               |
      CH2Br            H2C  /            CH2OSO2CH3

        L                 LI                 LII
```

Mannitol myleran (LII) was first synthesised as a water-soluble analogue of 1,4-dimethanesulphonyloxybutane (myleran, II) (Brown and Timmis, 1961). Myleran is the most effective compound in the series and is used for the treatment of chronic myeloid leukaemia.

```
      CH2Br                      H2C
       |                           \  O
   H—C—OH                      H—C /
       |                           |
  HO—C—H                     HO—C—H
       |                           |
  HO—C—H                     HO—C—H
       |                           |
   H—C—OH                      H—C
       |                           |  \ O
      CH2Br                     H2C  /

        LIII                       LIV
```

1,2:5,6-Dianhydrogalactitol (LIV) a more stable isomer of LI is of particular interest because it is structurally a compact low molecular weight molecule and, as such, is able to cross the blood–brain barrier, even though it is water-soluble. It is active against a range of experimental, including intracerebral, tumours (Elson *et al.*, 1968; Levin *et al.*, 1976). Phase I trials (Haas *et al.*, 1976) have revealed activity against lung tumours, and phase II studies are in progress.

Nitrosoureas

1-Methyl-3-nitro-1-nitrosoguanidine (LV) was discovered to have slight activity against L1210 leukaemia in random screening by the National Cancer Institute. As a result, a synthetic programme was initiated which led eventually to the introduction of 1,3-*bis*-(2-chloroethyl)-1-nitrosourea (BCNU, LVI) for general clinical use. A number of other nitrosourea derivatives are now undergoing clinical evaluation. These include especially 1-(2-chloroethyl)-3-cyclohexyl-1-nitrosourea (CCNU, LVII) and 1-(2-chloroethyl)-3-(*trans*-4-methylcyclohexyl)-1-nitrosourea (MeCCNU, LVIII).

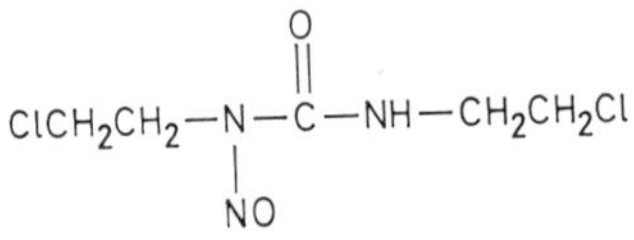

$$ClCH_2CH_2-N-\overset{\overset{\displaystyle O}{\|}}{C}-NH-CH_2CH_2Cl$$
$$\underset{NO}{|}$$

LVI

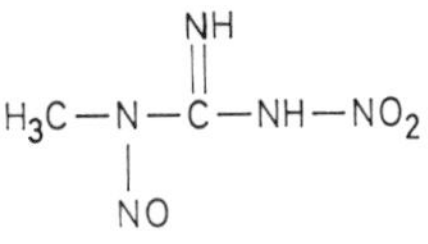

$$H_3C-N-\overset{\overset{\displaystyle NH}{\|}}{C}-NH-NO_2$$
$$\underset{NO}{|}$$

LV

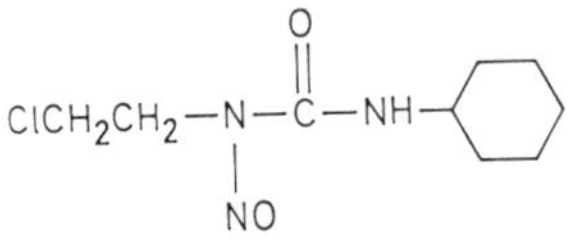

$$ClCH_2CH_2-N-\overset{\overset{\displaystyle O}{\|}}{C}-NH-\bigcirc$$
$$\underset{NO}{|}$$

LVII

$$ClCH_2CH_2-N-\overset{\overset{\displaystyle O}{\|}}{C}-NH$$
$$\underset{NO}{|}$$
CH₃

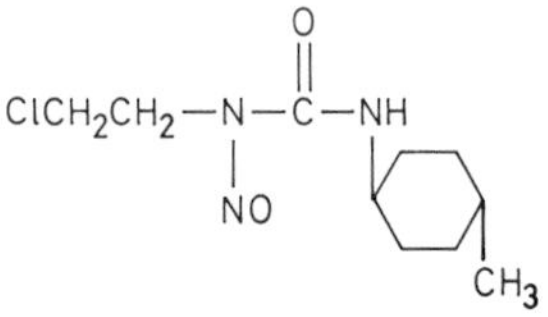

LVIII

Nitrosoureas resemble the difunctional alkylating agents in that they exhibit similar clinical side effects particularly toward the bone marrow and blood elements. The onset of toxicity is, however, delayed until 4 to 6 weeks after administration. In addition, nitrosoureas inhibit some animal tumours that are sensitive to alkylating agents, but are ineffective against other tumours with acquired resistance to alkylating agents.

These drugs are particularly useful against primary and metastatic tumours of the brain. Their ability to cross the blood-brain barrier results from their high lipid solubility. This characteristic has enabled nitrosoureas to find substantial use in the chemotherapy of acute lymphoid leukaemia with central nervous system involvement. They are also used in the treatment of Hodgkin's disease and other lymphomas as well as tumours of the lung. BCNU is of value against myeloma, melanoma, and breast cancer, whereas CCNU and MeCCNU have been given to patients with gastrointestinal adenocarcinoma and colorectal carcinoma respectively.

Although the experimental results tend to suggest that nitrosoureas are most likely just another type of alkylating agent, they have been the target of much research, probably because of their novel mechanism of antitumour activity. Despite obvious similarities to difunctional alkylating agents, they have a different spectrum of biological activity, and although little else is known about their mode of action they are known to break down chemically *in vivo* into a variety of monofunctional alkylating species.

The basic chemical features that a nitrosourea must have in order to exhibit antitumour activity against L1210 leukaemia are summarised in structure LIX, where X = Cl or F, and R = CH_2CH_2Cl, CH_2CH_2F or certain alicyclic or heterocyclic groups. A number of theories for the mechanism of decomposition of nitrosoureas have been put forward. In particular the work of Colvin and coworkers (Colvin *et al.*, 1976, Brundrett *et al.*, 1976) has suggested a pattern of product formation which adequately explains the origin of all the identified decomposition products of BCNU, CCNU and 1,3-*bis*(2-fluoroethyl)-1-nitro-

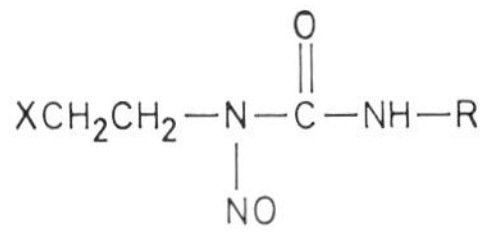

$$XCH_2CH_2-N-\overset{\overset{\displaystyle O}{\|}}{C}-NH-R$$

with N bearing NO below.

LIX

sourea in phosphate buffer at pH 7.4 and 37°C (see Chapter 9 for a more detailed account). Irrespective of the actual mechanism involved, it is clear that a number of reactive species are produced, any one or combination of which would be capable of reacting with biological macromolecules to bring about the observed antitumour activity and toxicity of the parent drug.

Other investigators (Hill *et al.*, 1975; Hilton and Walker, 1975; May *et al.*, 1975) have demonstrated that liver microsomal hydroxylation of the cyclohexyl ring in CCNU proceeds at a rate greater than the rate of chemical breakdown. Following the synthesis of all six possible hydroxylation products of CCNU (Johnston *et al.*, 1975), it was found that the *cis*- and *trans*-2-hydroxy and *cis*-3-hydroxy derivatives show virtually no carbamoylating activity. This is presumably due to an intramolecular carbamoylation reaction similar to that seen with chlorozotocin and streptozotocin.

Chlorozotocin (LX) and streptozotocin (LXI) are two antibiotics from *Streptomyces achromogenes* whose structure contains a nitrosourea moiety. Their unusually low haematological toxicity (Schein *et al.*, 1974) may be due to low carbamoylating activity. Chlorozotocin shows high activity against L1210 leukaemia but very little myelosuppression, which makes it a prime candidate for clinical trial. The glucosamine moiety may explain the affinity of this drug for certain cells of the pancreas, and may account for its activity against insulinomas.

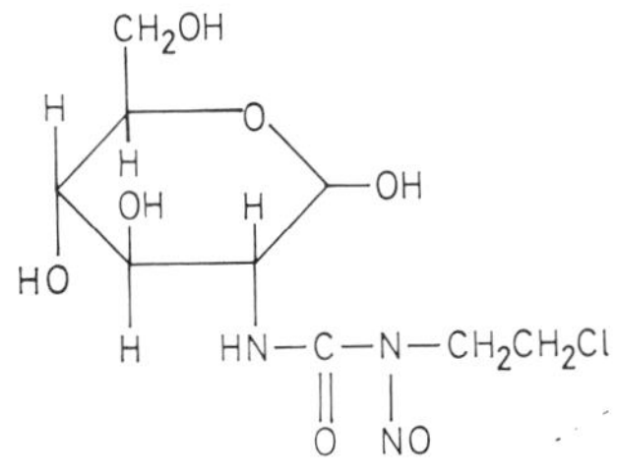

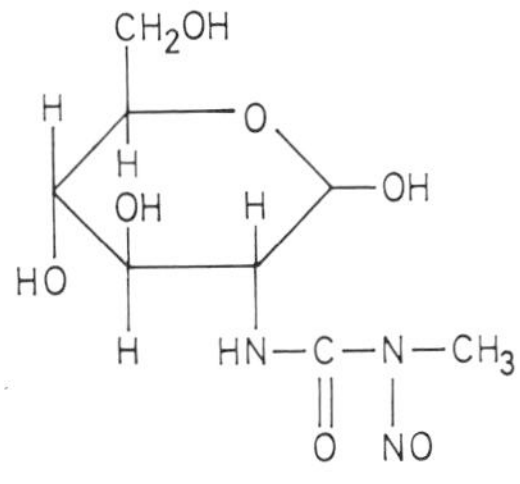

Triazenes

During an investigation of purine analogues with antineoplastic activity, Shealy and co-workers (Shealy *et al.*, 1961) showed that 5-diazoimidazole-4-carboxamide (LXIII), the diazonium salt derivative of 5-aminoimidazole-4-carboxamide

(AIC, LXII), is toxic to carcinoma cells in culture and inhibits the growth of the Walker 256 carcinoma in rats. Although stable under anhydrous conditions, LXIII readily undergoes intramolecular cyclisation in aqueous solution throughout the *p*H range to give 2-azahypoxanthine (LXIV). In the presence of primary or secondary amines the triazene LXV is formed (Shealy *et al.*, 1962). The *N*, *N*-disubstituted derivatives LXV (R and R' ≠ H) are stable except in the presence of light. Exposure to light results in the formation of 2-azahypoxanthine (LXIV) via the diazo derivative LXIII. The *N*-monosubstituted analogues LXV (R = H) are unstable in acid even in the dark.

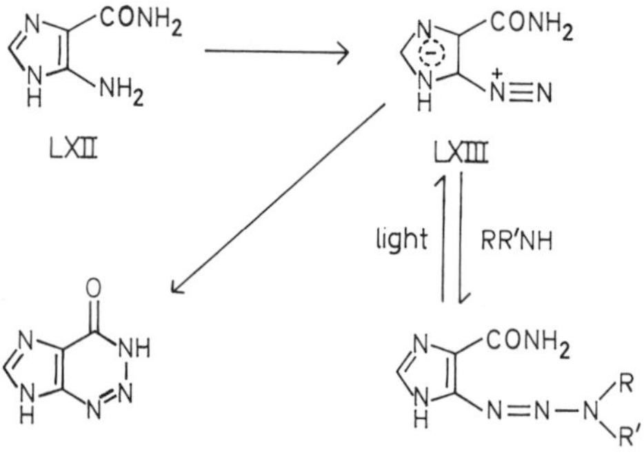

In order for triazenoimidazoles to be active against L1210 leukaemia, at least one methyl substituent is necessary (LXV, R' = CH$_3$) (Shealy and Krauth, 1966). On the basis of these results the most suitable compound for clinical investigation was judged to be 5-(3,3-dimethyl-1-triazeno)imidazole-4-carboxamide (DTIC, LXV, R = R' = CH$_3$). Preliminary clinical studies showed this agent to be particularly suitable for the treatment of malignant melanoma (Carter and Friedman, 1972), against which it is now the agent of choice, and also for the treatment of soft tissue sarcoma. Although other triazenoimidazoles such as 5-[3,3-*bis*-(2-chloroethyl)-1-triazeno]imidazole-4-carboxamide (LXV, R = R' = CH$_2$CH$_2$Cl) are more active than DTIC against L1210 leukaemia, none is as effective in the clinic.

Although the mechanism of action of triazenoimidazoles is not known, it seems likely that alkylation is involved. Methylation is most probably responsible for the antitumour activity, and alkylation in general for the carcinogenic and teratogenic effects. The accepted metabolism and subsequent chemical decomposition of triazenoimidazoles is shown in figure 10. Originally Skibba and co-workers (1970) suggested that the chemical species ultimately responsible for the antitumour activity of these compounds was diazomethane. In work on phenyltriazenes, Preussmann and co-workers (1969) concluded that a methyl carbonium ion is formed via methyl diazohydroxide which is released during decomposition of the monoalkyltriazene (figure 10). Final proof for the formation of the methylcarbonium ion rather than diazomethane came from an elegant experiment involving the decomposition of compound LXVI in deuterium oxide (D_2O). The methanol which formed was analysed by nuclear magnetic resonance spectroscopy and shown to have come from the carbonium ion (Nagasawa *et al.*, 1974).

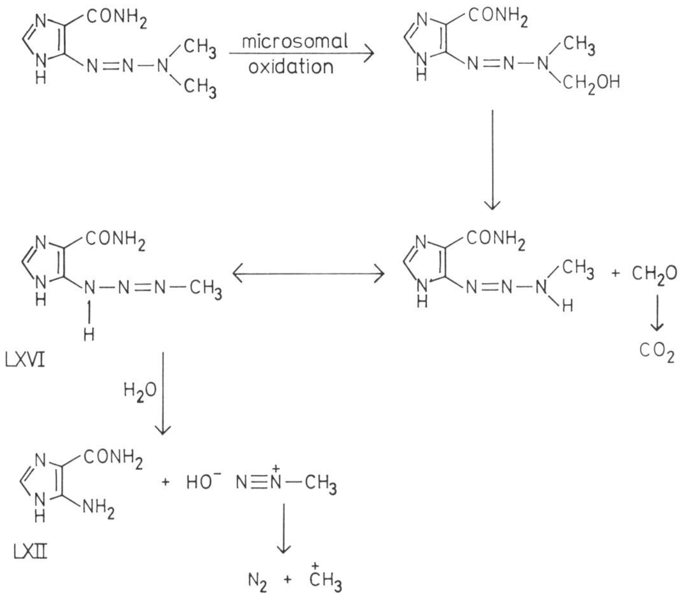

Figure 10 The metabolic pathway for triazenoimidazoles.

Many triazenes have been investigated in which the imidazole ring has been replaced by other heterocyclic systems (Shealy, 1970) and even by substituted phenyl derivatives (Audette *et al.*, 1973; Connors *et al.*, 1976). These studies have shown that the nature of the aromatic moiety does not affect the activity of N,N-dimethyltriazenes markedly. Substitution of one of the methyl groups, provided that the replacement group can undergo oxidative dealkylation to leave a monomethyltriazene, has no effect on antitumour activity. However, as the chain length of a replacement alkyl group increases so the antitumour activity reaches a maximum and then falls very rapidly (Wilman and Goddard, 1980). If this is not possible, as is the case with 1-(p-carbamoylphenyl)-3-*tert*-butyl-3-methyltriazene, the methyl group is lost and subsequent chemical breakdown produces the *tert*-butylcarbonium ion. Because of severe steric hindrance, this ion is not an alkylating agent and the compound shows no antitumour activity (Connors *et al.*, 1976). More recently the hydroxymethyl intermediate has been implicated as playing a significant role in the mechanism of action of the aryltri-azenes (Gescher *et al.*, 1981). That such species may be considered to be formed in mammals was shown by Kolar and Carubelli (1979), who isolated the O-glucuronide derivative of 3-hydroxymethyl-3-methyl-1-(2,4,6-trichlorophenyl)-triazene (LXVII). The N-hydroxymethyl metabolite itself has been isolated following treatment of rats with DTIC (Kolar, 1980).

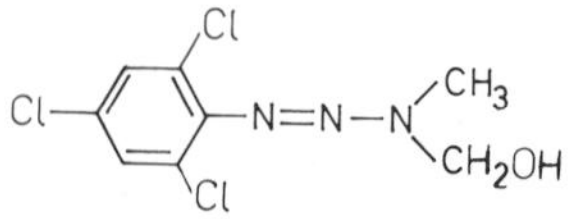

LXVII

The important chemical difference between aryl and imidazole triazenes is that the aryl derivatives are not affected by light either in the solid state or in solution. They are, however, susceptible to hydrolysis in acidic solution, the rate of decomposition being governed by the nature of the substituents in the aromatic ring and by the pH of the solution.

From the long-term point of view the most important aspect of the activity of the triazenes is not simply the fact that they are a new class of agents, but that the dialkyl compounds which contain no methyl groups and therefore give rise to higher alkyl carbonium ions are inactive. This difference between methyl and higher alkyl groups is also evident in other compounds such as hexamethyl-melamine and procarbazine. Although the reasons may be different in each case, an understanding of the mechanism which allows methyl substituents to pre-dispose so decisively toward activity will undoubtedly be very helpful in the design of other agents of this type.

Hexamethylmelamine

The antitumour activity of hexamethylmelamine (LXVIII) was discovered inde-pendently by Hendry and co-workers (1951) and by Buckley and co-workers (1950). Despite the fact that its activity against rodent tumours was at best marginal, it was introduced into the clinic (Wilson and de la Garza, 1965) and has proved to have a wide spectrum of activity against solid tumours (Legha *et al.*, 1976). Activity is most marked against ovarian cancer, lymphomas and car-cinoma of the cervix, and to a lesser extent bronchogenic carcinoma and carcin-oma of the breast.

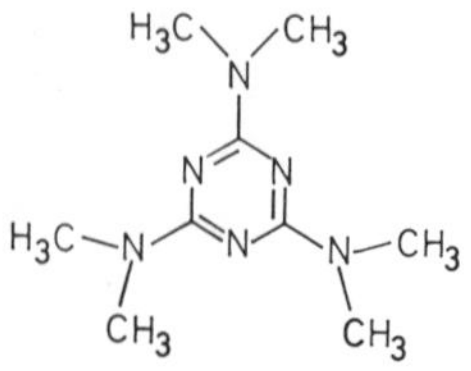

LXVIII

A number of reports on the metabolism of hexamethylmelamine in relation to its antitumour activity have been published (Worzalla *et al.*, 1973, 1974; Lake *et al.*, 1975). These indicate that enzymatic oxidative demethylation leads to loss of formaldehyde and the production of a series of lower homologues. One of the major problems associated with studies on the structure–activity relationship, metabolism and mechanism of action of hexamethylmelamine and its analogues is lack of response in the major tumour systems used for routine screening. Only one tumour, the mouse ADJ/PC6A plasma cell tumour, shows significant response, but even in this system the activity of hexamethylmelamine is in no way comparable to that of other agents to which the tumour is sensitive.

It is fortunate from many points of view that a completely new type of tumour test system has become available in the past 8 to 10 years. This method involves transplantation of human tumours into immunodeficient mice. The recipient animals used are either athymic nude mice or normal animals which have been thymectomised, lethally irradiated, and rescued with syngeneic bone marrow. A range of human tumour xenografts has been established by Cobb and Mitchley (1974). The response of these xenografts to a variety of drugs correlates well with clinical experience, but for reasons that are unknown at present some appear to be particularly sensitive to hexamethylmelamine (Mitchley *et al.*, 1975, 1976). A number of studies of the antitumour activity of compounds in this series have been carried out at the Institute of Cancer Research, London, with a human lung adenocarcinoma xenograft as the test tumour (Mitchley *et al.*, 1976).

Early theories concerning the mechanism of action of hexamethymelamine noted its superficial similarity to triethylenemelamine (TEM, XLII) and suggested that its role was that of an alkylating agent. However, the fact that hexamethylmelamine fails to react with 4-(*p*-nitrobenzyl)pyridine shows that this is not the case. Moreover its metabolism differs markedly from that of TEM (Worzalla *et al.*, 1973). From current evidence, it appears that the active metabolite formed after microsomal oxidation of hexamethylmelamine may be a methylol. This compound could condense subsequently with the amino groups of proteins and nucleic acids, or release formaldehyde (Rutty and Connors, 1977) which is mutagenic by virtue of its ability to interact with DNA. Again it should be noted that the analogous ethyl compound, hexaethylmelamine, is totally without activity.

The dose-limiting toxicity of hexamethylmelamine is neurologic, but gastrointestinal side effects and a moderate degree of myelosuppression are observed as well. Due to its very low aqueous solubility (< 0.1 mg ml^{-1}) hexamethylmelamine is usually administered orally. The only parenteral formulation available in early studies was an aqueous hydrochloric acid solution of *p*H 2. In an attempt to overcome this problem, gentisic acid complexes were prepared (Kreilgard *et al.*, 1975), with a *p*H range from 3.5 to 4.5. Unfortunately the complexes are unstable at physiological *p*H, so that injection resulted in immediate precipitation of the free drug.

A more logical approach to this problem is that followed by Cumber and Ross (1977). A series of hexamethylmelamine analogues of greater water-solubility were prepared and compared with the parent drug in respect to activity against the P246 lung tumour xenograft. Three promising compounds pentamethylmelamine (LXIX), hydroxymethylpentamethylmelamine (LXX) and *sym*-trihydroxymethyltrimethylmelamine (LXXI) were identified (Connors *et al.*, 1977). The hydroxymethyl derivatives are unstable in solution but compound LXIX is quite stable and its solubility of $2.16\,mg\,ml^{-1}$ is at least 20 times greater than that of hexamethylmelamine.

Phase I clinical trials of pentamethylmelamine (LXIX) have been undertaken in a number of centres (Ajani *et al.*, 1982; Benvenuto *et al.*, 1981; Casper *et al.*, 1981; Goldberg *et al.*, 1980; Ihde *et al.*, 1981; Muindi *et al.*, 1982). Whilst these trials have not demonstrated any useful clinical activity for the drug, the associated pharmacokinetic studies (Muindi *et al.*, 1982; Rutty *et al.*, 1982) have provided some insight into the reasons for this. Although in mice and to a lesser extent rats pentamethylmelamine is metabolised extensively to hydroxymethyl-metabolites, such metabolites being regarded as the proximally active species, in man they are not detectable.

Procarbazine

So far we have seen two classes of agents, the triazenes and hexamethylmelamine, in which replacement of the methyl substituents by higher alkyl groups results in total loss of activity. A further example of this situation is encountered with the hydrazine derivative, 1-methyl-2-(*p*-isopropylcarbamoyl)benzylhydrazine (procarbazine, LXXII). Like the triazenes and hexamethylmelamine, procarbazine is inactive against cells *in vitro* and requires enzymatic activation before it shows antitumour activity. Although *N*-demethylation is known to occur (Dewald *et al.*, 1969), a number of other pathways are implicated (Baggiolini *et al.*, 1969), especially oxidation of the hydrazine. The resultant azo linkage undergoes α-hydroxylation and carbon-nitrogen bond cleavage to yield a methyldiazene similar to the one obtained in the triazene series (figure 11).

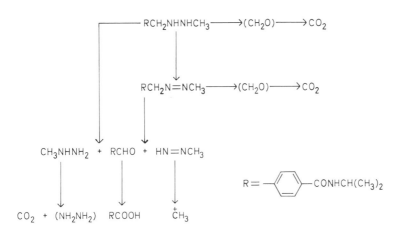

LXXII

Clinically, procarbazine has found use in the treatment of Hodgkin's disease and other lymphomas, myeloma, melanoma and lung cancer. Its side effects are those of a typical alkylating agent, but it also causes psychopharmacologic effects consistent with its ability to inhibit monoamine oxidase.

Figure 11

Platinum complexes

During an investigation of the effects of an electric field on bacterial growth, Rosenberg and co-workers (Rosenberg *et al.*, 1965) noticed that under certain conditions the bacteria grew into long filaments as a consequence of inhibition of cell division, although growth continued. They showed this to be due to a small amount of platinum dissolving away from the electrodes into the nutrient medium to form *cis*-tetrachlorodiammine platinum IV (LXXIII). Further investigation led to the discovery that compound LXXIII and *cis*-dichlorodiammine platinum II (LXXIV) were potent inhibitors of mouse sarcoma 180 (Rosenberg *et al.*, 1966). This finding has led to the synthesis and antitumour evaluation of many more compounds of similar structure (Connors *et al.*, 1972*b*; Clear and Hoeschele, 1973). (In Chapters 7 and 10 the mechanism of action of these compounds is discussed in detail.)

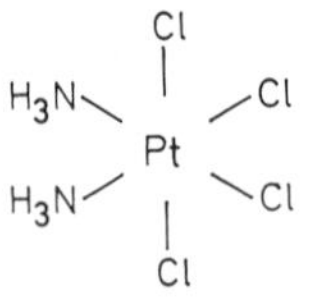

LXXIII

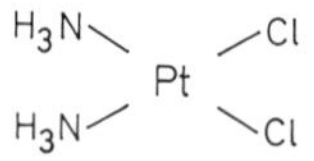

LXXIV

Activity against a wide range of experimental animal tumours has been observed for these compounds, but certain structural features have been found to be necessary. The *cis* configuration is an absolute requirement, all *trans* isomers being totally inactive (Connors *et al.*, 1972*b*). Variation of the ammine group can lead to improved activity (Connors *et al.*, 1972*b*), whereas replacement of the chlorine atoms (Cleare and Hoeschele, 1973), except by the bidentate malonate ligand, or of the amine ligand does not (Connors *et al.*, 1972*b*).

Some of the most interesting agents so far have been a series of N, N'-disubstituted *cis*-dichlorodiammineplatinum II compounds (LXXV, R = cycloalkyl) shown in table 2. Increase in the size of the alicyclic ring reduces the toxicity (LD_{50}) without a parallel loss of potency (ED_{90}) up to the N, N'-dicyclohexyl derivatives. However, as this compound is less potent that the lower homologues, the N, N'-dicyclopentyl derivative is of greatest interest as a candidate for clinical trial. One disadvantageous feature of compounds in this series is poor solubility, which makes formulation for intravenous injection difficult. Analogues containing various solubilising groups have been investigated in an attempt to overcome this problem.

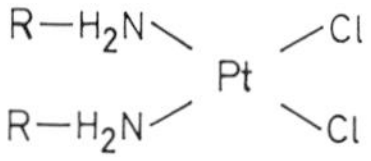

LXXV

Table 2   Antitumour activity of N, N'-disubstituted *cis*-dichloro-
diammine platinum II compounds against the ADJ/PC6A
(cf. structure LXXV)

R	LD_{50} (mg per kg)	ED_{90} (mg per kg)	Therapeutic index
Cyclopropyl	56.6	2.3	24.6
Cyclobutyl	90.0	2.9	31.0
Cyclopentyl	480	2.4	200
Cyclohexyl	> 3,200	12	> 267
Cyclooctyl	135	59	2.3

The mechanism of action of the platinum drugs is not clearly understood, but they do resemble the difunctional alkylating agents in a number of respects. For example, the N7 and 6-amino nitrogens of adenine are susceptible to attack by these platinum complexes. In the paired sequence A-T/T-A of double helical DNA the adenine moieties are 3 Å apart, an ideal situation for cross-linking by the *cis*-bidentate ligand (Thomson, 1974). Tumour lines with acquired resistance to a typical alkylating agent such as melphalan show cross-resistance not only to other alkylating agents but also to platinum complexes.

The observed similarities of the platinum complexes to traditional alkylating agents suggests that their clinical application may be limited. Clinical results with *cis*-dichlorodiammine platinum II (LXXIV), however, show this drug to be of use in the treatment of testicular tumours (Higby *et al.*, 1974; Merrin, 1979) and previously treated ovarian carcinoma. In particular, Wiltshaw and Carr (1974) have demonstrated responses in patients whose tumours had previously become unresponsive to alkylating agent therapy. Activity has also been seen against bladder cancer (Herr, 1980).

The nephrotoxicity observed with this drug has led to the development of a number of second-generation platinium analogues. One, *cis*-diammine-1, 1-cyclobutane dicarboxylate platinum II (CBDCA, JM8) has had a successful phase I clinical trial (Calvert *et al.*, 1982). The whole range of antitumour platinum analogues have recently been reviewed (Harrap, 1983).

ENHANCING THE SELECTIVITY OF ANTITUMOUR ALKYLATING AGENTS

The single characteristic which links alkylating agents together with all other antitumour agents is that they are generally cytotoxic and do not attack tumour cells preferentially. The basic aim of the chemotherapist is therefore to deliver the cytotoxic portion of the drug molecule selectively to the tumour by varying either the carrier moiety or the mode of administration. The latter approach is mainly clinical, but occasionally the medicinal chemist is asked to design agents with specific properties that will facilitate delivery to the tumour site. Examples of this are the water-soluble analogues of hexamethylmelamine.

The design of new agents is generally based on rational studies of the metabolism and mechanism of action of known antitumour agents, usually those that are already established in the clinic. This approach can often suggest improved treatment schedules for the clinical drug.

Although few of the known biochemical differences between normal and tumour cells have so far proved exploitable, there is reason to hope that these differences can lead to the successful design of new antitumour agents. Conversely, the activity of a particular drug may provide a clue to such a difference on occasion, even though the difference was not recognised when the drug was first prepared. As Cobb (1970) has pointed out, however, it may also be expedient in certain instances to destroy most of the normal tissue in a given site along with the tumour. This would apply especially in situations where the normal tissue is expected to regenerate at a rapid rate during the healing process, or where this tissue is not essential to the health of the patient in the first place.

Local administration

Because alkylating agents are general cytotoxic agents, the usual systemic mode of administration frequently leads to unacceptable damage to the bone marrow and gastrointestinal tract. Local administration of a drug to a tumour by intra-arterial infusion has been used for many years to confine cytotoxicity to the immediate area of the tumour. Any agent that does not require activation at a site distant from the tumour can be used for infusion if it is reasonably soluble under physiological conditions. Although infusion does not necessarily have to be in aqueous solution, its use is obviously limited if the drug precipitates at the site of administration. If the drug is not to affect more distant normal tissue, it must also have a very short chemical half life under physiological conditions. A number of workers have designed compounds with these points in mind.

Seligman and co-workers (1962) developed a series of sulphur mustard derivatives with very short half-lives. Of these, compound LXXVI (X = Cl, N = 3) had a half life of 1.3 minutes at pH 7.4 and 37°C and induced palliation in tumours of the head and pelvis when administered locally via the intraarterial route. Included in the series was a compound (LXXVI, X = I, n = 4) with the extremely short half life of 0.6 seconds. Unfortunately these compounds lacked the stable covalent linkage between alkylating centres which is probably necessary for carcinostatic activity.

Davis and Ross (1965) designed a short half life sulphur mustard derivative which is water-soluble and can bind tightly to proteins because it contains a carboxylic acid group that can form an anion at physiological pH. Additionally, the structure linking the two alkylating centres is chemically stable enough to allow permanent cross-links to be formed. The compound in question, p -[2, 3-*bis*(2-bromoethylthio)-n-propyloxy] benzoic acid (LXXVII), had a half life of about 4.8 seconds at 37°C and pH 8.3. Cobb (1967) showed that this drug, when administered intravenously, had no effect on the Walker 256 tumour in rats at a dose which was 100% curative when injected directly into the artery supplying the tumour. Its biological half life was of the same order as the *in vitro* value of 4.8 seconds. A clinical trial of the drug against a variety of tumours (Meyza and Cobb, 1971) indicated significant antitumour activity, especially against sarcoma. There was marked pain relief, no toxicity to the bone marrow, platelets, haemoglobin, or intestine, and only occasional effects on the white cell count. Although this preliminary human study demonstrate the advantages of using a chemically very reactive drug with a short biological half life, the results were not of sufficient interest to stimulate further clinical use.

$$XCH_2CH_2S-(CH_2)_n-CONHCH_2CH_2NHCO-(CH_2)_n-SCH_2CH_2X$$

LXXVI

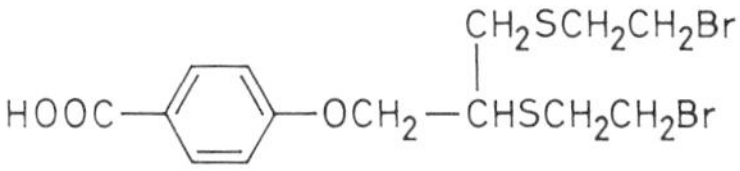

$$CH_2SCH_2CH_2Br$$

$$HOOC-\langle\bigcirc\rangle-OCH_2-\overset{|}{C}HSCH_2CH_2Br$$

LXXVII

Selectivity due to enzyme activation or detoxification

The simplest member of the series of aryl nitrogen mustards is *N*, *N-bis*(2-chloroethyl)aniline (aniline mustard, XV). Although first reported (Robinson and Watt, 1934) as a synthetic intermediate, its antitumour activity was not investigated until the late 1940s (Ross, 1949) and even then its full potential was not realised. Connors and Whisson (1965) showed that the ADJ/PC5 mouse plasma cell tumour is highly sensitive to aniline mustard. *p* -Hydroxyaniline mustard (LXXVIII) was some twelve times more toxic than aniline mustard itself, but the toxicity and antitumour effect of other *para* -substituted derivatives were greatly reduced. Connors and Whisson suggested a number of possible reasons for this selectivity:

(1) The tumour converts aniline mustard to the *p* -hydroxy derivative more effectively than the liver.

(2) The tumour *p* -hydroxylates aniline mustard just as effectively as the liver, but fails to convert the hydroxylated product to a glucuronide or sulphate conjugate. In this model conjugation is considered to be a detoxification reaction which, if it did not occur in the tumour, would enable *p* -hydroxyaniline mustard to exert its expected cytotoxic effect selectively at that site.

(3) *p* -Hydroxylation only occurs in the liver and is followed by conjugation, but the glucuronide or sulphate is taken up selectively by the tumour where it is converted to the cytotoxic *p* -hydroxy derivative if the tumour has a high level of glucuronidase or sulphatase.

Connors and Whisson (1966) later showed a correlation between the β-glucuronidase activity of tumours and their response to aniline mustard. Tumours whose β-glucuronidase activity was similar to normal liver (sarcoma 180 and NK lymphoma) were unaffected by aniline mustard, whereas those with elevated enzymes levels (ADJ/PC5 and ADJ/PC6A plasma cell tumours) underwent total regression. Double and Workman (1977) have shown another experimental tumour with high β-glucuronidase activity to be very susceptible to aniline mustard therapy. Recently Benckhuysen *et al.* (1981) have shown that 4 days after ovariectomy the β-glucuronidase levels of dimethylbenzanthracene-induced tumours increased considerably. At this time aniline mustard, which is ineffective against the tumour as a single agent, now acts synergistically with ovariectomy. These findings add further support to the hypothesis that the extreme sensitivity of the two plasma cell tumours to aniline mustard is a consequence of glucuronide formation in the liver, followed by breakdown of the conjugate inside the tumour to give the toxic *p* -hydroxy derivative *in situ*.

Subsequent synthesis of conjugates of *p* -hydroxyaniline mustard and testing against the ADJ/PC6A plasma cell tumour (Bukhari *et al.*, 1972) showed that various glucuronic acid derivatives and the sodium salt of the phosphate ester possess appreciable activity. The highest antitumour activity is shown by the methyl ester of the glucuronic acid derivative LXXXIX (R = CH$_3$), which has a somewhat higher activity than aniline mustard (XV) itself and six times the activity of the much more toxic unconjugated phenolic derivative LXXVIII.

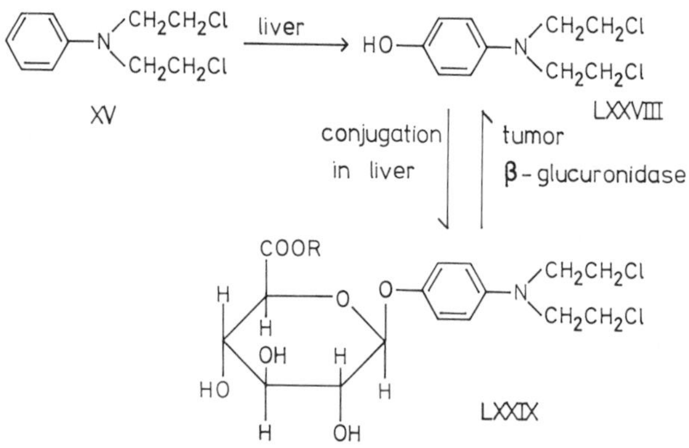

XV   liver   LXXVIII

conjugation in liver   tumor β-glucuronidase

LXXIX

Connors and co-workers (1973*a*) have studied the metabolism of aniline mustard in some detail. They showed that rat liver microsomes convert aniline mustard into the glucuronic acid derivative LXXIX (R = H), which was also identified in the serum of rats treated with aniline mustard. These findings validated the postulate that the glucuronide is a major metabolite of aniline mustard. This, and results obtained with the phosphate ester (Workman and Double, 1978) appear to recommend the use of the various conjugates against tumours with elevated levels of the appropriate catabolic enzyme (Bukhari *et al.*, 1972).

Young and co-workers (1976) undertook an extremely interesting clinical trial of aniline mustard. The novel feature of this investigation was that the β-glucuronidase activity of the tumour was determined by histochemical staining of preparations obtained by biopsy or aspiration. Significant antitumour activity with clinical benefit was demonstrated in 5 of 37 patients with cancer of the prostate, and in 1 of 8 patients with renal cancer. No marked response was seen in 4 patients with breast cancer, 8 with myeloma, and 2 with tumours of the bladder. A partial correlation appeared to exist between very intense glucuronidase staining and tumour regression in the kidney and prostate lesions; however, these high levels were observed only rarely. Perhaps the most significant observation was that two patients who showed high tumour β-glucuronidase activity and responded initially to aniline mustard therapy became unresponsive to further therapy on relapse and showed a concomitant loss of enzyme activity. A new trial of aniline mustard confined to those patients with a high level of tumour β-glucuronidase activity seems clearly indicated.

β-Glucuronidase has a *p*H optimum of 5.2, and since tumours are reported to have a lower *p*H than normal tissue (see later discussion in relation to tumour *p*H), it would be expected that the β-glucuronidase activity should be highest in the tumour. Glucose administration serves to further lower tumour *p*H and hence promote β-glucuronidase activity. That this is true was demonstrated by Bicker (1974) in respect of the β-D-glucuronide of 2-hydroxyquinoline. Vanquero and co-workers (1975) have shown that β-glucuronidase activity is greater in human malignant melanoma than surrounding skin, but it has not yet been possible to translate these findings into an effective clinical regimen against this very recalcitrant tumour.

A number of esters have been prepared from *p-bis*(2-chloroethyl)aminophenol (LXXXIII). The acetate and benzoate derivatives show a higher level of antitumour activity than the parent compound (Ross *et al.*, 1955*a*), and their activity correlates well with the presence of the appropriate esterase enzyme (Hebborn and Danielli, 1958). Urethane derivatives of LXXVIII have also been synthesised (Benn *et al.*, 1961). Of these, compound LXXX is of particular interest because of its relatively high therapeutic index of 17 against the Walker 256 rat tumour as compared to 2.3 for the parent phenol (Bardos *et al.*, 1965).

LXXX

The high toxicity of *p*-hydroxyaniline mustard (LXXXVIII) is due to the electron-donating character of the hydroxyl group, which is so strong that even the carbon-halogen bonds in the side chain become extremely reactive. Replacement of the hydroxy group by amino to give *p*-phenylenediamine mustard (LXXXI) enhances this effect. As with the hydroxyl derivative, substitution of the amino group results in reduced toxicity. This suggests that it may be possible to release the active drug from relatively non-toxic latent derivatives in the presence of suitable enzymes. A number of attempts have been made to prepare such derivatives, including the tetrazolium salt LXXXII and the formazan derivative LXXXIII which are presumably capable of undergoing bioreduction *in vivo* (Tsou and Su, 1963). Tumour inhibition tests against the Walker 256 tumour in rats showed the formazan derivative LXXXIII to have the higher activity. This could be due to poor diffusion of the cationic tetrazolium salt through the cell membrane. More recently the D-alanyl-leucyl-lysyl derivative of LXXXI has been prepared as a prodrug selectively activated by plasmin in tumours which produce large amounts of plasminogen activator (Carl *et al.*, 1980) and the γ-glutamyl derivative for activation in tumours high in γ-glutamyl transpeptidase (Manson *et al.*, 1981).

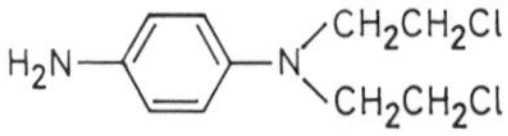

LXXXI

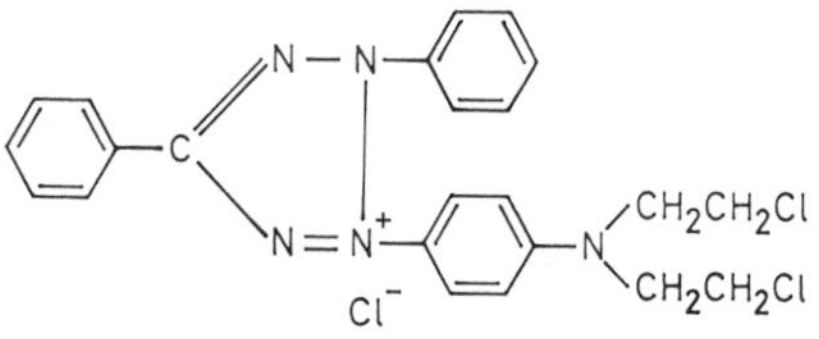

LXXXII

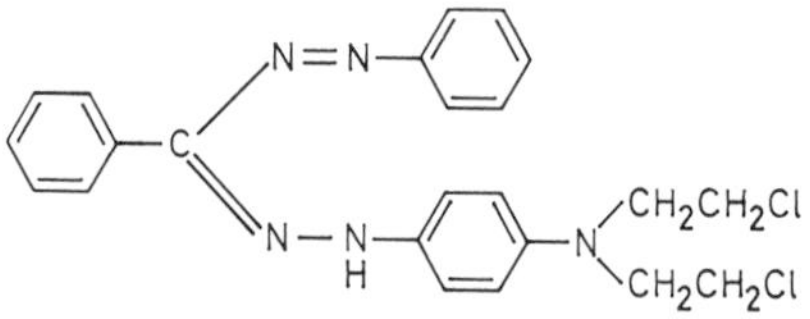

LXXXIII

For a prodrug to be truly selective, it must fulfill three criteria. First, the drug must be generally non-toxic to the host in the form in which it is administered. This means that when the drug becomes distributed throughout the body it must not severely damage normal sensitive tissues such as the bone marrow and gut epithelium. Second, when metabolised by the tumour, the active metabolite should be sufficiently toxic to kill the tumour cells; and third, the toxic metabolites should remain within the confines of the tumour if possible.

It is extremely unusual for a drug to be designed with a specific tumour as the target; more often, a compound shows selective activity against certain tumour types only after it reaches the clinic. A phenylenediamine mustard derivative has, however, been designed, with the foregoing criteria for selectivity in mind, specifically for the treatment of primary hepatocellular carcinoma. African and Far Eastern countries have a relatively high incidence of this type of liver tumour, which is particularly refractory to treatment. As a result of its rapid proliferation, patients have a life expectancy of only some 2 months from the time of diagnosis. This contrasts with the primary liver tumours seen in Europe and the United States, which grow much more slowly.

Enzyme activities in tumours are usually much lower than in normal tissue; however, in the case of primary hepatocellular carcinoma, a significant level of a specific enzyme called azoreductase is maintained (Autrup *et al.*, 1974). On this basis, a series of azomustards of general structure LXXXIV was investigated in an attempt to obtain a drug that would be specific for this tumour (Bukhari *et al.*, 1973). This investigation confirmed earlier work by Ross and Warwick

(1955*b*, 1956) in showing that, because of their highly conjugated structure, nitrogen mustards of this type tend to be chemically unreactive and therefore non-toxic. In this way the first criterion for selectivity was clearly fulfilled.

In the presence of azoreductase, compounds of type LXXIV are expected to be reduced to give two metabolites, the aniline derivative LXXV and the *p*-phenylenediamine derivative LXXVI (Bukhari *et al.*, 1973; Ross and Warwick, 1955*b*). Ross and Warwick (1955*b*) showed that this reduction occurs both chemically in the presence of hydrazine, and enzymatically under the influence of the xanthine–xanthine oxidase system. In particular they found a striking correlation between reducibility and antitumour activity against the Walker 256 rat carcinoma. From these investigations it was found that electron-donating substituents *ortho* to the azo linkage are necessary for activity. The most active compound of the series was 4-[*N*, *N*-*bis*(2-chloroethyl)amino]-2'-carboxy-2-methylazobenzene (CB 1414; LXXXIV, R = COOH, R' = CH$_3$, R" = H, X = Cl). This compound underwent clinical trial (Israels and Ritzmann, 1960) but was devoid of significant benefit, especially in comparison with chlorambucil; however, it should be noted that liver tumours were not investigated.

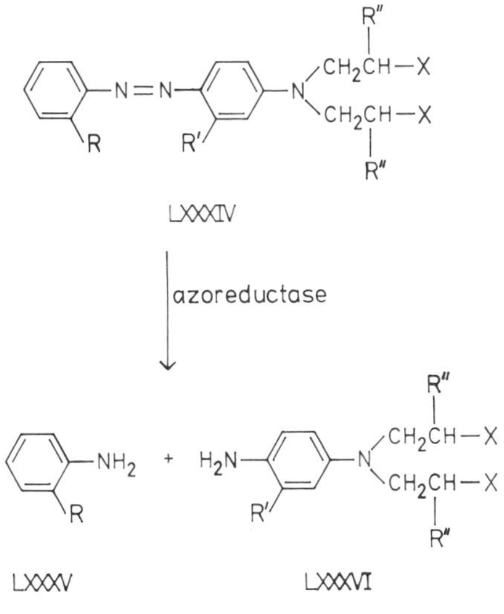

Some ten years later Connors and co-workers (1972*a*) confirmed the hypothesis that the enzyme system necessary for the reductive cleavage of azo mustards is actually present in liver homogenate, and that the expected metabolites are in fact formed. Of the two metabolites resulting from reduction of CB 1414, one is anthranilic acid (LXXXV, R = COOH) which has no antitumour activity and only a moderate host toxicity. The second metabolite (LXXXVI, R' = CH$_3$, R" = H, X = Cl) is a *p*-phenylenediamine mustard. Nitrogen mustards of this type are extremely potent alkylating agents, and are known to be clinically effective against tumours. Thus the second criterion of selectivity was fulfilled, that is the drug is metabolised to a potent cytotoxic agent by an enzyme which maintains its activity in a rapidly proliferating tumour. There remained the problem of retaining the metabolite within the tumour.

There is really no means, in this type of system, of ensuring that the active metabolite remains localised in the tumour. One way of overcoming the problem is to design a drug whose toxic metabolite will have a very short half life under physiological conditions and will have mostly decomposed to non-toxic products by the time it reaches sensitive normal tissues such as the bone marrow or gut epithelium. A series of azomustards LXXXIV (R = COOH, R' = CH$_3$) containing various substituents R'' and X, and also the corresponding phenylenediamine derivatives LXXXVI (R' = CH$_3$), were investigated by Bukhari and co-workers (1973) who found that the metabolite with the shortest half life was 4-*bis*(2-bromo-*n*-propyl)amino-2-methylaniline (LXXXVI, R' = R'' = CH$_3$, X = Br), which had a half life of 41 seconds at *p*H 7.5 and 37°C, approximately the 'circulation time' in man. Thus the compound 4-[*N,N-bis*(2-bromo-*n*-propyl)amino]-2'-carboxy-2-methylazobenzene (CB 10-252; LXXXIV, R = COOH, R' = R'' = CH$_3$, X = Br) is a virtually non-toxic drug, but it is reduced readily to a very toxic metabolite in tissues that contain high azoreductase activity. Although this compound exists in diasteroisomeric forms, because of the presence of an asymmetric centre in each alkyl chain, they are equally susceptible to enzymatic reduction (Wilman, 1980). The rapid hydrolysis of the metabolite prevents its toxic effect from spreading to other proliferative tissue. Although one could expect large amounts of the active metabolite to be released in normal liver, there should be little tissue damage since the cells are not dividing actively.

Confirmation of this selectivity was produced by Connors and co-workers (1973*b*). When the Walker 256 tumour was implanted in the liver of rats it was more susceptible to the action of CB 10-252 than to cyclophosphamide (LXXXVIII), whereas when the tumour was implanted in the flank the reverse was true.

These results were of sufficient interest for preliminary clinical trials to be undertaken in Africa, and initial data were encouraging. Unfortunately a clinical trial in England (Tattersall and Curt, 1976) yielded no useful results, although this was most probably due to differences between the African and European tumours, particularly in respect to proliferation rate. Additionally the azoreductase enzyme level in the tumour was only slightly higher than the level in other tissues, leading to dangerous myelosuppression.

The surprisingly high carcinostatic activity against a wide range of tumours (28 of 37 animal tumours tested were sensitive) of *N,N-bis*(2-methanesulphonyloxyethyl)-4-nitrosoaniline (LXXXVII, R = NO) is probably due to its *in vivo* reduction to the amino analogue LXXXVII (R = NH$_2$).

$$R-\!\!\left\langle\!\!\bigcirc\!\!\right\rangle\!\!-N\!\!\begin{array}{c}CH_2CH_2OSO_2CH_3\\[4pt]CH_2CH_2OSO_2CH_3\end{array}$$

LXXXVII

Probably the most clinically useful drug of the mustard type is cyclophosphamide (LXXXVIII). On the basis that tumours had been reported earlier to have a higher phosphoramidase content than normal tissue (Gomori, 1948), a large series of latent derivatives was prepared in the hope that they would be converted into active compounds *in vivo* by enzymatic hydrolysis. Following its synthesis in 1958 (Arnold *et al.*, 1958), cyclophosphamide was found to be the most active of these derivatives against experimental rodent tumours and was also very successful in clinical trials. It was soon demonstrated that simple enzymatic cleavage was not the reason for the drug's activity since it is inactive against tumour cells *in vitro*, and that in order to be active the drug requires oxidation by liver microsomes. Research by a large number of investigators has led to a vast body of information on the mechanism of action of cyclophosphamide (Montgomery and Struck, 1973; Hill, 1975). The currently favoured metabolic pathway (see also Chapter 9), deduced on the basis of actually isolated metabolites, involves initial microsomal oxidation of cyclophosphamide in the liver to 4-hydroxycyclophosphamide (LXXXIX) which appears to exist in equilibrium with its tautomer, aldophosphamide (XC) (figure 12). These intermediates give rise, on further enzymic reaction, to 4-ketocyclophosphamide (XCI) and carboxyphosphamide (XCII) respectively, both of which have low toxicity and are excreted in the urine. Aldophosphamide (XC) is chemically unstable, undergoing β-elimination to give acrolein (XCIV) and phosphoramide mustard (XCIII).

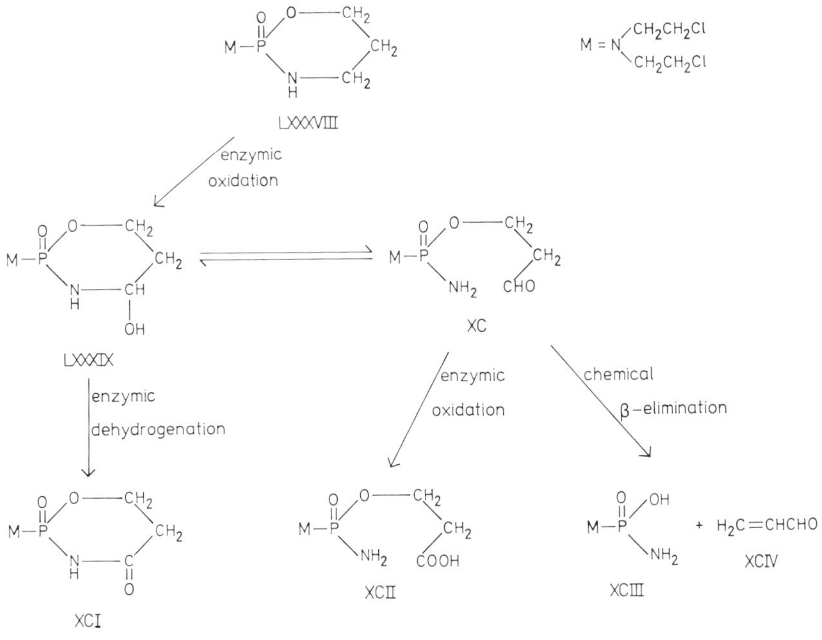

Figure 12   The metabolic pathways for cyclophosphamide

Of these ultimate metabolites, 4-ketocyclophosphamide (XCI) and carboxy-phosphamide (XCII) are less toxic than cyclophosphamide itself and phosphor-amide mustard (XCIII) and acrolein (XCIV) are several hundred times more toxic (Connors *et al.*, 1974). This has led to the conclusion that the activity of cyclophosphamide may well be due to differences in enzyme levels between normal and tumour cells. The observed selectivity of the drug would be explained if normal cells contained high levels of the enzymes required to form the non-toxic metabolites and if these enzymes were absent in the tumour (Connors *et al.*, 1974; Cox *et al.*, 1975). This theory has been criticised recently and it has been suggested that the selectivity may be the result of the binding of 4-hydroxy-cyclophosphamide (LXXXIX) to thiol-containing proteins to form an inactive 4-S-complex. The complex may then be acted on enzymatically to release the drug selectively in the tumour (Brock and Hohorst, 1977; Brock, 1983).

Physical distribution

Alkylating agents produce their cytotoxic effect by chemical reaction with a wide range of cellular components. They must therefore be designed in such a way that an adequate concentration of drug will reach the target site within the cell. There are a number of factors that affect the way in which a drug is dis-tributed in the body. The most important of these is probably its relative solu-bility in aqueous and lipid phases. For drugs to be administered by intravenous injection, water solubility at physiological *p*H is an obvious necessity. Chloram-bucil (IV) was designed with this purpose in mind. Other water-soluble compounds used in this manner include cyclophosphamide (LXXXVIII) and triethylenemela-mine (XLII), which have been used in treating tumours of the brain, lung, head and neck, liver, abdomen and extremities. This technique of administration called 'regional perfusion', owes its selectivity to the fact that the drug is intro-duced into the blood supply of a specific organ or region of the body which is isolated by physical means from the bulk of the circulation. This spares distant sites and also maintains a high drug concentration in the immediate vicinity of the tumour.

One of the criteria necessary for a compound to cross the blood-brain barrier is lipid solubility, a characteristic of BCNU (LVI). The partition coefficient of a drug between aqueous and lipid phases also influences its distribution through-out the whole organism as well as its diffusion across cell membranes. A number of studies have shown that a lipid–water partition coefficient close to unity is necessary for maximum biological activity. Series of compounds for which this applies include α, ω-*bis*(methanesulphonyloxy)hexanes (XXIX, R = H) by Hudson and co-workers (1958), analogues of 5-aziridino-2,4-dinitrobenzamide (XLVI) by Khan and Ross (1970), and the nitrosoureas by Hansch and co-workers (1972). It should be noted, however, that in the triazenoimidazole series Hansch and his colleagues (1972) observed activity to be associated with a wide range of partition coefficients.

For any series of tumours implanted at different sites the observed variation in antitumour activity of a drug with a particular partition coefficient must be viewed in the context of these studies (Cain, 1975). However, it is also possible that in some instances maximal deposition of a drug at various sites of the body will depend on physical properties other than the partition coefficient.

Tissue specificity

Since certain tissues of the body are not vital insofar as survival is concerned, it may be acceptable to destroy most of the normal tissue along with the tumour under some circumstances. As suggested by Cobb (1970), this might apply especially to such tissues as the breast, ovary, uterus, testis, prostate and pancreas.

Some classes of compounds have been shown to become localised in melanomatous tissue. For instance, analogues of chloroquine and quinacrine have been shown to have such an affinity. In this regard the work of Jones and co-workers (1968) on chloroquine mustard (XCV) and quinacrine mustard (XCVI) is particularly noteworthy. Similarly, L-dopa mustard (XCVII, R = $N(CH_2CH_2Cl)_2$) was synthesised (Vasil'eva *et al.*, 1972) on the basis that L-dopa (β-(3,4-dihydroxyphenyl)-L-alanine; XVCII, R = H) is incorporated into the insoluble melanin pigment which is deposited extensively in melanotic melanoma.

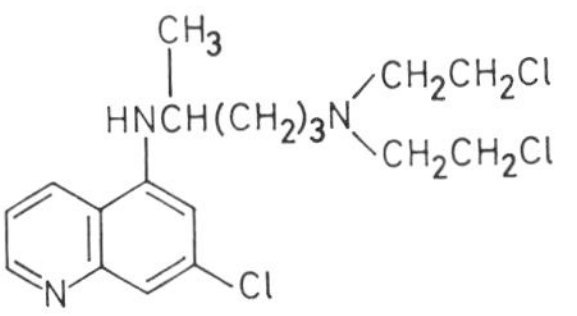

XCV 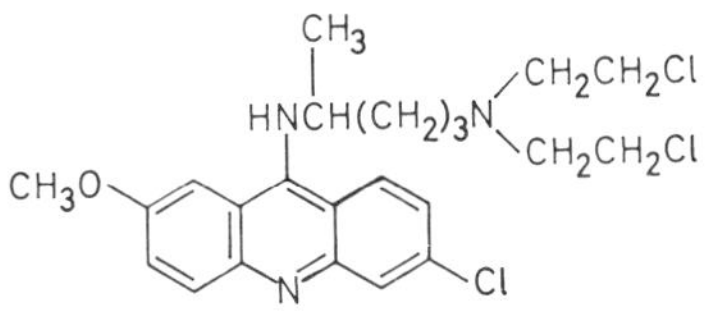   XCVI 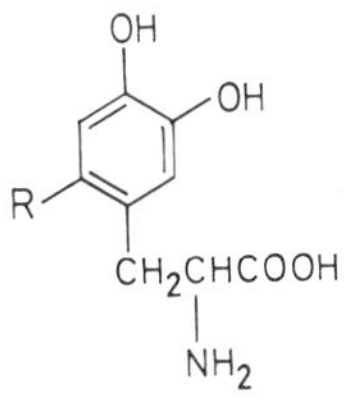   XCVII

The specific activity of the urethane mustard LXXX towards Walker tumour cells may be due to selective uptake by receptors unique to the tumour rather than to enzymatic activation (Bardos *et al.*, 1969; Hebborn *et al.*, 1970). Sieber and Adamson (1976) called attention to a number of compounds that are known to concentrate in specific tissues, and proposed ways of attaching cytotoxic groups to them. However, this must remain a theoretical exercise until evidence is put forward to show that the groups to be replaced are not those responsible for selective tissue concentration of the drug.

Tumour *pH*

Observations in patients with melanomatosis using a microelectrode technique (Ashby, 1966) have shown the mean *p*H of the tumour to be 6.80, compared with 7.45 for normal tissue, and that the tumour *p*H dropped to 6.54 following dextrose administration, confirming results already obtained in a number of experimental rat tumours. However, preliminary measurements using a non-invasive nuclear magnetic resonance technique to determine tumour *p*H has only shown a very slight *p*H difference and then only when compared with muscle (Griffiths *et al.*, 1981). The possible reason for an increased acidity of the tumour in relation to normal tissue is the high rate of production of lactic acid by anaerobic glycolysis in the tumour. It might therefore be expected that a drug with a pK_a close to the *p*H of the tumour would be deposited selectively at this site. Stevens and co-workers (1950) have shown this to be true for sulpha-pyrazine, even when the drug is administered at some distance from the tumour.

Ross and his colleagues have made a number of attempts to apply these observations to cancer chemotherapy. In assays against the Walker 256 rat tumour, Ross showed that prior glucose administration caused enhanced activity in the case of aromatic nitrogen mustards carrying a basic side chain (Ross, 1961). *p-Bis*(2-chloroethyl)aminophenethyldimethylamine (XCVII), with a pK_a of 8.9, is an example of this class of agent. Neutral (aniline mustard; XV), acidic (chlorambucil; IV) and zwitterionic (melphalan; XXXIX) aromatic nitrogen mustards would not be expected to be concentrated selectively in acid cells, and no enhancement of antitumour activity by glucose treatment was in fact observed.

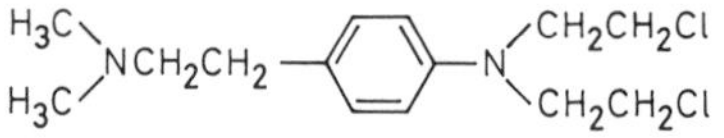

$$H_3C\diagdown NCH_2CH_2 \diagdown \diagup CH_2CH_2Cl$$
$$H_3C\diagup \qquad\qquad N\diagdown CH_2CH_2Cl$$

XCVIII

Following the work of Stevens and co-workers (Stevens *et al.*, 1950) a series of sulphonamide derivatives of nitrogen mustard were prepared and examined for antitumour activity (Calvert *et al.*, 1968). The most effective derivatives against the Yoshida sarcoma were those with a pK_a in the region of 6.5, that is, those derived from sulphadiazine. The most effective of the whole series was 2′-pyrimidinyl-*p*-[*N,N-bis*(2-chloropropyl)aminobenzene]sulphonamide (XCIX), which has a therapeutic index against the Yoshida sarcoma of 25.4 compared with 13.3 for melphalan and 24.5 against the Walker 256 tumour compared with 9.0 for melphalan. This activity was not confirmed clinically, even though it could be shown that sulphadiazine (C) is concentrated twofold more in the

Yoshida sarcoma than in normal liver. Later investigations using radioactively labelled materials by Abel and co-workers (1973) showed that despite the similarities in pK_a the mustard derivative XCIX was only poorly taken up by the Walker 256 tumour whereas the concentration of the parent sulphadiazine (C) was three times greater in the tumour than in the liver.

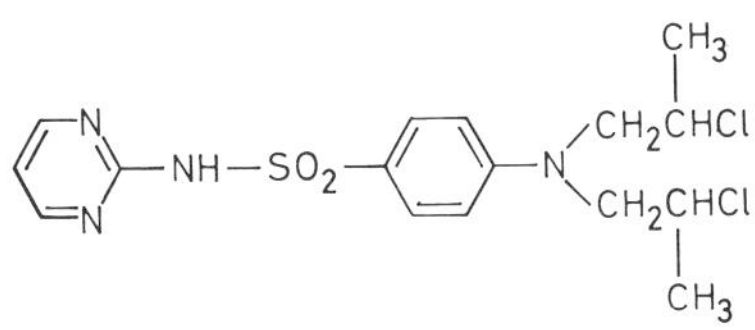

XCIX

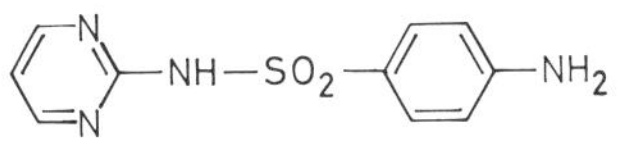

C

In a more extensive study on sulphonamides Abel and co-workers (1975) found that some compounds with pK_a values in the region of 6.5 showed high tumour to liver deposition ratios, whereas others showed no such selectivity. Selective deposition was sometimes also observed with drugs that had a pK_a higher than 6.5, as well as others that had a pK_a lower than 6.5. It would there-fore seem that there is no correlation between pK_a and selective accumulation of a sulphamide derivative in malignant cells. Moreover, there appears to be no correlation between the tissue accumulation of a sulphonamide and its solubility in 0.1 M phosphate buffer at pH 7.5 and 37°C or its partition coefficient between *n*-octanol and aqueous phosphate buffer. Replacement of the primary amino group in some of the sulphonamides by a 3,3-dimethyl-1-triazeno moiety led to activity against the TLX/5 tumour, though once again there was no cor-relation with pK_a.

The cytotoxic derivatives of sulphonamides seem to be a classic example of drugs that are active 'for the wrong reason'. Although the basis for their high activity is unknown, it seems not to be related to selective deposition in tumour tissues. Alteration of the primary amino group, as is true in every series of sub-stituted sulphonamides, can bring about dramatic changes in the physicochemical properties of the molecule as a whole. One should not be surprised, therefore, if the antitumour activity of any given series of sulphonamide derivatives fails to follow the predicted pattern. This need not mean that the sulphonamide carrier structure should be discarded in the search for improved drug selectivity. What it does mean is that, instead of merely making gross changes in the basic moiety, one ought to devise new and subtler means of introducing cytotoxic groups into the molecule.

Agents conjugated to tissue specific proteins

As this chapter has attempted to stress repeatedly, it is rather easy to design a cytotoxic agent, but extremely difficult, if not impossible, to ensure that activity is directed solely against tumour cells. A novel approach to the solution of this problem involves the use of the host's natural immune defence mechanism to carry the cytotoxic agent to the tumour. The rationale for this strategy is that a cytotoxic drug injected into a patient after first being linked chemically to a tumour-specific antibody might be delivered mainly to the tumour rather than to other tissues of the host.

Probably the first experiment of this kind was that of Mathé and co-workers (1958) who attached methotrexate to hamster antibodies specific for L1210 leukaemia by diazotisation and coupling. Leukaemic mice treated with the conjugate survived longer than those treated with methotrexate itself. The antibody alone was ineffective. A number of other macromolecular carriers have been examined in this way, including diphtheria toxin (Moolten *et al.*, 1976) and glucose oxidase. In the latter instance the conjugate is administered together with iodide and lactoperoxidase (Philpott *et al.*, 1973). The cytotoxic drug used most often in these types of conjugates has been chlorambucil (IV).

A chlorambucil–antibody complex, obtained by simple mixing of the drug and γ-globulin, was shown by Ghose and co-workers (1972) to produce a significant increase in survival of mice bearing the EL4 lymphoma in relation to animals treated with either component alone. However, the validity of the method used to prepare the complex is somewhat questionable (Israels and Linford, 1963), since one would expect the cytotoxic activity of the chlorambucil to be diminished considerably as a result of premature hydrolysis or reaction with the γ-globulin (Ross, 1974). Despite this objection, a number of other workers have used the same technique, or one that produces a physically bound complex instead of a covalently bonded structure.

Two groups have apparently succeeded in covalently attaching an alkylating agent to a carrier protein. Linfold and co-workers (1974) found that trenimon (XLIII) could be bound covalently to partially reduced proteins such as immunoglobulins to give four to ten trenimon residues per molecule of antibody. The conjugate retained antibody activity, was an active alkylating agent, and possessed immunologically specific cytotoxicity. The same group has also shown that a conjugate of trenimon with rabbit antibody to L5178Y mouse lymphoma cells has higher activity than non-conjugated controls (Froese *et al.*, 1976). Ross (1974) has demonstrated that there is little, if any, physical binding of chlorambucil to human γ-globulin in solution, though progressive chemical interaction does occur. In addition he has developed a method for conjugating chlorambucil to human γ-globulin through the carboxyl group (Ross, 1975*a*) which leaves the nitrogen mustard grouping intact (Ross, 1975*b*). By means of this method it is possible to covalently attach 30 molecules of chlorambucil to one molecule of γ-globulin.

Vennegoor and co-workers (1975) have reported the covalent attachment of chlorambucil to the γ-globulin fraction of antisera against human melanoma. However, although they found the molar ratio of active chlorambucil covalently bound to IgG to be in the range of 55 to 100, no therapeutic advantage was observed for the conjugate over a simple physical mixture of chlorambucil and γ-globulin. Although the authors stated that chlorambucil was still an active alkylating agent following covalent binding, this seems rather surprising in view of the high *p*H and temperature used for the reaction. The number of molecules of drug bound per molecule of γ-globulin also seems unusually high, since the addition of 30 molecules of chlorambucil is usually sufficient to cause γ-globulin to become denatured and precipitate from solution.

A way of overcoming this problem has been devised (Rowland *et al.*, 1975) in which many molecules of the cytotoxic agent are linked to an intermediate carrier molecule which is then conjugated to the antibody. The cytotoxic drug used in this instance was *p*-phenylenediamine mustard (LXXXI), and the carrier was poly-γ-L-glutamic acid. The mustard was linked through its *para*-amino group in the ratio of 45 molecules of drug per molecule of carrier. The complex was then coupled to a specific immunoglobulin against murine EL4 lymphoma. The final complex showed much greater antitumour activity against the EL4 tumour than any of its individual components.

Although chemical techniques for the conjugation of a wide range of cytotoxic agents to immunoglobulins are now available, the application of this valuable principle is hampered by the many practical difficulties that surround the isolation of pure antibodies against human tumours. In recent years due to the low potency of the alkylating agents they have tended to be superceded as the cytotoxic entity in this application by plant toxins such as abrin and ricin.

CONCLUSIONS

The foregoing discussion of the reactions of various classes of alkylating agents and their application to cancer chemotherapy is intended to give a glimpse of the very large number of compounds that have been investigated experimentally in relation to the very small number actually reaching the clinic. This enormous gap is due primarily to the fact that most compounds of the alkylating agent type turn out to be much less tumour-selective than one might hope. The major thrust in this field, therefore, is still the discovery of exploitable biochemical and pharmacological differences between tumour cells and normal cells, and the implementation of bold new synthetic programmes based on these findings.

REFERENCES

Abel, G., Connors, T. A., Ross, W. C. J., Nam, N-H., Hoellinger, H. and Pichat, L. (1973). *Eur. J. Cancer*, 9, 1

Abel, G., Connors, T. A., Goddard, P. M., Hoellinger, H., Nam, N-H., Pichat, L., Ross, W. C. J. and Wilman, D. E. V. (1975). *Eur. J. Cancer*, 11, 787

Ajani, J. A., Cabanillas, F. F. and Bodey, G. P. (1982). *Cancer Treat. Rep.*, 66, 1227

Arnold, H., Bourseaux, F. and Brock, N. (1958). *Nature, Lond.*, 181, 931

Ashby, B. S. (1966). *Lancet*, ii, 312

Audette, R. C. S., Connors, T. A., Mandel, H. G., Merai, K. and Ross, W. C. J. (1973). *Biochem. pharmac.*, 22, 1855

Autrup, H., Thurlow, B. J. and Warwick, G. P. (1974). *Biochem. Pharmac.*, 23, 2341

Baggiolini, M., Dewald, B. and Aebi, H. (1969). *Biochem. Pharmac.*, 18, 2187

Bardos, T. J., Datta-Gupta, N., Hebborn, P. and Triggle, D. J. (1965). *J. med. Chem.*, 8, 167

Bardos, T. J., Chmielewicz, Z. F. and Hebbron, P. (1969). *Ann. N. Y. Acad. Sci.*, 163, 1006

Benchkhuysen, C., Ter Hart, C. G. J. and Van Dijk, P. J. (1981). *Cancer Treat. Rep.*, 65, 567

Benn, M. H., Creighton, A. M., Owen, L. N. and White, G. R. (1961). *J. chem. Soc.*, 2365

Benn, M. H., Kazmaier, P., Watanatada, C. and Owen, L. N. (1970). *J. C. S. chem. Commun.*, 1685

Benvenuto, J. H., Stewart, D. J., Benjamin, R. S. and Loo, T. L. (1981). *Cancer Res.*, 41, 566

Bergel, F. and Stock, J. A. (1954). *J. chem. Soc.*, 2409

Bicker, U. (1974). *Nature, Lond.*, 252, 726

Brock, N. (1983). In *Structure-Activity Relationships of Antitumour Agents* (ed. D. N. Reinhoudt, T. A. Connors, H. M. Pinedo and K. W. van de Poll), Martinus Mijhoff B. V., Den Haag, Netherlands, p. 239

Brock, N. and Hohorst, H.-J. (1977). *Z. Krevsforsch.*, 88, 185

Brookes, P. and Lawley, P. D. (1960). *Biochem. J.*, 77, 478

Brown, D. M., Magrath, D. I. and Todd, A. R. (1955). *J. chem. Soc.*, 4396

Brown, S. S. and Timmis, G. M. (1961). *J. chem. Soc.*, 3656

Brundrett, R. B., Cowens, W., Colvin, M. and Jardine, I. (1976). *J. med. Chem.*, 19, 958

Buckley, S. M., Stock, C. C., Crossley, M. L. and Rhoads, C. P. (1950). *Cancer Res.*, 10, 207

Bukhari, M. A., Everett, J. L. and Ross, W. C. J. (1972). *Biochem. Pharmac.*, 21, 963

Bukhari, A., Connors, T. A., Gilsenan, A. M., Ross, W. C. J., Tisdale, M. J., Warwick, G. P. and Wilman, D. E. V. (1973). *J. natn. Cancer Inst.*, 50, 243

Cain, B. F. (1975). *Cancer Chemother. Rep.*, 59, 679

Calvert, N., Connors, T. A. and Ross, W. C. J. (1968). *Eur. J. Cancer*, 4, 627

Calvert, A. H., Harland, S. J., Newell, D. R., Siddik, Z. H., Jones, A. C., McElwain, T. J., Raju, S., Wiltshaw, E., Smith, I. E., Baker, J. M., Peckham, M. J. and Harrap, K. R. (1982). *Cancer Chemother. pharmac.*, 9, 140

Carl, P. L., Chakravarty, P. K., Katzenellenbogen, J. A. and Weber, M. J. (1980). *Proc. natn. Acad Sci. U.S.A.*, 77, 2224

Carter, S. K. and Friedman, M. A. (1972). *Eur. J. Cancer*, 8, 85

Casper, E. S., Gralla, R. J., Lynch, G. R., Jones, B. R., Woodcock, T. M., Gordon, C., Kelsen, D. P. and Young, C. W. (1981). *Cancer Res.*, 41, 1402

Chapman, N. B. and Jones, J. W. (1954). *J. chem. Soc.*, 2103

Cleare, M. J. and Hoeschele, J. D. (1973). *Plat. Metals Rev.*, 17, 2

Clifford, P. (1966). *East Afr. Med. J.*, 43, 179

Cobb, L. M. (1967). *Int. J. Cancer*, 2, 5

Cobb, L. M., Connors, T. A., Elson, L. A., Kahn, A. H., Mitchley, B. C. V., Ross, W. C. J. and Whisson, M. E. (1969). *Biochem. Pharmac.*, 18, 1519

Cobb, L. M. (1970). *Cancer Chemother. Rep.*, 54, 375

Cobb, L. M. and Mitchley, B. C. V. (1974). *Cancer Chemother. Rep.*, 58, 645

Colvin, M., Brundrett, R. M., Cowens, W., Jardine, I. and Ludlum, D. B. (1976). *Biochem. pharmac.*, 25, 695

Connors, T. A. and Ross, W. C. J. (1960). *Chem. Ind.*, 492
Connors, T. A. and Whisson, M. E. (1965). *Nature, Lond.*, 206, 689
Connors, T. A. and Whisson, M. E. (1966). *Nature, Lond.*, 210, 866
Connors, T. A. and Melzack, D. H. (1971). *Int. J. Cancer*, 7, 86
Connors, T. A., Foster, A. B., Gilsenan, A. M., Jarman, M. and Tisdale, M. J. (1972a). *Biochem. Pharmac.*, 21, 1309
Connors, T. A., Jones, M., Ross, W. C. J., Braddock, P. D., Khokhar, A. R. and Tobe, M. L. (1972b). *Chem.-Biol. Interact.*, 5, 415
Connors, T. A., Mandel, H. G. and Melzack, D. H. (1972c). *Int. J. Cancer*, 9, 126
Connors, T. A., Farmer, P. B., Foster, A. B., Gilsenan, A. M., Jarman, M. and Tisdale, M. J. (1973a). *Biochem. Pharmac.*, 22, 1971
Connors, T. A., Gilsenan, A. M., Ross, W. C. J., Bukhari, A., Tisdale, M. J. and Warwick, G. P. (1973b). In *Chemotherapy of Cancer Dissemination and Metastasis* (ed. S. Garattini and G. Franchi) Raven Press, New York, p. 367
Connors, T. A., Cox, P. J., Farmer, P. B., Foster, A. B. and Jarman, M. (1974). *Biochem. Pharmac.*, 23, 115
Connors, T. A., Goddard, P. M., Merai, K., Ross, W. C. J. and Wilman, D. E. V. (1976). *Biochem. Pharmac.*, 25, 241
Connors, T. A., Cumber, A. J., Ross, W. C. J., Clarke, S. A. and Mitchley, B. C. V. (1977). *Cancer Treat. Rep.*, 61, 927
Cox, P. J., Phillips, B. J. and Thomas, P. (1975). *Cancer Res.*, 35, 3755
Crick, F. H. C. and Watson, J. D. (1954). *Proc. R. Soc.*, A223, 80
Cumber, A. J. and Ross, W. C. J. (1977). *Chem.-Biol. Interact.*, 17, 349
Davis, W. and Ross, W. C. J. (1963). *Biochem. Pharmac.*, 12, 915
Davis, W. and Ross, W. C. J. (1965). *J. med. Chem.*, 8, 757
Dewald, B., Baggiolini, M. and Aebi, M. (1969). *Biochem. Pharmac.*, 18, 2179
Double, J. A. and Workman, P. (1977). *Cancer Treat. Rep.*, 61, 909
Elmore, D. T., Gulland, J. M., Jordan, D. O. and Taylor, H. F. W. (1948). *Biochem. J.* 42, 308
Elson, L. A., Jarman, M. and Ross, W. C. J. (1968). *Eur. J. Cancer*, 4, 617
Everett, J. L. and Ross, W. C. J. (1949). *J. chem. Soc.*, 1972
Everett, J. L., Roberts, J. J. and Ross, W. C. J. (1953). *J. chem. Soc.*, 2386
Fisher, B., Carbone, P. P., Economou, S. G., Frelick, R., Glass, A., Lerner, H., Redmond, C., Zelen, M., Katrych, D. L., Wolmark, N., Band, P. and Fisher, E. R. (1975). *New Engl. J. Med.*, 292, 117
Fox, M., Rees, R. W. M., Bennett, D. H. J. and Henry, L. (1971). *Lymphology*, 4, 35
Froese, G., Linfold, J. H., Berczi, I., Semon, A. H. and Israels, L. G. (1976). *Am. Ass. Cancer Res. Abs.*, 30
Gescher, A., Hickman, J. A., Simmonds, R. J., Stevens, M. F. G. and Vaughan, K. (1981). *Biochem. Pharmac.*, 30, 89
Chose, T., Norvell, S. T., Guclu, A., Cameron, D., Bodurha, A. and MacDonald, A. S. (1972). *Br. med. J.*, 3, 495
Gilman, A. (1963). *Am. J. Surg.*, 105, 574
Godeneche, D., Madelmont, J. C., Sauvezie, B. and Billaud, A. (1975). *Biochem. Pharmac.*, 24, 1303
Goldacre, R. J., Loveless, A. and Ross, W. C. J. (1949). *Nature, Lond.*, 163, 667
Goldenberg, R. S., Griffin, J. P., McSherry, J. W. and Krakoff, I. H. (1980). *Cancer Treat. Rep.*, 64, 1319
Golumbic, C., Fruton, J. S. and Bergmann, M. (1946). *J. org. Chem.*, 11, 518
Gomori, G. (1948). *Proc. Soc. exp. Biol. Med.*, 69, 407
Gram, H. F., Moscher, C. W. and Baker, B. R. (1959). *J. Am. Chem. Soc.*, 81, 3603
Griffiths, J. R., Stevens, A. N., Iles, R. A., Gordon, R. E. and Shaw, D. (1981). *Biosci. Rep.*, 1, 319
Haas, C. D., Stephens, R. L., Hollister, M. and Hoogstraten, B. (1976). *Cancer Treat. Rep.*, 60, 611
Haddow, A., Kon, G. A. R. and Ross, W. C. J. (1948). *Nature, Lond.*, 162, 824
Hansch, C., Smith, N., Engle, R. and Wood, H. (1972). *Cancer Chemother. Rep.*, 56, 443
Harrap, K. R., Riches, P. G., Gilby, E. D., Sellwood, S. M., Wilkinson, R. and Konyves, I. (1977). *Eur. J. Cancer*, 13, 873

Harrap, K. R. (1983). In *Cancer Chemotherapy* (ed. F. M. Muggia), Martinus Nijhoff, Massachusetts, in press

Hebborn, P. and Danielli, J. F. (1958). *Biochem. Pharmac.*, 1, 19

Hebborn, P., Simpson, C. L. and Mihich, E. (1970). *Cancer Chemother. Rep.*, 54, 23

Hendry, J. A., Homer, R. F., Rose, F. L. and Walpole, A. L. (1951). *Br. J. Pharmac.*, 6, 357

Herr, H. W. (1980). *J. Urol.*, 123, 853

Hickman, J. A. (1974). *Biochem. Pharmac.*, 23, 2833

Hickman, J. A. and Melzack, D. H. (1975). *Biochem. Pharmac.*, 24, 1947

Higby, D. J., Wallace, H. J., Albert, D. J. and Holland, J. F. (1974). *Cancer*, 33, 1219

Hill, D. L. (1975). *A Review of Cyclophosphamide*, Charles C. Thomas, Springfield, Illinois

Hill, D. L., Kirk, M. C. and Struck, R. F. (1975). *Cancer Res.*, 35, 296

Hilton, J. and Walker, M. D. (1975). *Biochem. Pharmac.*, 24, 2153

Hudson, R. F., Timmis, G. M. and Marshall, R. D. (1958). *Biochem. Pharmac.*, 1, 48

Ihde, D. C., Dutcher, J. S., Young, R. C., Cordes, R. S., Barlock, A. L., Hubbard, S. M., Jones, R. B. and Boyd, M. R. (1981). *Cancer Treat. Rep.*, 65, 755

Israels, L. G. and Linford, J. M. (1963). In *Proc. 5th Canadian Cancer Conf.* (ed. R. W. Begg), Academic Press, New York, p. 399

Israels, L. G. and Ritzmann, S. E. (1960). *Acta clin. int. Contra. Cancr.*, 16, 665

Izbicki, R., Al-Sarraf, M., Reed, M. L., Vaughan, C. B. and Vaitkevicius, V. K. (1972). *Cancer Chemother. Rep.*, 56, 615

Jarman, M., Melzack, D. H. and Ross, W. C. J. (1976). *Biochem. pharmac.*, 25, 2475

Johnston, T. P., McClab, G. S. and Montgomery, J. A. (1975). *J. med. Chem.*, 18, 634

Jones, R., Jonsson, U., Browning, M., Lessner, H., Price, C. C. and Sen, A. K. (1968). *Ann. N. Y. Acad. Sci.*, 68, 113

Kahn, A. H. and Ross, W. C. J. (1969/1970). *Chem. Biol. Interact.*, 1, 27

Kahn, A. H. and Ross, W. C. J. (1970/1972). *Chem. Biol. Interact.*, 4, 11

Kolar, G. F. and Carnbelli, R. (1979). *Cancer Lett.*, 7, 209

Kolar, G. F., Maurer, M. and Wildschutte, M. (1980). *Cancer Lett.*, 10, 235

Kreilgard, B., Higuchi, T. and Repta, A. J. (1975). *J. pharm. Sci.*, 64, 1850

Kupchan, S. M. and Schubert, R. M. (1974). *Science*, 185, 791

Lake, L. M., Grunden, E. E. and Johnson, B. M. (1975). *Cancer Res.*, 35, 2858

Larionov, L. F., Shkodinskaja, E. N., Troosheikina, V. I., Khokhlov, A. S., Vasina, O. S. and Novikova, M. A. (1955). *Lancet*, i, 169

Lawley, P. D. (1957). *Proc. chem. Soc.*, 290

Legha, S. S., Slavik, M. and Carter, S. K. (1976). *Cancer*, 38, 27

Levin, V. A., Freeman-Dove, M. A. and Meroten, C. E. (1976). *J. natn. Cancer Inst.*, 56, 535

Levins, P. L. and Papanastassiou, Z. B. (1965). *J. Am. Chem. Soc.*, 87, 826

Lin, A. J., Cosby, L. A. and Sartorelli, A. C. (1976). In *Cancer Chemotherapy* (ed. R. E. Gould), Americal Chemical Society, Washington, p. 71

Linford, J. H., Froese, G., Berczi, I. and Israels, L. G. (1974). *J. natn. Cancer Inst.*, 52, 1665

Loveless, A. (1969). *Nature, Lond.*, 223, 206

Manson, M. M., Legg, R. F., Watson, J. V., Green, J. A. and Neal, G. E. (1981). *Carcinogenesis*, 2, 661

Mathé, G., Loc, T. B. and Bernard, J. (1958). *C. r. hebd. Séanc. Acad. Sci. Paris*, 246, 1626

May, H. E., Boose, R. and Reed, D. J. (1975). *Biochemistry*, 14, 4723

McLean, A. (1975). *A. Rep. Marie Curie Mem. Found.*, 25

Merrin, C. E. (1979). *Cancer Treat. Rep.*, 63, 1579

Meyza, J. and Cobb, L. M. (1971). *Cancer*, 27, 369

Mitchley, B. C. V., Clarke, S. A., Connors, T. A. and Neville, A. M. (1975). *Cancer Res.*, 35, 1099

Mitchley, B. C. V., Clarke, S. A., Connors, T. A., Carter, S. M. and Neville, A. M. (1977). *Cancer Treat. Rep.*, 61, 451

Montgomery, J. A. and Struck, R. F. (1973). *Progr. Drug Res.*, 17, 320

Moolten, F., Zajdel, S. and Cooperband, S. (1976). *Ann. N. Y. Acad. Sci.*, 277, 690

Muindi, J. R. F., Newell, D. R., Smith, I. E. and Harrap, K. R. (1983). *Br. J. Cancer*, in press

Nagasawa, H. T., Shirota, F. N. and Mizuno, N. S. (1974). *Chem.-Biol. Interact.*, 8, 403
Philpott, G. W., Sherer, W. T., Bower, R. J. and Parker, C. W. (1973). *J. Immun.*, 111, 921
Preussman, R., von Hodenberg, A. and Hengy, H. (1969). *Biochem. Pharmac.*, 18, 1
Price, C. C. (1958). *Ann. N. Y. Acad. Sci.*, 68, 663
Price, C. C., Gaucher, G. M., Koneru, P., Shibakawa, R., Sowa, J. R. and Yamaguchi, M. (1969). *Ann. N. Y. Acad. Sci.*, 163, 593
Robinson, R. and Watt, J. S. (1934). *J. chem. Soc.*, 1536
Rosenberg, B., Van Camp, L. and Krigas, T. (1965). *Nature, Lond.*, 205, 698
Rosenberg, B., Van Camp, L., Trosko, J. E. and Mansour, V. M. (1969). *Nature, Lond.*, 222, 385
Ross, S. D. (1947). *J. Am. chem. Soc.*, 69, 2982
Ross, W. C. J. (1949). *J. chem. Soc.*, 183
Ross, W. C. J. (1950). *J. chem. Soc.*, 2257
Ross, W. C. J. (1953). *Adv. Cancer Res.*, 1, 397
Ross, W. C. J., Warwick, G. P. and Roberts, J. J. (1955a). *J. chem. Soc.*, 3110
Ross, W. C. J. and Warwick, G. P. (1955b). *Nature, Lond.*, 176, 298
Ross, W. C. J. and Warwick, G. P. (1956). *J. chem. Soc.*, 1364
Ross, W. C. J. (1961). *Biochem. Pharmac.*, 8, 235
Ross, W. C. J. (1962). *Biological Alkylating Agents*, Butterworths, London
Ross, W. C. J. and Mitchley B. C. V. (1964). *A. Rep. Br. Empire Cancer Campaign*, 42, 70
Ross, W. C. J. (1974). *Chem.-Biol. Interact.*, 8, 261
Ross, W. C. J. (1975a). *Chem.-Biol. Interact.*, 10, 169
Ross, W. C. J. (1975b). *Chem.-Biol. Interact.*, 11, 139
Rowland, G. F., O'Neill, G. J. and Davies, D. A. L. (1975). *Nature, Lond.*, 255, 487
Rutman, R. J., Chun, E. H. L. and Jones, J. (1969). *Biochim. biophys. Acta*, 174, 663
Rutty, C. J. and Connors, T. A. (1977). *Biochem. Pharmac.*, 26, 2385
Rutty, C. J., Newell, D. R., Muindi, J. R. F. and Harrap, K. R. (1982). *Cancer Chemother. Pharmac.*, 8, 105
Sartorelli, A. C. and Johns, D. G. (eds) (1975). *Antineoplastic and Immunosuppressive Agents*, Springer-Verlag, Berlin
Schein, P. S., O'Connell, M. J., Blom, J., Hubbard, S., Magrath, I. T., Bergevin, P., Wiernick, P. H., Ziegler, J. L. and DeVita, V. T. (1974). *Cancer*, 34, 993.
Seligman, A. M., Ulfohn, A., Gaby, S. D., Goodman, L. E., Aybar, O., Kramer, S. P., Bakal, D., Haber, S., Williamson, C. E., Miller, J. I., Sass, S. and Witten, B. (1962). *Ann. Surg.*, 156, 429
Shealy, Y. F., Struck, R. F., Holum, L. B. and Montgomery, J. A. (1961). *J. Org. Chem.*, 26, 2396
Shealy, Y. F., Krauth, C. A. and Montgomery, J. A. (1962). *J. org. Chem.*, 27, 2150
Shealy, Y. F. and Krauth, C. A. (1966). *J. med. Chem.*, 9, 34
Shealy, Y. F. (1970). *J. pharm. Sci.*, 59, 1533
Sieber, S. M. and Adamson, R. A. (1976). *Cancer Treat. Rep.*, 60, 217
Skibba, J. L., Beal, D. D., Ramirez, G. and Bryan, G. T. (1970). *Cancer Res.*, 30, 147
Stacey, K. A., Cobb, M., Cousens, S. F. and Alexander, P. (1958). *Ann. N. Y. Acad. Sci.*, 68, 682
Stevens, C. D., Quinlin, P. M., Meinken, M. A. and Kock, A. M. (1950). *Science*, 112, 561
Stratford, I. J., Sheldon, P. W., Williamson, C., Ahmed, I., Wilman, D. E. V. and Adams, G. E. (1982). *Br. J. Cancer*, 45, 629
Szybalski, W. and Iyer, V. N. (1967). In *Antibiotics*, Vol. 1 (ed. D. Gottlied and P. D. Shaw) Springer, Berlin
Tattersall, M. H. N. and Curt, G. A. (1976). *Br. J. Cancer*, 34, 320
Thomson, A. J. (1974). *Rec. Res. Cancer Res.*, 48, 38
Tisdale, M. J., Elson, L. A. and Ross, W. C. J. (1972). *Eur. J. Cancer*, 8, 255
Triggle, D. J. (1964). *J. theor. Biol.*, 7, 241
Tsou, K. C. and Su, H. C. F. (1963). *J. med. Chem.*, 6, 693
Vaquero, C., Masson, C., Guigon, M. and Hewitt, J. (1975). *Eur. J. Cancer*, 11, 739
Vasil'eva, M. N., Martynou, V. S. and Berlin, A. Y. (1970). *Zh. org. Khim.*, 6, 1677
Vennegoor, C., Smeerdijk, D. V. and Rumke, P. (1975). *Eur. J. Cancer*, 11, 725

Warwick, G. P. (1963). *Cancer Res.*, **23**, 1315
White, F. R. (1959). *Cancer Chemother. Rep.*, **4**, 52
Williamson, C. E. and Witten, B. (1967). *Cancer Res.*, **27**, 33
Wilman, D. E. V. (1980). *Biochem. Pharmac.*, **29**, 2919
Wilman, D. E. V. and Goddard, P. M. (1980). *J. med. Chem.*, **23**, 1052
Wilson, W. L. and de la Garza, J. G. (1965). *Cancer Chemother. Rep.*, **48**, 49
Wiltshaw, E. and Carr, B. (1974). *Rec. Res. Cancer Res.*, **48**, 178
Workman, P., Double J. A. and Wilman, D. E. V. (1976). *Biochem. Pharmac.*, **25**, 2347
Workman, P. and Double, J. A. (1978). *Biochem. Pharmac.*, **27**, 199
Worzalla, J. F., Johnson, B. M., Ramirez, G. and Bryan, G. T. (1973). *Cancer Res.*, **33**, 2810
Worzalla, J. F., Kaiman, B. D., Johnson, B. M., Ramirez, G. and Bryan, G. T. (1974). *Cancer Res.*, **34**, 2669
Young, C. W., Yagoda, A., Bittar, E. S., Smith, S. W., Grabstald, H. and Whitmore, W. (1976). *Cancer*, **38**, 1887

9

The Chemistry of DNA Damage by Antitumour Drugs

J. William Lown

INTRODUCTION

Much of present day cancer chemotherapy relies heavily on inhibition of nucleic acid synthesis or interference with mitosis as a means of discriminating selectively against fast growing tumour cells (Pratt and Ruddon, 1979; Cassady and Douros, 1980). There is much pharmacological evidence (antimitotic, cytological and mutagenic effects, and the inactivation of viruses) indicating that DNA is the principal cell target for many antitumour agents (Montgomery *et al.*, 1970; Ross, 1962; Gale *et al.*, 1981; Wheeler, 1962). The object of this chapter is to review the structure and chemical characteristics of antitumour drugs, the chemistry and topology of DNA as a drug target, and the nature of specific chemical changes that drugs induce in nucleic acids. For this purpose we may define 'damage' to DNA as any alteration involving chemical bond formation or breakage, permanent or transient, which affects the normal cellular function of DNA. The biology of DNA damage is considered in Chapter 10.

TYPES OF ANTICANCER AGENTS WHICH REACT CHEMICALLY WITH DNA

Chemical agents, both naturally-occurring and synthetic, which are employed clinically in treating cancer fall into six broad categories (1) alkylating agents (2) antibiotics (3) antimetabolites (4) specific mitotic inhibitors (5) steroidal hormones and (6) miscellaneous drugs such as *cis*-platinum complexes (Chapter 7). Of these types (1) and (2) act primarily on DNA while members of classes

Molecular Aspects of Anti-cancer Drug Action, ed. Neidle & Waring
0333–315561/83/283–314

(4) and (6) may do so. It should not be construed that the classification necessarily applies to mechanism of action for, as we shall see, most agents exhibit a variety of actions on even one cell target such as DNA.

Alkylating agents

Alkylating agents were among the first clinical agents of recognised value and the study of their mechanism of action has clarified our view of cancer chemotherapy (Ludlum, 1975; Pratt and Ruddon, 1979). It has led to increased understanding of DNA chemistry, structure and conformation and to investigation of mechanisms of repair of damaged DNA. In addition these studies have led to an awareness of the chemical origin of mutagenesis and carcinogenesis (Ludlum, 1977).

Some examples of alkylating agents are shown in figure 1. While sulphur mustard is too reactive for clinical use, three examples of nitrogen mustards, cyclophosphamide, chlorambucil and L-phenylalanine mustard have found widespread application (Schmidt *et al.*, 1965) (Chapter 8).

The $-N(CH_2CH_2Cl)_2$ grouping of nitrogen undergoes cyclisation in aqueous media to form aziridinium ions which also represent the active forms of the aziridine drugs like triethylenemelamine. The aziridinium ions attack electron rich centres in biological macromolecules (Ludlum, 1967). The drug cyclophosphamide was developed as a masked or transport form of a nitrogen mustard, and was designed in the expectation that it would be activated by the tumour itself (Bagley *et al.*, 1973; Connors *et al.*, 1974; Colvin *et al.*, 1975). The electron-withdrawing group attached to the mustard prevents cyclisation and activation until the ring structure is metabolised (Chapter 8). An alternative approach to tumour-activated 2-haloethylnitrosoureas will be discussed below. The 2-chloroethylnitrosoureas (CENUs) were developed as a result of the observation of promising antileukaemic properties of 1-methyl-1-nitrosourea in a routine screen. Examples including BCNU, CCNU, methylCCNU and chlorozotocin have an established role in clinical chemotherapy (Montgomery *et al.*, 1970; Wheeler, 1976; Johnston *et al.*, 1963, 1966). The CENUs are also of considerable theoretical interest and the study of their mechanism of action has contributed greatly to the understanding of anticancer action of drugs in general (Reed *et al.*, 1975; Kohn, 1977; Colvin *et al.*, 1976; Lown *et al.*, 1978*a*, 1979*b*, 1980). Their property of passing the blood–brain barrier (Schabel *et al.*, 1963) renders them especially useful in the treatment of central nervous system neoplasms, brain tumours and intracranial metastases (Walker, 1973).

While, as we shall see, the CENUs are subject to basic catalysis, the triazenes are complementary in the sense that they are activated specifically under acid conditions (Loo, 1975; Skibba and Bryan, 1971; Lown and Singh, 1981, 1982).

Alkylating agents are either electrophiles or generate electrophiles *in vivo* including carbonium ions or their kinetic equivalents (such as chloronium ions or alkyl diazotates) which may attack electron-rich sites in biological macromolecules, adding alkyl groups to nitrogen, sulphur or oxygen atoms (Ludlum,

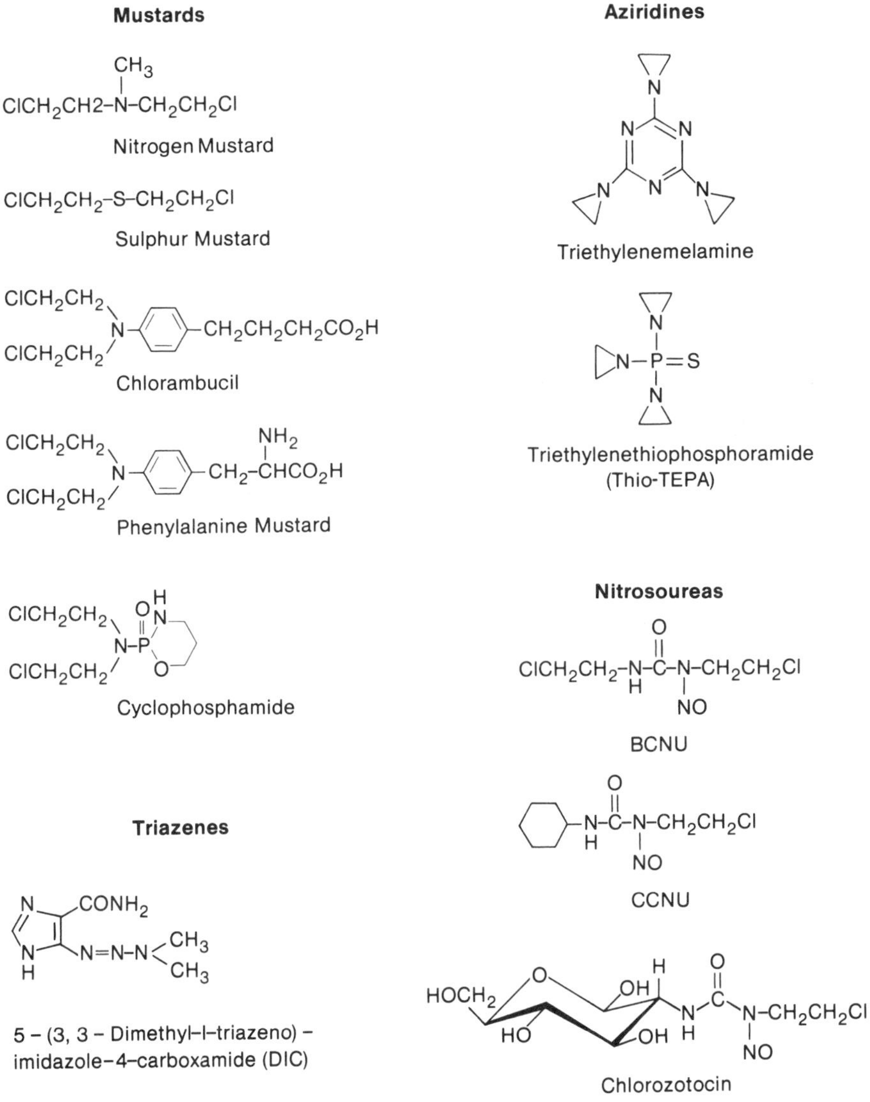

Figure 1   Structures of representative alkylating agents.

1975, 1977; Price, 1975; Roberts *et al.*, 1968). The attendant nucleophilic substitutions may proceed by S_N1 or S_N2 mechanisms. The distinction is significant since alkylating agents that adopt an S_N1 mechanism tend to substitute oxygen atoms in DNA and to esterify the phosphate groups in the sugar phosphate backbone (Lawley, 1972, Lawley *et al.*, 1971, 1972; Kusmierck and Singer, 1976).

While the origin of the alkylating species is evident in the case of mustards (Price *et al.*, 1969) or thioTEPA it is not immediately apparent how, for example, cyclophosphamide, CENUs or triazenes give rise to reactive species. Since cyclophosphamide requires metabolic activation its discussion will be postponed to a later section.

CENUs undergo spontaneous decomposition under physiological conditions without enzymatic activation to give rise to an isocyanate, products including 2-chloroethanol, 1, 2-dichloroethane, vinyl chloride and acetaldehyde which are consistent with the intermediacy of a chloroethyl cation, as well as carbamates and ureas (Reed *et al.*, 1975; Colvin *et al.*, 1975, 1976; Loo *et al.*, 1966).

Antibiotics

Quinone antibiotics

Probably the best-known antibiotics used for the treatment of cancer are the anthracyclines, which comprise a large family including Adriamycin, daunomycin, aclacinomycin and carminomycin. They are dealt with in Chapters 2-4. Rather less well known, but nonetheless important, are the mitomycin antibiotics first isolated from *Streptomyces caespitosus* (Hata *et al.*, 1956; Sugawara and Hata, 1956). Mitomycin C, the most thoroughly studied example, is effective against a range of neoplasms including chorioepithelioma, reticulum cell sarcoma and seminoma (Carter and Crooke, 1979). Mitomycin C and its methylated homologue porfiromycin are used clinically for treatment of malignancies of the breast, lung, colon and stomach, principally in Japan but increasingly nowadays in the United States (Carter and Crooke, 1979; Szybalski and Iyer, 1967). Mitomycin C contains three recognised carcinostatic groups: aziridine, carbamate and quinone. Its unique multifunctional nature is a model of structural economy (figure 2). In common with the anthracyclines, mitomycin C is subject to reductive metabolism at the quinone moiety. The concomitant release of the indole nitrogen lone pair causes expulsion of methanol activating the aziridine and carbamate moieties as potential alkylating groups (Iyer and Szybalski, 1963) (figure 2). Consequently the activated antibiotic (which has only a transient existence (Rao *et al.*, 1977)) can both alkylate and form inter-strand cross-links in DNA (Iyer and Szybalski, 1963). These effects appear to constitute the principal cytotoxic action but other effects are also evident. Concomitant oxidation of the hydroquinone moiety of the DNA-bound antibiotic generates the semiquinone, O_2^-, H_2O_2 and ultimately OH· which can cause single strand breaks in DNA (Lown *et al.*, 1976, 1978*b*). The semiquinones of both mitomycin C and mitomycin B although extremely short lived have been identified and characterised by epr spectroscopy (Lown *et al.*, 1978*b*). The formation of hydrogen peroxide has been observed from mitomycin C *in vivo* (Tomasz, 1976). The factors which may contribute to the preferential action of mitomycin C on neoplastic tissue will be discussed below.

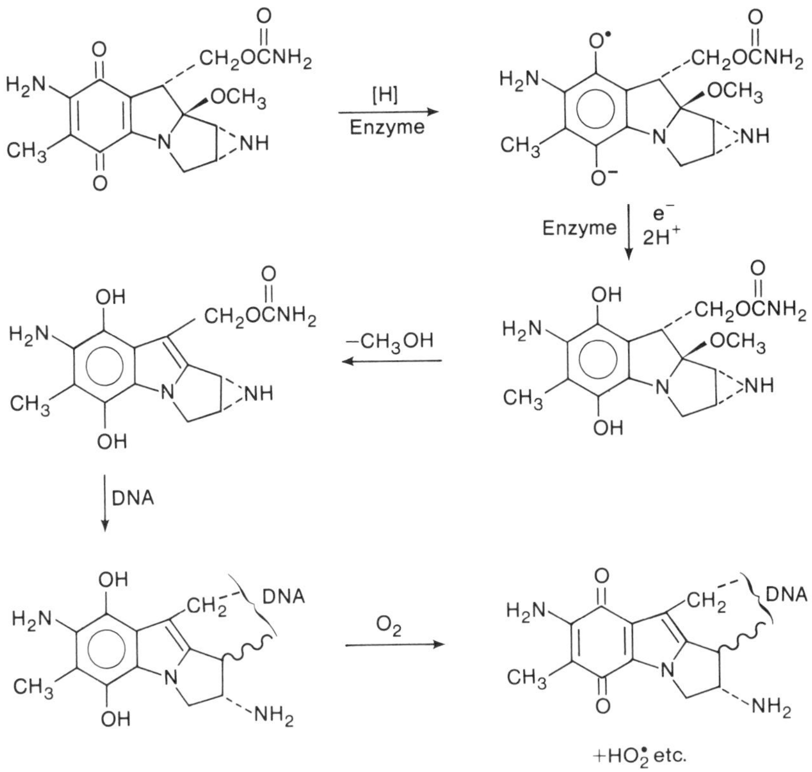

Figure 2   Reaction scheme for bioreductive activation of mitomycin C and the subsequent interstrand cross-linking of DNA.

Streptonigrin

Streptonigrin (figure 3) which is isolated from *Streptomyces flocculus* (Rao *et al.*, 1963; Rao and Cullen, 1959) has received only limited use as an anti-cancer agent owing to its severe toxicity (McBride *et al.*, 1966; Teller *et al.*, 1961; Ebert *et al.*, 1968). However, study of its mechanism of action on DNA has provided insight into the action of many other clinically important quinone antibiotics. Selective metal sequestration (Hadju and Armstrong, 1981) provides a charged complex which binds to DNA (White and White, 1966; White, 1977). Streptonigrin was the first antitumour antibiotic for which it was shown by the ethidium binding assay (described below) that once bound to DNA, is subject to an NADPH-mediated reductase action converting the quinone to a hydroquinone. The latter form of the antibiotic is unstable and is readily oxidised with the concomitant production of reactive oxygen species O_2^-, H_2O_2 and OH· the latter species of which causes single strand breaks in DNA (Cone *et al.*, 1976; Lown and Sim, 1976*a*,*b*).

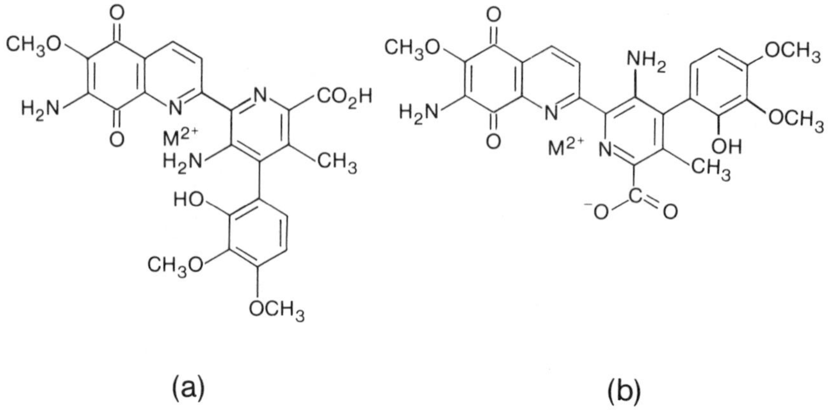

(a)   (b)

Figure 3   Streptonigrin forming two different metal chelation complexes.

Saframycins A, C and S

The saframycins, which are isolated from *Streptomyces laevendulae*, comprise a family of five heterocyclic *bis*-quinone antibiotics the structures of which have been established recently (Arai *et al.*, 1977, 1980, Ishiguro *et al.*, 1978). Two members of this group saframycin A (figure 4) and S show promising anticancer activity against Ehrlich ascites tumour, murine L1210 and P388 leukaemias and B16 melanoma (Arai *et al.*, 1977, Ishiguro *et al.*, 1978). These antibiotics are biosynthetically related to mimosamycin, naphthiridinomycin, chlorocarcins and streptonigrin via the shikimate-prephenate-tyrosine pathway (Arai *et al.*, 1977, 1980).

Like mitomycin C they compress within their relatively small structures a surprising range of specific chemical reactivities towards DNA. For example saframycin A exhibits acid-catalysed equilibrium binding to DNA, reversible covalent binding to DNA in a base and groove-specific manner and also bioreductive activation accompanied by DNA scission by a free radical pathway (Lown *et al.*, 1982).

Saframycins A and C have been shown by ^{1}H-NMR to be protonated at N—CH$_3$ (Lown *et al.*, 1981*b*) and to undergo binding to DNA preferentially at G+C-rich sites and in the minor groove of duplex DNA (Lown *et al.*, 1982). The latter property was indicated by their binding to T4 DNA in which the major goove is occluded by α-glycosylated cytidine residues. Also under acid catalysis or after bioreductive activation (figure 4) saframycin A loses the nitrile group and, in a sequence which plausibly involves a quinone methide species, generates an electrophilic cyclic immonium species. The latter intermediate may either add water and thereby form saframycin S or is subject to nucleophilic attack by DNA. The base preference noted above suggests attack by a guanine-2-NH$_2$ group resulting in formation of an aminal linkage similar to that suggested for anthramycin binding (Hurley, 1980). In addition, how-

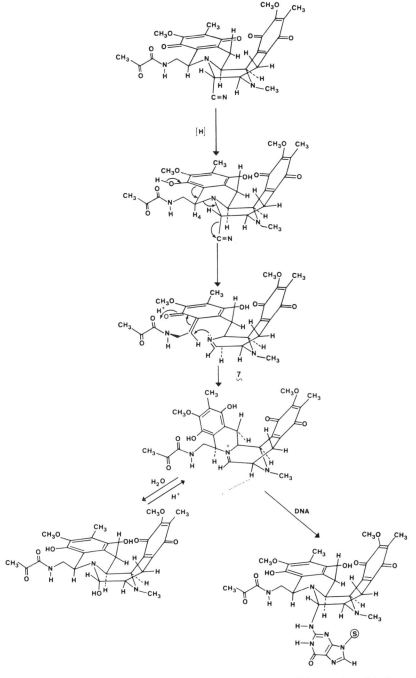

Figure 4 Reaction scheme for bioreductive-promoted reversible covalent binding of saframycin A to DNA.

ever, saframycins A and C reduced *in situ* with NADPH cause oxygen-dependent single-strand breaks (but not double-strand breaks) in supercoiled DNA. As in the case of the other quinone antibiotics the inhibition of this scission by superoxide dismutase, by catalase and by free radical scavengers implicate O_2^{-}, H_2O_2 and $OH^{\cdot}$ species in this process (Lown *et al.*, 1982). The intermediate semiquinone species was identified by EPR and an examination of the number and magnitude of the hyperfine splittings permitted an assignment of its conformation. The single-strand scission of DNA is pH-dependent and, unlike the binding of the antibiotics, is strongly inhibited by Mg^{2+} and Zn^{2+} ions (Lown *et al.*, 1982).

Peptide and glycopeptide antibiotics

This group includes bleomycin and tallysomycin (Chapter 6) as well as neocarzinostatin, a polypeptide antibiotic which has recently been shown to incorporate a chromophore (Povirk and Goldberg, 1980). Neocarzinostatin together with its chromophore, binds to DNA and causes an oxygen-dependent single-strand scission (Kappen and Goldberg, 1978; Burger *et al.*, 1978, Sim and Lown, 1978). While the structure of the chromophore has not yet been elucidated the evidence of selective DNA protection from scission by superoxide dismutase, catalase and free radical scavengers suggests a pathway mediated by O_2^{-}, H_2O_2 and $OH^{\cdot}$ species (Sim and Lown, 1978). It would appear therefore that neocarzinostatin, despite a structure very different from mitomycin C, streptonigrin or saframycin A, shares with these antibiotics a common mechanism for causing chemical damage to DNA.

Carzinophilin A is another antitumour antibiotic produced by *Streptomyces sahachiroi* whose molecular structure has recently been determined (Lown and Hanstock, 1982). It inhibits the growth of a wide spectrum of neoplasms in mammals and has been used clinically to treat human skin cancer, cancer of the jejunum, reticulosarcoma and certain cases of chronic leukaemia (Kuroyanagi *et al.*, 1956). Carzinophilin A rapidly produces interstrand cross-links in DNA (Terawaki and Greenberg, 1966; Lown and Majumdar, 1977) but, unlike mitomycin C, does not require enzymatic activation and shows a preference for binding to G, C rich sites. Its structure (figure 5) immediately suggests a plausible mechanism for its action on DNA, that is reversible binding by a process of *bis*-intercalation (Chapter 5) followed by acid-promoted opening of the aziridine rings to form covalent cross-links (Lown and Hanstock, 1982).

Pyrrolo(1,4)-benzodiazepine antibiotics

Anthramycin and tomaymycin have been isolated from *Streptomyces refuineus* var. *thermotolerans* (Tendler and Korman, 1963 Leimgruber *et al.*, 1965) and *Streptomyces achromogenes* var. *tomaymyceticus* respectively. Other members of this group are sibiromycin (Gause *et al.*, 1969) and the neothramycins A and B (Takeuchi *et al.*, 1976) (figure 6). A structural feature common to all these antibiotics is the pyrrolo(1,4)-benzodiazepinone nucleus. They all contain a

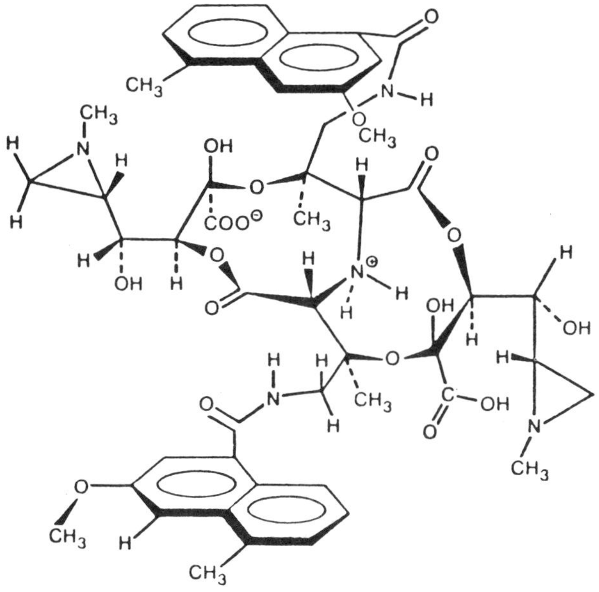

Figure 5   Structure of carzinophilin A.

carbinolamine function at positions 10, 11 (imine in the case of neothramycins) a phenolic hydroxy group and often an unsaturated side chain on the pyrrole ring. The increasingly important role of anthramycin in chemotherapy has generated further interest in the molecular mechanism of action of these agents. They bind reversibly in the minor groove of DNA and preferentially at GC-rich sites in an acid-promoted reaction. It has been demonstrated by employing modified anthramycins and model compounds that the active site on the molecule is the carbinolamine moiety (Hurley, 1977, 1980; Lown and Joshua, 1979). Reversible binding via an aminal linkage with the 2-NH_2 of guanine results in covalent linkage similar to that which obtains for saframycin A. Recent evidence supporting the site of binding has been obtained by [^{1}H] and [^{13}H]-NMR employing the [11-^{13}C]-labelled antibiotic (Ostrander *et al.*, 1983).

Camptothecin

Camptothecin (figure 7) is a cytotoxic alkaloid derived from the bark and stem wood of *Camptotheca acuminata* (family Nyssaceae), a tree native to China. Camptothecin inhibits the growth of numerous experimental tumours (Wall and Wani, 1980). At present it is used actively in the clinical treatment of cancer in the People's Republic of China. Study of the pharmacology of camptothecin reveals rapid inhibition of nucleic acid synthesis, fragmentation of DNA *in vivo* and lack of effect of campothecin on proteins and replicating enzymes, strongly

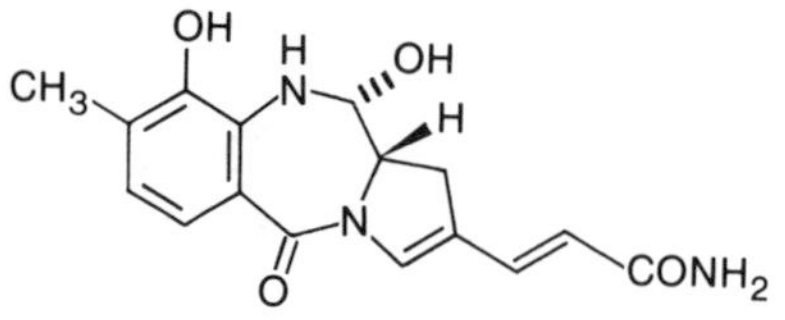

Anthramycin

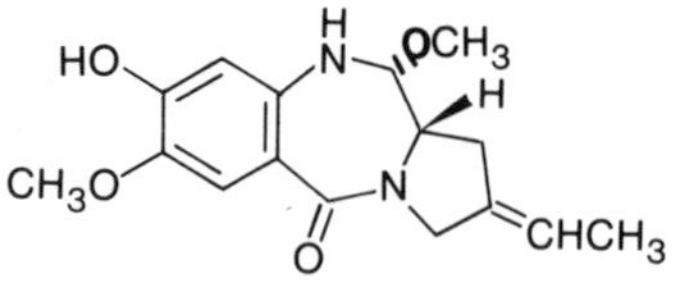

Tomaymycin

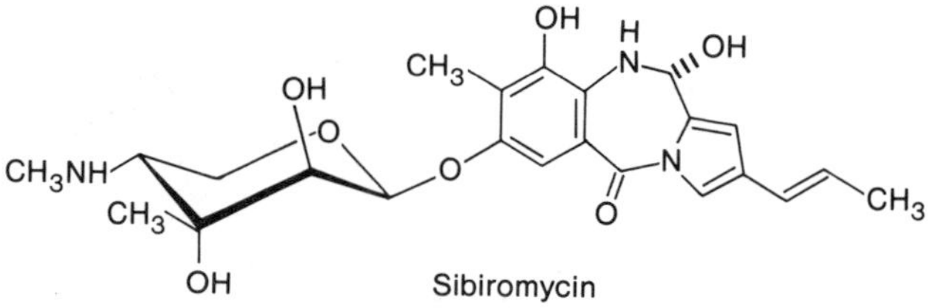

Sibiromycin

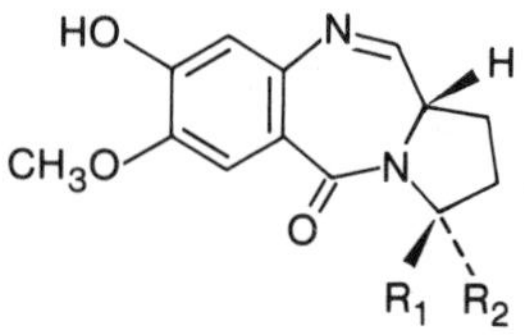

Neothramycin A $R_1 = H, R_2 = OH$

Neothramycin B $R_1 = OH, R_2 = H$

Figure 6   Structures of representative pyrrolo(1,4)-benzodiazepine antibiotics.

indicating that nucleic acids are the principal cell targets for the alkaloid (Wall and Wani, 1980).

Camptothecin causes an oxygen-dependent single strand scission of DNA when photoactivated by near-visible irradiation (Lown and Chen, 1980). The alkaloid also potentiates the bleomycin-induced cleavage of DNA by substituting for the reducing agent requirement of the antibiotic (Sartiano *et al.*, 1977). In the process camptothecin is oxidised to the corresponding hemiacetal (figure 7) (Lown *et al.*, 1981*a*).

MODIFICATION OF DRUGS PRIOR TO INTERACTION WITH DNA

Many of the drugs described above undergo chemical modifcation as a result of metabolism before or during their reaction with DNA. This can profoundly alter the structural and stereochemical characteristics as well as the chemical

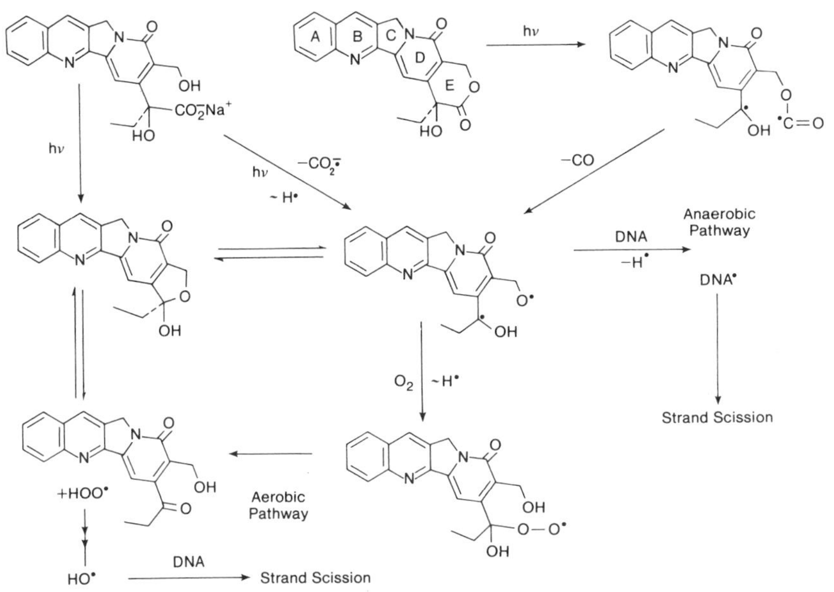

Figure 7 Reaction schemes for aerobic and anaerobic photoactivated generation of the reactive free radical species from the antitumour alkaloid camptothecin.

reactivity of the agents and often affects the specificity of their reactions with DNA.

Bioreductive activation

As we have seen, many drugs including mitomycin C, Adriamycin, daunomycin, actinomycin, streptonigrin, saframycin, bleomycin and tallysomycin exhibit a reducing requirement in their reactions with DNA. Depending on the redox characteristics of a particular drug this may be satisfied by NADPH or flavoprotein-mediated reduction. Since tumour tissue is characterised by having a somewhat higher reducing capacity than normal tissue, owing to metabolic differences (Cater and Phillips, 1954), this may lead to preferential activation of the drug. This property has given rise to the concept of bioreductive activation (Moore, 1977; Lin *et al.*, 1972, 1973, 1974) to assist in the rational design of antitumour drugs. Sartorelli and his coworkers have employed this concept to prepare benzoquinones, naphthoquinones, quinoline-quinones and naphthazarins that are mono- and disubstituted with CH_2-X (X a leaving group) substituents and have observed that many do show antineoplastic activity (Lin *et al.*, 1972, 1973, 1974).

Another example is provided by certain disulphide-linked nitrosocarbamates which were designed to be cleaved preferentially in tumour tissue so that the thiol nucleophile produced can trigger the release of the 2-chloroethyl cation by intramolecular nucleophilic attack (Lown *et al.*, 1980). Such compounds show promising antileukaemic activity in animal tests.

Selective protonation in tumour tissue

Tumour tissue also exhibits a slightly lower average pH than normal tissue owing to an increased rate of glycolysis and concomitant slight overproduction of lactic acid (Connors *et al.*, 1964; Ross, 1961). Consequently drugs which are specifically activated by protonation may exhibit a slight preferential effect. Examples include mitomycin C which will cross-link DNA at lower pHs without reductive activation (Lown and Sim, 1976*a*), and saframycin A which will bind reversibly and covalently to DNA under either acidic or reducing conditions as a result of selective protonation at the N-12 and loss of the cyano group (figure 4) (Lown *et al.*, 1982; Arai *et al.*, 1977, 1980). Aryl 2-haloethyltriazenes related to the clinically useful DTIC are also subject to preferential acid-promoted decomposition to give rise to electrophiles which attack DNA (Lown and Singh, 1982) (figure 8). Unlike the corresponding 2-chloroethylnitrosoureas the triazenes, being subject to specific acid catalysed decomposition, also exhibit a preference for reaction at the more acidic phosphate residues of DNA rather than at the bases.

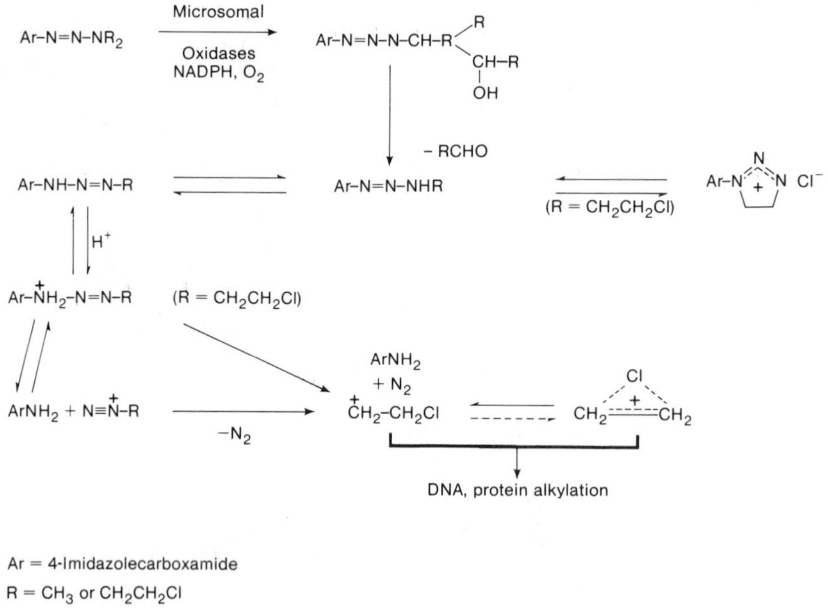

Ar = 4-Imidazolecarboxamide

R = CH_3 or CH_2CH_2Cl

Figure 8 Scheme for oxidative metabolism of aryldialkyltriazenes to arylalkyltriazenes and subsequent acid-promoted generation of electrophiles to attack DNA.

Metabolic modification of drugs

Many drugs are subject to extensive oxidative metabolism *in vivo* mediated by enzymes such as cytochrome P450 (Hill *et al.*, 1972; Colvin *et al.*, 1973). For example the cyclohexyl group in CCNU is extensively hydroxylated (May *et al.*,

1974; Farmer *et al.*, 1978). In this instance the portion of the molecule which gives rise to the electrophiles is not affected so that presumably the nature of the lesions induced in DNA is unaffected although the lipid solubility and transport properties of the drug are altered prior to attack on the cellular target.

In contrast the oxidative metabolism of cyclophosphamide constitutes an integral part of the mechanism by which the ultimate electrophiles are generated (Connors *et al.*, 1974; Cox *et al.*, 1975; Colvin *et al.*, 1973; Hill *et al.*, 1972). (See Chapter 8 for a more detailed account of cyclophosphamide metabolism.) In recent developments isophosphamide, having activity similar to cyclophosphamide, was shown to be converted to its active form *in vivo* by C_4-hydroxylation by hepatic cytochrome P-450 mixed-function oxidases. In an effort to study preactivated derivatives of cyclophosphamide, analogues were C_4-hydroperoxylated (Takamizawa *et al.*, 1978). These were found to be more stable and to be more effective against L1210 leukaemia. In the case of cyclophosphamide the initial metabolite is 4-hydroxycyclophosphamide (Hill *et al.*, 1972). From this the final alkylating species, most probably phosphoramide mustard, is generated intracellularly together with acrolein. Dialkyltriazenes are effective clinically in the treatment of cancer but pharmacological studies have shown that they are prodrugs in that one alkyl group is subject to enzymatic oxidative elimination so that the ultimate drug is the monoalkyltriazene (Loo, 1975) (figure 8).

Spontaneous decomposition under physiological conditions

The 2-chloroethylnitrosoureas (CENUs) and 2-chloroethylaryltriazenes are unique among anticancer agents which act principally on DNA in that they decompose spontaneously without enzymatic activation under physiological conditions to give rise to electrophilic species that modify DNA chemically. This property and the extraordinarily rich chemistry of CENUs in particular leads to a plethora of reactive species many of which attack DNA and proteins. Three major pathways of decomposition have been postulated (Colvin *et al.*, 1976; Ludlum *et al.*, 1975; Montgomery *et al.*, 1967; Chatterji, 1978). Among the most reactive electrophiles are the 2-chloroethyldiazohydroxides E and F (figure 9) leading to 2-chloroethyl cations. Recent evidence using specifically [^{15}N] and [^{13}C]-labelled compounds indicated the operation of strict stereoelectronic and conformational control in the decomposition of tetrahedral intermediates A-D (figure 9) (Lown and Chauhan, 1982). The relative contribution of each of the three pathways, examined by specific [^{18}O] labelling, is strongly dependent on the structure of the CENU (Lown and Chauhan, 1982).

Differential [^{14}C] labelling by CENU *in vivo* demonstrated that the released isocyanate becomes bound to protein by carbamoylation of $-NH_2$ groups while the electrophiles produced by pathways shown in figure 10 become attached to DNA (Cheng *et al.*, 1972).

Triazenes including DTIC (figure 8, AR = 4-imidazacarboxamide, R = CH_3) after the initial enzymatic dealkylation also decompose under physiological conditions but with enzymatic activation.

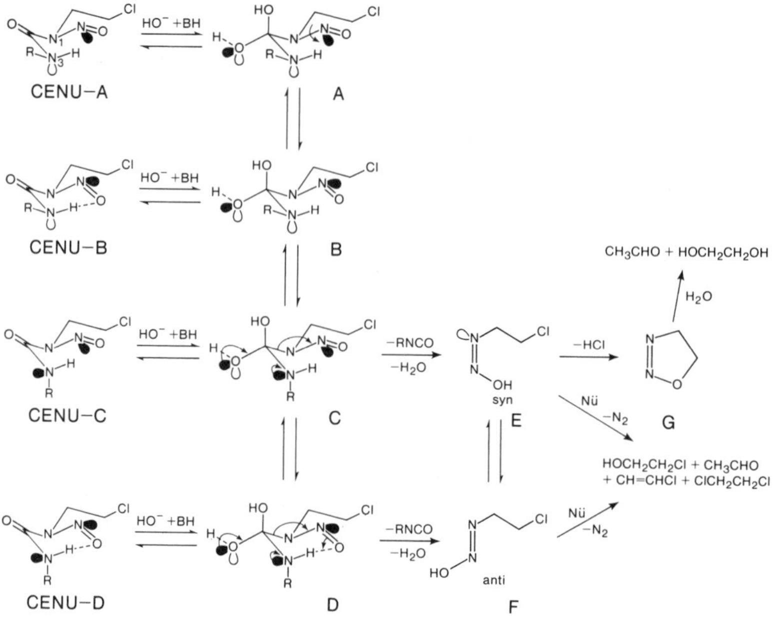

Figure 9 Operation of conformational and stereoelectronic control in the aqueous decomposition of chloroethylnitrosoureas leading to the generation of electrophiles.

Metal ion sequestration or exchange

Certain antitumour agents show specific metal ion requirements *in vivo* which need to be considered when attempts are made to rationalise the chemical damage they inflict on DNA as part of their cytotoxic action. For example it is evident that several clinically useful agents, especially the quinone antibiotics mitomycin C, streptonigrin, saframycin C, Adriamycin, daunomycin, actinomycin and the glycopeptide antibiotics bleomycin and tallysomycin cause, as part of their cytotoxic action, the generation of reactive oxygen species including O_2^- or HO_2^-, H_2O_2, OH^- and $Fe(III)-O_2^-$. The latter, which may be the reactive species ultimately responsible for damage to DNA in the case of the glycopeptide antibiotics, is generated via a Fe(II) ion complex interacting with oxygen (see Chapter 6).

In the case of the quinone antibiotics the evidence suggests that the Fe(II) required in the generation of OH^- is derived extramolecularly from iron complexed to protein or to ATP (Czapski, 1977) which has been 'decompartmentalized' in the view of Willson (1977).

Streptonigrin, the aminoquinone antibiotic shows additional specific metal ion requirements (White, 1977). Its unique bipyridyl-containing structure sequesters metal ions (Zn^{2+}, Mn^{2+} or Cu^{2+}) in two chemically distinct ways depending on the hardness or softness of the ion (figure 3) (Hadju and Armstrong,

1981). The resulting charged complex assists in the binding of the antibiotic to the DNA.

Another example of a specific metal ion requirement is provided by the aureolic acid group antibiotics, chromomycin A_3, olivomycin and mithramycin. These require an equivalent of magnesium in their non-intercalative interaction with DNA (Gause, 1975).

DETECTION AND QUANTITATION OF DNA LESIONS

Most DNA-damaging agents produce two or more types of lesions in their principal target which can pose severe analytical problems. In addition, special difficulties attend the detection of DNA lesions in intact cells or in large purified native DNAs owing to the fragility of large DNA molecules. The available methods for detecting and quantifying macromolecular lesions by drugs have hitherto been severely limited. Recently, however, new techniques have been developed which can provide quite detailed information on the different types of lesions and, in some cases, avoid DNA isolation steps.

Since many of these methods are reviewed in detail in Chapter 10, the present discussion will be confined to consideration of a group of related fluorescence techniques which have proved particularly suitable for detecting of individual lesions, concentrating on their sensitivity, versatility and experimental convenience.

Fluorescence techniques

Ethidium bromide is a planar dye which intercalates into duplex DNA apparently without any marked sequence preference (Waring, 1965). In the process there is enhancement of its fluorescence by a factor of 25 or 50 depending on the experimental conditions (Le Pecq and Paoletti, 1967). This property may therefore be employed to detect duplex DNA sensitivity and may be exploited in a number of convenient and versatile assays to detect and quantify a variety of chemical lesions in DNA produced by anticancer drugs which have decided advantages over alternative techniques (Morgan and Paetkau, 1972; Morgan and Pulleyblank, 1974).

Interstrand cross-linking may readily be detected by denaturing the DNA by heating at 96°C for 3 min followed by rapid cooling. Provided the pH is high (~11.8) this prevents the formation of regions of short self-complementarity so that potential ethidium intercalation sites in ordinary DNA are destroyed and the fluorescence falls to zero. By contrast, if a drug has introduced an interstrand cross-link this will serve as the nucleation site for rapid renaturation (Eigen and Porschke, 1970) following the heating and cooling cycle. The observation of ethidium fluorescence corresponding to the proportion of the DNA molecules present that contain at least one cross-link therefore permits quantification of this DNA lesion (Lown *et al.*, 1980; 1978*a*).

This assay may be refined by employing DNAs of different base composition to examine the base dependence of cross-link formation (Lown *et al.*, 1976). A second assay using supercoiled covalently closed circular PM2 phage DNA may be used to detect and quantify single strand scission. The assay is based on the different amounts of ethidium taken up by different forms of DNA. When the supercoiled DNA is nicked by a drug the resulting open-circular DNA accepts 30% more ethidium since it is subject to no topological constraints (Paoletti *et al.*, 1971). Correspondingly after heat denaturation the open circular DNA denatures in the *p*H 11.8 assay medium and the fluorescence falls to zero (figure 10). By contrast the supercoiled CCC–DNA used as control returns to

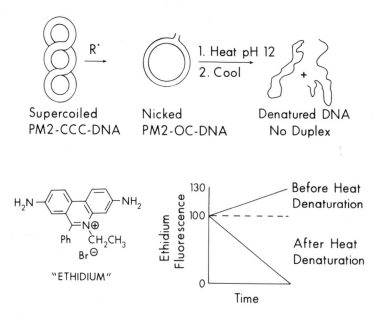

Figure 10   Ethidium fluorescence assay for determining single strand cleavage of supercoiled covalently closed circular DNA. The release of constraints upon nicking the DNA allows more dye to enter and gives a 30% rise in fluorescence. Denaturation at *p*H 12 prevents duplex formation and fluorescence falls to zero.

register after the heating and cooling cycle (Cone *et al.*, 1976; Lown and Sim, 1977). The fluorescence assay for an endonuclease using PM2 DNA is some 10^5 times more sensitive than the usual hyperchromicity assay (Morgan and Pulleyblank, 1974).

A further advantage of this technique for detecting single strand scission is that by adding selective enzyme and chemical inhibitors one may examine the chemical mechanism of the scission process (Cone *et al.*, 1976; Lown and Sim, 1977).

The sensitivity of the assay may be increased by first relaxing the supercoils of the DNA with a topoisomerase enzyme (Morgan *et al.*, 1979; Lown *et al.*,

1977*a*). The topological differences between the various forms of DNA reflected in the relative amounts of ethidium accepted is such that complete scission of the DNA results in a 100% increase in fluorescence (Morgan and Pulleybank, 1972). When using PM2-CCC-DNA the sensitivity of the assay is such that one chemical event may be detected in a DNA of the equivalent of 6×10^7 daltons (Morgan *et al.*, 1979).

The same sensitivity is obtained with an equivalent assay for detecting certain types of alkylation of DNA (Lown *et al.*, 1977*a*). The principle involved is that alkylated bases are susceptible to thermal *N*-glycosylic bond breakage under denaturing conditions leaving an apurinic or apyrimidinic site. The AP sites are readily cleaved in the alkaline *p*H 11.8 assay medium giving a strand break with the concomitant rise in fluorescence noted above. In favourable cases this assay may be adapted to detect alkylation at a particular site on the DNA base owing to the characteristic tendency of the N-7 alkylated guanine ring to open readily under basic conditions (Hsiung *et al.*, 1976).

It will be subsequently shown that these fluorescence assays also have the property of permitting discrimination between single strand scission resulting from base alkylation, depurination or depyrimidination and alkali cleavage of the AP site as described above, and that due to DNA backbone phosphotriester formation and hydrolysis.

TYPES OF DAMAGE TO DNA

Important DNA lesions include (1) alkylation of bases and/or phosphate residues, (2) inter- and intra-strand cross-linking (3) single strand breaks (4) double-strand breaks (5) DNA-protein cross-links. These are illustrated in figure 11. As indi-

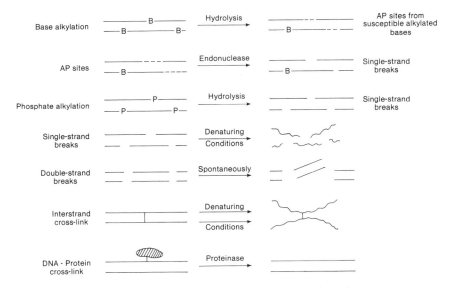

Figure 11   Different types of chemical lesions produced in DNA by antitumour agents.

cated above anticancer agents frequently produce more than one type of lesion which complicates their detection and analysis. Fluorescence techniques are particularly advantageous in that they can often be used in these situations to detect and discriminate different kinds of damage.

Alkylation of DNA bases

Antitumour drugs of the alkylating class fall into roughly two categories depending on the nature of the electrophiles generated (1) those that produce carbonium ions or their precursors (or kinetic equivalents like diazohydroxides) and are anticipated to react by an S_N1 mechanism, such as nitrosoureas and triazenes, and (2) substances which generate strained ring aziridinium or thiiranium ions or oxirane rings as intermediates and which will react by S_N2 mechanisms such as nitrogen and sulphur mustards (Chapter 8). The reactions are generally measured by incubating the nucleic acid with the reagent and then hydrolysing the product and measuring the proportions of alkylated bases (Lawley *et al.*, 1971, 1972; Brookes and Lawley, 1964; Brown, 1974). The main products are 7-alkyl-guanines which can be identified directly (Ludlum, 1975). Less common products are investigated by performing the reaction with radioactive anticancer agents and identifying the products by co-chromatography with authentic samples (Lawley, 1973). If the agent is of the nitrosourea type the alkylated bases will include O^6-alkylguanine. Since such compounds are acid-labile, the product is digested enzymatically to nucleosides and the O^6-alkylguanosine is identified (Loveless, 1966). The most reactive base for this type of reaction is guanine followed by adenosine and cytosine. The products of such alkylations are 7-alkyldeoxyguanosine, 1-alkyldeoxyadenosine and 3-alkyldeoxycytidine (Brookes and Lawley, 1961; Lawley and Brookes, 1963; Lawley, 1966). Alkylation of cellular DNA by ethylnitrosourea results in the formation of 7-ethyldeoxyguanosine, O^6-ethyldeoxyguanosine, 3-ethyldeoxyadenosine, O^4-ethylthymidine, O^2-ethylthymidine, and O^2-ethyldeoxycytidine (Singer *et al.*, 1978; Bodell *et al.*, 1979). There is no reaction with uracil or deoxyuracil (Lawley, 1961).

The DNA alkylation products formed in cells treated with 2-chloroethyl-nitrosoureas are base alkylation products, DNA interstrand cross-links and DNA-protein cross-links. Ludlum and coworkers have identified the following alkylation products: 7-(2-hydroxyethyl)deoxyguanosine; 7-(2-chloroethyl)deoxy-guanosine; 3-(2-hydroxyethyl)deoxycytidine; 3,N^4-ethanodeoxycytidine and 1,2-(didedoxyguanosin-7-yl)-ethane (Tong and Ludlum, 1981; Gombar *et al.*, 1980).

It may be noted that the classical depiction of the purines with sp^3 hybridised exocyclic nitrogens is misleading since this view leads to the incorrect prediction of protonation or electrophilic attack preferentially at these positions. In fact these NH_2 nitrogens are sp^2 hybridised and the nitrogen lone pairs are delocalised into an extended π orbital system and therefore do not normally serve as sites for electrophilic attack. An apparent exception is the nucleophilic addition of the guanine 2-NH$_2$ to the carbinolamide (or carbinolamine) moiety of anthramycin (Hurley, 1977) and saframycin A (Lown *et al.*, 1982) which may be due to stereochemical constraints incurred in the molecular recognition process. By contrast those nitrogens in the purines which were already sp^2 hybridised in the classical structures, that is N$_1$, N$_3$ and N$_7$ in adenine and N$_3$ and N$_7$ in guanine have a lone pair of electrons in a σ orbital in the plane of the ring. It is these lone pairs which are the sites of potential alkylation. In adenine the N-1 nitrogen would normally be expected to be the site for preferential protonation or electrophilic attack. This prediction is confirmed when studying the alkylation products of adenine (Brookes and Lawley, 1961; Lawley and Brookes, 1963; Lawley, 1966) or the adenine residues in RNA or poly rA (Ludlum, 1965). However, N-1 is involved in hydrogen bonding in a Watson-Crick base-pair. Therefore in native DNA or in poly rA·poly rU the major alkylated adenine is 3-methyladenine (Lawley and Brookes, 1963). Methylation of DNA always produces some 7-methyladenine in addition (Lawley and Brookes, 1964).

The predicted reactive site for electrophilic attack in the pyrimidines based on, for example, calculated electron density distribution (Berthod *et al.*, 1967) is N-3 and this is found to be the case for cytidine with alkylating agents at neutral *p*H. However, since this position is also involved in Watson-Crick hydrogen bonding (as in the case of the N-1 position of adenine noted above) there is very little pyrimidine alkylation in DNA owing to the generally greater reactivity of the purines.

The other types of alkylating agents are the sulphur and nitrogen mustards, for example *bis*-β-chloroethyl methylamine and derivatives (see Chapter 8 for further discussion of these compounds). The reactive intermediates in such cases are the thiiranium and aziridinium ions respectively. In the case of bifunctional alkylating agents, once a 7-alkylguanosine has been produced it still possesses a reactive group (Cl—CH$_2$—CH$_2$—X—) for which either water or another DNA base are competing nucleophiles.

The application of the ethidium binding assay provides further insight into the DNA alkylation phenomena. For example the extent of alkylation of DNA by nitrogen mustard, dimethyl sulphate and mitomycin C is proportional to the decrease in the fluorescence of intercalated ethidium (figure 12) (Hsiung *et al.*, 1976). The fluorescence losses due to the first two types of reagents show a marked *p*H dependence, with greater losses of fluorescence being observed at alkaline *p*H values. At *p*H 11.8 the fluorescence shows a slow recovery so that with low levels of methylation ($\sim$4% deoxyguanosine residues modified as determined by radiolabelling techniques) one observes complete return of fluorescence. These phenomena are attributed to the conversion of 7-methyldeoxyguanosine to the zwitterionic form and partial denaturation of the DNA duplex with loss of ethidium binding sites. Base-catalysed imidazole ring opening, and the removal of the positive charge, permits reannealing with concomitant return of ethidium intercalation sites. This conclusion was substantiated by enzymatic hydrolysis of ^{14}C-labelled methylated DNA and identification of the

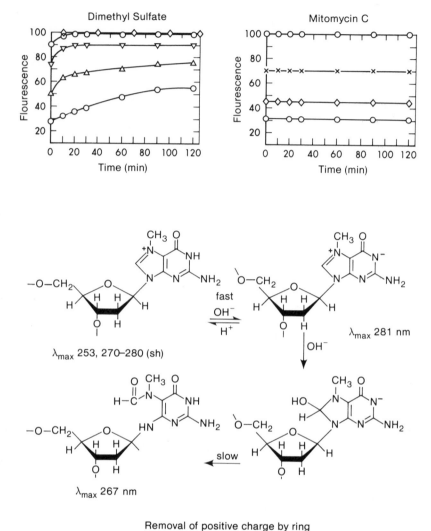

Figure 12 Fluorescence assay for distinguishing 7-guanosine alkylation from reaction at other base sites.

two types of deoxyguanosine residues formed under the different conditions of the ethidium assay (Hsiung *et al.*, 1976).

The distinctly different behaviour of mitomycin C supports previous conclusions that its alkylation of DNA, preferentially on guanine, does not take place at the N-7 position (Tomasz *et al.*, 1974). Studies such as these provide a convenient insight into some aspects of the base preference and chemoselectivity of DNA alkylation by anticancer drugs.

Interstrand cross-linking

Bifunctional alkylating agents can react with two nucleoside residues to link them together. Depending on the relative positions of the two bases this may result in either intrastrand or interstrand cross-linking. Examination of space-filling models of duplex DNA indicates stereochemical constraints on the possible reactions of a bifunctional alkylating anticancer agent. This may in part account for the fact that alkylation of DNA is a much more common event than interstrand cross-linking, typically by a factor of 10 to 20.

7-Alkylation of guanine in this way renders the base susceptible to hydrolytic elimination so that prolonged heating can result in a non-renaturable product by depurination (Lawley and Brookes, 1967). Interstrand cross-links in DNA are formed by such diverse anticancer drugs as nitrogen and sulphur mustards, butane diepoxide, 2-chloroethylnitrosoureas (CENUs), reduced mitomycin C, and carzinophilin A. Although the N-7 position of guanine is the most readily alkylated site, evidence suggests that this is not the position of binding selected by all bifunctional agents. In the case of mitomycin C for example [³H]-exchange procedure has been employed to establish that the N-7 position of guanine can not be involved and in this case the O-6 position of the base is implicated (Tomasz *et al*., 1974).

Similarly CENUs in addition to alkylating the N-7 position of guanine also alkylate the N_3 and N_4 nitrogens of cytidine (Kramer *et al*., 1974; Ludlum *et al*., 1975). By using fluorescence assays one may follow the kinetics of the cross-linking reaction which was shown to take place by a distinct two step process (Lown *et al*., 1978*a*).

While carzinophilin A shows a preference for cross-linking G+C rich sites in DNA the two bonds formed are of unequal lability (Lown and Majumdar, 1977) and the positions of attachment have yet to be established.

The dearth of hard chemical evidence in this area is due to two factors (1) bifunctional anticancer agents frequently will not react with model mononucleotides which hampers the chemical characterisation of adducts and (2) the inherent difficulties of sequential and selective enzymatic degradation of cross-linked DNA (Dubelman and Shapiro, 1977) which may disrupt the original cross-link. It is evident, however, that DNA can accommodate bifunctional linkers of quite disparate lengths from a two-carbon bridge with CENUs to a five-atom bridge for the mustard side-chain CENUs (Lown *et al*., 1980) to a nine-atom bridge spanning the two N-7 positions of G-residues in the case of triethylenemelamine (figure 1) (Lawley and Brookes, 1967). Some distortion of the three-dimensional structure of DNA is implied for the shortest of these links. There is evidence that bifunctional alkylating agents yield interstrand cross-links *in vivo*, for example CENUs (Ewig and Kohn, 1977, 1978) *bis*-methanesulphonylbutane-1,4-diol, myleran, mitomycin C. A consequence of cross-linking is that operation of DNA polymerase must be impeded, so that the lesion may be considered lethal unless the damaged region is either repaired or alternatively deleted. Verly and Brakier

(1969) established that excisions were produced in *Chlamydomonas* after treatment with myleran and that the extent of the excisions was consistent with the levels of alkylation and cross-linking determined.

Most agents that produce interstrand cross-links including bifunctional alkylating agents, nitrosoureas and *cis*-DDP also produce DNA-protein cross-links which are revealed by a reduction of the cross-linking effect by proteinase treatment (Chapter 10).

Single-strand breaks

Anticancer agents of a wide range of structural types have been observed to produce strand breaks in DNA *in vivo* and/or *in vitro*. These include 2-chloroethylnitrosoureas (Ludlum, 1967, 1969; Lown and McLaughlin, 1979*a*,*b*), aryltriazenes (Lown and Singh, 1982), streptonigrin (Cone *et al.*, 1976), mitomycin C (Lown *et al.*, 1976), Adriamycin, daunomycin (Lown *et al.*, 1977*b*, 1981*a*), bleomycin (Umezawa, 1975), tallysomycin (Lown and Joshua, 1980), saframycins A and C (Lown *et al.*, 1982) neocarzinostatin (Kappen and Goldberg, 1978) and camptothecin (Horwitz and Horwitz, 1971). This type of lesion frequently accompanies other types of damage like cross-linking and double strand breaks and whereas it may not always be the primary cytotoxic lesion it appears to be so in the cases of streptonigrin (Cone *et al.*, 1976) and bleomycin (Chapter 6). It is only comparatively recently that sufficiently sensitive techniques have been developed to permit discrimination between different mechanisms of strand scission. There are three main types based on quite distinct chemical origins.

Type I single strand scission (SSS) due to DNA phosphotriester formation
Selective enzyme inhibitors such as superoxide dismutase or catalase, or free radical scavengers, have no effect on the cleavage produced by nitrosoureas. Other possibilities therefore are (1) depurination or depyrimidination and cleavage of an AP site or (2) phosphotriester formation and subsequent hydrolysis.

DNA strand scission can be detected by various methods (Chapter 10) but we have found that fluorescence methods based on the DNA nicking assay employing covalently closed circular PM2 DNA are especially informative (Cone *et al.*, 1976; Morgan and Pulleyblank, 1974). These assays are similar to the one outlined in figure 11. A modification introduced to increase the sensitivity of the assay was to employ calf thymus topoisomerase to relax the native supercoiling of PM2 DNA; this alters the topological differences between the different forms of DNA such that 100% increase in fluorescence is observed for complete single-strand cleavage relative to the control (Morgan *et al.*, 1979). Employing this assay one can readily detect two different types of single strand scission (1) Type 1 scission which occurs relatively rapidly (and without heating) when assayed at *p*H 11.8 (figure 13) and (2) a much slower Type II scission (Lown and McLaughlin, 1979*a*,*b*) (figure 14). The hydroxy substituted nitrosourea CHNU (which is a

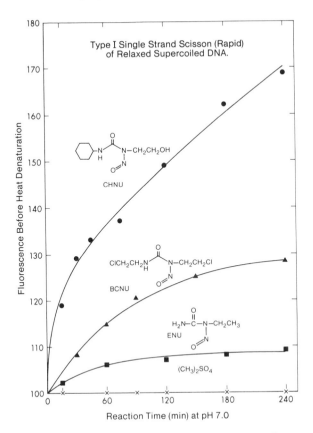

Figure 13　Fluorescence behaviour of covalently closed circular PM2 DNA as a result of Type I single strand scission (relatively rapid and due to phosphate residue alkylation and subsequent hydrolysis of the phosphotriesters) by different alkylating agents.

product of CCNU) is more efficient at cleaving DNA than CCNU or BCNU (Lown and McLaughlin, 1979*b*). Dimethyl sulphate, which is known to alkylate DNA bases but not phosphates (Lawley, 1966; Singer, 1975) has little or no effect in this assay. If aliquots of the first assay are incubated at *p*H 11.8 and 37°C one sees the second and much slower Type II scission, which is caused by dimethyl sulphate. The rate of the latter process is closely similar to the hydrolysis of AP DNA under similar conditions.

We now address the problem of the more rapid Type I scission. Many alkylating agents are known to attack the phosphate groups of DNA (Ludlum, 1967, 1969; Lawley, 1973; Shooter *et al.*, 1974). The resulting phosphotriesters, while known to be stable under neutral conditions undergo quite rapid basic hydrolysis. RNA can serve as a model to test whether Type I scission and phosphate alkylation are related because RNA internucleotide linkages are less stable and the *N*-glycosylic bonds are more stable than those in DNA (Brown *et al.*, 1955).

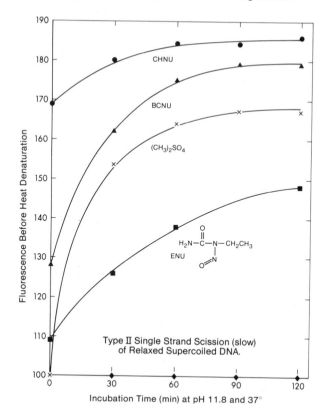

Figure 14 Fluorescence behaviour of relaxed PM2 DNA as a result of Type II single strand scission (relatively slow and due to base alkylation, depurination or depyrimidination and cleavage of the AP site) by different alkylating agents.

Phosphotriesters of ribonucleotides are unstable over the entire *p*H range presumably owing to participation in the hydrolysis step of the β-hydroxyl group on the sugar moiety (Brown, 1974). Alkylation of the base residues of RNA produces a much more stable system than in DNA and therefore depurination of alkylated bases followed by hydrolytic cleavage of the AP site is less likely to interfere. Reaction of poly(rA) with ethylnitrosourea, BCNU and CHNU at 37°C and *p*H 7 was followed by molecular weight analysis using sedimentation velocity measurements. The RNA is readily cleaved and CHNU is the most effective agent. The rates of hydrolysis of the three corresponding phosphotriesters showed that the β-hydroxyethyl diethylphosphate hydrolysed most readily. Further evidence demonstrating that an OH function is necessary to assist DNA cleavage is obtained by comparing with the results obtained for 3-cyclohexyl-1-(2-methoxyethyl)-1-nitrosourea. Upon methylating the OH group, much less Type I scission is observed (Lown and McLaughlin, 1979*a,b*).

Type II scission due to base alkylation, depurination or depyrimidination and cleavage of the AP site

The results suggested that the Type II scission is due to alkylation of the bases followed by depurination or depyrimidination and subsequent strand cleavage at the AP site. This was confirmed by the use of endonuclease VI which is specific for AP sites (Verly and Rassart, 1975). Addition of the endonuclease after 2 hours of treatment of PM2 DNA with BCNU resulted in a characteristic very rapid rise in fluorescence. This was paralleled by incubating the DNA at pH 11.8 which also resulted in strand scission (Lown and McLaughlin, 1979*a,b*; Lown and Singh, 1982). The chemical reactions occurring in a typical example of this process are shown in figure 15.

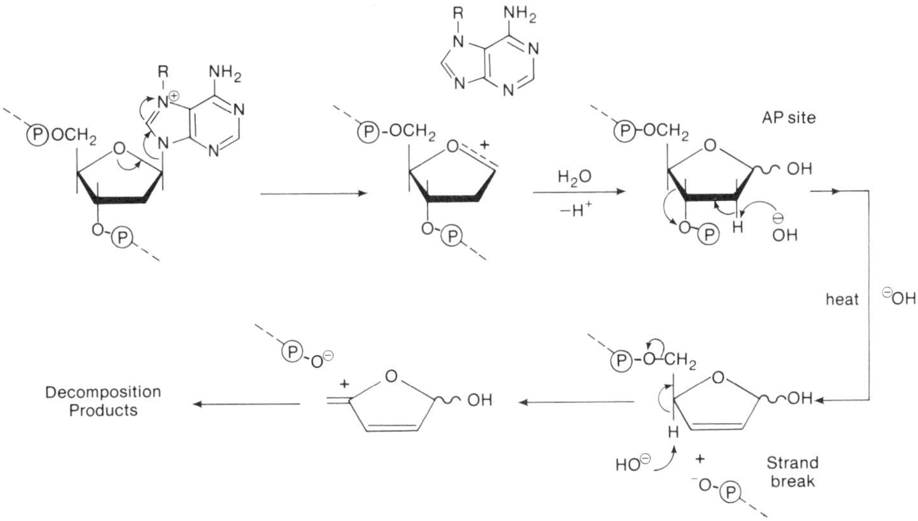

Figure 15   Spontaneous loss of alkylated purine with generation of AP site and subsequent hydrolysis to form a DNA strand break.

Aqueous decomposition of *N*-substituted nitrosoureas gives rise to amines. If aliphatic amines such as cyclohexylamine are generated they are unlikely to affect the AP sites. However, aromatic amines, which are formed in the hydrolysis of the corresponding aryltriazenes, are much more effective in accelerating the cleavage of AP-DNA. This is attributed to Schiff base formation which is more efficient for aromatic amines than for aliphatic amines (Peterson and Burton, 1964). The efficiency of cleavage of AP-DNA for a series of aromatic amines follows the stability of the Schiff base. This process may therefore be significant for the mode of action of aryl-nitrosoureas or aryltriazenes.

Free radical single-strand breaks

Many antitumour agents including mitomycin C, streptonigrin, daunomycin, Adriamycin, bleomycin, tallysomycin, neocarzinostatin and saframycins A and

C when reduced *in situ* cause different kinds of single-strand breaks (and in some cases double strand breaks) in DNA which are different in origin from those previously discussed. In contrast to the nitrosoureas these processes show a definite oxygen requirement. It is only comparatively recently that the free radical mechanism of action common to these widely different structures was established. For example 45 μM saframycin A at pH 7.2 and 38°C in the presence of NADPH causes nicks in covalently closed circular PM2 DNA. The nicking process may be inhibited (and the DNA thereby protected) by (1) superoxide dismutase or (2) catalase or (3) more efficiently by a combination of the two enzymes or (4) by free radical scavengers such as isopropyl alcohol or mannitol or sodium benzoate (figure 16). This confirms the involvement respectively of O_2^- and H_2O_2, and strongly implicates OH˙ in the scission processes (Cone *et al.*, 1976). The participation of OH˙ is confirmed by spin-trapping experiments using N-butylphenylnitrone in conjunction with electron paramagnetic resonance

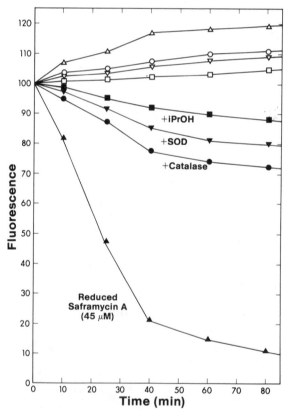

Figure 16   Single strand scission of PM2 DNA by the antibiotic saframycin A reduced *in situ* with NADPH (▲) and its selective inhibition by (●) catalase, (▼) superoxide dismutase, (■) isopropyl alcohol, using the ethidium fluorescence assay. Open symbols are fluorescence readings before heat denaturation and closed symbols are readings after heat denaturation.

spectroscopy (Lown *et al.*, 1978*b*). Strictly speaking detectionof the OH˙-spin-trapped nitroxide adduct simply indicates that the spin-trap has been oxidised. However, independent kinetic criteria confirm the generation of OH˙ radicals in these examples (Lown and Chen, 1981). A general mechanism for its formation may be formulated:

$$A + NADPH \xrightarrow{\text{H}^+} AH_2 + NADP^+ \tag{1}$$

$$AH_2 + O_2 \longrightarrow AH^\cdot + HO_2 \tag{2}$$

$$HO_2^\cdot + HO_2^\cdot \xrightarrow{\text{SOD}} H_2O_2 + O_2 \tag{3}$$

$$2H_2O_2 \xrightarrow{\text{catalase}} 2H_2O + O_2 \tag{4}$$

$$CMX \cdot Fe(III) + O_2^{\cdot-} \longrightarrow CMX \cdot Fe(II) + O_2 \tag{5}$$

$$CMX \cdot Fe(II) + H_2O_2 \longrightarrow CMX \cdot Fe(III) + OH^\cdot + OH^- \tag{6}$$

$$OH^\cdot + DNA \longrightarrow \text{strand scission} \tag{7}$$

[where A = antibiotics; SOD = superoxide dismutase; CMX·Fe(III) = iron complexed with ATP or protein.]

The evidence suggests that for antibiotics including mitomycin C, streptonigrin, the saframycins and anthracyclines, free OH˙ radicals are the result of the reaction of hydrogen peroxide with adventitous cellular iron complexed with protein or ATP (Czapski and Ilan, 1977) [designed CMX·Fe(II) in the reaction scheme.]

The generation of reactive oxygen species from bleomycin or tallysomycin differs from the above reaction scheme in that the required iron is first complexed with the antibiotic (Chapter 6).

While the precise nature of the oxygen species which causes damage of DNA is at present unclear (Kuramochi *et al.*, 1981) it has been shown with the aid of spin-traps that these antibiotics generate $O_2^{\cdot-}$ and OH˙ (Oberley and Buettner, 1979; Lown and Joshua, 1980) which are obvious candidates for the role.

The significance of the OH˙ radical is that it is the most reactive oxygen radical encountered in biological systems and reacts indiscriminately to degrade any molecule within diffusion distance (Fridovich, 1977) and is often the ultimate species responsible for degrading DNA (Dizdaroglu *et al.*, 1975). Neither an antibiotic semiquinone (AH˙) nor the superoxide anion $O_2^{\cdot-}$ (which are relatively stable and belong to the same family as the semidiones) is sufficiently reactive to abstract a hydrogen atom from DNA.

There is a great deal of evidence from radiation chemistry as to the types of damage produced in DNA by OH˙ radicals (Hutterman *et al.*, 1978). Vulnerable sites are the bases and the 4'- and 5'-positions of the deoxyribose moiety. Some of the reaction pathways suggested for sugar damage by OH˙ and consequent backbone scission are given in figure 17. The fact that the action of such antibiotics as bleomycin, tallysomycin and streptonigrin on DNA leads to dialdehyde

DNA Strand Breakage and Base Release by Hydroxyl Radicals

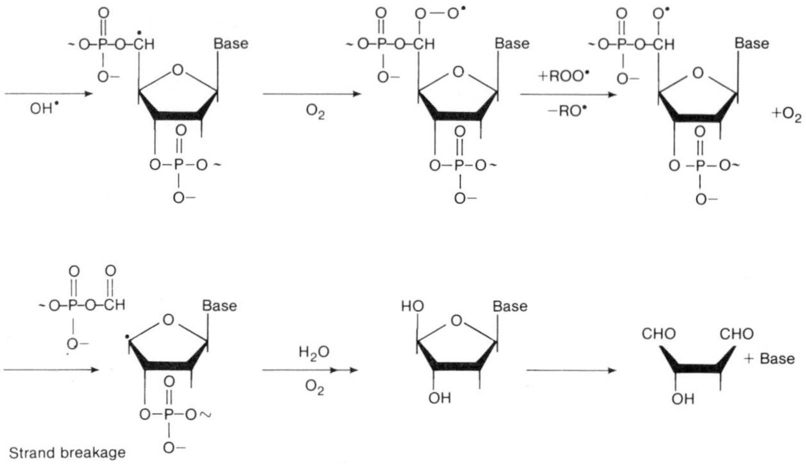

Figure 17 Suggested pathway for hydroxyl radical-induced attack on DNA deoxyribose moiety with subsequent single strand breakage, sugar fragmentation and base release.

sugar fragments supports the analogy with radiation damage (Umezawa, 1974, 1975) with, however, one important difference. Indiscriminately generated OH˙ damages both DNA bases and the sugar residues. However bleomycin and tally-somycin, while degrading the deoxyribose, release the bases (predominantly thymine) undamaged (Chapter 6). This is compatible with a 'site-specific' gener-ation of an oxygen radical close to a susceptible site on the deoxyribose moiety and remote from the bases.

Double-strand breaks

A double strand break is formed when both strands of a DNA helix are broken and the opposing breaks are sufficiently close that the entire molecule is severed under non-denaturing conditions. Such a lesion is presumed always to be lethal to the cell. Double-strand breaks may be most conveniently detected by agarose gel electrophoresis using a circular DNA such as SV40 or PM2 since the covalently closed circular, nicked circular and linear forms (corresponding to a double-strand break) are readily separable.

Bleomycin and tallysomycin induce double strand breaks whose frequency exceeds that anticipated on a statistical basis considering the number of single strand breaks. This suggests a superimposed stereochemical factor or more likely intercalative binding mediated via the bithiazole moiety at sites in the duplex DNA which are especially susceptible to double-strand breaks (Chapter 6).

CONCLUSIONS AND PROSPECTS

Considerable progress has been made towards understanding the molecular mechanism of action of antitumour drugs especially as regards specific types of chemical damage produced in their principal cell targets the nucleic acids. In some instances a fairly complete picture is emerging of the complex series of events preceding and following the generation of reactive species, such as from nitrosoureas and their characteristic DNA alkylation sites. It is now becoming possible to place the synthesis of more effective nitrosoureas on a rational basis in terms of their reactivity towards DNA. There has also been some success in separating the anticancer effects of anthracyclines from the undesired cardio-toxic effects by a consideration of their chemical reactivity towards macro-molecules including DNA.

Many problems remain however. More rapid, selective, reproducible and less tedious methods are needed to analyse, separate, identify and quantify the various alkylation sites in DNA produced by anticancer drugs, and to determine how specific sites relate to repair mechanisms, resistance and to anticancer efficacy. Methods are needed to isolate DNA cross-linked sites so that the chemical bonds are not disrupted during the isolation process. A thorough study of the chemical degradation of DNA by reactive oxygen species, including OH˙, is needed since at present we rely heavily on the analogy with damage produced by ionising radiation. These problems represent a fertile area for further studies.

REFERENCES

Arai, T., Takahashi, K., Ishiguro, K. and Yazawa, K. (1980). *J. Antibiot.*, **33**, 951

Arai, T., Takahashi, K. and Kubo, A. (1977). *J. Antibiot.*, **30**, 1015

Bagley, C. M., Bostick, F. W. and De Vita, V. T. (1973). *Cancer Res.*, **33**, 226

Berthod, H., Giessner-Prettre, C. and Pullman, A. (1967). *Int. J. Quantum Chem.*, **1**, 123.

Bodell, W. J., Singer, B., Thomas, G. H. and Cleaver, J. E. (1979). *Nucleic Acids Res.*, **6**, 2819

Brookes, P. and Lawley, P. D. (1961). *Biochem. J.*, **80**, 496

Brookes, P. and Lawley, P. D. (1964). *J. cell. comp. Physiol.*, **64**, Suppl. 1, 111

Brown, D. M. (1974). In *Basic Principles in Nucleic Acid Chemistry*, Vol. II (ed. P. O. P. Ts'O), p. 1

Burger, R. M., Perisach, J. and Horwitz, S. B. (1978). *J. biol. Chem.*, **253**, 4830

Brown, D. M., Magrath, D. J. and Todd, A. R. (1955). *J. chem. Soc.*, 4396

Cassady, J. M. and Douros, J. D. (ed.) (1980). *Anticancer Agents Based on Natural Product Models*, Academic Press, New York

Carter, S. K. and Crooke, S. T. (1979). *Mitomycin C–Current Status and New Developments*, Academic Press, New York

Cater, D. B. and Phillips, A. F. (1954). *Nature, Lond.*, **174**, 121

Chatterji, D. C., Green, R. F. and Gallelli, J. F. (1978). *J. pharm. Sci.*, **67**, 1527

Cheng, C. J., Fujimara, S., Grunberger, D. and Weinstein, I. B. (1972). *Cancer Res.*, **32**, 22

Colvin, M., Brundrett, R. B., Cowens, W., Jardine, I. and Ludlum, D. B. (1976). *Biochem. Pharmac.*, **25**, 695

Colvin, M., Cowens, J. W., Brundrett, R. B., Kramer, B. S. and Ludlum, D. B. (1975). *Biochem. biophys. Res. Commun.*, **60**, 515

Colvin, M., Padgett, C. A. and Fenselau, C. (1973). *Cancer Res.*, **33**, 915

Cone, R., Hasan, S. K., Lown, J. W. and Morgan, A. R. (1976). *Can. J. Biochem.*, **54**, 219

Connors, T. A., Cox, P. J., Farmer, P. B., Foster, A. B. and Jarman, M. (1974). *Biochem. Pharmac.*, 23, 115

Connors, T. A., Mitchley, B. C. V., Rosenoer, V. M. and Ross, W. C. J. (1974). *Biochem. Pharmac.*, 13, 395

Cox, P. J., Phillips, B. J. and Thomas, P. (1979). *Cancer Res.*, 35, 3755

Czapski, G. and Ilan, Y. A. (1977). Abstract C-11, *International Conference on Singlet Oxygen and Related Species in Chemistry and Biology*, August 21-26, Pinawa, Manitoba

Dizdaroglu, M., von Sonntag, C. and Schulte-Frohlinde, D. (1975). *J. Am. chem. Soc.*, 97, 2277

Dubelman, S. and Shapiro, R., (1977). *Nucleic Acids Res.*, 4, 1815

Ebert, P. S., Chirigos, M. A. and Ellsworth, P. A. (1968). *Cancer Res.*, 28, 363

Eigen, M. and Porschke, D. (1970). *J. molec. Biol.*, 53, 123

Ewig, R. A. G. and Kohn, K. W. (1977). *Cancer Res.*, 37, 2114

Farmer, P. B., Foster, A. B., Jarman, M., Oddy, M. R. and Reed, D. J. (1978). *J. med. Chem.*, 21, 514

Fridovich, I. (1977). In *Biochemical and Medical Aspects of Active Oxygen* (ed. O. Hayaishi and K. Asada), University Park Press, Baltimore, p. 171

Gale, E. F., Cundliffe, E., Reynolds, P. E., Richmond, M. H. and Waring, M. J. (1981). In *The Molecular Basis of Antibiotic Action* 2nd edn, Wiley Interscience, New York, pp 258-401

Gause, G. F. (1975). In *Antibiotics III. Mechanism of Action of Antimicrobial and Antitumour Agents* (ed. J. W. Corcoran and F. E. Hahn), Springer-Verlag, Berlin, p. 197

Gause, G. F., Preobrazhenskaya, T. P., Ivanitskaya, L. P. and Sveshnikova, M. A. (1969). *Antibiotiki*, 14, 963

Gombor, C. T., Tong, W. D., Ludlum, D. B. (1980). *Biochem. Pharmac.*, 29, 2643

Hadju, J. and Armstrong, E. C. (1981). *J. Am. chem. Soc.*, 103, 232

Hata, T., Sano, Y., Sugawara, R., Matsumae, A., Kanamori, K., Shima, T. and Hoshi, T. (1956). *J. Antibiot. Ser. A.*, 9, 141

Horwitz, M. S. and Horwitz, S. B. (1971). *Biochem. biophys. Res. Commun.*, 45, 723

Hill, D. L., Laster, W. R. and Struck, R. F. (1972). *Cancer Res.*, 32, 658

Hurley, L. H. (1980). *Acc. Chem. Res.*, 13, 263

Hurley, L. H. (1977). *J. Antibiot.*, 30, 349

Hsiung, H., Lown, J. W. and Johnson, D. (1976). *Can. J. Biochem.*, 54, 1047

Huttermann, J., Kohnlein, W. and Teoule, R. (ed.) (1978). In *Effects of Ionizing Radiation on DNA, Physical, Chemical and Biological Aspects*, Springer-Verlag, Berlin

Ishiguro, K., Sakiyama, S., Takahashi, K. and Arai, T. (1978). *Biochemistry*, 17, 2545

Iyer, V. N. and Szybalski, W., (1963). *Proc. natn. Acad. Sci. U.S.A.*, 50, 355

Johnston, T. P., McCaleb, G. S. and Montgomery, J. A. (1963). *J. med. Chem.*, 6, 669

Johnston, T. P., McCaleb, G. S., Opliger, P. S. and Montgomery, J. A. (1966). *J. med. Chem.*, 9, 892

Kappen, L. S. and Goldberg, I. H. (1978). *Biochemistry*, 17, 729

Kohn, K. W. (1977). *Cancer Res.*, 37, 1450

Kramer, B. S., Fenselau, C. C. and Ludlum, D. B. (1974). *Biochem. biophys. Res. Commun.*, 56, 783

Kuramochi, H., Takahashi, K., Takita, T. and Umezawa, H., (1981). *J. Antibiot.*, 34, 576

Kusmierek, J. T. and Singer, B. (1976). *Nucleic Acids Res.*, 3, 989

Kuroyanagi, S., Miyajima, S., Hirota, M. and Yamana, T. (1956). *Gann*, 47, 359

Lawley, P. D. (1961). *J. chem. Soc.*, 1011

Lawley, P. D. (1966). *Proc. Nucleic Acid Res. molec. Biol.*, 5, 89

Lawley, P. D. (1973). *Chem.-biol. Interact.*, 7, 127

Lawley, P. D. and Brookes, P. (1967). *J. molec. Biol.*, 25, 143

Lawley, P. D. and Brookes, P. (1963). *Biochem. J.*, 89, 127

Lawley, P. D. and Brookes, P. (1964). *Biochem. J.*, 92, 19c

Lawley, P. D., Orr, D. J. and Shah, S. A. (1971-1972). *Chem.-biol. Interact.*, 4, 431

Lawley, P. D. (1972). In *Topics in Chemical Carcinogenesis* (ed. W. Nakahara, S. Takayama, T. Sugimura and S. Odashima), University Park Press, Baltimore, p. 237-258

Leimgruber, W., Stefanovic, V., Schenker, F., Karr, A. and Berger, J. (1965). *J. Am. chem. Soc.*, 87, 5791

Le Pecq, J. B. and Paoletti, C. (1967). *J. molec. Biol.*, 27, 87

Lin, A. G., Cosby, L. A. and Sartorelli, A. C. (1974). *Cancer Chemother. Rep.*, 4, 23

Lin, A. J., Cosby, L. A., Shansky, C. W. and Sartorelli, A. C. (1972). *J. med. Chem.*, 15, 1247

Lin, A. J., Pardini, R. S., Cosby, L. A., Lillis, B. J., Shansky, C. W. and Sartorelli, A. C. (1973). *J. med. Chem.*, 16, 1268

Loo, T. L. (1975). In *Antineoplastic Immunosuppressive Agents, Part II* (ed. A. C. Sartorelli and D. G. Johns), Springer-Verlag, Berlin, pp. 544-553

Loo, T. L., Dion, R. L., Dixon, R. L. and Rall, D. P. (1966). *J. pharm. Sci.*, 55, 492

Loveless, A. (1966). In *Genetic and Allied Effects of Alkylating Agents*, Pennsylvania State University Press, University Park

Lown, J. W., Begleiter, A., Johnston, D. and Morgan, A. R. (1976). *Can. J. Biochem.*, 54, 110

Lown, J. W. and Chauhan, S. M. S. (1982). *J. org. Chem.*, 47, 85

Lown, J. W. and Chen, H. H. (1981). *Can. J. Chem.*, 59, 390

Lown, J. W. and Chen, H. H. (1980). *Biochem. Pharmac.*, 29, 905

Lown, J. W., Chen, H. H. and Plambeck, J. A. (1981*a*). *Chem.-biol. Interact.*, 35, 55

Lown, J. W. and Hanstock, C. C. (1982). *J. Am. chem. Soc.*, 104, 3213

Lown, J. W. and Joshua, A. V. (1979). *Biochem. Pharmac.*, 28, 2017

Lown, J. W. and Joshua, A. V. (1980). *Biochem. Pharmac.*, 29, 521

Lown, J. W., Joshua, A. V. and Chen, H. H. (1981*b*). *Can. J. Chem.*, 59, 2945

Lown, J. W., Joshua, A. V. and Lee, J. S. (1982). *Biochemistry*, 3, 419

Lown, J. W., Joshua, A. V. and McLaughlin, L. W. (1980). *J. med. Chem.*, 23, 798

Lown, J. W. and Majumdar, K. C. (1977). *Can. J. Biochem.*, 55, 630

Lown, J. W., Majumdar, K. C., Meyers, A. I. and Hecht, A. (1977*a*). *Bioorg. Chem.*, 6, 463

Lown, J. W., McLaughlin, L. W. and Chang, Y. M. (1978*a*). *Bioorg. Chem.*, 7, 97

Lown, J. W. and McLaughlin, L. W. (1979*a*). *Biochem. Pharmac.*, 28, 2123

Lown, J. W. and McLaughlin, L. W. (1979*b*). *Biochem. Pharmac.*, 28, 1631

Lown, J. W. and Sim, S. K. (1976*a*). *Can. J. Chem.*, 54, 2563

Lown, J. W., Sim, S. K., Majumdar, K. C. and Chang, R. Y. (1977*b*). *Biochem. biophys. Res. Commun.*, 76, 705

Lown, J. W. and Sim, S. K. (1976*b*). *Can. J. Chem.*, 54, 2563

Lown, J. W. and Sim, S. K. (1977). *Biochem. biophys. Res. Commun.*, 77, 1150

Lown, J. W., Sim, S. K. and Chen, H. H. (1978*a*). *Can. J. Biochem.*, 56, 1042

Lown, J. W. and Singh, R. (1982). *Biochem. Pharmac.*, 31, 1257

Lown, J. W. and Singh, R. (1981). *Can. J. Chem.*, 59, 1347

Ludlum, D. B. (1965). *Biochim. biophys. Acta.*, 95, 674

Ludlum, D. B. (1967). *Biochim. biophys. Acta.*, 142, 282

Ludlum, D. B. (1969). *Biochim. biophys. Acta.*, 174, 773

Ludlum, D. B. (1975). In *Handbook of Experimental Pharmacology* (ed. A. C. Sartorelli and D. G. Johns), Springer-Verlag, Berlin, pp. 6-17

Ludlum, D. B. (1977). In *Cancer Vol. 5 A Comprehensive Treatise* (ed. F. F. Becker), Plenum Press, New York, p. 285-305

Ludlum, D. B., Kramer, B. S., Wang, J. and Fenselau, C. (1975). *Biochemistry*, 14, 5480

McBride, T. J., Oleson, J. J. and Wolff, D. (1966). *Cancer Res.*, 26, 727

May, H. E., Boose, R. and Reed, D. J. (1974). *Biochem. biophys. Res. Commun.*, 57, 426

Montgomery, J. A., James, R., McCaleb, G. S. and Johnston, T. P. (1967). *J. med. Chem.*, 10, 668

Montgomery, J. A., Johnston, T. P. and Shealy, Y. F. (1970). In *Medicinal Chemistry*, 4th edn (ed. M. E. Wolff), Wiley Interscience, New York, p. 630

Moore, H. W., (1977). *Science*, 197, 527

Morgan, A. R., Lee, J. S., Pulleyblank, D. E., Murray, N. L. and Evans, D. H. (1979). *Nucleic Acids Res.*, 7, 547

Morgan, A. R. and Paetkau, V. (1972). *Can. J. Biochem.*, 50, 210

Morgan, A. R. and Pulleyblank, D. E. (1974). *Biochem. biophys. Res. Commun.*, 61, 396

Oberley, L. W. and Buettner, G. R. (1979). *FEBS Lett.*, 97, 47

Ostrander, J. M., Hurley, L. H., Balakrishnan, M. S. and Krugh, T. R. (1983). *Biochemistry*, in press

Paoletti, C., Le Pecq, J. B. and Lehman, J. R. (1971). *J. molec. Biol.*, 55, 75

Peterson, G. B. and Burton, K. (1964). *Biochem. J.*, 92, 666

Povirk, L. F. and Goldberg, I. H. (1980). *Biochemistry*, 19, 4773
Pratt, W. B. and Ruddon, R. W. (1979). In *The Anticancer Drugs*, Oxford, New York
Price, C. C. (1975). In *Antineoplastic and Immunosuppressive Agents Part II* (ed. A. C. Sartorelli and D. G. Johns), Springer-Verlag, Berlin, 1–5
Price, C. C., Gaucher, G. M., Koneru, D. Shibakawa, R., Sowa, J. R. and Yamaguchi, M. (1969). *Ann. N. Y. Acad. Sci.*, 163, 593
Rao, G. M., Begleiter, A., Lown, J. W. and Plambeck, J. A. (1977). *J. Electrochem. Soc.*, 124, 199
Rao, G. M., Lown, J. W. and Plambeck, J. A. (1977). *J. electrochem. Soc.*, 124, 195–199
Rao, K. V. and Cullen, W. P. (1959). *Antibiot. Annu.*, 950
Rao, K. V., Biemann, K. and Woodward, R. B. (1963). *J. Am. Chem. Soc.*, 85, 2532
Reed, D. J., May, H. E., Boose, R. B., Gregory, K. M and Beilstein, M. A. (1975). *Cancer Res.*, 568
Roberts, J. J., Brent, T. P. and Crathorn, A. P. (1968). In *The Interaction of Drugs and Subcellular Components in Animal Cells* (ed. P. M. Campbell), Churchill, London, pp. 5–27
Ross, W. C. J. (1961). *Biochem. Pharmac.*, 8, 235
Ross, W. C. J. (1962). *Biological Alkylating Agents*, Butterworths, London
Sartiano, G. D., Coetzea, M. L., Klein, K. and Ove, P. (1977). *J. natn. Cancer Inst.*, 58, 1357
Schabel, F. M., Johnston, T. P., McCaleb, G. S., Montgomery, J. A., Laster, W. R. and Skipper, H. E. (1963). *Cancer Res.*, 23, 725
Schmidt, L. H., Fradkin, R., Sullivan, R. and Flowers, A. (1965). *Cancer Chemother. Rep. Suppl.*, 2, 1
Shooter, K. V., Howse, R. and Merrifield, R. K. (1974). *Biochem. J.*, 137, 313
Sim, S. K. and Lown, J. W. (1981). *Biochem. biophys. Res. Commun.*, 81, 99
Singer, B., Bodell, W. J., Cleaver, J. E., Thomas, G. H., Rajewsky, M. F. and Thon, W. (1978). *Nature, Lond.*, 276, 85
Singer, B. (1975). *Prog. Nucleic Acid Res. molec. Biol.*, 15, 219
Skibba, J. L., and Bryan, G. T. (1971). *Tox. appl. Pharmac.*, 18, 707
Sobell, H. M. (1973). *Prog. Nucleic Acid Res. molec. Biol.*, 13, 153
Sugawara, R. and Hata, T. (1956). *J. Antibiot. Ser. A.*, 9, 147
Szybalski, W. and Iyer, V. N. (1967). In *Antibiotics I. Mechanism of Action of Antimicrobial and Antitumour Agents*, Springer-Verlag, Berlin, pp. 211–245
Takamizawa, A., Matsumoto, S., Iwata, T., Makino, I., Yamaghuchi, K., Uchida, N., Kasai, H., Shiratori, O. and Takase, S. (1978). *J. med. Chem.*, 21, 208
Takeuchi, T., Miyamoto, M., Ishizuka, M., Naganawa, H., Kondo, S., Hamada, M. and Umezawa, H. (1976). *J. Antibiot.*, 29, 93
Teller, M. N., Wagshul, S. F. and Woolley, G. W. (1961). *Antibiot. Chemother.*, 11, 165
Tendler, M. D., and Korman, S. (1963). *Nature, Lond.*, 199, 501
Terawaki, A. and Greenberg, J. (1966). *Biochim. biophys. Acta*, 119, 59
Tomasz, M., Mercado, C. M., Olson, J., and Chatterjee, N. (1974). *Biochemistry*, 13, 4878
Tomasz, M. (1976). *Chem.-biol. Interact.*, 13, 89
Tong, W. P. and Ludlum, D. B. (1981). *Cancer Res.*, 41, 380
Umezawa, H. (1974). *Fedn Proc.*, 33, 2296
Umezawa, H. (1975). In *Antibiotics III Mechanism of Action of Antimicrobial and Antitumor Agents* (ed. J. W. Corcoran, and F. E. Hahn), Springer-Verlag, Berlin, p. 21
Verly, W. A. and Rassart, E. (1975). *J. biol. Chem.*, 250, 8214
Walker, M. D. (1973). *Cancer Chemother. Rep. Suppl.*, 4, 21
Wall, M. E. and Wani, M. C. (1980). In *Anticancer Agents Based on Natural Produce Models* (ed. J. M. Cassady and J. D. Douros), Academic Press, New York, p. 417
Waring, M. J. (1965). *J. molec. Biol.*, 13, 269
Wheeler, G. P. (1962). *Cancer Res.*, 22, 651
Wheeler, G. P. (1976). *Am. chem. Soc. Symp. Ser.*, No. 30
White, J. R. (1977). *Biochem. biophys. Res. Commun.*, 77, 387
White, H. L. and White, J. R. (1966). *Biochim. biophys. Acta*, 123, 648
Willson, R. L. (1977). *Iron Metabolism*, 51, 331

10
Biological Aspects of DNA Damage by Crosslinking Agents

Kurt W. Kohn

INTRODUCTION

Since the earliest successes in cancer chemotherapy, which began with the clinical trial of nitrogen mustard (Gilman and Philips, 1946), it was realised that there is something special about compounds having two or more functional groups. As the chemical mechanism of alkylation became clear, it was surmised that the antitumour alkylating agents produce toxic effects due to the formation of covalent crosslinks. By the 1950s, it was suspected that DNA might be the crucial target of crosslinking (Goldacre et al., 1949; Stacey et al., 1958). The model of DNA as a double helix and its role in replication, together with the finding that alkylation can occur on the ring nitrogens of the bases (Brookes and Lawley, 1960, 1961), suggested interstrand crosslinking as a possibly crucial effect, and this phenomenon was soon demonstrated in purified DNA and in cells (Geiduschek, 1961; Iyer and Szybalski, 1963; Kohn et al., 1965, 1966).

The antitumour action of these drugs almost certainly involves selective killing of tumour cells, and structure–activity data indicate that bifunctionality, and hence crosslinking, plays a crucial role. Alkylation reactions, however, occur at a wide variety of biomolecular sites, and only a minority of the total alkylations produced in a cell are of the crucial crosslinking type. A major current problem is to identify the crucial molecular lesions and eventually to develop drugs that would specifically produce only those lesions.

A second major problem is to determine the mechanisms by which tumour cells might be selectively killed. In general, selective cytotoxicity might arise

Molecular Aspects of Anti-cancer Drug Action, ed. Neidle & Waring
0333-315561/83/315-361

from differences between tumour and critical normal cells in (1) drug uptake, (2) drug inactivation, (3) accessibility of critical intracellular targets, (4) repair of DNA (or other) damage, and (5) ability of cells to withstand a period of metabolic stress brought about by the DNA (or other) damage until it is repaired.

There can be marked differences in drug sensitivities among tumours, even of the same histological type. Sensitive tumours may have a biochemical derangement that makes them more sensitive than normal cells to certain kinds of insult. The optimal use of chemotherapy may require that a selected treatment is optimised to take advantage of biochemical derangements that may exist in the cell population of an individual malignancy.

A third major problem, then, is to develop methods to determine key biochemical characteristics in individual tumours, and to develop an armamentarium of agents to take best advantage of any specific vulnerabilities which may be found.

The over-all problem outlined above is particularly pertinent to DNA-damaging drugs, especially crosslinking agents. This chapter reviews some recent experimental approaches to these problems.

MEASUREMENT OF DNA CROSSLINKING IN CELLS

Several methods are now available to measure DNA interstrand crosslinking with sufficient sensitivity to permit studies of the effects of treatment of cells with pharmacologically relevant doses of bifunctional drugs. These methods include alkaline elution, alkaline sedimentation, and strand reassociation. A somewhat more direct procedure is the density-hybrid method which however has a relatively low sensitivity and requires that the cells incorporate a large quantity of 5-bromo-2'-deoxyuridine (BrdUrd) into one DNA strand prior to treatment with a crosslinking drug. Some essential characteristics of these methods are summarised below.

Density-hybrid method

This method directly demonstrates crosslinkage between complementary DNA strands (Roberts and Pascoe, 1972; Roberts and Friedlos, 1981). In order to allow the strands to be distinguished, cells are grown for several hours in the presence of BrdUrd which becomes incorporated into the strand being synthesised at that time. The BrdUrd increases the density of the strand into which it is incorporated. In addition, a radioactive precursor (such as ^{3}H-thymidine) is included, so that this strand will be radioactive as well as of increased density. The cells are then treated with the drug whose crosslinking activity is to be studied. If interstrand crosslinks are formed, some of them will link a radioactive density-labelled strand with its parental strand which will be non-radioactive and of normal density. The DNA is then isolated; the mechanical shear during the isolation procedure breaks the DNA double-strands into segments of average size approximately 2×10^7 daltons. The DNA is then sedimented to equilibrium in alkaline CsCl at a pH high enough to break the hydrogen bonding between paired strands. The density-labelled ('heavy') strands form a band that is separated from the unlabelled ('light') strands (figure 1). Crosslinks between heavy and light strands cause DNA molecules to band in an intermediate position.

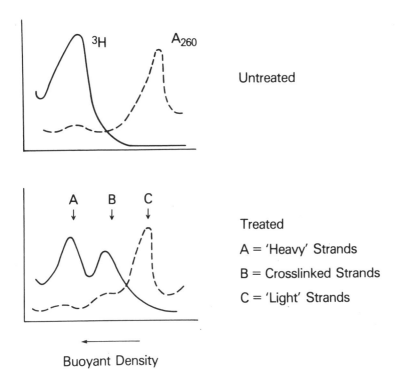

Figure 1 Density-hybrid method for determination of interstrand crosslinks. In a typical procedure (Roberts and Friedlos, 1981), cells are grown for 7 hr in the presence of BrdUrd (5 μg ml^{-1}); 3H-Thd is also present for some time within this period (for example between hours 1 and 4). The cells are then either treated with a crosslinking drug (lower panels) or are left untreated (upper panel). The DNA is then isolated and sedimented to equilibrium in alkaline CsCl at a pH high enough to eliminate hydrogen bonding between complementary DNA strands. Strands labelled with ^{3}H have an increased buoyant density due to incorporation of BrdUrd, whereas the bulk of the DNA, measured by absorbance at 260 nm (A_{260}), is largely of normal density. Crosslinks between 'heavy' and 'light' strands show up as material of intermediate density (lower panel, peak B).

The limited sensitivity of this method is due to the cleavage of the DNA molecules during the isolation step. In order to be measurable, the crosslink frequency must be high enough so that a substantial fraction of the cleaved DNA segments will contain one or more crosslinks.

Alkaline elution

This relatively new method has now been used extensively in several laboratories to measure interstrand crosslinking and DNA–protein crosslinking in mammalian cells (Kohn, 1979, 1981a,b; Kohn et al., 1981).

Essentially, the procedure is to deposit treated cells on a membrane filter, and then to lyse the cells on the filter using a detergent solution. The lysis solution is allowed to flow out through the filter and removes most of the non-DNA material of the cell. The high molecular weight nuclear DNA remains quantitatively on the filter. The procedure is carried out with minimal mechanical shear, so that the cell DNA remains unbroken. The DNA is then slowly eluted from the filter by means of an alkaline solution ($pH \sim 12$) which disrupts the hydrogen bonding between paired strands. The normal intact cell DNA elutes only to a small extent (typically 10-20%) under the conditions used. However, if single-strand breaks have been introduced, for example by X-rays, then the DNA elutes at a rate that increases with the X-ray dose. When the single-strand breaks are randomly distributed, as is the case with X-irradiated cells, the DNA elution rate is first-order with respect to time and with respect to break frequency (figure 2). The sensitivity is high enough to allow quantitation of single-strand break frequencies as low as 1 per 10^{10} daltons of DNA.

For the measurement of interstrand crosslinks, it is necessary to introduce a known frequency of randomly distributed single-strand breaks in order to make the DNA from control cells elute at a defined rate. This is usually accomplished by exposing the cells to an appropriate dose of X-rays at $0°C$. Interstrand crosslinks reduce the elution rate of strand segments due to the effective increase in their size (figure 3).

The apparent crosslink frequency can be quantitated by the following formula (Ewig and Kohn, 1978);

$$p_c = \left[\left(\frac{1 - R_0'}{1 - R_1} \right)^{\frac{1}{2}} - 1 \right] (p_{br} + p_{bd}) \qquad (1)$$

This formula includes a correction that can be applied to take into account drug-induced strand breaks, provided that the frequency of these is low. p_c, p_{br} and p_{bd} are the frequencies of interstrand crosslinks, X-ray-induced single-strand breaks and drug-induced single-strand breaks, respectively. R_1 is the fraction of the DNA from drug-treated cells that is retained on the filter in a standard assay. R_0' is the fraction of the DNA from control cells retained under the same conditions; the prime indicates the correction that may be applied when the drug also produces strand breaks. R_0' then is the fraction of DNA that would be retained if only the calculated drug-induced strand breaks, but no crosslinks, were present.

The values obtained by this computation procedure have usually been proportional to drug concentration, indicating that the values give a proportional measure of crosslink frequencies. The values obtained are fairly independent of

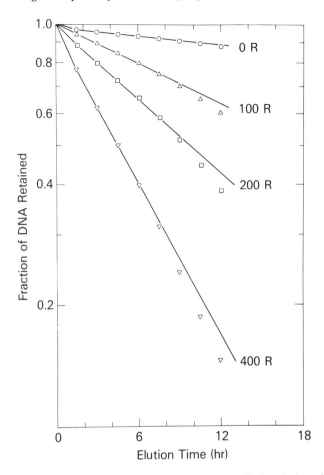

Figure 2   Effect of X-ray-induced single-strand breaks on the alkaline elution of DNA from mouse leukaemia L1210 cells. Cells were exposed to the indicated doses of X-ray, and kept cold to prevent DNA repair.

the exact elution time end point used. For further details of the computational procedure and its conceptual basis, see Ewig and Kohn (1978).

In addition to interstrand crosslinks, bifunctional drugs usually produce DNA-protein crosslinks. In assays for interstrand crosslinks, it is necessary to eliminate DNA-protein crosslinks, because proteins may adsorb to filters and cause retention of DNA strands to which they may be linked. The effects of DNA-protein crosslinks can be eliminated by inserting a proteinase digestion step into the assay and by using appropriate filters and detergents to minimise adsorption. These factors have little effect on DNA elution kinetics in the absence of DNA-protein crosslinks. The elution kinetics apparently depend on the purely mechanical process by which DNA strands of various lengths

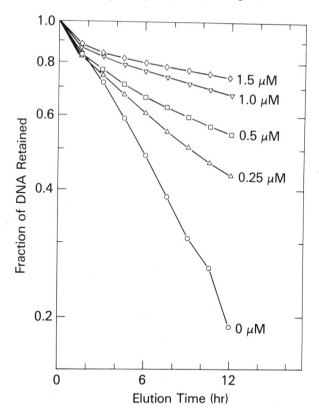

Figure 3 Effect of nitrogen mustard-induced interstrand crosslinks on alkaline elution of DNA from mouse leukaemia L1210 cells. After treatment with the indicated concentrations of nitrogen mustard for 30 min, the cells were chilled and exposed to 300 rads of X-rays. (The filter retention of DNA crosslinked to proteins was eliminated by the combined use of proteinase K, non-adsorptive polycarbonate filters, and sodium dodecyl sulphate).

can flow through a network of filter pores. This process does not involve adsorption of DNA to the filters, and is insensitive to the geometry of the filter pores.

DNA–protein crosslinking is measured by using filters that adsorb proteins with high efficiencies and solutions that do not interfere with this adsorption. Quantitation can be obtained by using a high X-ray dose, so that DNA single-strands that are not linked to protein will elute very rapidly. Those strands that are linked to protein elute very slowly, so that the mass fraction of DNA existing as protein-linked strands can readily be determined (figure 4).

Assuming random and independent distributions of DNA–protein crosslinks and of X-ray induced single-strand breaks, the DNA–protein crosslink frequency is given by

$$p_c = [(1 - r_0)^{-\frac{1}{2}} - 1] \, p_b \tag{2}$$

where r_0 is the fraction of the DNA found to be linked to protein (that is,

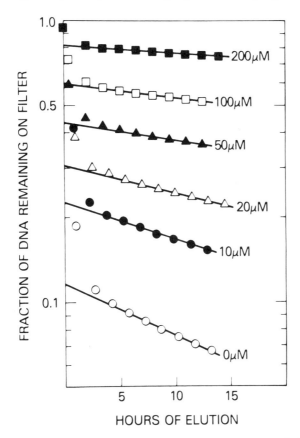

Figure 4   Determination of DNA-protein crosslinks induced by trans-Pt(II)-diaminedichloride in mouse leukaemia L1210 cells. This compound produces little or no interstrand crosslinking or single-strand breakage (Zwelling *et al.*, 1979*a*). Cells were treated with the indicated concentrations of the agent for 1 hr and then chilled and exposed to 3000 R of X-rays in order to introduce a relatively large frequency of strand breaks.

retained on the filter), and p_b is the single-strand break frequency (Kohn and Ewig, 1979). The values obtained by this computation are independent of the X-ray dose used in the assay.

Strand reassociation

This method is based on the fact that the hydrogen bonding between complementary strands is disrupted in alkaline solution ($pH > 12$), so that the strands gradually separate and move apart; an interstrand crosslink prevents complementary strands from moving apart and, upon neutralisation, allows the strands to reassociate rapidly (Kohn, Steigbigel and Spears, 1965; Kohn, Spears and Doty, 1966; Jolley and Ormerod, 1973).

Pera *et al.* (1981) recently showed that this method can be made sensitive and quantitative in assays with mammalian cells. The cells are exposed to 1,500 rad of X-rays at 0°C in order to introduce random strand breaks. They are added to an alkaline solution and sufficient time is allowed for complete unwinding of the double helices (Rydberg, 1975). The solution is then slowly neutralised, so as to allow crosslinked complementary strands to reassociate without mechanically inducing strand breakage. The fraction of the DNA that reassociates to form double helices can be determined by means of S_1 nuclease, an enzyme that preferentially digests DNA single strands.

Assuming that the interstrand crosslinks and X-ray-induced single-strand breaks are randomly and independently distributed, the crosslink frequency is given by

$$p_c = (\sqrt{f^0_{ss}/f_{ss}} - 1)\,(2p_{br}) \tag{3}$$

where f^0_{ss} and f_{ss} are the fraction of the DNA that is single stranded in assays of control and drug-treated cells, respectively; p_{br} is the break frequency introduced by X-ray. (It is assumed that the drug does not itself cause significant strand breakage.) This formula is a simple statistical statement and, in contrast to the case of alkaline elution, gives an absolute rather than a proportional estimate of interstrand crosslinking.

This method is promising but has not yet been widely used to estimate crosslink frequencies.

Alkaline sedimentation

In this procedure, as in the preceding two methods, treated and control cells are exposed to X-rays in order to introduce an appropriate frequency of single-strand breaks, and the cells are placed in an alkaline solution which disrupts hydrogen bonds and allows DNA double helices to unwind. The DNA which is released from the cells is then centrifuged in an alkaline sucrose gradient. Interstrand crosslinks between different single-strands increase the effective size of the linked DNA strands and cause these strands to sediment more rapidly. It is difficult to calculate crosslink frequencies from changes in the sedimentation pattern, but Pera *et al.* (1981) have applied computer modelling to this problem.

Comparison of methods

Pera *et al.* (1981) have recently compared alkaline elution, DNA renaturation and alkaline sedimentation methods in a study of the formation and loss of interstrand crosslinks following treatment of Chinese hamster cells with the platinum drug *cis*-platinum (II) diammine dichloride (*cis*-DDP). The three methods were found to be sufficiently sensitive to measure interstrand crosslinking following pharmacologically realistic doses. Mutually consistent results were obtained among these methods, and also with those obtained at higher doses by the more direct, but less sensitive, hybrid density method (Roberts and Friedlos, 1981). The three sensitive methods showed interstrand crosslinking to increase several-fold during several hours following treatment, and then to decrease with a half-time of 12 to 24 hr. The delay in interstrand crosslink formation is attributed to the time required for DNA monoadducts to react with the complementary strand (Zwelling *et al.*, 1979*a*). The density-hybrid method did not show this delay, probably because the DNA purification procedure used in this assay needed an incubation period during which most monoadducts capable of forming interstrand crosslinks could complete this reaction.

CROSSLINKING AND LETHALITY

Inactivation of viral DNA

Bacteriophage and viruses provide useful systems for the study of the biological effects of DNA damage. The phage or viruses can be treated with DNA-damaging agents and the nature of the damage can be characterised prior to infection. Since the host cells remain untreated, the damage is restricted to the infecting genomes.

In early studies with T2 phage, Loveless and Stock (1959*a,b*) observed a delayed inactivation of the phage following brief treatments with either HN2 or with di(2:3-epoxypropyl)ether. The continued inactivation during the post-incubation period was attributed to the progressive conversion of monoalkylations to dialkylations upon reaction of the second functional group with the phage material. The progressive inactivation due to the reaction of the second alkylating arm could be prevented by sodium thiosulphate which rapidly neutralises any unreacted alkylating groups.

The inactivation by bifunctional agents could be due in part to blocked injection of the phage DNA into the cell, because of crosslinks within the DNA or between the DNA and the phage capsid (Loveless and Stock, 1959*c*). This factor must be taken into account in attempting to deduce the biological activity of damaged DNA within the cell.

More recent work has correlated phage inactivation with measurements of DNA alkylation and crosslinking. Verly and Brakier (1969) found that treatment of T2 phage with nitrogen mustard inactivated the phage to an extent similar to the fraction of phage DNA molecules that would sustain one or more interstrand crosslinks when the purified DNA was treated under similar conditions. Thus a single interstrand crosslink in a phage DNA molecule appeared to be lethal, and the infected *Escherichia coli* B apparently did not repair the crosslinks in the phage DNA. A possibility not excluded, however, was that intra-strand crosslinks or monoalkylations also contributed to lethality, but that repair led to a fortuitous agreement between interstrand crosslinking and survival. Since *E. coli* B does repair interstrand crosslinks in its own DNA (Kohn *et al.*, 1965), a failure to repair phage T2 DNA might be a function of the infected state.

Lawley *et al.* (1969) found that T7 phage were much more sensitive to inactivation by mustard gas (di-(2-chloroethyl)sulphide) than by monohydrolysed derivative which has only one alkylating arm. This points to bifunctional attack as the major cause of lethality. The inactivation of an average phage by mustard gas occurred when there were 7 monoalkylations per phage and 1.3 equivalents of di-(guanin-7-yl-ethyl)sulphide. The latter represented the sum of inter- and intra-strand crosslinks at a ratio of $1:2$ to $1:3$. By contrast, the one-armed mustard required 280 alkylations to inactivate an average phage. One can conclude that intra-strand crosslinks were the major cause of inactivation by mustard gas in this system.

When phage that have been treated either with a bifunctional or with a monofunctional alkylating agent are incubated at $37°C$, there is a delayed inactivation which can be attributed to the spontaneous loss of alkylated purines (Verly, 1974). The resulting base-free sites would be cleaved by AP-endonuclease which is specific for these sites. In principle, such cleavage might lead either to inactivation or to repair. It was estimated that an inactivating dose corresponded to 7–8 depurinations per phage; hence depurinations are not necessarily lethal, and presumably can be repaired. Verly (1974) proposed that inactivation occurs when a base-free site is formed in certain critical parts of the DNA molecule.

Thus two factors might contribute to delayed inactivation by bifunctional drugs: (1) delayed alkylation by the second functional group, and (2) formation of base-free sites consequent to the elimination of alkylated purines. Base-free sites that form from bifunctional alkylations would differ from those that form from monoalkylations in that, apposed (on the opposite strand) or adjacent (on the same strand) to the base-free site, there would still be another lesion, such as an alkylation or another base-free site. The systems which repair simple base-free sites may have difficulty in dealing with such neighbouring lesions.

In the case of *cis*-DDP complexes, Shooter *et al.* (1972) reported that 5 platinations per phage were required per lethal hit, but that 35 platinations were required per crosslink. This would seem to argue against crosslinking as the major lethal lesion. However, it is possible that more crosslinks could have formed than was apparent in these experiments, because the delayed increase in interstrand crosslinking (Zwelling *et al.*, 1979*a*) was not taken into account.

The relationship between Pt-induced DNA crosslinks and DNA inactivation was studied by Filipski *et al.* (1980) in phage λ DNA. Biological activity was measured by the ability of the DNA to transfect *E. coli*. Both *cis*- and *trans*-DDP complexes were found to produce interstrand crosslinks in the purified DNA. Interstrand crosslinking was measured by loss of alkaline denaturability of the DNA. In order to determine the loss of biological activity corresponding to a given crosslink frequency, it was important to exclude possible continued conversion of monoadducts to crosslinks during the transfection assay. This was done either by reacting the DNA long enough to assure complete reaction of monoadducts, or by quenching any remaining monoadducts with thiourea (Filipski *et al.*, 1979). In the case of *trans*-DDP, the survival of transfectivity approximated the fraction of the DNA remaining uncrosslinked; thus only the uncrosslinked molecules remained transfective. In the case of *cis*-DDP, however, the fraction of the DNA that remained transfective was much less than the fraction of the DNA that remained uncrosslinked. This suggested that both *trans*-DDP and *cis*-DDP produce interstrand crosslinks which inactivate phage DNA but that only *cis*-DDP also produces other inactivating lesions. Support for this hypothesis was obtained by isolating the non-crosslinked DNA fraction by alkaline sedimentation and reannealing the single-stranded fraction. In the case of *trans*-DDP, the reannealed non-crosslinked DNA retained transfectivity, whereas, in the case of *cis*-DDP, the DNA prepared in this way was devoid of transfectivity. In the latter case, transfectivity could be restored by treatment with thiourea. The results suggested that *cis*-DDP, but not *trans*-DDP, forms intra-strand crosslinks which are inactivating lesions.

Furocoumarins (psoralens)

The most selective way presently available to introduce interstrand crosslinks in cells is by the photochemical reaction of furocoumarins (psoralens) with the double-bonds of pyrimidine bases (reviewed by Song and Tapley, 1979). Psoralen binds reversibly to DNA by intercalation between base-pairs. The binding constant is of the order of $0.021 \, mol^{-1}$ and there is a preference for AT-containing sequences (Ou *et al*., 1978). Upon irradiation with near-UV light (such as 360 nm), the intercalated molecules react with pyrimidine bases to form cyclobutane-bridged adducts. Initially, a psoralen molecule reacts with a pyrimidine base to form a monoadduct. Some monoadducts can then undergo a second photochemical reaction of similar type in which the opposite side of the psoralen molecule reacts with a pyrimidine base in the complementary DNA strand. The result is a diadduct which is an interstrand crosslink. Because of the mode and geometry of DNA binding, psoralens probably do not produce intra-strand or DNA–protein crosslinks.

Interstrand crosslinks probably are the main lethal lesions produced by psoralen plus light, both in bacteria and in mammalian cells (Cole, 1971; Ben-Hur and Elkind, 1973). A particularly clear demonstration of the lethality of interstrand crosslinks, in contrast to monoadducts, was recently obtained by Grover *et al*. (1981) in *E. coli*. The relative frequency of interstrand crosslinks to monoadducts was maximised by delivering a short pulse of light in the presence of psoralen, washing the psoralen away, and then delivering a second pulse. The first pulse generates mainly monoadducts, many of which are converted to interstrand crosslinks by the second pulse. This regimen produces a much higher interstrand crosslink to monoadduct ratio than does prolonged exposure to light in the presence of the psoralen. Cell survival was dependent only on the interstrand crosslink frequency, and was independent of the regimen employed. In this system, therefore lethality resulted exclusively from interstrand crosslinking.

Two steps in crosslink formation

Crosslink formation by a bifunctional agent requires a sequence of two steps: the reaction of the first functional group to form a monoadduct followed by reaction of the second functional group to form a crosslink. If the second step is relatively slow, crosslink formation will exhibit a lag.

There is a lag of several hours in the formation of interstrand crosslinks by chloroethylnitrosoureas in purified DNA (Kohn, 1977; Lown *et al.*, 1978) and in mammalian cells (Ewig and Kohn, 1977, 1978; Erickson *et al.*, 1980*a*). In mammalian cells, the post-incubation time required to reach maximum cross-linking is about 6 hr. Delays of 4-12 hr in peak interstrand crosslinking in mammalian cells have also been observed with L-phenylalanine mustard (Ross, *et al.*, 1978), *cis*-DDP (Zwelling *et al.*, 1978, 1979*a*), an activated cyclophosphamide derivative (Erickson *et al.*, 1980*c*), and a bifunctional aziridinyl quinone (R. A. Ewig, L. Szmigiero and K. W. Kohn, unpublished). Similar delays in interstrand crosslinking have been reported in yeast treated at 36°C with an activated cyclophosphamide derivative (Fleer and Brendel, 1981) and with triaziquone (Trenimon, a trifunctional aziridinyl quinone) (Fleer and Brendel, 1979).

In the case of nitrogen mustard, the delay in the reaction of the second arm is less pronounced. Rutman *et al.* (1969) were able to detect a lag of about 10 minutes in the crosslinking of purified DNA at 37°C. At 0°C, the monoalkylation and crosslinking steps could be clearly separated. In mammalian cells at 37°C, no lag (of < 1 hr) in crosslinking was detected (Ewig and Kohn, 1977; Ross *et al.*, 1978). In yeast at 23°C, a lag of about 3 hr was observed in the development of maximum interstrand crosslinking (Fleer and Brendel, 1979).

In relating DNA crosslinking to cytotoxicity, it is important to take into account the possible delays in crosslink formation. At later times following treatment, crosslink levels often decline, possibly due to the action of repair mechanisms. An approximate balance of crosslink formation (by continued second-arm reaction) and crosslink removal may persist for many hours during the post-treatment period; during this time, the crosslink frequency may remain nearly constant.

In the case of *cis*-DDP, the further conversion of monoadducts to crosslinks can be blocked by treatment with thiourea (Filipski *et al.*, 1979; Zwelling *et al.*, 1979*c*). Thiourea reacts rapidly with available Pt sites, and can quench these sites at sub-toxic doses. In a typical protocol (figure 5), L1210 cells were treated with *cis*-DDP for 1 hr, and then with various concentrations of thiourea for 1 h; interstrand crosslinking was measured 5 hours or more later, when near its peak. Colony formation also was determined. It is seen that thiourea reduced interstrand crosslinking and simultaneously increased colony formation. It is plausible, although not entirely compelling, to suppose that these effects of thiourea are due to the quenching of monoadducts which otherwise would generate potentially lethal interstrand crosslinks.

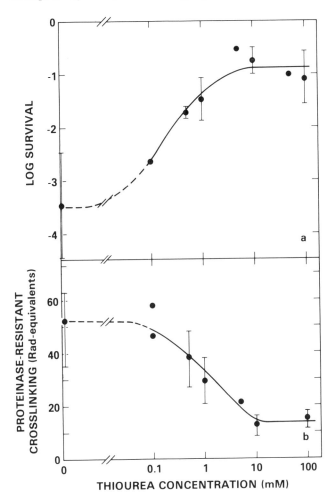

Figure 5   Effect of thiourea on *cis*-DDP-induced cell killing and interstrand crosslinking in L1210 cells. Cells were treated with 20 μM *cis*-DDP for 1 hr. The cells were then washed and treated with thiourea for 1 hr. Survival of colony formation in soft agar was determined. Protease-resistant crosslinking was measured by alkaline elution 6–12 hr after removal of *cis*-DDP. The effects of thiourea are attributed to the quenching of Pt-DNA monoadducts. (From Zwelling *et al.*, 1979c)

INHIBITION OF DNA REPLICATION

Since strand separation is necessary in the semi-conservative replication of DNA, it is unlikely that replication could proceed past an interstrand crosslink. Indeed, a variety of crosslinking agents are known to selectively inhibit cellular DNA synthesis, as well as DNA polymerase enzymes.

In an illuminating study *in vitro*, Tanaka and Yoshida (1981) found that low concentrations of mitomycin-crosslinked DNA (but not free mitomycin) inhibit

DNA polymerase enzymes, even in the presence of a large excess of template-primer DNA. Thus the addition of a small amount of crosslinked DNA appeared to poison the enzyme. Polymerase beta from calf thymus was slightly more sensitive than polymerase alpha, and terminal deoxynucleotidyl transferase was inhibited to an extent similar to polymerase beta. The inhibitions were competitively reversed by increasing the concentration of template-primer DNA, but were not affected by increasing the concentrations of enzyme or deoxyribonucleoside triphosphates. The competitive inhibition kinetics show that the polymerase enzymes are not irreversibly bound to the DNA lesions, but have the ability to transfer to undamaged DNA template-primer. Tanaka and Yoshida suggest that the DNA conformation around an interstrand crosslink may in some way resemble a replication fork and thus constitute a site of unusually strong binding for polymerases. This is plausible if one considers that the DNA strands may become separated at the immediate site of polymerase binding, but that the crosslink prevents further strand separation down-stream.

Ou *et al.* (1978) found DNA polymerase reactions to be inhibited by the photobinding to DNA of 8-methoxy-psoralen or 5,7-demethoxycoumarin. The inhibition of DNA template activity was noted to be greater than expected from the fraction of the DNA molecules crosslinked. Although the authors concluded that monoadducts contributed a major part of the inhibition, it is possible, in view of the work of Tanaka and Yoshida mentioned above, that the crosslinked DNA sites could have bound polymerase molecules so as to make the enzyme unavailable to non-crosslinked DNA molecules.

Pohl and Christophers (1979) reported that the inhibition of DNA synthesis in skin fibroblasts treated with furocoumarin plus light was enhanced if, after removal of free furocoumarin, the cells were again exposed to light. The second light exposure converts monoadducts, which were produced by the first light exposure, to interstrand crosslinks. The result supports the premise that interstrand crosslinks are a major cause of DNA synthesis inhibition.

Varga *et al.* (1982) found that treatment of murine melanoma cells with psoralen (trioxsalen) plus light at doses which produced about one photoadduct per 10^6 base pairs tended to arrest the cells in the S and G2 phases of the cell cycle. The monofunctional compound, angelicin, which can produce monoadducts but not crosslinks, did not have this effect. Hence very low frequencies of interstrand crosslinks can markedly perturb the cell cycle, at least in part by inhibiting DNA replication.

INTERSTRAND CROSSLINK REPAIR

Possible mechanisms

In the light of present knowledge, two types of excision mechanisms can be proposed for the early steps in the repair of interstrand crosslinks. The better known of the two is nucleotide excision repair, in which a damage-specific endonuclease introduces two breaks into the same strand, one on each side of the crosslink (figure 6). This would allow a strand segment containing the crosslink to swing out of the way, and would set the stage for later steps in the repair sequence.

The second possible mechanism would involve excision of one of the cross-linked bases by cleavage of its bond to the deoxyribose moiety of the DNA backbone. This could be accomplished by a glycosylase. Several enzymes of this type have been found to remove specific types of altered bases (Lindahl, 1982). The removal of crosslinks by a glycosylase however has not been demonstrated and is still hypothetical. Following base excision, a base-free site would remain which would be susceptible to strand cleavage by an AP-endonuclease (figure 6). The residual base-free deoxyribose moiety might then be removed by an exonu-clease. The gap in the strand could be filled by the DNA repair polymerase, and

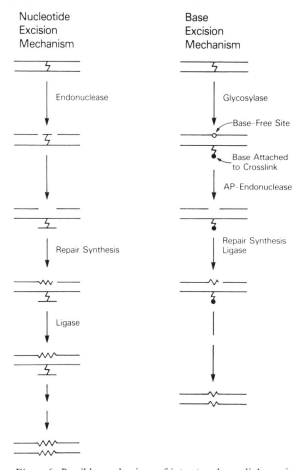

Figure 6   Possible mechanisms of interstrand crosslink repair.

the final strand ends could be joined by DNA ligase; these late steps are similar to the steps in classical nucleotide excision repair.

In both of these mechanisms, the transient occurrence of single-strand breaks would be expected. In the case of nucleotide excision repair, the gap which must be filled by repair polymerase may be several hundred nucleotides long, whereas the gap formed following AP-endonuclease would probably be only a few nucleotides.

In both mechanisms, however, the repair synthesis would encounter the opposite end of the crosslink which would still be attached to the template strand. The problem thus encountered is how a replication process can cope with a defect in the template. An incompletely repaired interstrand crosslink may block repair synthesis in the gap and may stimulate genetic recombination. The repair of crosslinks following the incision step may also occur by recombination with intact sister DNA duplexes. Recombinational repair has been demonstrated in bacteria and may also occur in yeast.

A repair process is thought to exist in mammalian cells which can replicate through template defects by an uncertain mechanism, apparently not involving recombination. This mechanism, commonly (but unfortunately) termed 'post-replication repair', appears to be deficient in certain xeroderma 'variant' cells and to be sensitive to caffeine (Cleaver, 1972; Lehmann *et al.*, 1975, 1977).

Bacteria

Nitrogen and sulphur mustards

The repair of interstrand crosslinks was first noted in *E. coli* treated with nitrogen mustard (Kohn *et al.*, 1965). In order to measure DNA crosslinking, DNA isolated from treated organisms was exposed to alkali. Single and double stranded DNA were then separated on the basis of buoyant density difference by equilibrium sedimentation in CsCl. Upon further incubation after removal of drug, wild type *E. coli* B eliminated crosslinks from their DNA. The elimination of crosslinks was probably a repair process, because the UV-excision repair-defective mutant, Bs-1, failed to remove crosslinks and had an increased sensitivity to the drug. Similar findings were reported by Venitt (1968) using sulphur mustard.

Since *E. coli uvr⁻* mutants, which are defective in a damage-specific endonuclease, fail to remove nitrogen mustard-induced interstrand crosslinks, the repair in this case is probably by a nucleotide excision mechanism, rather than by a glycosylase mechanism.

Psoralens

The repair of interstrand crosslinks produced by psoralen plus light in *E. coli* was studied in detail by Cole and his coworkers. The presence of covalent links between complementary strands was shown by the ability of crosslinked strands to reassociate rapidly after exposure of the isolated DNA to denaturing conditions (in this case, heat). The resulting single and double strands were separated on hydroxyapatite. In wild-type *E. coli*, crosslinks disappeared at the first-order rate with a half-time of about 10 min at 30°C (Cole *et al.*, 1976). Concurrently, single-strand breaks were observed in alkaline sedimentation studies (Cole, 1973). Immediately after treatment with psoralen plus light, the sedimentation of the DNA strands was increased as expected for crosslinked molecules. After 30 min, however, the single-strand lengths were reduced well below the original sizes, indicating the introduction of large numbers of single-strand breaks. By 90 min, the DNA sedimentation had reverted to normal. Thus, while crosslinks were removed, strand breaks appeared transiently and then rejoined. So far this is consistent with an excision-repair mechanism, although the breaks could have been associated with the repair of monoadducts or DNA-protein crosslinks as well as interstrand crosslinks.

E. coli mutants defective in endonuclease specific for UV-type base damage failed to remove psoralen crosslinks, indicating that this repair involves a nucleotide excision mechanism (Cole *et al.*, 1976).

In analogous studies, Ben-Hur *et al.* (1979) calculated that there were approximately 14 psoralen crosslinks per *E. coli* genome per lethal hit. The crosslinks were removed with a half-time of 8 min at 37°C in these bacteria. In an excision repair-deficient *E. coli uvr*B mutant, there were only about two crosslinks per genome per lethal hit, and the crosslinks were not removed. Thus most crosslinks in wild-type cells were removed by excision-repair. The excision repair-deficient mutant might however be able to by-pass some crosslinks by recombinational repair.

The evidence cited so far is consistent with an excision and repair synthesis mechanism. Studies using density labelling however indicated the covalent joining of old to newly synthesised DNA strands (Cole, 1973). This unexpected finding for classical excision-repair was observed 90 min after treatment of *E. coli* with psoralen plus light, at which time the crosslinks and strand breaks appeared to have been removed. The result indicated that strands of sister DNA duplexes were joined to each other, as occurs in genetic recombination. This mechanism was supported by studies using *rec⁻* mutants (Sinden and Cole, 1978). These mutants removed interstrand crosslinks and accumulated single-strand breaks but did not reseal them. The ability of mutant strains to carry out genetic recombination was found to be correlated with the rate of rejoining of these breaks and the formation of viable progeny after treatment with psoralen plus light. (This evidence is not conclusive however since the *rec*A protein is now known to act not only in recombination, but also to be involved in the control of a variety of repair functions induced by the SOS system (Little and Mount, 1982).)

Yoakum and Cole (1977) demonstrated the removal of psoralen crosslinks in permeabilised *E. coli*. The organisms were permeabilised by treatment with toluene. The removal of interstrand crosslinks required ATP, Mg^{2+} and the *uvr*A/*uvr*B gene products, but deoxyribonucleoside triphosphates were not obligatory. It was concluded that a *uvr* endonuclease cleaves a strand on the

5' side of one arm of the crosslink, producing a 3'-hydroxyl terminus, and that a second strand cut on the 3' side of the crosslink is made by an exonuclease activity associated with DNA polymerase I.

E. coli cells having a defect either in the *uvr*AB endonuclease gene or in the *rec*A gene are killed by one or by very few psoralen crosslinks (Seki *et al.*, 1978; Bridges and Wright, 1981; Sinden and Cole, 1978). When both *uvr*AB and *rec*A are defective, the organisms are even killed by one or a few psoralen–DNA mono-adducts. Bridges and von Wright (1981) find that in addition to the *uvr* and *rec* genes, crosslink repair can involve the *rep* gene which codes for a DNA unwinding enzyme that promotes strand separation during DNA replication.

The multiplicity of repair functions is again shown in the recent report by Bridges and Stannard (1982) that psoralen monoadduct repair can occur in *uvr*⁻ mutants. *E. coli uvr*A⁻ were exposed to 8-methoxypsoralen plus near-UV light. This exposure produced mainly monoadducts, some of which could be converted to crosslinks by a subsequent light exposure in the absence of drug. If a period of incubation was inserted between the two light exposures, the lethal effect was reduced, suggesting that monoadducts were repaired. That the reduction in lethality was due to a repair process was indicated by the finding that, in organisms defective in *pol*A as well as in *uvr*B, the lethality did not decrease with time between light exposure. Hence the DNA repair polymerase, coded by *pol*A, was involved in the repair of psoralen monoadducts in excision-defective organisms.

Platinum complexes

Effects of *cis* and *trans*-platinum compounds in wild type and repair deficient *E. coli* were recently studied by Alazard (1982). Both *uvr*A/B and *rec*A mutants were found to have markedly increased sensitivities to *cis* and to *trans*-Pt, even when normalised with respect to the quantity of Pt-DNA adducts formed. In combination, the *uvr* and *rec*A defects produced additive effects on sensitivity, suggesting that there may be at least 2 different repair paths for Pt-DNA lesions. Loss of viability was correlated with inhibition of DNA synthesis immediately after treatment. Compared to *cis*-Pt DNA adducts, *trans*-Pt adducts were much less lethal and much less inhibitory towards DNA replication and DNA repair synthesis. In mammalian cells, *trans*-Pt adducts have a much lower probability of forming interstrand crosslinks than do *cis*-Pt adducts (Zwelling *et al.*, 1978). If this is true also in *E. coli*, it could relate to the low potency of the *trans*-Pt adducts.

Nitrosoureas

The *uvr*-dependent excision mechanism in *E. coli* is not required for the repair of lethal and mutagenic damage introduced by CCNU or by MNU, according to Morimoto *et al.* (1980) who found that a *uvr*A mutant exhibited normal survival and mutagenesis following treatment with these drugs.

Yeast

Yeasts are useful organisms for DNA repair studies because they are eukaryotes in which DNA repair-defective mutations have been well defined genetically.

Excision-defective strains of yeast (*rad* 1, 2 or 3 mutants, corresponding to *uvr*A, B of *E. coli*) exhibit markedly increased sensitivities to interstrand crosslinking by trimethylpsoralen plus light (Averbeck and Moustacchi, 1975; Grant *et al.*, 1979; Jachymczyk *et al.*, 1981). Jachymczyk *et al.* (1981) found that wild type yeast (*Saccharomyces cerevisiae*) remove trimethylpsoralen crosslinks completely within 8 hr. Mutants at the *rad* 3 locus showed marked impairment or total inability to remove these crosslinks. Hence the removal of these crosslinks, as in *E. coli*, probably requires an endonuclease step.

Defects in recombination (*rad* 51 mutants) also increased the sensitivity to trimethylpsoralen plus light, but did not interfere with the loss of crosslinking. Hence the repair of the lesions left after strand incision near the crosslinks, as in *E. coli*, may involve recombination.

Whereas the endonuclease step is required for the repair of both furocoumarin crosslinks and monoadducts, the recombination step appears to be required only for the repair of crosslinks. This was indicated by the finding by Jachymczyk *et al.* (1981) that angelicin (a furococumarin that can generate monoadducts but not crosslinks) killed *rad* 3 mutants almost as effectively as did trimethylpsoralen, but had little effect on *rad* 51 mutants.

Jachymczyk *et al.* (1981) observed the appearance of single-strand and double-strand breaks during the repair of interstrand crosslinks. The precise mechanisms involved however do not seem clear. Cole *et al.* (1976) did not detect double-strand breaks during the repair of psoralen crosslinks in *E. coli*, suggesting that there are differences in mechanism between *E. coli* and yeast.

Fleer and Brendel (1979, 1981) observed some removal of interstrand crosslinks induced by triaziquone (Trenimon), nitrogen mustard or activated cyclophosphamide in an excision-defective *rad* 2 strain. The loss of crosslinks was slow and may have been due to spontaneous depurination.

Mammalian cells

Mustards

Mammalian cells are capable of removing both interstrand and intrastrand crosslinks of nitrogen or sulphur mustards between guanine-N7 positions. Reid and Walker (1969) found di-guaninyl and single guanine alkylations by sulphur mustard to be removed in mouse L-cells at rates of about 20% in 5 hr. By 24 hr, most of the guanine alkylations were eliminated. Yin *et al.* (1973) reported similar results for nitrogen mustard in Ehrlich ascites tumour cells of mice. In both cases, the loss of renaturable (crosslinked) DNA occurred more rapidly than loss of adducts. This may not, however, indicate a selective removal of interstrand crosslinks, because (1) DNA strand breaks might have reduced the renaturability for a given frequency of interstrand crosslinks, and (2) an inter-strand crosslink can be broken by eliminating either of two guanines, thus making the loss of interstrand crosslinks inherently more sensitive than the loss of adducts.

Walker and Reid (1971) reported that sub-lines of mouse L-cells selected for resistance to sulphur mustard removed alkylations somewhat faster than did the parental cells. Yin *et al.* (1973) on the other hand, found no difference in removal of nitrogen mustard-induced alkylations or interstrand crosslinks between a sensitive and a resistant line of Ehrlich ascites cells, and Ball and Roberts (1970) found no difference in removal of sulphur mustard alkylations to account for a 20-fold difference in sensitivity between two sub-lines of Yoshida sarcoma. Clearly, there are several possible mechanisms of resistance to crosslinking agents, but resistance on the basis of enhanced repair has yet to be demonstrated convincingly.

Despite the fact that adult mouse cells have very low capacity for nucleotide excision-repair, mouse cells are quite capable of removing interstrand crosslinks produced by nitrogen mustard (Ewig and Kohn, 1977, 1978), as well as by *cis*-DDP (Zwelling *et al.*, 1979*a*). The repair of these crosslinks was not accompanied by any observable single-strand breakage, suggesting that a mechanism other than nucleotide excision-repair may be involved. It is conceivable that mouse cells remove crosslinks by a base excision mechanism mediated by a glycosylase enzyme capable of removing an abnormal base (figure 6, right). Enzyme activities that remove N7-alkyl-guanines from DNA have been detected in extracts of mammalian cells (Singer and Brent, 1981; Margison and Pegg, 1981).

Alkylation at guanine-N7 by phosphoramide mustard has recently been found to lead to spontaneous cleavage of the guanine imidazole ring (Mehta *et al.*, 1980; Chetsanga *et al.*, 1982), producing several ring-opened guanine alkylation products, some of which are removed by a specific glycosylase activity that is present in *E. coli* (Chetsanga *et al.*, 1981), and that has been detected in rodent liver extracts (Margison and Pegg, 1981). This sequence of reactions—imidazole ring-opening of N7-alkylated guanines followed by glycosylase action—could cleave crosslinks. One can speculate that imidazole ring opening, which ordin-arily occurs at a substantial rate only at high *p*H, is catalysed at neutral *p*H by the cationic phosphamide group in the case of phosphoramide mustard alkylation.

The subsequent steps in this repair path would involve incision of the base-free site created by the removal of the alkylated guanine, and closure of the gap by repair synthesis (perhaps only one or two nucleotides need be added and ligated). Another possibility is that a new base is directly inserted at the base-free site by an 'insertase' enzyme (Deutsch and Linn, 1979*a,b*). With either type of mechanism, the vestige of the interstrand crosslink may still remain attached to the template strand, and could cause the repair to be error prone.

Alkaline elution studies of mouse L1210 cells treated with low doses of nitro-gen mustard have shown that loss of crosslinks begins with little or no delay (not more than about 1 hr) (Ewig and Kohn, 1977, 1978). On the other hand, L-phenylalanine mustard gives rise to delayed crosslink formation which overlaps the phase of crosslink removal (Ross *et al.*, 1978).

Platinum complexes

As in the case of L-phenylalanine mustard, study of the removal of *cis*-DDP-induced interstrand crosslinks is complicated by delayed crosslink formation which overlaps the phase of crosslink removal (Zwelling *et al.*, 1978, 1979*a*). In the case of *cis*-DDP, however, delayed interstrand crosslink formation can be aborted by using thiourea to quench monoadducts (Zwelling *et al.*, 1979*c*; Micetich *et al.*, 1981). The true kinetics of removal of interstrand crosslinks can then be measured.

Pera *et al.* (1981) measured interstrand crosslinking as a function of time after treatment of Chinese hamster cells with *cis*-DDP (Chapter 7). The delayed formation of interstrand crosslinks, followed by their removal was verified using three different methods of measurement.

Psoralens

Ben-Hur and Elkind (1973*b*) studied DNA photoadducts of 4, 5′, 8-trimethylpsoralen in Chinese hamster cells. It was estimated that 2,000 interstrand crosslinks were produced per lethal hit, suggesting that the vast majority of crosslinks are either repaired or by-passed. Using [^{3}H] labelled trimethylpsoralen, it was estimated that 11% of the DNA-bound psoralen was in the form of interstrand crosslinks, and that about 90% of the bound psoralen was removed with a half-time of approximately 5 hours. Some evidence of crosslink removal was also noted.

Interstrand crosslinking of DNA in mammalian cells treated with psoralen plus light can be studied using the alkaline elution technique (Cohen *et al.*, 1980). This technique was recently employed by Bredberg *et al.* (1982) who reported excision repair of psoralen-induced crosslinks in normal human fibroblasts. During a period of 24 hr after treatment of cells with 8-methoxypsoralen and light, the reduced elution signifying crosslinkage was replaced by an increase in elution which indicates the presence of strand breaks. In xeroderma cells, the phase of increased elution was absent. This suggests that the strand breakage is a manifestation of excision repair. However, the strand breakage could have been due to excision repair of monoadducts (which are present in excess) rather than of crosslinks.

Kay *et al.* (1980) studied repair replication in normal human embryo cells treated either with 8-methoxypsoralen or with the monofunctional congener, angelicin. The cells were exposed to light in the presence of either of these photosensitizers. Repair replication was measured by combined density (BrdUrd) and radioactivity ([^{3}H] thymidine) labelling. The magnitude and time-course of repair replication were found to be the same for the two compounds. Since crosslinks are produced by 8-methoxypsoralen but not by angelicin, the replication could be accounted for by repair of monoadducts.

Repair-deficient cells

The use of repair-deficient mammalian cells to elucidate repair mechanisms has not yielded results as clear as in the case of microorganisms, because the exact nature of the repair deficiencies available in mammalian cells is not known and the defects often appear to be multiple. Investigation of the mechanisms of crosslink repair, in particular, is still at an early stage.

Xeroderma pigmentosum (classical type) and related repair defects. The most extensively studied mammalian repair deficiency is xeroderma pigmentosum, a genetic defect in man. The main defect in the complementation group A variety, in which the classical form of this disease is clearly expressed, appears to be in a damage-specific endonuclease which acts at sites of ultraviolet light-induced pyrimidine dimers and at sites of some, but not all, types of chemical adducts to the DNA molecule. The first question that may be asked about crosslink repair is whether this endonuclease is involved. The enzyme might, of course, act on some types of crosslinks and not others. Fujiwara and Tatsumi (1977) found that xeroderma cells remove mitomycin-induced interstrand crosslinks at a normal rate and survive treatment with this drug similarly to normal cells. The induced rate of unscheduled DNA synthesis was also normal. However, the xeroderma cells were deficient in strand scission in response to treatment with decarbamoyl-mitomycin, a monofunctional analogue. Hence the cross-links might be removed by a mechanism which is not associated with high levels of strand scission; glycosylase involvement would be one such possibility.

Kaye *et al.* (1980), however, found that xeroderma group A cells are deficient in the ability to remove psoralen crosslinks. Furthermore, neither 8-methoxy-psoralen nor angelicin stimulated repair replication in these cells. Hence the damage-specific endonuclease in which these cells are deficient appears to be required for the repair of psoralen monoadducts and crosslinks, but not for the repair of mitomycin crosslinks.

Meyn *et al.* (1982) recently studied the effects of *cis*-DDP and mitomycin C on a repair-deficient mutant line of Chinese hamster ovary cells. The line had been selected for increased sensitivity to ultraviolet light, and resembled human xeroderma pigmentosum cells in that there was an absence of the endonuclease step of excision repair (Thompson *et al.*, 1982). Meyn *et al.* found the cells also to have markedly enhanced sensitivity to *cis*-DDP and to mitomycin C. Alkaline elution assays indicated that the line was deficient in the removal of interstrand and/or DNA–protein crosslinks. This is as would be expected if crosslinks were the major cause of cytotoxicity and were repaired by a nucleotide excision mechanism. This mechanism may repair monoadducts as well as crosslinks, and there was some suggestion in the data that the mutant cells might also be defective in monoadduct repair. The role of nucleotide excision-repair is not proven however, because, as in the case of xeroderma cells, it may be that the mutant CHO cells have defects in several different repair processes. Indeed, the question must be considered whether DNA strand breaks occur during the repair, as would be expected for an excision mechanism. Roberts and Friedlos (1981) failed to detect any single-strand breaks or alkali-labile sites for up to 6 hours after treatment of Chinese hamster V79 cells with high doses of *cis*-DDP. The possible role of a base excision mechanism in this story is not excluded.

Fanconi's anaemia. A defect in the ability to repair DNA crosslinks is suspected to exist in cells from patients with Fanconi's anaemia (Fujiwara and Tatsumi, 1975; Fujiwara *et al.*, 1977; Sasaki, 1975; 1978). A variety of DNA crosslinking agents have been noted to have an enhanced effect on these cells as regards the production of chromosome aberrations (Auerbach and Wolman, 1976; Berger *et al.*, 1980; Latt *et al.*, 1975; Sasaki and Tonomura, 1973) and cytotoxicity (Weksberg *et al.*, 1979; Ishida and Buchwald, 1982).

Ishida and Buchwald (1982) recently extended the list of crosslinking agents studied in Fanconi's anaemia cells to include mitomycin C, nitrogen mustard, L-phenylalanine mustard, butadiene diepoxide and *cis*-DDP. Lymphoblastoid cells were used and cytotoxicity was measured by inhibition of proliferation rate. The dose-modification factor for these drugs in four Fanconi's anaemia cell lines compared with four lines derived from normal individuals was usually greater than 10. Sensitivity to chloroethyl-nitrosoureas varied markedly among cell lines from either normal or Fanconi's anaemia subjects, although the Fanconi's anaemia cells tended to be more sensitive than the normal cells. Methylating agents had variable cytotoxic effects, and there was no difference between Fanconi's anaemia and normal cells. There was also no significant difference in sensitivity to X-rays, bleomycin or 4-nitroquinoline oxide.

Fujiwara, Sasaki and their coworkers have consistently observed a deficiency in the removal of mitomycin crosslinks in Fanconi's anaemia cells, using a strand reassociation method. Fornace *et al.* (1979), however, failed to detect any deficiency in crosslink removal in one Fanconi's anaemia line, studied by alkaline elution. It is conceivable that the discrepancy could be dose-related, since the two methods differ in sensitivity. Kaye *et al.* (1980) found that Fanconi's anaemia cells remove psoralen crosslinks normally. Thus the story about crosslink removal in Fanconi's anaemia cells is still confused.

Xeroderma pigmentosum (variant form) and inhibition of post-replication repair by caffeine. Xeroderma pigmentosum variant cells differ from cells of the classical form of the disorder in that their repair replication response to ultraviolet light is normal, indicating that damage-specific endonuclease function is normal (Cleaver, 1972). The cells appear to have a defect in 'post-replication repair', that is, the rate at which gaps or strand breaks are resealed (Lehmann *et al.*, 1975, 1977). The defect is probably in the ability to replicate past sites of template damage. The process is noted for its sensitivity to inhibition by caffeine, and caffeine interferes with the ability of these cells to survive DNA damage.

In this regard, it is interesting to note the finding by Hartley-Asp (1978) that caffeine enhances the frequency of chromatid breaks and exchanges produced by the crosslinking agent, mitomycin C. The frequency of mitomycin-induced chromosome aberrations was higher in group A xeroderma cells, which are defective in the incision step of excision repair, than in normal human cells. Xeroderma variant cells produced an intermediate level of aberrations. The enhancement by caffeine occurred in all three cell types, but was significantly greater in the variant cells than in either the normal or the group A cells. These results are consistent with the idea that interstrand crosslink repair involves replication through a template defect (an inherent assumption being that improperly repaired interstrand crosslinks can produce chromosome aberrations).

Caffeine, when included in the post-treatment medium, also enhances the killing effect of mitomycin C, as measured by colony formation in mouse L-cells (Rauth, 1970). Again this might be due to inhibition of a critical step in crosslink repair. Caffeine, however, also enhances the lethality of ultraviolet light (Domon and Rauth, 1969), which does not produce significant interstrand crosslinking. In both situations, caffeine may act by interfering with replication through a template defect.

Szumiel (1979) compared two strains of L5178Y murine lymphoma, one of which was relatively sensitive to X-rays, UV light and a *cis*-Pt (II) complex, whereas the other was relatively resistant to all three agents. The survival of the resistant strain following treatment with the *cis*-Pt (II) compound was substantially reduced by including 0.75 mM caffeine in the agar colony growth medium. The survival of the sensitive strain in similar experiments was unaffected by caffeine. It is possible that the sensitive cells were deficient in a caffeine-sensitive post-replication repair process. This kind of deficiency could be responsible for the sensitivity of some rodent cell types to DNA damaging agents.

Normal human cells, however, do not exhibit this type of caffeine sensitivity. In contrast to rodent cells, human cells appear to deal with DNA damage mainly by excision repair. This difference between the species may be a reason for the poor correspondence between the sensitivity of experimental rodent tumours to a variety of DNA damaging agents and their effects in human cancer.

DNA-PROTEIN CROSSLINKS

DNA–protein crosslinking is a prominent feature in the action of bifunctional alkylating agents and Pt (II) complexes in mammalian cells (Ewig and Kohn, 1978; Kohn and Ewig, 1979; Zwelling *et al.*, 1979a; Erickson *et al.*, 1980a,b,c; Laurent *et al.*, 1981; Strandberg *et al.*, 1982). DNA–protein crosslinks may be of a wide variety of types, depending on the identity of the bound protein and on the DNA site of binding. These chemical details, however, have not yet been elucidated, because of the difficulties of analysing the small number of protein molecules that may be linked to DNA when pharmacologically reasonable drug doses are used. The alkaline elution technique allows sensitive quantitative of the frequency of covalently linked proteins along DNA strands, but it does not distinguish between different proteins or between different DNA binding sites.

DNA-protein crosslinks are not necessarily highly lethal to cells. This is most clearly seen with agents that produce DNA-protein crosslinks as the predominant DNA lesion, namely with *trans*-DDP (Zwelling *et al.*, 1979*a*; Kohn and Ewig, 1979) and with formaldehyde (Ross and Shipley, 1980). Treatment of L1210 cells with 100 μM *trans*-DDP for 1 hr was estimated to produce 1.4 DNA-protein crosslinks per 10^6 nucleotides (Kohn and Ewig, 1979). Cell survival (colony formation in soft agar) following this treatment was over 50% (Zwelling *et al.*, 1979*a*). The cells removed and presumably repaired most of these crosslinks over a 24 hr period.

From the data of Ross and Shipley (1980), it can be estimated that L1210 cells treated with 100 μM formaldehyde for 2.5 hr acquired 0.5-1.0 DNA-protein crosslink per 10^6 nucleotides. Following this treatment, more than 50% of the cells were still able to form colonies. The cells removed most of the crosslinks within 6 hr and essentially all of them within 24 hr.

Little is known about the mechanism of removal and repair of DNA-protein crosslinks. No strand breaks are ordinarily observed during the repair of *trans*-DDP or formaldehyde-induced DNA-protein crosslinks. However, when the repair was allowed to take place in the presence of the such DNA polymerase inhibitors as arabinosylcytosine and hydroxyurea, single-strand breaks were observed to accumulate (Fornace, 1982). The accumulation of these breaks was markedly less in excision-deficient xeroderma pigmentosum cells, and these cells failed to remove the crosslinks (Fornace, 1981). Xeroderma pigmentosum cells also are unable to remove DNA-protein crosslinks introduced by UV light (Fornace and Kohn, 1976). The results suggest that these types of DNA-protein crosslinks may be removed by the nucleotide-excision mechanism in mammalian cells.

Excision-deficient yeast cells (*rad* 1-3), on the other hand, have been reported to retain the ability to remove formaldehyde-induced DNA-protein crosslinks (Magana-Schwenke and Moustacchi, 1980).

DETERMINANTS OF TUMOUR CELL SENSITIVITY

Methodologic considerations

Since DNA damage and cytotoxicity are measured by quite different experimental procedures and are not necessarily linear relationships, the problem of quantitating the relation between the two requires attention. DNA damage is essentially a chemical measurement, whereas cytotoxicity is a biological measurement, and the two are expressed in entirely different dimensional units. A quantitative comparison requires that the measurements be expressed in consistent dimensions. This can be conveniently accomplished, for example, by comparing two cell types on the basis of a dose modification factor for equal actions. An example is shown in figure 7 where cell survival curves are compared for a sensitive (K25) and a resistant (ZCR9) line of mouse leukaemia

L1210 cells treated with *cis*-DDP (Zwelling *et al.*, 1981). The survival curves for the two cell types are plotted on the same graph, but using separate dose scales, chosen so as to make the curves coincide. This requires that the survival curves are similar in shape. An advantage of this method is that the survival curves do not have to be linear. The dose modification factor is the ratio of the dose scales that gives best superimposition of the two survival curves. For the case shown in figure 7, this ratio is 2.4.

When survival curves differ markedly in shape, the comparison requires a more detailed analysis. For example, if the curves can be analysed into a

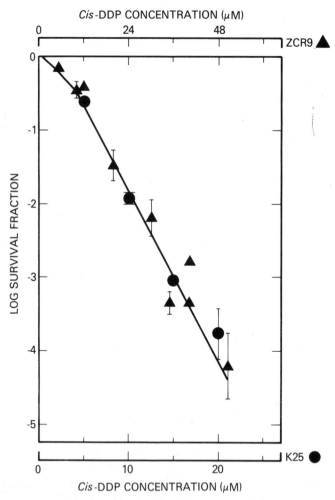

Figure 7 Quantitative comparison of two lines of L1210 cells for sensitivity to *cis*-DDP. Survival of colony-forming ability is plotted against separate drug concentration scales, chosen to make the survival curves coincide. Drug exposure time was 1 hr. K25, parent line; ZCR9, line selected for resistance to *cis*-DDP. (From Zwelling *et al.*, 1981.)

shoulder region and an exponential portion, then dose modification factors may be evaluated for each of these components separately.

DNA damage can be evaluated in a similar manner, that is, in terms of dose ratio for equal effect. However, attention must be given to the dependence of damage present as a function of time after treatment; following drug treatment, DNA crosslinking frequently rises for some time, reaches a peak, and then falls. If the time-course of DNA damage in cells of different drug sensitivities is similar, it is appropriate to use values obtained near the time of maximum DNA damage. If the time courses differ substantially, then it may be necessary to consider the area under the damage-versus-time curve.

DNA damage at an appropriate time following treatment is often found to be a linear function of dosage. The dose modification ratio is then equal to the ratio of the slopes of the linear dose–response curves. An example of this is shown in figure 8 for the formation of DNA–protein and interstrand crosslinks using the same cells and the same drug treatments as in figure 7. The DNA–protein crosslinking ratio for the two cell types was 3.0, which was slightly greater than the dose modification factor, 2.4, for survival. In the case of interstrand crosslinking, the ratio of slopes was 2.5, very close to the dose-modification ratio. Table 1 summarises the comparison between the K25 and ZCR9 lines with respect to interstrand crosslinking, DNA-protein crosslinking, and two measures of cytotoxicity colony survival and growth inhibition. In the case of *cis*-DDP), the dose modification ratios for all four measurements fell within the range 2.4–3.0. Thus the biological difference between these cell types in response to *cis*-DDP was similar in magnitude to the difference in the formation of bifunctional DNA lesions. In the case of treatment with L-PAM, the cytotoxicity differential was similar in magnitude to the potency ratio for interstrand crosslinking, but not for DNA–protein crosslinking. Except for the unchanged DNA–protein crosslinking in the resistant ZCR9 line treated with L-PAM, the results could be explained by reduced drug uptake into the cells or increased drug inactivation within the cells.

Table 1   Relative cytotoxicity and DNA crosslinking in a parent and a resistant line of L1210 cells treated with cis-DDP or L-phenylalanine mustard (L-PAM).

Effect measured	cis-DDP	L-PAM
Colony survival	2.4	1.8
Growth inhibition	2.7	1.6
Interstrand crosslinking	2.5	2.0
DNA–protein crosslinking	3.0	1.0

The figures express drug concentration ratios for equal effects in the two cell lines, resistant/parent or the inverse as appropriate. (From Zwelling *et al.*, 1981)

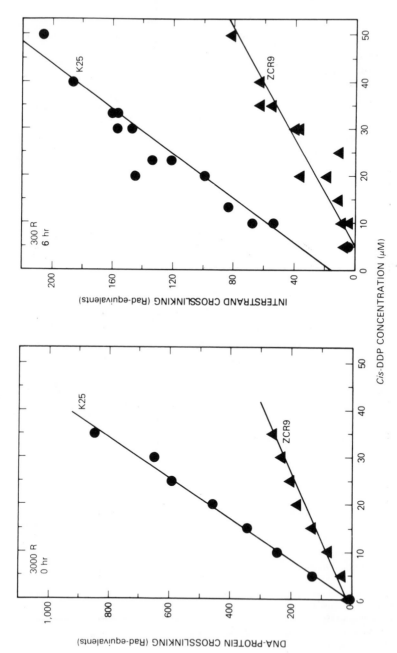

Figure 8 Quantitative comparison of sensitive (K25) and resistant (ZCR9) lines for DNA-protein and interstrand crosslinking following treatment with various concentrations of *cis*-DDP for 1 hr. DNA-protein crosslinking was measured immediately after drug removal, at which time the frequency of these lesions was maximal. Interstrand crosslinking was measured after 6 hr, near its peak. In order to introduce the DNA single-strand breaks required in the assays, the cells were exposed to X-rays (3,000 rad for the DNA-protein crosslink assay and 300 rad for the interstrand crosslink assay). (From Zwelling *et al.*, 1981.)

Mouse leukaemia cells

Zwelling *et al.* (1981) compared the drug sensitivities of L1210 ascites tumour lines *in vivo* with interstrand crosslink or DNA–protein crosslink formation in the ascites cells treated *in vitro*. A line which had been selected for resistance to *cis*-DDP *in vivo* (Burchenal *et al.*, 1977) and a line selected for resistance to L-PAM (Schabel *et al.*, 1978) were each compared with their respective parent lines (table 2). DNA crosslinking was measured as a function of time after 1 hr drug treatments. Interstrand crosslinking increased for approximately 6 hr, and then began to decline at 6–12 hr. The *cis*-DDP-resistant line (L1210/DDP) showed markedly less interstrand crosslinking and DNA–protein crosslinking following treatment with *cis*-DDP than did its parent line. The L1210/DDP line was not cross-resistant to L-PAM, and showed no significant reduction in either type of crosslink in response to L-PAM. The L-PAM-resistant line (L1210/PAM) showed the expected reduction in both types of crosslinks following treatment with L-PAM. However, although this line was strongly cross-resistant to *cis*-DDP, it exhibited only modest reductions in crosslinking.

The latter discrepancy was investigated further by isolating cell lines from the L1210/PAM tumour line *in vitro*, as well as from its parent line (Micetich *et al.*, 1981). The cell lines were compared on the basis of dose-modification factors for equal survival and equal crosslinking. In accord with the results *in vivo*, the resistant cell line exhibited higher peaks of crosslinking (both types) than would be expected if the changes in survival and crosslinking were proportionate. The kinetics of interstrand crosslink formation and removal suggested that the resistant line had a greater ability to quench DNA-drug monoadducts so as to reduce the probability of monoadduct-to-crosslink conversion. Quenching of monoadducts might be due to the presence in the resistant cells of a relatively high concentration of the thiol which could react with the second

Table 2 Sensitivity *in vivo* and crosslinking *in vitro* by cis-DDP and L-PAM in various L1210 tumour lines (Zwelling *et al.*, 1981)

| Tumour line | Cis-DDP | | | L-PAM | | |
| | | Crosslinking | | | Crosslinking | |
	%ILS	Total	Interstrand	%ILS	Total	Interstrand
L1210 (MSKI)	151	1.45	0.21	94	0.60	0.22
L1210/DDP	11	0.19	0.01	76	0.55	0.18
L1210 (NCI)	66	1.64	0.28	96	0.49	0.18
L1210/PAM	6	1.12	0.18	25	0.15	0.08

Crosslinking coefficients were determined 6–12 hr after 1 hr drug treatment *in vitro*; %ILS records the increase in survival time of treated mice. L1210/DDP: selected for resistance to cis-DDP derived from L1210 (MSKI); L1210/PAM: selected for resistance to L-PAM derived from L1210 (NCI).

reactive arm of bifunctional monoadducts so as to prevent alkylation by the second arm to form a crosslink.

Strandberg *et al.* (1982) pursued this problem further by examining three cell lines derived from Burchenal's resistant L1210/DDP tumour. The dose-modification ratios for the resistant lines in growth assays ranged from 20 to 50. This high degree of resistance was specific for *cis*-DDP, and did not carry over to the trans isomer or to L-PAM (Eastman and Bresnik, 1981). Using the alkaline elution technique, these workers confirmed the findings of Zwelling *et al.* (1981) that interstrand crosslinks rise to a peak at approximately 12 hr after treatment, whereas DNA-protein crosslinks appear promptly. This was true both in the parent and in the resistant lines. Although the resistant lines showed reduced crosslinking of both types, the dose modification ratios were only about 5 for DNA-protein crosslinks and 3 for interstrand crosslinks. This is substantially less than the dose-modification ratios of 20-50 for cytotoxicity, and indicates that peak crosslinking levels cannot be the sole determinants of cytotoxicity. Thus the resistant lines were able to survive much higher peak crosslink frequencies than was the parent line. In agreement with Zwelling *et al.*, the resistant cells removed most of the crosslinks within 24 hr. However, the possibility is not excluded that the discrepancy could be due to differences in the persistence of interstrand crosslinks beyond 24 hr.

Another notable finding by Strandberg *et al.* (1982) was that DNA-protein crosslinks persisted in the parent line for over 72 hr, whereas these lesions were completely removed by this time in resistant cells. It is possible that DNA-protein crosslinking sometimes could be a late secondary phenomenon in dying cells. On the other hand, it is quite possible that persistant DNA-protein crosslinking could inhibit cell proliferation.

DNA-protein crosslinks could be of a variety of types, depending on the identity of the protein and on the site of DNA binding. Hence these lesions may vary in ability to be repaired and in cytotoxic significance.

Burkitt's lymphoma cells

Ducore *et al.* (1982) studied DNA crosslinking and cytotoxicity in three human Burkitt's lymphoma cell lines having different sensitivities to *cis*-DDP (table 3). Quantitative comparison of two of the lines revealed dose-modification ratios of approximately 3 for survival and 2-4 for interstrand or DNA-protein crosslinking. Thus there were proportionate differences in cytotoxicity and DNA lesions for this pair of lines. The third line, however, exhibited survival similar to the more sensitive of the first pair, but crosslinking similar to the more resistant of the first pair, that is, there was high sensitivity with relatively little crosslinking. This anomalous line differed from both lines of the first pair in that the cells tended to undergo morphological lysis with release of DNA and protein into the medium 12-24 hr following the treatment (for 1 hr) with *cis*-DDP. This line also exhibited a similar anomaly following treatment with L-PAM.

Table 3 DNA crosslinking and cytotoxicity produced by cis-DDP in Burkitt's lymphoma cells (Ducore *et al.*, 1982)

Cell strain	Relative cytotoxicity		Relative crosslinking	
	Cell number	Colony formation	Total	Interstrand
BHM	1.0	1.0	1.0	1.0
W-1	2.8	3.5	3.4	2.6
WS	2.6	4.1	1.6	1.4

Relative cytotoxicity is based on the ratio of drug concentrations for equal effect on cell number (3 days after treatment) or on colony formation (e.g., in order to produce an equal reduction in cell number in BHM compared to W-1, a 2.8-fold higher drug concentration was needed.) Relative crosslinking was measured 6–12 hr after treatment with 12 μM cis-DDP for 1 hr.

At 24 hr, crosslinking levels in all three cell lines were still greater than 50% of their peak values. In order to survive, cells would have to remain physiologically intact during the period of metabolic stress following treatment until potentially lethal DNA lesions are repaired. The anomalous cell type apparently could not survive intact during the post-treatment period of metabolic stress induced by DNA damage.

Human tumour cells

Chloroethylnitrosoureas

The 1-(2-chloroethyl)-1-nitrosoureas are highly reactive compounds that have extraordinary anti-tumour activity in mice (Chapter 8). Their degree of effectiveness against clinical cancer in man, however, is less than might have been expected. The compounds are alkylating agents which produce chloroethyl and hydroxyethyl adducts at various DNA sites. The chloroethyl adducts can undergo a second alkylation reaction with the elimination of chloride. This results in a crosslink consisting of an ethano ($-CH_2CH_2-$) bridge between two nucleophilic atoms. Two types of DNA crosslinks have been chemically identified: (1) an ethano bridge between two guanine-N7 positions, and (2) an ethano bridge between a guanine-N1 and a cytosine-N3 position (Tong and Ludlum, 1981, 1982). The diguaninyl crosslink is almost certainly restricted to adjacent guanines in the same DNA strand, and very little can be said as yet about its biological significance. The GC dimer, however, could be an interstrand crosslink of key significance.

The crosslinkage between guanine-N1 and cytosine-N3 at first appears puzzling because these atoms are inaccessible in the DNA helix. Tong and Ludlum (1982), however, have recently provided evidence for a sequence of reactions that could explain the formation of this crosslink. According to this proposal, the first reaction is a chlorethylation of the guanine-O6 position. This disrupts the hydrogen bonding to the paired cytosine and makes the guanine-N1 position accessible. The chloroethyl group at O6 then reacts intramolecularly with N1, eliminating chloride and forming an ethano bridge. The resulting N1, O6-ethano-guanine moiety is a cation which is susceptible to nucleophilic attack at the ethano group, the O6 being the leaving group. If the nucleophile in this reaction is a cytosine-N3, then the result would be an interstrand crosslink and the observed ethano-GC moiety.

Chloroethylnitrosoureas have been observed to crosslink DNA strands, both in reaction with purified DNA (Kohn, 1977; Lown *et al.*, 1978) and in cells (Ewig and Kohn, 1977, 1978; Erickson *et al.*, 1980). In both cases, the reaction is separable into two steps: (1) an initial alkylation step which does not by itself produce crosslinks, and (2) conversion of monoadducts to crosslinks, which can take place after removal of free drug. If the initial alkylation (at sites potentially leading to interstrand crosslinks) is at guanine-O6, then a repair mechanism that removes alkyl groups from this site could prevent crosslink formation. A guanine-O6 alkyltransferase is known to remove methyl and ethyl groups from the guanine-O6 position and to transfer them to a thiol group of a protein cysteine moiety (Mehta *et al.*, 1981; Olsson and Lindahl, 1980). There is as yet only indirect evidence that this alkyltransferase can handle chloroethyl groups.

The guanine-O6 alkyltransferase mechanism is a rapidly acting repair system with a limited capacity. The limited capacity seems to be due to the consumption of the enzyme in the reaction; thus the enzyme reacts stoichiometrically. Shiloh and Becker (1981) estimated the capacity for the removal of O6-methylguanine by repair-proficient human cells to be 3–5 fmol per μg DNA. This corresponds to one or two O6-methylguanines repairable per million nucleotides.

Human cell strains derived from tumours or by viral transformation generally fall into two classes, differing in the efficiency with which they are able to repair O6-methylguanines. Most cell strains have a relatively high efficiency, but about 25% are deficient in this repair (Day *et al.*, 1980; Sklar and Strauss, 1981). Cells that repair these lesions efficiently are designated Mer$^+$ or Mex$^+$, and cells that are deficient are designated Mer$^-$ or Mex$^-$. This phenotypic distinction is justified because cell strains fall into a bimodal distribution of repair efficiencies.

The distinction between the two phenotypes (Mer$^+$ versus Mer$^-$) is observed in the extent to which cells remove O6-methylguanines from their nuclear DNA after treatment of cells with a methylating agent. The ability to restore or remove an altered base from DNA does not, however, guarantee that the DNA is thereby repaired. The conclusion that the DNA is actually repaired must be shown by the recovery of DNA function. Recovery of DNA function after inactivation by treatment with a methylating agent has been most clearly shown using viruses. Day and his coworkers (1979, 1980*a,b*, 1981) treated adenovirus 5 with methylating agents (MNNG or MNU) and then plated the virions on monolayer cells whose repair efficiency was to be tested. Human tumour cell strains were found to segregate into two types of survival curves for viral plaque formation (figure 9). The cells exhibiting the steeper survival curve had a reduced

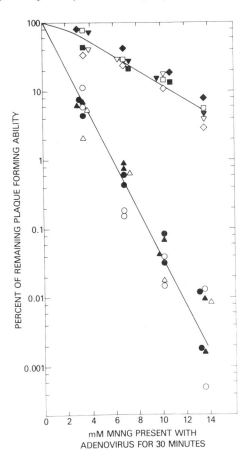

Figure 9 Survival curves for MNNG-treated adenovirus 5 plated on various human tumor cell strains (different symbols). The cell strains segregate into two types of survival curves, depending on the ability of the cells to repair O6-methylguanines in DNA. (From Day *et al.*, 1980*b*).

ability to remove O6-methylguanines, but not N7-methylguanines from DNA (Day, 1981). All of these cell strains exhibited the same survival curves for virus irradiated with ultraviolet light. Thus Mer⁻ cell strains were characterized by a reduced ability to support the proliferation of methylated adenovirus due to a deficiency in a specific DNA repair mechanism.

This type of experiment, that is, host cell reactivation, has the advantage that it avoids exposure of the cells to the chemical agent which could cause a variety of direct and indirect effects. Nevertheless, treatment of the cells themselves did show that Mer⁺ cells were better able to survive methylation damage than were Mer⁻ cells (Day *et al.*, 1980*b*). These studies were carried out using 1-methyl-1-nitrosourea (MNU) and 1-methyl-1-nitroso-3-nitroguanidine

(MNNG) both of which are capable of methylating the guanine-O6 position to a significant degree.

If the Mer system were capable of removing chloroethyl groups from the guanine-O6 position, then it is possible that this system could prevent interstrand crosslinking by chloroethyl-nitrosoureas. The removal of the monoadducts, if fast enough, could prevent the delayed formation of interstrand crosslinks. Indeed, the removal of methyl groups from O6-methylguanines following treatment of Raji lymphoma cells with MNNG was observed to proceed with a half-time of less than 1 hr (Sklar and Strauss, 1981). Shiloh and Becker (1981) noted that loss of O6-methyl-guanines was completely expressed within 1 hr at 37°C by human lymphoblastoid cells, and the removal half-time was estimated to be 10–15 min. Thus the removal of alkyl groups from the guanine-O6 position can, in principle, prevent the conversion of chloroethyl monoadducts to interstrand crosslinks.

In order to determine whether there is a relationship between interstrand crosslinking by chloroethylnitrosoureas and the Mer repair system, these properties were compared in a series of human tumour cell strains (Erickson *et al.*, 1980*b*). As shown in table 4, the correlation was striking. The cell strains are listed in the table in order of increasing interstrand crosslinking. Except for two intermediate cases, the Mer$^+$ strains clearly show less interstrand crosslinking than do the Mer$^-$ strains. DNA-protein crosslinking, on the other hand, showed no correlation with Mer phenotype. Hence the difference between the Mer$^+$ and Mer$^-$ strains cannot be attributed to a difference in drug uptake or intracellular reactivity. The DNA–protein crosslinking measurement can, in fact, be used as an estimate of intracellular drug reaction. A plot of interstrand crosslinking against DNA–protein crosslinking (figure 10) then normalises the data with respect to extent of intracellular drug reaction. Figure 10 shows that the cell strains are bimodally distributed with respect to their capacity to show interstrand crosslinking, and exhibit a segregation of the strains according to their Mer status, as was the case in the host-cell reactivation assay with adenovirus. This further supports the existence of a discrete phenotype, Mer, characterised by the ability of cells to (1) repair methylated viral DNA, (2) remove or dealkylate DNA O6-methylguanines, and (3) prevent the formation of chloroethylnitrosourea-induced interstrand crosslinks.

The Mer$^+$ character also enhances the ability of cells to survive treatment with MNU (Day, 1980*b*, 1981) or with chloroethylnitrosoureas (Erickson, 1981) (figure 11). The improved survival to MNU suggests, but does not prove, that methylation at guanine-O6 is a major contributor to cell killing; the possibility not excluded is that the Mer phenotype may govern the function of several different repair processes. Similarly, the improved survival to chloroethylnitrosourea suggests, but does not prove, that interstrand crosslink formation is a major contributor to cell killing by this drug. The main possibility not excluded in this case is that the cytotoxicity prevented by the Mer function stems from particularly cytotoxic monoadducts generated by chloroethylnitrosourea but

Table 4 Alkaline elution assays of DNA interstrand and DNA-protein crosslinking by 1-(2-chloroethyl)-1-nitrosourea (CNU) (crosslink index $\times 10^3$)

	Cell strain	Origin	Mer phenotype	Interstrand crosslinking (Proteinase assay)		Total crosslinking (No proteinase)	
				50 μM CNU	100 μM CNU	50 μM CNU	100 μM CNU
◼	HT29	Colon carcinoma	+	0	7	126, 135	268, 410
●	IMR90	Embryonic lung	+	3	23	210, 216	485
◇	A2182	Lung carcinoma	+	14	31	154, 218	788
◻	A673	Rhabdomyosarcoma	+	16	16	186, 261	523
◆	A549	Lung carcinoma	+	33	52	203, 216	504, 545
◁	HT1080	Fibrosarcoma	+	62	127	221, 251	437, 462
◐	A431	Epidermoid carcinoma (vulva)	+	69	143	190, 190	385, 446
⬡	A1336	Ovarian carcinoma	−	69	221	135, 152	692
▷	A427	Lung carcinoma	−	72	149	122, 208	295, 471
☐	BE	Colon carcinoma	−	81	201	166, 207	438, 530
◇	A172	Astrocytoma (grade IV)	−	84	217	193, 204	432, 457
◁	VA13	SV40-transformed WI-38	−	88	232	148, 152	384
◯	A875	Melanoma	−	112	240	250, 281	540

From Erickson *et al.*, 1980b

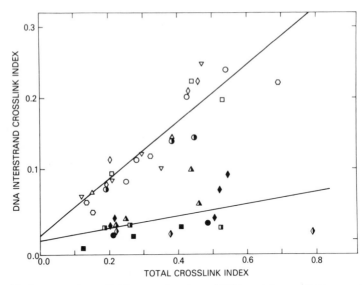

Figure 10 Interstrand crosslinking as a function of DNA-protein crosslinking produced by chloroethylnitrosourea in various human cell strains. The symbols for different cell strains are the same as in table 4; open symbols, Mer⁻ strains; closed and half-closed symbols, Mer⁺ strains. 'Total crosslink index' (horizontal axis) is mainly a measure of DNA–protein crosslinks in these assays. (From Erickson *et al.*, 1980*b*.)

not by methylnitrosourea. Since chloroethylnitrosoureas are much more cytotoxic than simple alkylnitrosoureas, there must be something related to chloroethylation that is especially cytotoxic. Chloroethylnitrosoureas also generate hydroxyethylations, but these products are unlikely to be major contributors, because hydroxyethylnitrosoureas are less toxic than chloroethylnitrosoureas and do not have a high degree of antitumour activity.

It has been found however that fluoroethylnitrosoureas are as effective as chloroethylnitrosoureas in the treatment of several mouse tumours (Johnson *et al.*, 1982 and R. K. Johnson, personal communication), even though the fluoro compounds produce much less interstrand crosslinking, or produce these lesions only after a delay of 24–48 hours (Sharkey *et al.*, 1982). The fluoroethyl compounds are as cytotoxic as the chloroethyl compounds and are selectively effective against Mer⁻ cell strains. The effectiveness of the fluoro compounds might be due to a highly delayed crosslink formation or to the formation of especially cytotoxic monoadducts.

A difference in crosslinking between a sensitive ('BE', Mer⁻) and a resistant ('HT29', Mer⁺) human colon tumour was demonstrable *in vivo* in mice bearing these tumours as xenografts (Thomas *et al.*, 1978). The resistant tumour exhibited less crosslinking than the sensitive tumour following treatment of the mice with methyl-1-(2-chloroethyl)-1-nitroso-3-cyclohexylurea (MeCCNU). Working with cell strains derived from these tumour lines, Erickson *et al.* (1978*c*) reported a loss of MeCCNU-induced crosslinks in the resistant cells but not in the sensitive

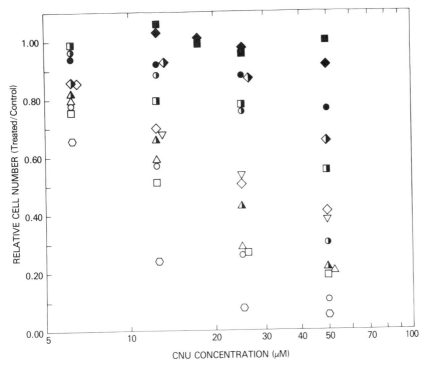

Figure 11 Cytotoxicity of chloroethylnitrosourea against various human cell strains. Cells were exposed to drug for 1 hr. After drug removal, the cells were allowed to proliferate in fresh medium for 3 days (3–5 doubling times for control cells). Symbols as in table 4 and figure 10. (After Erickson *et al.*, 1980*b*.)

cells. The loss of crosslinks was only seen after 48 hr of post-treatment incubation. The results suggested a repair of potentially lethal crosslinks in the resistant cells. The possibility could not be excluded, however, that by 48 hr a selective loss of a highly crosslinked cell population might have taken place.

Ahlgren *et al.* (1982) recently studied the fate of DNA adducts generated by chloroethylnitrosoureas in mouse L1210 leukaemia and bone marrow cells (which gave similar results). Cells were treated with [^{14}C-chloroethyl]-labelled drug; at various times thereafter, DNA was extracted and its specific activity determined. When cells were treated with chlorozotocin (a chloroethylnitrosourea which lacks carbamoylating activity), about 40% of the adducts were lost from the DNA within a few hours; the loss was limited and did not continue upon more prolonged incubation. The limited removal suggests that chloroethylations or hydroxyethylations are removed from some DNA sites but not from others (the frequency of crosslinks is too small to contribute appreciably to this measurement). Chloroethylnitrosoureas such as 1-(2-chloroethyl)-1-nitroso-3-cyclo-hexylurea that have carbamoylation activity, however, were not removed or removed only after a long delay. Since carbamoylation can inhibit a variety of enzymatic processes, including the ligase step in DNA repair (Kann *et al.*, 1974, 1980*b*; Fornace *et al.*, 1978; Erickson *et al.*, 1978*b*), it is possible that the carbamoylating function of compounds such as CCNU can inhibit the repair of some DNA lesions produced by these drugs. Kann *et al.* (1980*a*) have reported that repair-inhibiting nitrosoureas enhance the lethal effect of ionizing radiation on L1210 cells.

Platinum complexes

The question can be raised as to whether the ability of Mer$^+$ cells to prevent interstrand crosslink formation is specific for chloroethylating agents, or whether it would apply also to other crosslinking agents. *Cis*-DDP is a suitable agent for this test because, despite an entirely different chemistry, this agent resembles chloroethylnitrosoureas in that both interstrand and DNA–protein crosslinks are formed and interstrand crosslink formation is delayed a few hours whereas DNA–protein crosslink formation is relatively prompt.

Several human tumour cell strains having the Mer$^+$ or Mer$^-$ character were treated with *cis*-DDP, and were compared in regard to crosslinking and cyto-toxicity (Laurent *et al.*, 1981). The experiments were similar to the previous study of chloroethylnitrosoureas. As was the case with the latter drugs, cyto-toxicity was generally proportional to interstrand crosslinking. Unlike the case of the chloroethylnitrosoureas, however, DNA–protein crosslinking was pro-portional to interstrand crosslinking. More importantly, the extents of cyto-toxicity or crosslinking were independent of whether the cells were Mer$^+$ or Mer$^-$. Hence the susceptibility of human tumour cells to *cis*-DDP, although perhaps related to crosslinking, depends on cell properties (such as DNA repair capacities) that are different from the factors that govern susceptibility to chloroethylnitrosoureas.

MUTAGENESIS

The importance of the mutagenic and hence potentially carcinogenic effects of DNA damaging agents used in cytotoxic chemotherapy is shown by recent clinical experience indicating that perhaps 10% of cancer patients who are cured by chemotherapy will subsequently develop new tumours that are probably a late consequence of the treatment. As a first question, one may ask whether the relative magnitudes of the mutagenicity and cytotoxicity can vary independently of each other when different DNA damaging agents are used. The answer is clearly yes if one includes the entire range of alkylating agents, since the mono-functional methylating agents, such as MNU, can produce high mutation fre-quencies with relatively low cell killing, compared to bifunctional alkylating agents such as chloroethylnitrosoureas which have high cytotoxicities at low concentrations, at which mutation frequencies are relatively low. In order to achieve high cell kill with methylating agents, it is necessary to go to highly mutagenic concentrations. The more pertinent question is whether it is possible to reduce mutagenic and carcinogenic hazards through modification of the highly cytotoxic classes of drugs used in chemotherapy.

The chemotherapeutic crosslinking agents produce an excess of monoadducts in addition to crosslinks, and these monoadducts may be the main source of the mutagenic and carcinogenic side-effects. There has not yet been any systematic effort to improve these agents by minimising the ratio of monoadducts to cross-links. There is evidence however that mutagenicity and cytotoxicity can be dissociated from each other within the group of monofunctional and within the group of bifunctional alkylating agents. There is also some tentative information on the relation of these effects to specific types of DNA lesions.

Peterson *et al.* (1979) compared the potencies of several methylating and ethylating agents for production of cytotoxicity, mutation and DNA damage in Chinese hamster V79 cells. Cytotoxicity was measured by inhibition of colony formation; DNA strand lesions (strand breaks or alkali-labile sites) were measured by alkaline sedimentation; the methylating agents used were methyl-methane sulphonate (MMS) and MNNG; the ethylating agents were the ethyl homologues (EMS and ENNG). The ratio of DNA strand lesions to cytotoxicity varied widely, and was much higher for MNNG than for MMS. The frequency of DNA strand lesions did however seem to correlate linearly with mutagenesis. The authors argued that the major DNA lesion measured was the phosphotriester group (formed by alkylation of DNA phosphates). DNA phosphotriester groups are known to be alkali-labile in that they are converted to single-strand breaks at the high *p*H used in alkaline sedimentation assays (Shooter and Merrifield, 1978). Since guanine-O6 and phosphodiester groups are both among the less nucleophilic DNA sites, a high reactivity at phosphodiester groups would go along with a high reactivity at guanine-O6 positions. The result then would be consistent with guanine-O6 being the major mutagenic lesion, as generally believed, but would suggest that some other factors may have a major role in determining cytotoxicity. Regardless of the specific mechanism involved, the important observation is that the cytotoxicity and mutagenesis produced by this class of agents are not uniformly correlated with each other.

Suter *et al.* (1980) found marked variations in mutation frequencies at equi-toxic doses of alkylating agents in V79 cells. The induced mutation frequencies for cells treated at the D_{37} dose of MNU, MNNG and MMS were 400, 70 and 15 thioguanine-resistant mutants induced per 10^5 cells, respectively. Nitrogen mustard produced too few mutations to be detected at the D_{37} dose. Mutation frequency increased linearly with increasing dose, whereas cytotoxicity was disproportionately low at low doses (that is, the survival curves had shoulders). Measurements of O6-methylguanine and N7-methylguanine supported a quantitative dependence of mutations on O6-methylguanine lesions.

Similar conclusions were reached by Newbold *et al.* (1980). For a given extent of survival of colony forming ability of V79 cells, MNU was much more mutagenic than were MMS or dimethyl sulphate. MNU produced a much higher frequency of O6-methylguanine than did dimethyl sulphate, while the frequencies of N7-methylguanine were comparable for the two compounds.

Friedman and Huberman (1980) selected a line of V79 cells for resistance to MNNG and found the line to be much more resistant to cytotoxicity than to mutagenesis by this agent. The lines apparently did not differ in ^{14}C-MNNG binding or in the removal of ^{14}C-methyl adducts from DNA. Like other DNA-damaging agents, MNNG reduced the rate at which newly replicated DNA strands grow or join to form very long strands. The resistant line was less inhibited in this regard than was the parent line. The authors hypothesised that replication forks in the resistant cells have an increased capability of passing through methylation sites in the template. This may increase cell survival. The increased survival however may be at the cost of an increased probability of errors in base matching which may occur during replication at sites of alkylated template.

Peterson (1980) subsequently extended his study to include the response of DNA synthesis inhibition. The same four alkylating agents were used as in the previous study. The order of potency was MNNG > ENNG > MMS > EMS for inhibition of DNA synthesis, as well as for cytotoxicity, DNA strand lesions (breaks and/or alkali-labile sites) and mutagenesis. DNA synthesis inhibition however was found to correlate much better with cytotoxicity than with either mutagenesis or DNA strand lesions. The quantitative relation between inhibition of DNA synthesis and cytotoxicity was coincident for two cell types: Chinese hamster V79 cells and mouse C3H/10T1/2 cells. There was no such agreement, however, between DNA synthesis and DNA strand lesions. It seems therefore that, at least for methylating and ethylating agents, mutagenesis goes along with DNA strand lesions (breaks and/or alkali-labile lesions), whereas cytotoxicity goes along with inhibition of DNA synthesis.

Since MNNG can react to a greater extent with weaker nucleophilic sites than can MMS, the two agents generate a different pattern of DNA methylation sites. It may be that, at least for small alkyl groups, some relatively weak nucleophilic sites are major factors in mutagenesis whereas some relatively strong nucleophilic sites are important contributors to cytotoxicity.

Peterson and Peterson (1982) measured the formation of methyl purines in the DNA of V79 cells treated with MNNG or MMS. At equitoxic doses, the formation of N7-methylguanine and of N3-methylguanine by MNNG was about one-half that produced by MMS, but O6-methylguanine formation by MNNG was more than 10 times as great as by MMS.

There is no precise correspondence between any single type of DNA lesion and the magnitude of cytotoxicity or mutagenicity. It is possible, and even likely, that more than one type of lesion generally contributes to each type of biological effect. The possible role of pyrimidine alkylations (Singer, 1976, 1979; Singer *et al.*, 1978), in particular, still remains to be determined.

Alkylation of guanine-O6 has also been implicated in the production of sister chromatid exchanges (SCE) (Swenson *et al.*, 1980). V79 cells were treated with doses of MMS, EMS, MNU or ENU that produced the same frequency of SCEs

(30 per cell). A variety of DNA alkylation sites were measured, including sites on all four bases and at phosphates. If one alkylation site were predominantly responsible for SCE production, the extent of alkylation at that site should be the same for all four compounds. Of the twelve different alkylation sites measured, none obeyed this criterion. However, alkylation at guanine-O6 gave good agreement (3.3–3.5 alkylations 10^6 bases) for all of the compounds except MMS. MMS produced only 1/10 of this frequency of guanine-O6 alkylations. Hence, if it is accepted that guanine-O6 alkylation is the main cause of SCEs in the case of the other three compounds, then another alkylation site must emerge as predominant in the case of MMS. A possible candidate for this alternative site was identified as the adenine-N3 position; MMS at the equi-effective dose produced considerably more alkylation at this site than did the other compounds. (These inferences assume that any removal or repair of alkylation sites is similar for methylations as for ethylations.)

Aside from repair by the alkytransferase mechanism, O6-alkyl-guanines may be repaired by an excision mechanism. Goth-Goldstein (1977) has reported that xeroderma pigmentosum fibroblasts which are deficient in excision repair have greatly reduced abilities to remove O6-methylguanine or O6-ethylguanine lesions from DNA. (The removal was studied at 24 and 48 hr after treatment; the more rapid alkyltransferase repair was not evident in these experiments.)

Simon *et al.* (1981) observed an increased lethality of excision repair-deficient xeroderma pigmentosum cells treated with ENU. There was also a commensurate increase in mutagenicity. However, MNU or MNNG each produced the same lethality in xeroderma pigmentosum as in normal cells. Assuming that O6-alkylguanine lesions are the major cause of cytotoxicity and mutagenicity, the results can be explained on the basis of two repair mechanisms. It may be that O6-methylguanine is mainly repaired by alkyltransferase whereas O6-ethylguanine is mainly repaired by excision. Then loss of excision-repair would be more consequential for ethylating than for methylating agents (assuming that alkyltransferase repair is intact). On the other hand, one might suppose that the selective loss of alkyltransferase repair would have a greater impact on methylating than on ethylating agents.

There are fewer studies attempting to dissociate mutagenicity from cytotoxicity for bifunctional agents. Such a dissociation has been observed in a comparison between *cis*- and *trans*-DDP complexes (Zwelling, 1979*b*; Johnson, 1980). Although the two Pt (II) isomers are similarly efficient in crosslinking DNA in chemical systems, in cells *trans*-DDP produces much less interstrand crosslinking than does *cis*-DDP (Zwelling *et al.*, 1979*b*). *Trans*-DDP is much less cytotoxic than is *cis*-DDP, and it is possible that the magnitude of the cytotoxicity depends on interstrand crosslinking. The difference in mutagenicity of the two isomers however is even greater than the cytotoxicity difference, as *trans*-DDP is clearly less mutagenic than *cis*-DDP even at equitoxic doses. Johnson (1980) found that the number of mutants obtained per DNA-bound Pt was almost 1,000-fold greater for *cis*-DDP than for *trans*-DDP. This is an interesting result, although unfortunately not therapeutically useful, because *trans*-DDP is devoid of antitumour activity. Both *cis*- and *trans*-DDP also produce DNA-protein crosslinks; these lesions are repairable and apparently do not contribute either to mutagenicity or to cytotoxicity (Zwelling *et al.*, 1979*a,b*).

Although it is not detectably mutagenic in either mammalian cells or bacteria, *trans*-DDP does induce high levels of sister chromatid exchanges in rodent cells and induces malignant transformations in mouse C3H 10T1/2 and 3T3 cells (Fornace, 1980). This fits with the idea that malignant transformation can arise from genome rearrangements occurring by processes other than those responsible for point mutations.

The relative mutagenicity and cytotoxicity of several nitrosoureas was studied by Erickson *et al.* (1978) in Chinese hamster V79 cells (table 5). The methylnitroso compounds, MNU and streptozotocin, were found to have much higher mutagenicity/cytotoxicity ratios than did the other compounds, which were all 2-haloethylnitrosoureas. The chloroethylnitrosoureas, 1-(2-chloroethyl)-1-nitrosourea (CNU), CCNU and 1,3-*bis*(2-chloroethyl)nitrosourea (BCNU), exhibited similar low ratios, whereas the ratios for chlorozotocin and 1-(2-fluoroethyl)-1-nitroso-3-cyclohexylurea (FCNU) were somewhat higher.

Table 5   Cytotoxicity and mutagenicity of nitrosoureas in V79 Chinese hamster cells (Erickson *et al.*, 1978*a*)

Compound	Drug concentration giving 1 log kill (D_{37})	Drug concentration giving 1–2 mutants per 10^5 cells	Ratio
	(μM)	(μM)	
MNU	400	50	8.0
Streptozotocin	544	58	9.4
CNU	13.2	33.1	0.40
Chlorozotocin	12.8	12.8	1.0
CCNU	5.5	15	0.37
FCNU	21	21	1.0
BCNU	8.5	18	0.47

When methylnitrosoureas are administered at the high concentrations required to produce substantial cytotoxicity, high frequencies of DNA strand breaks (or alkali-labile lesions) are produced (table 6). At equitoxic doses, the haloethylnitrosoureas produced much less strand breakage than did the methylnitrosoureas. When the comparison was made at equimutagenic doses, all of the compounds except chlorozotocin produced break frequencies of a similar order of magnitude; the mean values ranged over a factor of <3. This is consistent with the idea that strand breaks are mechanistically related to mutagenesis. The ability to form crosslinks, as with haloethylnitrosoureas, seems to enhance cytotoxicity markedly without enhancing mutagenicity.

Table 6 DNA strand break frequencies measured by alkaline elution after 2 hr exposure of V79 cells to equitoxic and equimutagenic doses of nitrosoureas (Break frequencies expressed in number of single-strand breaks per 10^{10} daltons of DNA.)

	Equitoxic doses		Equimutagenic doses	
	μM	Break frequency	μM	Break frequency
MNU	400	10.55 ± 0.86	50	$0.94 + 0.24$
Streptotozotin	544	7.02 ± 0.63	58	0.53 ± 0.31
CNU	13.2	0.18 ± 0.07	33.1	0.43 ± 0.09
Chlorozotocin	12.8	0.12 ± 0.03	12.8	0.12 ± 0.03
CCNU	5.5	0.27 ± 0.23	15	0.39 ± 0.07
FCNU	21	0.32 ± 0.11	21	0.38 ± 0.11
BCNU	8.5	0.32 ± 0.04	18	0.52 ± 0.12

From Erickson *et al.* (1978*a*).

The idea that crosslinks tend to be lethal whereas monoadducts are mainly mutagenic has also received experimental support in the case of furocoumarin photoadducts, in which the ratio of crosslinks to monoadducts introduced into the DNA can be experimentally controlled (Bridges *et al.*, 1979; Seki, 1978). In an excision repair-defective strain of bacteria, the conversion of monoadducts to crosslinks greatly enhanced the lethality, but reduced the mutation frequency. On the other hand, Ruhland *et al.* (1979) found in haploid yeast that a mono-functional nitrogen mustard was, if anything, less mutagenic than equitoxic doses of bifunctional mustards.

In haploid yeast, Ruhland *et al.* (1978, 1979) found that certain repair functions, most notably *rev*, reduce killing but increase mutation rates in organisms treated with nitrogen mustard or other mono- and bi-functional alkylating agents. Organisms bearing a *rev* mutation were highly sensitive to alkylating agents, but there was no mutagenesis among the survivors. The *rev* mutation is thought to block an error-prone repair process. It appears that mutations produced by alkylating agents may mostly result from error-prone repair. Hence mutagenesis may be the cost of certain types of repair.

The chemotherapeutic crosslinking agents produce an excess of monoadducts in addition to crosslinks, and these monoadducts may be the main source of the mutagenic and carcinogenic side effects. There has not yet been any systematic effort of improve these agents by minimising the ratio of monoadducts to crosslinks. It is encouraging however that mutagenicity and cytotoxicity in mammalian cells can be dissociated from each other within the class of monofunctional alkylating agents and within the class of bifunctional alkylating agents, and that information is beginning to emerge about the relation of these effects to specific DNA lesions.

REFERENCES

Ahlgren, J. D., Green, D. C., Tew, K. D. and Schein, P. S. (1982). *Cancer Res.*, **42**, 2605

Alazard, R., Germanier, M. and Johnson, N. P. (1982). *Mutat. Res.*, **93**, 327

Auerbach, A. D. and Wolman, S. R. (1976). *Nature, Lond.*, **261**, 494

Berger, R., Bernheim, A., Gluckman, E. and Gisselbrecht, C. (1980). *Br. J. Haemat.*, **45**, 565

Ball, C. R. and Roberts, J. J. (1970). *Chem.-biol. Interact.*, **2**, 321

Ball, C. R. and Roberts, J. J. (1972). *Chem.-biol. Interact.*, **4**, 297

Ben-Hur, E. and Elkind, M. M. (1973). *Mutat. Res.*, **18**, 315

Ben-Hur, E. and Elkind, M. M. (1973). *Biochim. biophys. Acta*, **331**, 181

Ben-Hur, E., Prager, A. and Riklis, E. (1979). *Photochem. Photobiol.*, **29**, 921

Berger, R., Bernheim, A., Gluckman, E. and Gisselbrecht, C. (1980). *Br. J. Haemat.*, **45**, 565

Bredburg, A., Lambert, B. and Söderhall, S. (1982). *Mutat. Res.*, **93**, 221

Bridges, B. A. and Stannard, M. (1982). *Mutat. Res.*, **92**, 9

Bridges, B. A., Mottershead, R. P. and Knowles, A. (1979). *Chem.-biol. Interact.*, **27**, 221

Bridges, B. A. and von Wright, A. (1981). *Mutat. Res.*, **82**, 229

Brookes, P. and Lawley, P. D. (1960). *Biochem. J.*, **77**, 478

Brookes, P. and Lawley, P. D. (1961). *Biochem. J.*, **80**, 496

Burchenal, J. H., Kalaher, K., O'Toole, T. and Chisholm, J. (1977). *Cancer Res.*, **37**, 3455

Chetsanga, C. J., Lozon, M., Makaroff, C. and Savage, L. (1981). *Biochemistry*, **20**, 5201

Chetsanga, C. J., Polidori, G. and Mainwaring, M. (1982). *Cancer Res.*, **42**, 2616

Cleager, J. E. (1972). *J. invest. Dermat.*, **58**, 124

Cohen, L. F., Ewig, R. A. G., Kohn, K. W. and Glaubiger, D. (1980). *Biochim. biophys. Acta*, **610**, 56

Cole, R. S. (1971). *J. Bact.*, **107**, 846

Cole, R. S. (1973). *Proc. natn. Acad. Sci. U.S.A.*, **70**, 1064

Cole, R. S., Levitan, D. and Sinden, R. R. (1976). *J. molec. Biol.*, **103**, 39

Day, R. S., III. (1981). *Biosciences*, **31**, 807

Day, R. S., III and Ziolkowski, C. H. J. (1979). *Nature, Lond.*, **279**, 797

Day, R. S., III, Ziolkowski, C. H. J., Scudiero, D. A., Meyer, S. A., Lubiniecki, A., Girardi, A., Galloway, S. M. and Bynum, G. D. (1980a). *Nature, Lond.*, **288**, 724

Day, R. S., III, Ziolkowski, C. H., Scudiero, D. A., Meyer, S. A. and Mattern, M. R. (1980b). *Carcinogenesis*, **1**, 21

Deutsch, W. A. and Linn, S. (1979a). *Proc. natn. Acad. Sci.*, **76**, 141

Deutsch, W. A. and Linn, S. (1979b). *J. biol. Chem.*, **25**, 12099

Domon, M. and Rauth, A. M. (1969). *Radiat. Res.*, **40**, 414

Ducore, J. M., Erickson, L. C., Zwelling, L. A., Laurent, G. and Kohn, K. W. (1982). *Cancer Res.*, **42**, 897

Eastman, A. and Bresnick, E. (1981). *Biochem. Pharmac.*, **30**, 2721

Erickson, L. C., Bradley, M. O. and Kohn, K. W. (1978a). *Cancer Res.*, **38**, 3379

Erickson, L. C., Osieka, R. and Kohn, K. W. (1978c). *Cancer Res.*, **38**, 802

Erickson, L. C., Bradley, M. O. and Kohn, K. W. (1978). *Cancer Res.*, **38**, 672

Erickson, L. C., Bradley, M. O., Ducore, J. M., Ewig, R. A. G. and Kohn, K. W. (1980a). *Proc. natn. Acad. Sci. U.S.A.*, **77**, 467

Erickson, L. C., Laurent, G., Sharkey, N. A. and Kohn, K. W. (1980b). *Nature, Lond.*, **288**, 727

Erickson, L. C., Ramonas, L. M., Zaharko, D. S. and Kohn, K. W. (1980c). *Cancer Res.*, **40**, 4216

Ewig, R. A. G. and Kohn, K. W. (1977). *Cancer Res.*, **37**, 2114

Ewig, R. A. G. and Kohn, K. W. (1978). *Cancer Res.*, **38**, 3197

Filipski, J., Kohn, K. W. and Bonner, W. M. (1980). *Chem.-biol. Interact.*, **32**, 321

Filipski, J., Kohn, K. W., Prather, R. and Bonner, W. M. (1979). *Science*, **204**, 181

Fleer, R. and Brendel, M. (1979). *Molec. gen. Genet.*, **176**, 41

Fleer, R. and Brendel, M. (1981). *Chem.-biol. Interact.*, **37**, 123

Fornace, A. J. (1980). *Proc. Am. Ass. Cancer Res.*, **21**, 118

Fornace, A. J. (1981). *Proc. Am. Ass. Cancer Res.*, **22**, 80

Fornace, A. J. (1982). *Cancer Res.*, **42**, 145

Fornace, A. J. and Kohn, K. W. (1976). *Biochim. biophys, Acta*, **435**, 95

Fornace, A. J., Kohn, K. W. and Kann, H. E. (1978). *Cancer Res.*, **38**, 1064

Fornace, A. J., Little, J. B. and Weichselbaum, R. R. (1979). *Biochim. biophys. Acta*, **561**, 99

Friedman, J. and Huberman, E. (1980). *Proc. natn. Acad. Sci. U.S.A.*, **77**, 6072

Fujiwara, Y. and Tatsumi, M. (1975). *Biochem. biophys. Res. Commun.*, **66**, 592

Fujiwara, Y., Tatsumi, M., and Sasaki, M. S. (1977). *J. molec. Biol.*, **113**, 635

Fujiwara, Y., and Tatsumi, M. (1977). *J. molec. Biol.*, **15**, 635

Geiduschek, E. P. (1961). *Proc. natn. Acad. Sci. U.S.A.*, **47**, 950

Gilman, A. and Philips, F. S. (1946). *Science*, **103**, 409

Goldacre, R. J., Loveless, A. and Ross, W. C. J. (1949). *Nature, Lond.*, **163**, 667

Goth-Goldstein, R. (1977). *Nature, Lond.*, **267**, 81

Grant, E. L., von Borstel, R. C. and Ashwood-Smith, M. J. (1979). *Eviron. Mutag.*, **1**, 55

Grover, N. B., Margalit, A., Zaritsky, A., Ben-Hur, E. and Hansen, M. T. (1981). *Biophys. J.*, **33**, 93

Hartley-Asp, B. (1978). *Mutat. Res.*, **49**, 117

Ishida, R. and Buchwald, M. (1982). *Cancer Res.*, **42**, 4000

Iyer, V. N., and Szybalski, W. (1963). *Proc. natn. Acad. Sci. U.S.A.*, **50**, 355

Jachymcyzk, W. J., von Borstel, R. C., Mowat, M. R. and Hastings, P. J. (1981). *Molec. gen. genet.*, **182**, 196

Johnson, N. P., Hoeschele, J. D., Rahn, R. O., O'Neill, J. P. and Hsie, A. W. (1980). *Cancer Res.*, **40**, 1463

Johnson, R. K., Faucette, L. F., Wodinsky, I., and Clement, J. J. (1982). *Proc. Am. Ass. Cancer Res.*, **23**, 116

Jolley, G. M. and Ormerod, M. G. (1973). *Biochim. biophys. Acta*, **308**, 242

Kann, H. E., Blumenstein, B. A., Petkas, A. and Schoot, M. A. (1980*a*). *Cancer Res.*, **40**, 771

Kann, H. E., Kohn, K. W. and Lyles, J. M. (1974). *Cancer Res.*, **34**, 398

Kann, H. E., Schott, M. A. and Petkas, A. (1980*b*). *Cancer Res.*, **40**, 50

Kay, J., Smith, C. A. and Hanawalt, P. C. (1980). *Cancer Res.*, **40**, 696

Kohn, K. W. (1977). *Cancer Res.*, **37**, 1450

Kohn, K. W. (1979). In *Methods in Cancer Research*, Vol. 16 (ed. V. T. DeVita and H. Busch) Academic Press, New York, p. 291

Kohn, K. W. (1981*a*). *Biosciences*, **31**, 593

Kohn, K. W. (1981*b*). In *Molecular Actions and Targets for Cancer Chemotherapeutic Agents*, (ed. A. C. Sartorelli) Academic Press, New York, p. 3

Kohn, K. W., Ewig, R. A. G., Erickson, L. C., and Zwelling, L. A. (1981). In *DNA Repair a Laboratory Manual of Research Procedures*, Vol. 1 (ed. E. C. Friedberg and P. C. Hanawalt), Marcel Dekker, New York, p. 379

Kohn, K. W., Steigbigel, N., and Spears, C. L. (1965). *Proc. natn. Acad. Sci. U.S.A.*, **53**, 1154

Kohn, K. W., Spears, C. L., and Doty, P. (1966). *J. molec. Biol.*, **19**, 266

Kohn, K. W., and Ewig, R. A. G. (1979). *Biochim. biophys. Acta*, **562**, 32

Latt, S. A., Stetten, G., Juergens, L. A., Buchanan, G. R. and Gerald, P. S. (1975). *Proc. natn. Acad. Sci. U.S.A.*, **72**, 4066

Laurent, G., Erickson, L. C., Sharkey, N. A. and Kohn, K. W. (1981). *Cancer Res.*, **41**, 3347

Lawley, P. D., Lethbridge, J. H., Edwards, P. A. and Shooter, K. V. (1969). *J. molec. Biol.*, **39**, 181

Lehmann, A. R., Kirk-Bell, S., Arlett, C. F., Paterson, M. C., Lohman, P. H. M., DeWeerd-Kastelein, E. A. and Bootsma, D. (1975). *Proc. natn. Acad. Sci. U.S.A.*, **72**, 219

Lehmann, A. R., Kirk-Bell, S., Arlett, C. F., Harcourt, S. A., DeWeerd-Kastelein, E. A., Keijzer, W. and Hall-Smith, P. (1977). *Cancer Res.*, **37**, 904

Lindahl, T. (1982). *A. Rev. Biochem.*, **51**, 61

Little, J. W. and Mount, D. W. (1982). *Cell*, **29**, 11

Loveless, A. and Stock, J. C. (1959*a*). *Proc. R. Soc.*, **B150**, 486

Loveless, A. and Stock, J. C. (1959*b*). *Proc. R. Soc.*, **B150**, 423

Loveless, A. and Stock, J. C. (1959*c*). *Proc. R. Soc.*, **B151**, 148

Lown, J. W., McLaughlin, L. W. and Chang, Y-M. (1978). *Bioorgan. Chem.*, 7, 97

Magana-Schwenke, N and Moutacchi, E. (1980). *Mutat. Res.*, 29

Margison, G. P. and Pegg, A. E. (1981). *Proc. natn. Acad. Sci. U.S.A.*, 78, 861

Mehta, J. R., Ludlum, D. B., Renard, A. and Verly, W. G. (1981). *Proc. natln. Acad. Sci. U.S.A.*, 78, 6766

Mhta, J. R., Przybylski, M. and Ludlum, D. B. (1980). *Cancer Res.*, 40, 4183

Meyn, R. E., Jenkins, S. F. and Thompson, L. H. (1982). *Cancer Res.*, 42, 3106

Micetich, K., Michaels, S., Jude, G., Kohn, K., and Zwelling, L. (1981). *Proc. Am. Ass. Cancer Res.*, 22, 252

Morimoto, K., Yoshikawa, K. and Yamaha, T. (1980). *Gann*, 71, 674

Newbold, R. F., Warren, W., Medcalf, A. S. and Amos, J. (1980). *Nature, Lond.*, 283, 596

Olsson, M. and Lindahl, T. (1980). *J. biol. Chem.*, 255, 10569

Ou, C-N., Tsai, C-H., Tapley, K. J. and Song, P-S. (1978). *Biochemistry*, 17, 1047

Pera, M. F., Jr,., Rawlings, C. J., Shackleton, J. and Roberts, J. J. (1981). *Biochim. biophys. Acta*, 655, 152

Peterson, A. R., Peterson, H. and Heidelberger, C. (1979). *Cancer Res.*, 39, 131

Peterson, A. R. (1980). *Cancer Res.*, 40, 682

Peterson, A. R. and Peterson, H. (1982). *Proc. natn. Acad. Sci. U.S.A.*, 79, 1643

Pohl, J. and Christophers, E. (1979). *J. inv. Dermat.*, 73, 176

Rauth, A. M., Barton, B. and Lee, C. P. (1970). *Cancer Res.*, 30, 2724

Reid, B. D. and Walker, I. G. (1969). *Biochem. biophys. Acta*, 179, 179

Roberts, J. J. and Friedlos, F. (1981). *Biochim. biophys. Acta*, 655, 146

Roberts, J. J. and Pascoe, J. M. (1972). *Nature, Lond.*, 235, 282

Ross, W. E. and Shipley, N. (1980). *Mutat. Res.*, 79, 277

Ross, W. E., Ewig, R. A. G., and Kohn, K. W. (1978). *Cancer Res.*, 38, 1502

Ruhland, A., Fleer, R. and Brendel, M. (1978). *Mutat Res.*, 58, 241

Ruhland, A. and Brendel, M. (1979). *Genetics*, 92, 83

Rutman, R. J., Chun, E. H. L., and Jones, J. (1969). *Biochim. biophys. Acta*, 174, 663

Rydberg, B. (1975). *Radiat. Res.*, 61, 274

Sasaki, M. S. (1975). *Nature, Lond.*, 257, 501

Sasaki, M. S. and Tonomura, A. (1973). *Cancer Res.*, 33, 1829

Schabel, F. M., Trader, M. W., Laster, L. R., Wheeler, G. P. and Witt, M. H. (1978). *Antibiol. Chemother.*, 23, 200

Seki, T., Nozu, K. and Kondo, S. (1978). *Photochem. Photobiol.*, 27, 19

Sharkey, N. A., Erickson, L. C. and Kohn, K. W. (1982). *Proc. Am. Ass. Cancer Res.*, 23, 164

Shiloh, Y. and Becker, Y. (1981). *Cancer Res.*, 41, 5114

Shooter, K. V., Howse, R., Merrifield, R. K. and Robins, A. B. (1972). *Chem.-biol. Interact.*, 5, 289

Shooter, K. V. and Merrifield, R. K. (1978). *Biochim. biophys. Acta*, 521, 155

Simon, L., Hazard, R. M., Maher, V. M. and McCormick, J. J. (1981). *Carcinogenesis*, 2, 567

Sinden, R. R. and Cole, R. S. (1978). *J. Bact.*, 136, 538

Singer, B. (1976). *Nature, Lond.*, 264, 333

Singer, B. (1979). *J. natn. Cancer Inst.*, 62, 1329

Singer, B., Bodell, W. J., Cleaver, J. E., Thomas, G. H., Rajewsky, M. F. and Thon, W. (1978). *Nature, Lond.*, 276, 85

Singer, B. and Brent, T. P. (1981). *Proc. natn. Acad. Sci. U.S.A.*, 78, 856

Sklar, R. and Strauss, B. (1981). *Nature, Lond.*, 289, 417

Song, P.-S. and Tapley, K. J. Jr (1979). *Photochem. Photobiol.*, 29, 1177

Stacey, K. A., Cobb, M., Cousens, S. F. and Alexander, P. (1958). *Ann. N. Y. Acad. Sci.*, 68, 682

Strandberg, M. C., Bresnick, E. and Eastman, A. (1982). *Chem.-biol. Interact.*, 15, 169

Suter, W., Brennand, J., McMillan, S. and Fox, M. (1980). *Mutat. Res.*, 73, 171

Swenson, D. H., Harbach, P. R. and Trzos, R. J. (1980). *Carcinogenesis*, 1, 931

Szumiel, I. (1979). *Chem.-biol. Interact.*, 24, 73

Tanka, M. and Yoshida, S. (1981). *Biochem. Pharmac.*, 30, 299

Thomas, C. B., Osieka, R. and Kohn, K. W. (1978). *Cancer Res.*, 38, 2448

Thompson, L. H., Brookman, K. W., Carrano, A. V. and Dillehay, L. E. (1982). *Proc. natn. Acad. Sci. U.S.A.*, **79**, 534

Tong, W. P. and Ludlum, D. B. (1981). *Cancer Res.*, **41**, 380

Tong, W. P., Kirk, M. C. and Ludlum, D. B. (1982). *Cancer Res.*, **42**, 3102

Varga, J. M., Wiesehahn, G., Bartholomew, J. C. and Hearst, J. E. (1982). *Cancer Res.*, **42**, 2223

Venitt, S. (1968). *Biochem. biophys. Res. Commun.*, **31**, 355

Verly, W. G. and Brakier, L. (1969). *Biochim. biophys. Acta*, **174**, 674

Walker, I. G. and Reid, B. D. (1971). *Cancer Res.*, **31**, 510

Weksberg, R., Buchwald, M., Sargent, P., Thompson, M. W. and Siminovitch, L. (1979). *J. Cell Physiol.*, **101**, 311

Yoakum, G. H. and Cole, R. W. (1977). *J. biol. Chem.*, **252**, 7023

Yin, L., Chun, E. H. L. and Rutman, R. J. (1973). *Biochim. biophys. Acta.*, **324**, 472

Zwelling, L. A., Anderson, T., and Kohn, K. W. (1979*a*). *Cancer Res.*, **39**, 365

Zwelling, L. A., Bradley, M. O., Sharkey, N. S., Anderson, T. and Kohn, K. W. (1979*b*). *Mutat. Res.*, **67**, 271

Zwelling, L. A., Filipski, J., and Kohn, K. W. (1979*c*). *Cancer Res.*, **39**, 4989

Zwelling, L. A., Kohn, K. W., Ross, W. E., Ewig, R. A. G. and Anderson, T. (1978). *Cancer Res.*, **38**, 1762

Zwelling, L. A., Michaels, S., Schwartz, H., Dobson, P. P. and Kohn, K. W. (1981). *Cancer Res.*, **41**, 640

11
Inhibitors of Dihydrofolate Reductase

B. Roth, E. Bliss and C. R. Beddell

INTRODUCTION

Methotrexate (*1b*), which remains the single most important antifol anticancer agent after 30 years of clinical use, owes its activity to its stoichiometric inhibition of the enzyme dihydrofolate reductase (EC 1.5.1.3.). Methotrexate was introduced as an antileukaemic agent 10 years before dihydrofolate reductase was demonstrated to be its prime intracellular target, although it was designed to inhibit the utilisation of the vitamin folic acid (*2*). Its close structural relationship to folic acid, and the fact that it was found to be a competitive inhibitor of dihydrofolate reductase, led to the seemingly obvious conclusion that it binds to the active site of the reductase in the manner of folic acid. Only very recently has there appeared evidence that methotrexate actually binds to dihydrofolate reductase in a conformation substantially different from that of the vitamin.

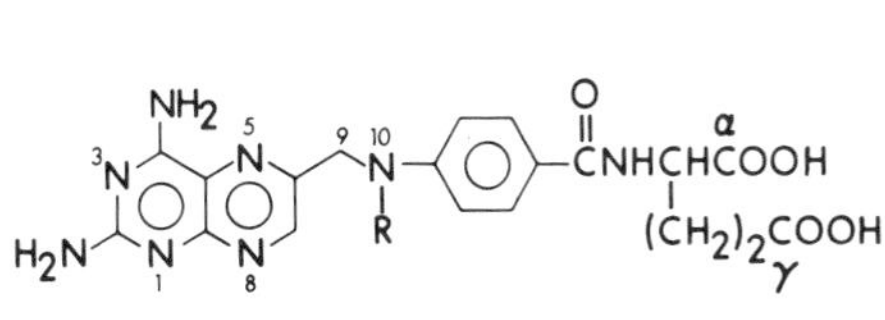

(*a*) R = H (aminopterin)
(*b*) R = CH$_3$ (methotrexate)

1

Molecular Aspects of Anti-cancer Drug Action, ed. Neidle & Waring
0333-315561/83/363-393   © The Contributors 1983

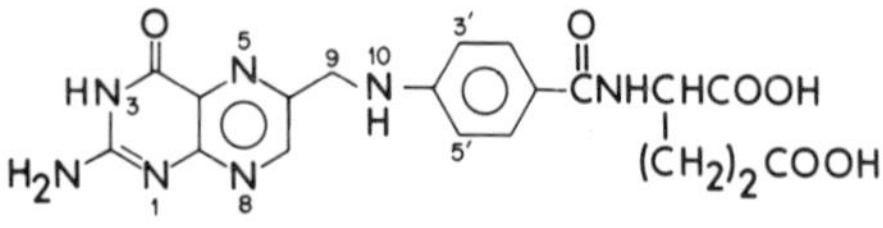

(folic acid)

2

The enzyme dihydrofolate reductase has fascinated medicinal chemists and biochemists since the early realization that certain types of small molecule inhibitor show remarkable species specificity for dihydrofolate reductase from various sources; particularly evident is selectivity for enzymes from prokaryotic as opposed to eukaryotic cells. This has led to the purification and amino acid sequencing of many dihydrofolate reductases during the past decade, followed by the crystallisation of the enzyme from several sources with methotrexate and with other inhibitors, as well as with the cofactor, NADPH. The three-dimensional structures of three dihydrofolate reductases have now been solved, at resolutions better than 2 Å.

This chapter describes the enzymatic basis for the activity of methotrexate and other dihydrofolate reductase inhibitors at the molecular level, and also illustrates other important aspects of the activity of various anticancer antifols, including physicochemical properties, pharmacokinetics and metabolism, selectivity and means for achieving it, and other possible sites of action for such inhibitors.

BIOCHEMICAL PERSPECTIVES

Methotrexate and the biochemical role of dihydrofolate reductase

The site of biochemical action of methotrexate was demonstrated in the late 1950's to be its stoichiometric binding to dihydrofolate reductase (Futterman, 1957; Zakrzewski and Nichol, 1958; Osborn *et al.*, 1958; Werkheiser, 1961). In the years which have followed, a picture of much greater complexity has emerged, however. Figure 1 illustrates the biochemical pathways involving this enzyme. Originally it was considered that the major role for dihydrofolate reductase was to reduce folic acid in two steps to its 5, 6, 7, 8-tetrahydro derivative, in which active form it transfers single carbon moieties. These in part form the building blocks for certain amino acids and pyrimidine and purine bases (see for example reactions 1–6, figure 1), which lead ultimately to the synthesis of DNA. A total of at least 17 functions are performed by the reduced folates, only a few of which are illustrated here.

The vitamin is now known to be supplied to cells in its tetrahydro stage of reduction, particularly as its stabilised 5-methyl and 5-formyl derivatives. These are conveyed across cell membranes by active transport (Kessel *et al.*, 1965;

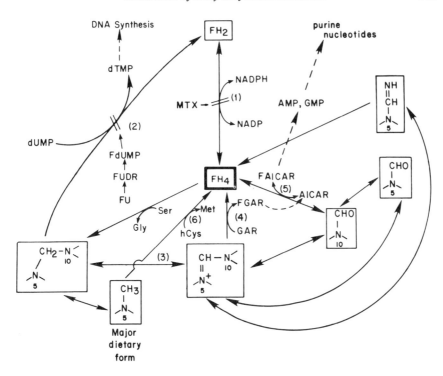

Figure 1 Some important folate metabolic pathways, illustrating the role of dihydrofolate reductase (1). (1) Dihydrofolate reductase (tetrahydrofolate dehydrogenase, EC 1.5.1.3; (2) thymidylate synthetase (EC 2.1.1.45); (3) methylene tetrahydrofolate dehydrogenase (EC 1.5.1.5); (4) phosphoribosylglycinamide formyltransferase (EC 2.1.2.2); (5) phospho-ribosylaminoimidazolecarboxamide formyltransferase (EC 2.1.2.3); (6) tetrahydropteroyl-glutamate methyltransferase (EC 2.1.1.1.3); (5>N–CH$_2$–N<)=5,10-methylenetetrahydro-folate; (5>N–CH$_3$–=5-methyltetrahydrofolate; (5>N$^+$=CH–N <)=5,10-methenyltetrahy-drofolate; (5>N–CHO)=5-formyltetrahydrofolate (citrovorum factor, CF); 5>N–CH=NH)= 5-formiminotetrahydrofolate; (10>N–CHO) = 10-formyltetrahydrofolate; hCys = homo-cysteine; FGAR = formylglycinamide ribonucleotide; GAR = glycinamide ribonucleotide; AICAR = aminoimidazolecarboxamide ribonucleotide, FAICAR = 5-formamidoimidazole-carboxamide ribonucleotide; dUMP = deoxyuridylate; AMP = adenylic acid; GMP = guanylic acid; dTMP = thymidylate; FU = 5-fluorouracil, FUDR = 5-fluorodeoxyuridine; FdUMP = 5-fluorodeoxyuridylate

Goldman *et al.*, 1968; Goldman 1971; Nahas *et al.*, 1972; Sirotnak and Donsbach, 1972; Bender, 1975). Folic acid might then be considered as a mere artefact, caused by the ready oxidation of reduced folates which are not stabilised by 5-substitution; however, it is found to some extent in foods as well as in vitamin supplements. Dihydrofolate reductase regenerates the active cofactor.

In the presence of sufficient reduced folate, it can be seen from figure 1 that the primary role for dihydrofolate reductase would be to reduce the dihydro-folic acid created from the thymidylate synthetase-catalysed conversion of deoxyuridylate (dUMP) to thymidylate (dTMP) (Werkheiser, 1969). This is made necessary by the fact that dTMP synthesis (pathway 2, figure 1) depletes the tetrahydrofolic acid pool more rapidly than it can be replenished from 5-methyltetrahydrofolate (pathway 6). Cells which are rapidly synthesising DNA will of course be depleted of tetrahydrofolic acid most rapidly. For a more complete understanding of the action of dihydrofolate reductase inhibitors, it is important to look at the binding of inhibitors to thymidylate synthetase at the molecular level, particularly since some compounds inhibit both enzymes, or are substrates for one, and inhibitors of the other.

The thymidylate reaction and thymidylate synthetase catalysis

The reaction controlled by thymidylate synthetase is unusual, in that the cofactor, methylenetetrahydrofolate, not only provides the one carbon source for forming dTMP but also reduces this moiety to a methyl group, thereby becoming self-oxidised to the dihydro level. This biochemical reaction is the only known *de novo* source of thymidine. Blockade of this event causes disruption of DNA synthesis, despite continued protein and RNA formation, a condition which Cohen (1971) has called 'thymineless' death. A number of recent reviews have covered various aspects of the biochemistry of this enzyme (Santi, 1980; 1981; Danenberg, 1977; Pogolotti and Santi, 1977; Friedkin, 1973).

The catalytic mechanism of the one-carbon transfer effected by thymidylate synthetase is thought to proceed via the addition of a nucleophilic group on the enzyme (very likely a sulphydryl group) to the 6-position of dUMP. The coenzyme then becomes added to the 5-position of dUMP through its 5, 10-methylene group, which then becomes cleaved from the coenzyme. This leaves a 5-methylene group on dUMP which is reduced by transfer of a hydrogen from the coenzyme, thus producing dTMP and dihydrofolic acid (Pogolotti and Santi, 1974; 1977). A prototype inhibitor, 5-fluorodeoxyuridine monophosphate (FdUMP) and the cofactor, react with the enzyme to give the covalent derivative 3.

3

In contrast with the substrate, which then loses a C_5 proton, the 5-CF bond apparently remains unbroken (Danenberg, 1977), leading to the accumulation of an analogue of a steady-state intermediate, and blockade of dTMP synthesis. In the absence of the cofactor, FdUMP binds poorly to the enzyme.

Certain folic acid analogues also inhibit the action of thymidylate synthetase. The development of this type of inhibitor followed the observation that successive resistance to methotrexate and dichloromethotrexate in murine L1210 leukaemia was accompanied by a tremendous increase in the dihydrofolate reductase content of the tumour (Misra *et al.*, 1961). The authors suggested that a fraudulent substrate for dihydrofolate reductase might conceivably produce a product which would inhibit subsequent enzymes on the folate pathway, such as thymidylate synthetase. This approach would take advantage of a unique property of the resistant tumour cell. Dihydrohomofolate and certain analogues were in fact found to behave in this manner. Another point to note is that thymidylate synthetase is present at a very low level in normal mammalian cells, but occurs to a greater extent in rapidly growing (tumour) cells (Blakley, 1969). Despite the inherent interest in such an approach, the preparation of inhibitors to date has been fraught with problems of synthesis and instability, and the inhibitors have been unsuitable for practical use.

The effect of combined therapy with methotrexate and thymidylate synthetase inhibitors

There have been many conflicting reports on the effects of combining methotrexate with 5-fluorouracil or its deoxyriboside; some have reported synergy (Bareham *et al.*, 1974; Kline *et al.*, 1966), others have reported antagonism (Tattersall *et al.*, 1973; Bowen *et al.*, 1978). Bertino *et al.* (1977) found that the result was dependent on the dosage schedule; pretreatment with methotrexate enhanced the antitumour activity of 5-fluorouracil but combined or subsequent treatment produced antagonism.

Borsa and Whitmore (1969*a,b*) concluded that methotrexate owed its activity in living cells to the inhibition of thymidylate synthetase rather than dihydrofolate reductase from studies on incorporation of labelled deoxyuridine in the presence of methotrexate. Moran *et al.* (1979) disagreed with this conclusion; they found that methotrexate inhibition of mouse or human leukaemia cell growth was partially reversed in the presence of either hypoxanthine or thymidine; 5-fluorodeoxyuridine in the presence of small amounts of 5-formyltetrahydrofolic acid and thymidine had similar effects. They concluded that a mechanism invoking thymidylate synthetase as a rate-limiting step was insufficient to explain methotrexate cytotoxicity, since purines, as well as thymidine, produce the effect, and both cannot be rate-limiting. These authors proposed (1) that the availability of dUMP consequent to feedback inhibition determines the rate of the thymidylate synthetase reaction *in vivo*, and (2) that direct sites of inhibition other than dihydrofolate reductase need not be invoked to understand the mechanism of methotrexate cytotoxicity. There is no doubt that all of these studies are complicated by the variation in enzyme content in various types of cells, and by the dynamics of cell influx and efflux of inhibitors under varying conditions (see the next section).

The transport of methotrexate and other dihydrofolate reductase inhibitors: selectivity effects

The clinical response to methotrexate is now established to be closely related to cellular uptake of the drug (Kessel *et al.*, 1965; Bertino, 1979; Fernandes *et al.*, 1981). Transport of this polar ionised molecule is accomplished by a carrier mechanism shared by the naturally occurring folates, including 5-methyl- and 5-formyltetrahydrofolate (Goldman *et al.*, 1968; Goldman, 1971). There is a marked temperature dependence on the unidirectional influx, which demonstrates Michaelis–Menten kinetics, with competitive inhibition by analogues. 5-Formyltetrahydrofolic acid, for example, competitively inhibits methotrexate influx, leading to the conclusion that it may be bound to the carrier like methotrexate, which is not the case with dihydrofolate reductase, where 5-formyltetrahydrofolic acid is bound poorly. Methotrexate is less well transported than is aminopterin (*1a*) into normal murine cells, but comparably well into murine tumour cells (Sirotnak and Donsbach, 1976, Sirotnak *et al.*, 1979). Methotrexate thus shows a relative selectivity for tumour tissues which originates from differences in transport rather than in intrinsic inhibitory potency against dihydrofolate reductase. Regulation of transport seems to depend on the intracellular level of cyclic-3′, 5′-adenosine monophosphate (cAMP) (Henderson) *et al.*, 1978).

The importance of the carrier protein for the delivery of methotrexate to target tissues, and the intriguing possibility for achieving selectivity with dihydrofolate reductase inhibitors of tumour cells via a second membrane-bound protein, have promoted extensive efforts to isolate and purify the carrier. Henderson and co-workers (1977) have isolated and purified a protein of this type from *Lactobacillus casei* cells which has broad specificity for folates. A parallel programme is aimed at isolation of the corresponding binding protein from L1210 cells (Huennekens *et al.*, 1981). This has required the development of procedures for covalently attaching a marker ligand to the protein prior to its release from the membrane. Photoaffinity labelling with 8-azido-AMP, or better still interaction with a carbodiimide-activated methotrexate derivative has proved effective in irreversibly inhibiting the methotrexate transport system (Henderson *et al.*, 1979; 1980*b*); a binding component with high affinity for 5-methyltetrahydrofolic acid has been found on the external surface of L1210 cells (Henderson *et al.*, 1980*a*).

In contrast with the 'classical' folic acid antagonists, small-molecule inhibitors such as pyrimethamine, metoprine or 2,4-diamino-5-adamantyl-6-methylpyrimidine are transported by passive diffusion (Wood *et al.*, 1961; Cavallito *et al.*, 1978; Greco and Hakala, 1980). Such compounds are normally quite lipophilic, and are therefore able to cross the blood/brain barrier and reach target tissues not affected by methotrexate and analogues.

Biochemical transformations of methotrexate

Methotrexate is rapidly converted into poly-gamma-glutamyl derivatives intracellularly, and these metabolites have been reported to have a very high affinity for dihydrofolate reductase and to replace previously bound methotrexate (Fry *et al.*, 1982). These peptides are retained in the cell more effectively than is methotrexate and build up a chemical gradient (inside:outside) over 200:1. This recent finding may have important implications for the activity of methotrexate as an anticancer agent. Fry *et al.* (1982) postulate that the capacity of the parent methotrexate to achieve sustained suppression of dihydrofolate reductase activity in the cell is limited, since only a small fraction of the enzyme is required to maintain tetrahydrofolic acid synthesis, and the high levels of dihydrofolate that then accumulate compete effectively with methotrexate. However, if the poly-glutamates of methotrexate build up in the cell, prolonged suppression of dihydrofolate reductase activity may result.

The metabolism of methotrexate in various species has been studied extensively, and differs from species to species. As a generalization, methotrexate can be cleaved to a pteroate and glutamate in rodents and man (Henderson *et al.*, 1965*a,b*; Zaharko *et al.*, 1969; Zaharko and Oliverio, 1970). It can also undergo oxidation by means of liver aldehyde oxidase to give 7-hydroxymethotrexate in man, rabbits, and guinea pigs (Henderson *et al.*, 1965*b*; Johns *et al.*, 1965; Johns and Loo, 1967). The 7-hydroxy metabolite is less than one hundredth as active as methotrexate (Johns and Valerino, 1971).

'Rescue' techniques

The use of 5-formyltetrahydrofolic acid (citrovorum factor, CF, leucovorin) to protect the host against the toxicity of methotrexate or aminopterin by replenishing pools of reduced folic acid ('leucovorin rescue') has been employed since the early 1950s (Goldin *et al.*, 1953). The use of leucovorin permitted the administration of larger doses of methotrexate without many of the side effects (Vogler and Jacobs, 1971; Jacobs and Santicky, 1978). The delay of leucovorin administration after treatment with methotrexate has been found to enhance the drug's antineoplastic effect and minimise toxicity (Goldin *et al.*, 1954; 1955; 1966; Sandberg and Goldin, 1970). After 12–24 hours, the action of methotrexate on leukaemic cells was essentially complete, but leucovorin was well able to protect the host against toxicity. When leucovorin is administered orally, it is almost quantitatively converted to 5-methyltetrahydrofolic acid (Nixon and Bertino, 1972).

Combinations of methotrexate with thymidine, or with thymidine-inosine together, have also been studied for the rescue of normal cells. The latter combination might possibly be more effective than leucovorin in providing selective protection (Nederbragt *et al.*, 1981).

Carboxypeptidase G_1, which hydrolyses the Glu moiety from folic acid, and therefore can serve as an antineoplastic agent, has also been used for rescue from methotrexate toxicity, since it can also cleave methotrexate (Chabner *et al.*, 1972*a,b*).

Resistance

The most common cause of resistance to methotrexate and other dihydrofolate reductase inhibitors is the development of elevated levels of the enzyme. This has been found to be associated with an increase in messenger ribonucleic acid (mRNA) levels for dihydrofolate reductase (Chang and Littlefield, 1976; Kellems *et al.*, 1976). These events result from a proportional increased number of gene copies for the enzyme (Alt *et al.*, 1978). Such gene amplification, and natural selection of such mutants might be circumvented by the avoidance of situations in which cells are exposed to relatively low levels of drug for long periods of time, using instead brief schedules with maximally tolerated doses of drug, and by alternately using drugs with other modes of action.

A second cause of resistance results from an impairment in the methotrexate transport mechanism, which causes decreased permeability (Sirotnak *et al.*, 1968; Lindquist *et al.*, 1978; Hakala, 1965). The use of 'non-classical' inhibitors, for example those which enter the cell by passive diffusion, should obviate the problem (Hill *et al.*, 1975; 1978; Hill and Price, 1977; Sedwick *et al.*, 1979; Grivsky *et al.*, 1980).

A third, and less likely, cause of resistance can derive from alterations in the dihydrofolate reductase molecule (Blumenthal and Greenberg, 1970; Goldie *et al.*, 1981). With the present detailed structural information available concerning dihydrofolate reductase and given the altered amino acid sequence, it should be possible in the future to design inhibitors which are tailor-made for such a situation.

STRUCTURAL STUDIES

Dihydrofolate reductase is the target of several drugs used to combat bacteria, protozoa and cancer and there has been much interest in establishing how some of these drugs are selective inhibitors of certain dihydrofolate reductases. This has led in the last decade to structural studies on the enzyme from a variety of sources (see reviews by Blakley, 1981; Jardetzky and Roberts, 1982). The amino acid sequences of several vertebrate, bacterial and plasmid dihydrofolate reductases are now known (summarised by Stone, 1981) and the elucidation of the sequence of human dihydrofolate reductase is in progress in several laboratories. The crystallographic investigation of dihydrofolate reductase has been conducted by two groups, Matthews *et al.* at La Jolla, USA and Beddell *et al.* at Beckenham, UK. The crystal structure of the binary complex of methotrexate with the *Escherichia coli* enzyme (Matthews *et al.* 1977) was followed shortly thereafter by that of the ternary complex of methotrexate, reduced nicotinamide adenine dinucleotide phosphate (NADPH) and *L. casei* dihydrofolate reductase (Matthews *et al.*, 1978). Further crystallographic work with the *E. coli* enzyme has shown how the antibacterial drug trimethoprim (Baker *et al.*, 1981*a*) and certain analogues (Baker *et al.*, 1981*b*) interact in the binary complex. This group has work in progress on the trimethoprim-NADPH-*E. coli* dihydrofolate reductase ternary complex. Crystallographic studies on the vertebrate enzymes were initiated after those on the bacterial enzymes. The structure of the ternary complex between a dihydrotriazine inhibitor, NADPH and chicken liver dihydrofolate reductase (Volz *et al.*, 1982) and of certain other inhibitor complexes (Matthews, 1981) are reported, and crystallographic work is in progress on the mouse enzyme in the UK.

Of particular interest in the present context would be the structure of the complex between methotrexate and vertebrate enzyme, but this structure is not yet established. However, the homology between bacterial and vertebrate dihydrofolate reductases in both amino acid sequence and in three-dimensional folding enables us to postulate a mode of binding for methotrexate and a relationship between the binding of this inhibitor and of the structurally related substrate *2*.

Enzyme architecture

The amino acid sequence of chicken dihydrofolate reductase is shown in figure 2. The single peptide chain of 189 residues contains a single cysteine at position 11 and therefore there can be no internal disulphide. The amino acid sequence is closely homologous (72–89%) with those of other vertebrate DHFRs sequenced to date (bovine liver, porcine liver, L1210) as detailed in figure 2. Significant homology also exists between the dihydrofolate reductases of chickens and those of bacteria of known amino acid sequence isolated from two strains of *E. coli* (RT500, MB1428), *Streptococcus faecium* and *L. casei*. Figure 2 shows those residues which are strictly conserved between dihydrofolate reductases from chicken and bacteria in the amino acid sequence alignment of Volz *et al.* (1982) which is based on the three-dimensional superposition of the structures of the enzyme from *E. coli*, *L. casei* and chicken. The overall architecture of chicken dihydrofolate reductase is shown in figure 3. The chain participates in the formation of an eight-stranded β-sheet, with one strand, H, inserted between F and G and antiparallel. Four α-helices, denoted B, C, E and F, are involved in the interstrand connections which lead to strands, B, C, E and F respectively. There is a crescent-shaped cleft backed by strands A, B, C, E and F and the carboxyl end of αF, and flanked on one side by the B helix and the preceding connecting strand from βA and on the other side by the C helix and subsequent strand connecting to βC. The cofactor occupies one half of this cleft, in an extended conformation, with the adenine outermost, and reduced nicotinamide near the middle of the cleft. The diamino inhibitors which have been studied are found to bind close to the nicotinamide in the other half of the cleft.

Dihydrotriazine binding

The dihydrotriazine inhibitor used for the crystallographic study, 2,4-diamino-5,

Molecular Aspects of Anti-cancer Drug Action

Figure 2 Comparison of dihydrofolate reductase sequences. A, chicken liver (Kumar *et al.*, 1980); B, mouse L1210 (Stone *et al.*, 1979); C, porcine liver (Smith *et al.*, 1979); D, bovine liver (Lai *et al.*, 1979); E, *Escherichia coli* RT500 (Stone *et al.*, 1977, Baccanari *et al.*, 1981); F, *Escherichia coli* MB1428 (Bennett *et al.*, 1978); G, *Streptococcus faecium* (Gleisner *et al.*, 1974); H, *Lactobacillus casei* (Bitar *et al.*, 1977). The residue numbering shown is for chicken liver (top), *E. coli* (middle) and *L. casei* enzymes and the alignment of the sequences, together with insertion of gaps where appropriate, follows that of Volz *et al.* (1982) and is based on three-dimensional structural superposition of the chicken, *E. coli* and *L. casei* enzymes. Identity with the corresponding residue in the chicken liver enzyme is shown by means of a dot. The high degree of homology between vertebrate enzymes (A–D) is clearly evident and also the conservation for all enzymes shown at positions 9, 16–17, 23–24, 27, 34, 38, 53, 56, 67, 75, 116–117, 136, 143 and 145 (chicken numbering).

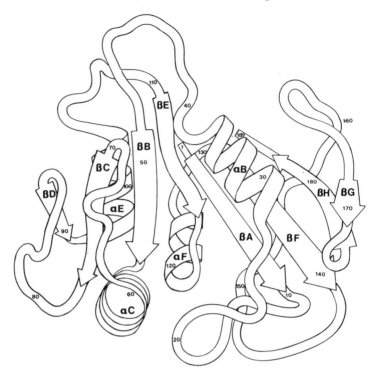

Figure 3 Schematic illustration of the folding of chicken liver dihydrofolate reductase. The overall folding of the polypeptide chain is dominated by an eight-stranded β-sheet composed of seven parallel strands and a single antiparallel strand at the C-terminus. The sheet, viewed along the strands, shows the usual right-handed twist. The molecule contains four helical regions. Cofactor and inhibitors bind in a cleft some 15Å wide which cuts across one face of the molecule and gives the structure a bi-lobed appearance.

6-dihydro-6,6-dimethyl-5-(4'-methoxyphenyl)-*s*-triazine (MPT) has a K_i of 3.7×10^{-7} M for chicken liver enzyme (Volz *et al.*, 1982). Figure 4 shows the binding of this inhibitor. The bound MPT has its principal inter-ring C–N torsion angle $\theta = 88°$ close to that observed in the X-ray structure ($\theta = 82°$) for the *p*-chloro analogue *(4)* (Hunt *et al.*, 1980) and to the minimum energy value ($\theta = 70°$) calculated by consistent force-field methods (Hopfinger, 1980). MPT

4

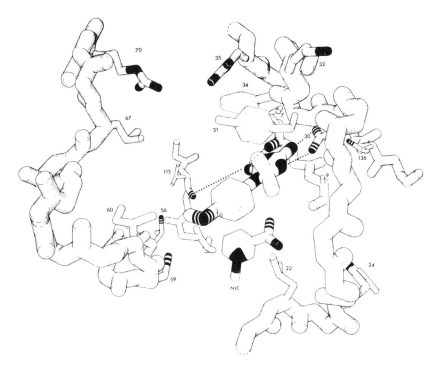

Figure 4 Schematic illustration of the active site of chicken liver dihydrofolate reductase showing the binding of a dihydrotriazine inhibitor, and the nicotinamide moiety of the reduced cofactor (NADPH). Selected nitrogen and oxygen atoms are emphasised by solid or striped shading respectively. The dotted lines show hydrogen bonding of the N_1 and the 2-amino group in the triazine with the carboxyl group in Glu-30, and also the hydrogen bonding of the 4-amino group to the carboxyl oxygens of residues 7 and 115.

binds with the side chain carboxyl of Glu-30 hydrogen bonded to the N_1 and 2-amino groups of the inhibitor. Modest (1956) reported that the pK_a for the p-chloro analogue of MPT is 11.2. Therefore MPT itself will have a high pK_a and will be protonated in the complex with enzyme. Thus the interaction between the carboxyl and N_1 comprises ionic attraction between the carboxylate anion and a protonated N_1. Lining the binding pocket and making hydrophobic contacts with the dihydrotriazine ring or its attached methyl groups are the side chain of Ile-7, the peptide bond between Val-8 and Ala-9, the side chains of Ala-9, Leu-22, Tyr-31, Phe-34, Val-115 and the carboxamide of NADPH. The hydroxyl of Thr-136 hydrogen-bonds to a carboxylate oxygen of Glu-30 and to the 2-amino group of the inhibitor via an intervening water molecule. Hydrogen bonds are donated by the 4-amino groups of the dihydrotriazine to the carbonyls of Ile-7 and Val-115.

The inhibitor's methoxyphenyl group makes van der Waals contacts with the side chains of Leu-22, Phe-34, Thr-56, Ser-59, Ile-60, Val-115 and the nicotinamide ring of NADPH. In conclusion, substantial hydrophobic, and also ionic and hydrogen-bonding interactions contribute to the tight binding of the dihydrotriazine inhibitor.

Methotrexate binding

The structure of the corresponding methotrexate complex with the chicken liver enzyme is not yet known, but extrapolation is possible from the methotrexate–bacterial enzyme complexes because of the close structural homology. Aside from certain restricted regions, the superposition of the main peptide chain of the chicken enzyme with that of the *L. casei* or *E. coli* enzyme is very close, and so are the positions and interactions involving the cofactor present in the *L. casei* and chicken complexes.

Figure 5 shows methotrexate binding to the *L. casei* enzyme determined by Matthews *et al.* (1978). The methotrexate in the *L. casei* structure is bound in an extended conformation, with the diaminopyrimidine moiety of the pteridine similarly placed to the diaminotriazine part of MPT. Cocco *et al.* (1981) have shown that the effective pK_a of enzyme-bound methotrexate is greater than 10, so again N_1 will be protonated. The corresponding pK_a for unbound methotrexate is 5.7. The pK_a increase accompanying the binding of methotrexate indicates that compound protonated at N_1 binds to dihydrofolate reductase at least 2×10^4 times more tightly than does unprotonated compound; the interaction between N_1 on methotrexate and dihydrofolate reductase can therefore contribute significantly to the overall binding energy. Other similar contacts are made between the protein and pyrimidine, but in the *L. casei* structure Phe-30 is more inclined to the pyrimidine than the corresponding residue in the MPT complex. The pteridine binding pocket is lined in addition by side chains from Leu-4, Ala-6, Leu-19, Leu-27 and the main chain atoms of Ala-97. The backbone at Trp-5 and Ala-6 approaches closely the pteridine ring at N_1, C_2, the 2-amino group and N_3. The 2- and 4-amino groups hydrogen bond with $O\gamma$ of Thr-116 (via a water molecule) and with the backbone carbonyls of Leu-4 and Ala-97 respectively. The side chain of Leu-19 is tucked up under the pyrazine part of the pteridine which is also close to the nicotinamide ring of NADPH. This latter interaction may in part produce the 670-fold co-operative increase in binding between methotrexate and NADPH (Birdsall *et al.*, 1980; 1981). The side chain of Asp-26 closely approaches N_1 and the 2-amino group. The aromatic ring of the *p*-aminobenzoyl portion of methotrexate interacts with the side chains of Leu-27, Phe-49 and Leu-54. The aromatic side chain of Phe-49 is nearly parallel to and in van der Waals contact with the benzene ring in the inhibitor. The α-carbon of Ser-48 is in contact with the methyl group of N_{10} in methotrexate. The glutamate portion of the inhibitor is hydrogen-bonded between the α-carboxyl and the guanidinium group of Arg-57 and the γ-carboxyl interacts with $N\epsilon_2$ of His-28. A similar binding of methotrexate to the *E. coli*

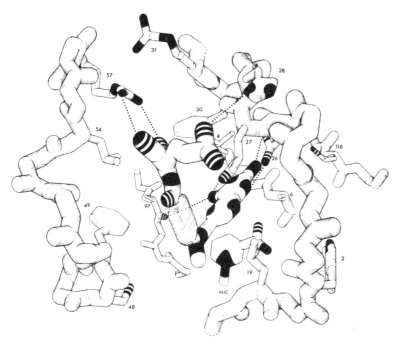

Figure 5 Schematic illustration of the active site of *L. casei* dihydrofolate reductase showing the binding methotrexate and NADPH. Details of representation are as described in the legend to Figure 4.

enzyme (no cofactor being present) has been found, with a counterpart for every interaction between *L. casei* enzyme and methotrexate in the *E. coli* enzyme and methotrexate complex, with but two small differences. The counterpart of Leu-19, that is, Met-20, in *E. coli* is not orientated to provide contact with the pteridine ring. There is no basic residue homologous to His-28 and the γ-carboxyl is hydrogen-bonded to a water molecule which in turn is hydrogen-bonded to Lys-32.

The search function of Rossman and Argos (1976) for alignment of homologous protein structures has been used (D. K. Stammers, unpublished data) to superimpose separately the molecules of the *L. casei* dihydrofolate reductase-methotrexate-NADPH complex and the two observed molecules of the *E. coli* dihydrofolate reductase-methotrexate complex with the chicken dihydrofolate reductase structure, so that it may then be observed how the methotrexate itself *might* interact with the chicken reductase. In the putative binding mode for methotrexate in chicken enzyme, figure 6, the interactions found are analogous to those seen in the observed complex with *L. casei* dihydrofolate reductase, and indeed a very similar picture is obtained if instead the methotrexate is aligned in

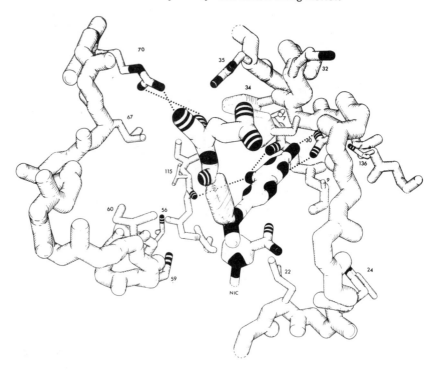

Figure 6 Schematic illustration of the active site of chicken liver dihydrofolate reductase and a postulated fit for methotrexate derived by superposition. The atomic coordinates of the *L. casei* dihydrofolate reductase–methotrexate complex were superimposed on the coordinates of the chicken dihydrofolate reductase so as to optimise the alignment of structurally homologous segments of protein chain. From this resultant fit, the coordinates for the methotrexate were combined with these for the chicken dihydrofolate reductase. The comparison of this illustration with the previous reveals that the interaction between methotrexate and chicken dihydrofolate reductase in this postulated fit is generally similar to that observed between methotrexate and *L. casei* dihydrofolate reductase. The impossibly close contacts between the inhibitor and the side-chains of Tyr-31 and Gln-35 might readily be relieved by movements of the side chains. Substantial movement of the Tyr-31 side-chain is induced by trimethoprim binding (Matthews, 1981).

the chicken enzyme by superposition of either of the two molecules of the *E. coli* dihydrofolate reductase–methotrexate complex with the chicken enzyme.

The pteridine binding pocket is lined by Ile-7, Ala-9, Leu-22, Leu-27 and the main chain atoms of Val-115, these being closely similar interactions to those seen for *L. casei* dihydrofolate reductase. N_1, C_2, N_3 and the 2-amino group approach the main chain at Val-8 and Ala-9. The 2- and 4-amino groups hydrogen bond with Oγ of Thr-136 (via a water molecule) and the carbonyls of Ile-7 and Val-115 respectively. The side chain of Leu-22 and the nicotinamide ring of NADPH are close to the pyrazine ring. The side chain of Glu-30 bonds to N_1 and the 2-amino group. The aromatic ring of the *p*-aminobenzoyl portion of methotrexate interacts with the side chains of Tyr-31, Ile-60 and Leu-54. Tyr-31 replaces leucine and the side chain conformation is different and causes close contacts with the glutamate of methotrexate. However, it is known (Matthews, 1981) that in some complexes between chicken dihydrofolate reductase and diamino inhibitors (for example, trimethoprim) the tyrosine side chain swings downwards, to an orientation closer to that adopted by Leu-27 in the *L. casei* enzyme. The side chain of Ile-60 occupies a similar position to that of the Phe-49 in *L. casei*. Ser-59 is in contact with the methyl at N_{10}. The glutamate α-carboxyl interacts with the side chain of Arg-70 and the γ-carboxyl could potentially interact with the side chain of Lys-32. The close contacts between the glutamate of methotrexate and the side chain of Gln-35 could be relieved by rotation upwards of this side chain, to an orientation closer to that adopted by the homologous residue Arg-31 in *L. casei* dihydrofolate reductase.

There is therefore in the postulated interactions with methotrexate a counterpart in the chicken enzyme for the interatomic contacts described with the *L. casei* enzyme.

Relationship between substrates and inhibitors

Crystallographic information is not yet available on the complex between dihydrofolate reductase and substrate or product (folic acid, dihydrofolate or tetrahydrofolate). Methotrexate itself is close in structure to folic acid and indeed was originally made as a substrate analogue. Therefore, one might hope to use the observed binding of methotrexate to dihydrofolate reductase as a model for substrate binding. The crystallographically observed structure of the ternary complex predicts, assuming the pteridine ring of dihydrofolate is bound in the same orientation as that of methotrexate, that the absolute configuration at C_6 in the product would be R according to the Cahn–Ingold–Prelog convention (Cahn *et al.*, 1966). In fact, however, the X-ray-crystallographic study of the absolute configuration of the natural diastereoisomer of 5, 10-methenyltetrahydrofolic acid (Fontecilla-Camps and Bugg, 1979) has shown that the configuration at C_6 in tetrahydrofolate is S. Thus the orientation of the dihydropteridine ring in the productive enzyme–substrate complex must differ from that observed in enzyme-methotrexate complexes.

This is also true for the reduction of folate by *L. casei* DHFR (Charlton *et al.*, 1979). Folate was reduced enzymatically with cofactor deuterated in the 4-pro-R position, which is on the A-face of the reduced nicotinamide ring. The ^{1}H NMR spectrum for the product lacked the multiplets corresponding in tetrahydrofolic acid itself to ^{1}H at position 6 and *cis*-^{1}H at position 7, and the product was therefore [*cis*-6,7-[2]H_2]-5,6,7,8-tetrahydrofolic acid. The ^{2}H which is transferred in both reductions, folic acid to dihydrofolic acid and dihydrofolic acid to tetrahydrofolic acid, has therefore originated from the A-face of the reduced nicotinamide ring and has been transferred to the same face of the bicyclic system.

An orientation that appears to be required for catalysis can be achieved by rotations of the pteridine around the C_6-C_9, C_9-N_{10} and $N_{10}-C_{11}$ bonds (D. A. Matthews, personal communication). The plane of the pteridine in the substrate is then inverted relative to that observed for methotrexate. It follows, therefore, that in the design of a drug closely similar to the natural substrate in general features, the small differences in chemical structure may alter the physico-chemical properties in a way which significantly modifies the way in which the compound binds to its receptor. Indeed, in this instance methotrexate binds several orders of magnitude more strongly to enzyme than does substrate and the increase of basicity at N_1 contributes significantly to the tight binding in the alternative conformation adopted by methotrexate.

STRUCTURE–ACTIVITY RELATIONSHIPS

Direct analogues of folic acid

MTX and aminopterin

The announcement of the structure of the vitamin folic acid (Angier *et al.*, 1946) was followed just one year later by the discovery of a very potent inhibitor, aminopterin, (Seeger *et al.*, 1947). Its 10-methyl derivative, methotrexate described shortly thereafter (Seeger *et al.*, 1949) remains today the single leading antifol for cancer chemotherapy, with no serious rivals. Many reviewers have found cause to comment on this fact, considering that the number of analogues with potent dihydrofolate reductase inhibitory activity which have been synthesised since then now runs into many thousands. However, in the intervening years a great deal has been learned about the specificity of methotrexate for neoplasms, and about how to use the drug most effectively (Bertino, 1963, 1975, 1979).

What perhaps has not been clear, as a consequence of the tremendous literature explosion in recent years, is that the basic structure–activity relationships among folic acid inhibitors had been very well worked out by the original team (Angier *et al.*, 1946, and colleagues), despite the fact that the nature of the enzyme inhibition was not understood until considerably later. By 1949 nearly 200 analogues of folic acid had been prepared in an analytically pure state (Smith, 1949) and the list grew rapidly in subsequent years (Williams, 1952). Many modifications of the pyrimidine and pyrazine rings, of the bridge atoms, the benzene moiety, and of the glutamate fragment had been made. At a 1951 conference which reviewed these antagonists for the treatment of leukaemia, Dr Sidney Farber stated, "May I bring what might be a long discussion to an abrupt end by saying that no results have been achieved in children with acute leukaemia with any of these compounds that are any better than those obtained with aminopterin and methotrexate" (Farber, 1952). Examples of the types of variations studied, both then and in more recent years, will be considered below in the light of current knowledge.

The complexity of the interaction of methotrexate with living systems which was noted early in this review makes difficult a structure–activity comparison among dihydrofolate reductase inhibitors which is based on reductase inhibition alone. In making such comparisons, the question should be raised as to what qualities are desirable in new dihydrofolate reductase inhibitors. The problems of resistance, toxicity, inability to reach the site of action – for example, the central nervous system (CNS) – the need for improved pharmacokinetics, inactivity against certain neoplasms and other factors, all dictate that the search continue.

Variations in the pteridine moiety

Retention of the 2-amino-4-oxo substitution pattern in the pteridine ring of folic acid analogues, coupled with modification in other parts of the molecule, produces variable biological results which often do not involve inhibition of dihydrofolate reductase. For example 9- and 10-methylfolic acid (Hultquist *et al.*, 1949; Cosulich and Smith, 1948) are substrates, rather than inhibitors, of dihydrofolate reductase (Bertino *et al.*, 1965) but the reduction products are inhibitors of thymidylate synthetase. N_{10}-formylfolic acid, on the other hand, is a potent inhibitor of dihydrofolate reductase (Bertino *et al.*, 1965). Dihydrohomofolic acid, which contains an extra methylene group in the bridge atoms, is also a substrate for dihydrofolate reductase, which forms a potent thymidylate synthetase inhibitor upon reduction (Goodman *et al.*, 1964). This section will be mainly concerned with diamino derivatives of folic acid, which presumably all bind to dihydrofolate reductase with a protonated ring nitrogen atom interacting with an active site glutamic acid residue, and with the hetero rings interacting as an approximately 180° rotamer of dihydrofolic acid.

Replacement of any of the hydrogen atoms of the 2- or 4-amino groups in *1a* or *1b* by methyl or other alkyl substituents leads to virtually complete loss of activity (Roth *et al.*, 1950, 1951; Johns *et al.*, 1964). Such substitution would lead to loss of important hydrogen-bonds to dihydrofolate reductase, and to a poor fit in the active site due to crowding. A 2,4-dioxo derivative was likewise found inactive (Oleson, 1950; Bertino *et al.*, 1965).

7-Methyl analogues of *1a* and *1b* were synthesised in the belief that they might prevent biological oxidation to the inactive 7-hydroxy analogues (Farquhar *et al.*, 1972). Although very active as dihydrofolate reductase inhibitors, both compounds were inactive against leukaemia L1210. The 7-aza analogue of *1a* was inactive against pigeon liver dihydrofolate reductase (Temple *et al.*, 1975), possibly as a result of lowered basicity due to electron withdrawal by the added ring nitrogen (Roth and Strelitz, 1969).

Important information concerning the requirements for dihydrofolate reductase binding was obtained with 1- and 3-deazamethotrexate (Temple *et al.*, 1971). The former had only about 1/400 the activity of methotrexate against dihydrofolate reductase. This compound would be expected to bind to the enzyme as the folic acid rotamer, since its protonation site will necessarily be at N_3, with N_1 missing (the resultant pyridine ring is considerably more basic than the pyrazine ring). With a different combination of proton donor and acceptor atoms in the pyrimidine ring than with folic acid, it probably would not be expected to act as a substrate, however. The 3-deaza analogue of methotrexate experienced a fifteen-fold loss of its original potency with substitution of CH for N. This is to be compared with a 300-fold loss in activity by a small molecule analogue, 3-deazatrimethoprim compared to trimethoprim, with *E. coli* dihydro-folate reductase (Rauckman and Roth, 1980). The 3-deazamethotrexate analogues should bind to dihydrofolate reductase as the methotrexate rotamer, and may exhibit the increased thermal vibration at the 3-position due to crowding by the CH- function that is observed with 3-deazatrimethoprim (Baker *et al.*, 1981*b*). The reason for the smaller loss of inhibitory potency for the metho-trexate analogue relative to the trimethoprim analogue is not known.

2, 6-Diaminopurine analogues of *1a* and *1b* were found to be devoid of activity for L1210 leukaemia (Weinstock *et al.*, 1968; 1970).

Replacement of the 5- and 8-nitrogen atoms in methotrexate and analogues by carbon resulted in quinazoline analogues of high potency against dihydrofolate reductase (Davoll and Johnson, 1970). Two compounds chosen for further study were methasquin and chlorasquin (*5a* and *5b*), both of which contain small lipo-philic 5-substituents, and Asp rather than Glu as the amino acid moiety (Hutchison, 1968; Etcubanas *et al.*, 1972). These compounds were not only potent inhibitors of dihydrofolate reductase from leukaemia and neuroblastoma cells, but also inhibited thymidylate synthetase (Scanlon *et al.*, 1979), although not as efficiently as the 4-oxo analogues (Bird *et al.*, 1970; Scanlon *et al.*, 1979). Furthermore, they retained antileukaemic activity against several methotrexate-resistant L1210 sublines. Methasquin was the most effective of several quinazolines in leukaemic mice; however, the compounds were found to be quite toxic in several animal models, and not well absorbed from the intestinal tract (Philips *et al.*, 1971; Hutchison *et al.*, 1971).

(a) R = CH_3, methasquin
(b) R = Cl, chlorasquin

5

Small hydrophobic substituents at the 5-position on both quinazolines and pyrido (2, 3-d) pyrimidine analogues have been found to have dramatic effects in many cases in enhancing inhibitory activity towards dihydrofolate reductase

(Hurlbert *et al.*, 1968; Elslager and Davoll, 1974). Models of dihydrolate reductase show that a small substituent of this type will fit very well into a small hydrophobic pocket in the enzyme cavity.

Variations in the bridge atoms

The 10-nitrogen of *1a* has been replaced by carbon, sulphur and oxygen, with retention of high activity in the first two cases (DeGraw *et al.*, 1974; Nair *et al.*, 1975); the oxo analogue, although less active, showed excellent inhibition of dichloromethotrexate-resistant *L. casei* dihydrofolic acid reductase (Nair and Campbell, 1976).

Exchange of the C_9 and N_{10} atoms usually lowers activity to some extent (Nair *et al.*, 1974; Elslager and Davoll, 1974). Insertion of an additional CH_2 linkage between the amino group and the benzene ring decreases activity (Montgomery *et al.*, 1979). Other 3-atom bridges were found to give poor activity (Elslager and Davoll, 1974). However, with dihydrohomofolic acid, it will be recalled that the substance was a substrate for the enzyme, so with this rotamer at least, a proper fit to the enzyme can be achieved with the extra atoms in the bridge.

Increasing the bulk of N_{10} substitution beyond ethyl on methotrexate analogues also lowers activity (Montgomery *et al.*, 1979). Sirotnak *et al.* (1981) have reported that appropriate modification of the bridge area may have more important ramifications for transport than for enzyme binding.

Substitution on the benzene ring

Halogenation of the benzene ring of methotrexate to give 3'- or 3', 5'-dihalo derivatives (Cosulich *et al.*, 1951; 1953; Angier and Curran, 1959) results in highly effective anticancer agents, with greater lipophilicity than methotrexate. 3'-Bromo-5'-chloro- and 3', 5'-dichloromethotrexate have produced a fairly high number of long-term survivors (Goldin *et al.*, 1959). However, with lymphosarcoma and Hodgkin's disease the tumour regressions were generally short-lasting (Frei *et al.*, 1965). The compounds are less toxic than is methotrexate (Motycka, 1971) possibly due to greater ease of metabolism to the inactive 7-hydroxy derivatives (Misra *et al.*, 1963; Loo and Adamson, 1965).

Modification of the glutamate moiety

A very large number of variations have been made in this portion of the folic acid and methotrexate molecules, in an exercise which has proved largely futile. The Glu of folic acid and aminopterin has been replaced by most of the common amino acid residues (Hutchings *et al.*, 1947; Wright *et al.*, 1949; Mead *et al.*, 1965; Montgomery *et al.*, 1979), and in addition, many combinations of di- and tripeptides were synthesised (Boothe *et al.*, 1949; Semb *et al.*, 1949; Mowat *et al.*, 1949; Oleson, 1950; Williams, 1952). Most of these changes, other than Asp, have led to a considerable loss of activity *in vitro* and *in vivo*, probably as a consequence of poor transport. It would appear that a dicarboxylic amino acid residue is required for active transport of such molecules.

A vast number of potentially labile 'prodrug' derivatives of methotrexate have been prepared, with the object of improving membrane transport by increasing lipid solubility. These include mono- and diesters, amides, hydrazides and similar derivatives (Rosowsky 1973; Rosowsky *et al.*, 1977, 1978; Sirotnak *et al.*, 1979; Piper and Montgomery, 1979). The results have been disappointing here as well; most derivatives are less active than the parent amino acid derivatives, or only slightly more active at high doses.

Small-molecule analogues

Simple pyrimidines

Probably the reason why small molecule analogues of folic acid did not receive much attention as anticancer agents until the past decade is that testing methods, including pharmacokinetic studies, were not sufficiently well advanced, and neither were techniques for combating toxicity. It certainly was well recognised that more lipophilic compounds than methotrexate were required in order to cross the blood/brain barrier.

2, 4-Diamino-5-(3, 4-dichlorophenyl)-6-methylpyrimidine (*6a*, DDMP or metoprine) and its 6-ethyl analogue (*6b*), had been reported as antimalarial agents, along with pyrimethamine (*7*) (Russell and Hitchings, 1951). As a consequence of the cytotoxicity noted for *6a* and *6b*, the compounds were submitted for tumour screening, with responses to the effect that "no final conclusions can be drawn – all compounds give identical results – a broad diversity of screens is needed" (Gellhorn and Hirschberg, 1955; Sugiura, 1956). However, one 1953 report which has been largely neglected stated that combined therapy of leukaemia in mice with methotrexate and metoprine retarded local tumour growth and increased survival time at dose levels that were ineffective for each drug given alone (Nadel and Greenberg, 1953). Whereas this does not represent synergy, it nevertheless seems clear that if the lipophilic small molecule enters the cell by passive diffusion (Wood *et al.*, 1961), whereas methotrexate enters by carrier transport and cannot reach all target cells, a combination might well prove more effective in knocking out all malignant cells.

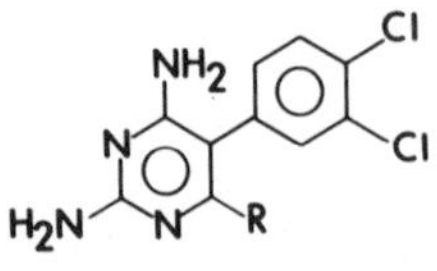

(*a*) R = CH₃, metoprine
(*b*) R = C₂H₅, etoprine

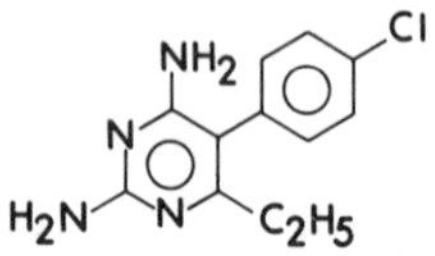

pyrimethamine

6

7

Compounds *6a* and *6b* were resurrected in about 1970, with the hope that these lipophilic molecules might cross the blood/brain barrier and be useful for neoplasms with cerebral involvement. Both compounds actually readily enter cerebrospinal fluid and penetrate into animal brain tumours (Nichol, 1977; Denlinger *et al.*, 1976; Nichol *et al.*, 1977). Clinical studies of *6a* demonstrated that its potential was limited by its long half-life (10 days in man) (Cavallito *et al.*, 1978) and its limited efficacy at tolerated doses. It is bound to human plasma protein at 87% or higher level at therapeutic concentrations. Furthermore, it produces neurological, cutaneous and gastrointestinal toxicities, as well as haematological effects (Price *et al.*, 1975; Price and Hill, 1976; Miller *et al.*, 1976; Alberto *et al.*, 1978). Its inhibition of histamine *N*-methyltransferase may conceivably cause some of the neurological side effects (Duch *et al.*, 1980a).

Among a number of 2,4-diaminopyrimidines containing adamantyl substituents, derivatives with the adamantyl function attached directly to the pyrimidine 5-position, and with methyl or ethyl at the 6-position (*8a*, *8b*) were good inhibitors of mammalian dihydrofolate reductase (100 times less active than methotrexate) (Jonak *et al.*, 1970, Ho *et al.*, 1972); however, they were more inhibitory than methotrexate in cell culture, indicating that the very lipophilic 5-substituent greatly facilitated passage of pyrimidines through the cell plasma membrane. These derivatives are very similar in structure to *6a* and *6b*, and presumably occupy the same position on dihydrofolate reductase, as a consequence of their relative rigidity. They also have similar antitumour effects (Zakrzewski *et al.*, 1978). Compound *8a* is currently in clinical trials. The compound is extensively metabolized to polar and non-polar metabolites; neurological effects have been noted (Zakrzewski *et al.*, 1982). Possibly improved regimens or use of drug combinations may obviate such effects. The crystal structure of one of these inhibitors has been carefully studied and analysed in regard to dihydrofolate reductase binding, and compared to related compounds (Cody and Zakrzewski, 1982).

(*a*) R = CH$_3$, DAMP
(*b*) R = C$_2$H$_5$, DAEP

8

Dihydrotriazines

The late B. R. Baker synthesised thousands of compounds for cancer chemotherapy, many of which were dihydrofolate reductase inhibitors. His rationale was based on the concept of specificity for tumour versus normal mammalian enzymes, which might arise from small differences in amino acids sequences. In 1960, after hearing a paper on diaminodihydrotriazines which had high specificity for intestinal helminths and virtually no toxicity (Roth *et al.*, 1960), he concluded that dihydrotriazines were highly species-specific dihydrofolate reductase inhibitors, and initiated an investigation of this class of compounds (Baker *et al.*, 1960) which was to last until his death in 1971. He particularly sought specificity by the empirical principle of seeking irreversible drug binding to non-conserved amino acid residues near the active site (Baker, 1967; 1969). Among his many dihydrotriazines designed as reversible or irreversible inhibitors, two have aroused current interest: *9* (Baker's triazinate, TZT, Baker's soluble antifol; Baker and Ashton, 1973) and *10*, an irreversible inhibitor (Baker and Vermeulen, 1970).

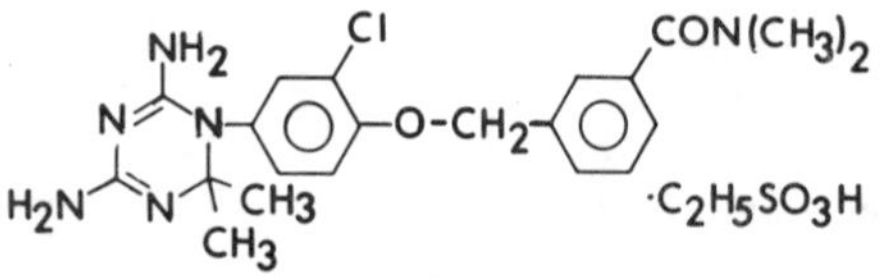

" Baker's soluble antifol "

9

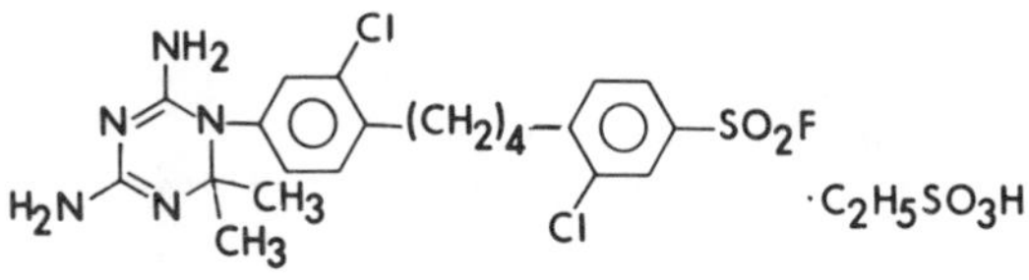

10

Dihydrotriazine *9* had very high activity against Dunning leukaemia ascites and against Walker 256 carcinosarcoma in the rat; it was found to enter the latter cells by facilitated diffusion using a different carrier than that used by methotrexate (Baker, 1971; Skeel *et al.*, 1973; Bertino and Lindquist, 1978).

More than 30 patients with solid tumours or acute leukaemia were evaluated in Phase I and II studies with *9*. A Phase II trial demonstrated objective regression in lung cancer, particularly adenocarcinoma (Rodriguez *et al.*, 1977). The compound proved to have toxicity related to liver dysfunction; the severity of gastrointestinal disturbances, dermatitis, and other complications might possibly compromise prolonged usage of the drug except at low doses (Creagan *et al.*, 1980). On the other hand, single-dose or a 3–5 day course has shown the drug to be relatively well tolerated; a 20% response rate with adenocarcinoma of the colon has been recorded (Bertino and Lindquist, 1978).

Recently compound *10* was chosen for studies of irreversible binding with chicken liver dihydrofolate reductase (Kumar *et al.*, 1981) and found to inactivate the enzyme irreversibly. The binary complex of *10* and dihydrofolate reductase was sequenced, and Tyr-31 was identified as the specific site of covalent attachment. Thus Baker's postulate has achieved fruition, in the sense that active-site-directed irreversible inhibition has been shown to occur with dihydrofolate reductase. Whether an active anticancer drug may ever result from this approach remains to be seen.

Bicyclic derivatives

Studies with the potent dihydrofolate reductase inhibitors *5a* and *5b* suggested the preparation of analogues lacking the acidic amino acid moiety, which might be transported by passive diffusion into malarial parasites (Elslager and Davoll, 1974). Many derivatives of this type were made and investigated for this purpose. The quinazolines *11* and *12* were found to be very potent inhibitors of the purified reductase from human acute lymphocytic leukaemia and from a methotrexate-resistant subline of L1210 murine leukaemia (Bertino *et al.*, 1979), and were therefore screened in animal tumour systems. Compound *11* was found more potent than methotrexate against several cell cultures and active in four murine tumours, where methotrexate was effective against only one. It was found to be a more potent inhibitor of DNA synthesis than methotrexate in human leukaemic cells, which suggests improved transport properties. It is also a potent inhibitor of histamine N-methyltransferase (Duch *et al.*, 1980a). No clinical studies have been reported.

11 *12*

As a sequel to the clinical studies with *6a*, a series of pyridopyrimidines (Hurlbert *et al.*, 1968) which had shown high antibacterial potency but almost equally high inhibitory activity against mammalian dihydrofolate reductase were reinvestigated, and new compounds synthesised. Some of these were found to have (low activity against histamine N-methyltransferase (Grivsky *et al.*, 1980; Duch *et al.*, 1980a). Compound *13* was chosen for further investigation. Its inhibitory activity towards dihydrofolate reductase from human chronic granulocytic leukaemia cells was equivalent to that of methotrexate, and it caused regression of several tumours in rodents (Duch *et al.*, 1980b). In contrast to *6a*, it had a short half-life in rodents and dogs, due to metabolism. It was less lipophilic than *6a*, as well. The compound is now undergoing clinical evaluation as an antitumour agent (C. A. Nichol, personal communication).

13

ACKNOWLEDGEMENTS

We are grateful to Dr D. Matthews, Dr D. Henry and Dr C. A. Nichol for their helpful comments and to Miss D. J. Baker for the schematic illustrations of the enzyme.

REFERENCES

Alberto, P., Peytremann, R., Medenica, R. and Beretta-Piccoli, M. (1978). *Cancer Chemother. Pharmac.*, **1**, 101

Alt, F. W., Kellems, R. E., Bertino, J. R. and Schimke, R. T. (1978). *J. biol. Chem.*, **253**, 1357

Angier, R. B., Boothe, J. H., Hutchings, B. L., Mowat, J. H., Semb, J., Stokstad, E. L. R., SubbaRow, Y., Walker, C. W., Cosulich, D. B., Fahrenbach, M. J., Seeger, D. R., Sickels, J. P. and Smith, J. M. Jr (1946). *Science*, **103**, 667

Angier, R. B. and Curran, W. G. (1959). *J. Am. Chem. Soc.*, **81**, 2814

Baccanari, D. P., Stone, D. and Kuyper, L. (1981). *J. biol. Chem.*, **256**, 1738

Baker, B. R. (1967). *The Design of Active-Site-Directed Irreversible Enzyme Inhibitors* John Wiley and Sons, New York, 192

Baker, B. R. (1969). *Acc. Chem. Res.*, **2**, 129

Baker, B. R. (1971). *Ann. N. Y. Acad. Sci.*, **186**, 214

Baker, B. R. and Ashton, W. T. (1973). *J. med. Chem.*, **16**, 209

Baker, B. R., Lee, W. W., Skinner, W. A., Martinez, A. P. and Tong, E. (1960). *J. pharm. Sci.*, **53**, 1137

Baker, B. R. and Vermeulen, N. M. J. (1970). *J. med. chem.*, **13**, 1154

Baker, D. J., Beddell, C. R., Champness, J. N., Goodford, P. J., Norrington, F. E. A., Smith, D. R. and Stammers, D. K. (1981a). *FEBS Lett.*, **126**, 49

Baker, D. J., Beddell, C. R., Champness, J. N., Goodford, P. J., Norrington, F. E. A., Roth, B. and Stammers, D. K. (1981b). *12th Gen. Assembly, Int. Union of Crystallography, Ottawa*, C58

Bareham, C. R., Griswold, D. E., Calabresi, P. (1974). *Cancer Res.*, **34**, 571

Bender, R. A. (1975). *Cancer Chemother. Rep.* (Part 3) **6**, 73

Bennett, C. D., Rodkey, J. A., Sondey, J. M. and Hirschmann, R. (1978). *Biochemistry*, **17**, 1328

Bertino, J. R. (1963). *Cancer Res.*, **23**, 1286

Bertino, J. R. (1975). *Handb. Exp. Pharmakol.*, **38**. *Antineoplast. Immunosuppr. Agents* Part 2, Springer-Verlag, Berlin, 468

Bertino, J. R. (1979). *Cancer Res.*, **39**, 293

Bertino, J. R. and Lindquist, C. (1978). *Advances in Cancer Chemotherapy*. (ed. H. Umezawa) Japan Sci. Soc. Press, Tokyo/Univ. Park Press, Baltimore. 155

Bertino, J. R., Perkins, J. P. and Johns, D. G. (1965). *Biochemistry*, **4**, 839

Bertino, J. R., Sawicki, W. L., Lindquist, C. A. and Gupta, V. S. (1977). *Cancer Res.*, **37**, 327

Bertino, J. R., Sawicki, W. L., Moroson, B. A., Cashmore, A. R., Elslager, E. F. (1979). *Biochem. Pharmac.*, 28, 1983

Bird, O. D., Vaitkus, J. W. and Clarke, J. (1970). *Molec. Pharmac.*, 6, 573

Birdsall, B., Burgen, A. S. V., Hyde, E. I., Roberts, G. C. K. and Feeney, J. (1981). *Biochemistry*, 20, 7186

Birdsall, B., Burgen, A. S. V. and Roberts, G. C. K. (1980). *Biochemistry*, 19, 3723

Bitar, K. G., Blankenship, D. T., Walsh, K. A., Dunlap, R. B., Reddy, A. V. and Freisheim, J. H. (1977). *FEBS Lett.*, 80, 119

Blakley, R. L. (1969). *The Biochemistry of Folic Acid and Related Pteridines*, North-Holland, Amsterdam

Blakley, R. L. (1981). *Molecular Actions and Targets for Cancer Chemotherapeutic Agents*, Academic Press, New York, 303

Blumenthal, G. and Greenberg, D. M. (1970). *Oncology*, 24, 223

Boothe, J. H., Semb, J., Waller, C. W., Angier, R. B., Mowat, J. H., Hutchings, B. L., Stokstad, E. L. R., SubbaRow, Y. (1949). *J. Am. Chem. Soc.*, 71, 2304

Borsa, J. and Whitmore, G. F. (1969a). *Molec. Pharmac.*, 5, 503

Borsa, J. and Whitmore, G. F. (1969b). *Molec. Pharmac.*, 5, 318

Bowen, D., White, J. C. and Goldman, I. D. (1978). *Cancer Res.*, 38, 219

Cahn, R. S., Ingold, C. K. and Prelog, V. (1966). *Angew. Chem. int. Edn Engl.*, 5, 385

Cavallito, J. C., Nichol, C. A., Brenckman, W. D., DeAngelis, P. L., Stickney, D. R., Simons, W. S. and Sigel, C. W. (1978). *Drug Metab. Disp.*, 6, 329

Chabner, B. A., Chello, P. L. and Bertino, J. R. (1972a) *Cancer Res.*, 32, 2114

Chabner, B. A., Johns, D. G. and Bertino, J. R. (1972b) *Nature, Lond.* 239, 395

Chang, S. and Littlefield, J. W. (1976). *Cell*, 7, 391

Charlton, P. A., Young, D. W., Birdsall, B., Feeney, J. and Roberts, G. C. K. (1979) *J. C. S. Chem. Commun.* 922

Cocco, L., Carroll, T. Jr., Montgomery, J. A., London, R. E. and Blakley, R. L. (1981). *Biochem. biophys. Res. Commun.*, 100, 413

Cody, V. and Zakrzewski, S. F. (1982). *J. med. Chem.* 25, 427

Cohen, S. S. (1971) *Ann. N. Y. Acad. Sci.*, 186, 292

Cosulich, D. B., Seeger, D. R., Fahrenbach, M. J., Collins, K. H., Roth, B., Hultquist, M. E. and Smith, J. M. (1953). *J. Am. Chem. Soc.*, 75, 4675

Cosulich, D. B., Seeger, D. R., Fahrenbach, M. J., Roth, B., Mowat, J. H., Smith, J. M. Jr and Hultquist, M. E. (1951). *J. Am. Chem. Soc.*, 73, 2554

Cosulich, D. B. and Smith, J. M. Jr. (1948). *J. Am. Chem. Soc.*, 70, 1922

Creagan, E. T., Eagan, R. T., Fleming, T. R., Frytak, S., Ingle, J. N., Kvols, K. and Nichols, W. C. (1980). *Cancer Treat. Rep.*, 64, 1057

Danenberg, P. V. (1977). *Biochim. biophys Acta*, 473, 73

Davoll, J. and Johnson, A. M. (1970). *J. Chem. Soc.*, (C), 977

DeGraw, J. I., Kisliuk, R. L., Gaumont, Y., Baugh, C. M. and Nair, M. G. (1974). *J. med. Chem.*, 17, 552

Denlinger, R. H., Nichol, C. A., Cavallito, J. C. and Sigel, C. W. (1976). *Proc. Am. Ass. Cancer Res.*, 17, 95

Duch, D. S., Edelstein, M. P. and Nichol, C. A. (1980a). *Molec. Pharmac.*, 18, 100

Duch, D. A., Sigel, C. W., Bowers, S. W., Edelstein, M. P., Cavallito, J. C., Foss, R. G. and Nichol, C. A. (1980b). *Am. Soc. Microbiol.*, 1597

Elslager, E. F. and Davoll, J. (1974). *Lectures in Heterocyclic Chemistry* (eds R. N. Castle and L. B. Townsend) Hetero Corp., Orem., Utah, 97

Etcubanas, E., Tan, C., Go, S. C. and Krakoff, I. H. (1972). *Proc. Am. Ass. Cancer Res.*, 13, 48

Farber, S. (1952). *Blood*, 7, 107

Farquhar, D., Loo, T. L. and Vadlamudi, S. (1972). *J. med. Chem.*, 15, 567

Fernandes, D. J., Moroson, B. A. and Bertino, J. R. (1981). *Cancer Treat. Rep. Suppl. 1*, 65, 29

Fontecilla-Camps, J. C. and Bugg, C. E. (1979). *Chemistry and Biology of Pteridines*, (eds R. L. Kisliuk and G. M. Brown) North-Holland, Amsterdam, 235

Frei, E., Spurr, C. L., Brindley, C. O., Selaway, O., Holland, J. F., Rall, D. P., Wasserman, L. R., Hoogstraten, B., Schnider, B. I., McIntyre, O. R., Matthews, L. B. and Miller,

S. P. (1965). *Clin. Pharmac. Ther.*, **6**, 160

Friedkin, M. (1973). *Adv. Enzym.*, **38**, 235

Fry, D. W., Yalowich, J. C. and Goldman, I. D. (1982). *J. biol. Chem.*, **257**, 1890

Futterman, S. (1957). *J. biol. Chem.*, **228**, 1031

Gellhorn, A. and Hirschberg, E. (1955). *Cancer Res.* Suppl. 3, **125**, 1

Gleisner, J. M., Peterson, D. L. and Blakley, R. L. (1974). *Proc. natn. Acad. Sci. U.S.A.*, **71**, 3001

Goldie, J. H., Dedhar, S. and Krystal, G. (1981). *J. biol. Chem.*, **256**, 11629

Goldin, A., Mantel, N., Greenhouse, S. W., Venditti, J. M. and Humphreys, S. R. (1953). *Cancer Res.*, **13**, 843

Goldin, A., Mantel, N., Greenhouse, S. W., Venditti, J. M. and Humphreys, S. R. (1954). *Cancer Res.*, **14**, 43

Goldin, A., Humphreys, S. R., Venditti, J. M. and Mantel, N. (1959). *J. natn. Cancer Inst.*, **22**, 811

Goldin, A., Venditti, J. M., Humphreys, S. R., Dennis, D., Mantel, N. and Greenhouse, S. M. (1955). *Cancer Res.*, **15**, 57

Goldin, A., Venditti, J. M., Kline, I. and Mantel, N. (1966). *Nature, Lond.*, **212**, 1548

Goldman, I. D. (1971). *Trans. N. Y. Acad. Sci.*, **186**, 400

Goldman, I. D., Lichenstein, N. S., Oliverio, V. T. (1968). *J. biol. Chem.*, **243**, 5007

Goodman, L., DeGraw, J., Kisliuk, R. L., Friedkin, M., Pastore, E. G., Crawford, E. J., Plante, L. T., Al-Nahas, A., Morningslar, J. F. Jr., Kwok, G., Wilson, L., Donovan, E. F. and Ratzan, J. (1964). *J. Am. chem. Soc.*, **86**, 308

Greco, W. R. and Hakala, M. T. (1980). *J. Pharm. exp. Ther.*, **212**, 39

Grivsky, E. M., Lee, S., Sigel, C. W., Duch, D. S. and Nichol, C. A. (1980). *J. med. Chem.*, **23**, 327

Hakala, M. T. (1965). *Biochem. biophys. Acta*, **102**, 198

Henderson, E. S., Adamson, R. H., Denham, C. and Oliverio, V. T. (1965a). *Cancer Res.*, **25**, 1008

Henderson, E. S., Adamson, R. H. and Oliverio, V. T. (1965b). *Cancer Res.*, **25**, 1018

Henderson, G. B., Grzelakowska-Sztabert, B., Zevely, E. M., Huennekens, F. M. (1980a). *Archs. biochem. Biophys.*, **202**, 144

Henderson, G. B., Zevely, E. M. and Huennekens, F. M. (1977). *J. biol. Chem.* **252**, 3760

Henderson, G. B., Zevely, E. M. and Huennekens, F. M. (1978). *Cancer Res.*, **38**, 359

Henderson, G. B., Zevely, E. M., Huennekens, F. M. (1980b). *J. biol. Chem.*, **255**, 4829

Henderson, G. B., Zevely, E. M., Huennekens, F. M. (1979). *J. biol. Chem.*, **254**, 9973

Hill, B. T., Bailey, B. D. and Goldman, I. D. (1978). *Proc. Am. Ass. Cancer. Res.*, **19**, 49

Hill, B. T. and Price, L. A. (1977). *Cancer Topics*, **1**(8), 2.

Hill, B. T., Price, L. A., Harrison, S. I. and Goldie, J. H. (1975). *Biochem. Pharmac.*, **24**, 535

Ho, Y. K., Hakala, M. T. and Zakrzewski, S. F. (1972). *Cancer Res.*, **32**, 1023

Hopfinger, A. J. (1980). *J. Am. Chem. Soc.*, **102**, 7196

Huennekens, F. M., Vitols, K. S., Suresh, M. R. and Henderson, G. B. (1981). *Molecular Actions and Targets for Cancer Chemotherapeutic Agents* (ed. A. C. Sartorelli, J. S. Lazo, and J. R. Bertino) Academic Press, New York

Hultquist, M. E., Smith, J. M. Jr., Seeger, D. R., Gosulich, D. B., Kuh, E. (1949). *J. Am. chem. Soc.*, **71**, 619

Hunt, W. E., Schwalbe, C. H., Bird, K. and Mallinson, P. D. (1980). *Biochem. J.*, **187**, 533

Hurlbert, B. S., Ferone, R., Herrmann, T. A. Hitchings, G. H., Barnett, M. and Bushby, S. R. M. (1968). *J. med. Chem.*, **11**, 711

Hutchings, B. L., Mowat, J. H., Oleson, J. J., Stokstad, E. R. L., Boothe, J. H., Waller, C. W., Angier, R. S., Semb, J. and SubbaRow, Y. (1947). *J. biol. Chem.*, **170**, 323

Hutchison, D. J. (1968). *Cancer Chemother. Rep.*, **52**, 697

Hutchison, D. J., Shimoyama, M. and Schmid, F. A. (1971). *Cancer Chemother. Rep.*, **55**, 123

Jacobs, S. A. and Santicky, M. J. (1978). *Cancer Treat. Rep.*, **62**, 397

Jardetzky, O. and Roberts, G. C. K. (1982). *NMR in Molecular Biology*, Academic Press, New York

Johns, D. G., Iannotti, A. T., Sartorelli, A. C., Booth, B. A. and Bertino, J. R. (1965).

Biochim. biophys. Acta, **105**, 380

Johns, D. G., and Loo, T. L. (1967). *J. pharm. Sci.*, **56**, 356

Johns, D. G., Sartorelli, A. C., Iannotti, A. T. and Bertino, J. R. (1964). *Proc. Am. Ass. Cancer Res.*, **5**, 32

Johns, D. G. and Valerino, D. M. (1971). *Ann. N. Y. Acad. Sci.*, **186**, 378

Jonak, J. P., Zakrzewski, S. F., Mead, L. H. and Hakala, M. T. (1970). *J. med. Chem.*, **13**, 1170

Kellems, R. E., Alt, F. W. and Schimke, R. T. (1976). *J. biol. Chem.*, **251**, 6987

Kessel, D., Hall, T. C., Roberts, D., Wodinsky, I. (1965). *Science*, **150**, 752

Kline, I., Venditti, J. M., Mead, J. A. R., Tyrer, D. D. and Goldin, A. (1966). *Cancer Res.*, **26**, 848

Kumar, A. A., Blankenship, D. T., Kaufman, B. T. and Friesheim, J. H. (1980). *Biochemistry*, **19**, 667

Kumar, A. A., Mangum, J. H., Blankenship, D. T. and Freisheim, J. H. (1981). *J. biol. Chem.*, **256**, 8970

Lai, P-H., Pan, Y-C., Gleisner, J. M., Paterson, D. L. and Blakley, R. L. (1979). *Chemistry and Biology of Pteridines* (eds R. L. Kisliuk and G. M. Brown) North-Holland, Amsterdam, 437

Lindquist, C. A., Moroson, B. A. and Bertino, J. R. (1978). *Proc. Am. Ass. Cancer Res.*, **19**, 165

Loo, T. L. and Adamson, R. H. (1965). *J. med. Chem.*, **8**, 513

Matthews, D. A. (1981). *Molecular Structure and Biological Activity* (eds J. F. Griffin and W. L. Duax) Elsevier, New York, 13

Matthews, D. A., Alden, R. A., Bolin, J. T., Filman, D. J., Freer, S. T., Hamlin, R., Hol, W. G. J., Kisliuk, R. L., Pastore, E. J., Plante, L. T., Xuong, N. H. and Kraut, J. (1978). *J. biol. Chem.*, **253**, 6946

Matthews, D. A., Alden, R. A., Bolin, J. T., Freer, S. T., Hamlin, R., Xuong, N., Kraut, J., Poe, M., Williams, M. and Hoogsteen, K. (1977). *Science*, **197**, 452

Mead, J. A. R., Greenberg, N. H., Schrecker, A. W., Seeger, D. R. (1965). *Biochem. Pharmac.*, **14**, 105

Miller, D. S., Rundles, R. W., Nichol, C. A., Woolley, J. L. and Sigel, C. W. (1976). *Proc. Am. Soc. clin. Oncol.*, **17**, 263

Misra, D. K., Adamson, R. H., Loo, T. L. and Oliverio, V. T. (1963). *Life Sci.*, **2**, 407

Misra, D. K., Humphreys, S. R., Friedkin, M., Goldin, A., Crawford, E. J. (1961). *Nature, Lond.*, **189**, 39

Modest, E. J. (1956). *J. org. Chem.*, **21**, 1

Montgomery, J. A., Piper, J. R., Elliott, R. D., Temple, C., Roberts, E. C., Shealy, Y. F. (1979). *J. med. Chem.*, **22**, 862

Moran, R. G., Mulkins, M. and Heidelberger, C. (1979). *Proc. natn Acad. Sci. U.S.A.*, **76**, 5924

Motycka, K. (1971). *Natn. Cancer Inst. Monogr.*, **34**, 167

Mowat, J. H., Gazzola, A. L., Hutchings, B. L., Boothe, J. H., Waller, C. W., Angier, R. B., Semb, J. and SubbaRow, Y. (1949). *J. Am. chem. Soc.*, **71**, 2308

Nahas, A., Nixon, P. F. and Bertino, J. R. (1972). *Cancer Res.*, **32**, 1416

Nadel, E. N. and Greenberg, J. (1953). *Cancer Res.*, **13**, 865

Nair, M. G. and Campbell, P. T. (1976). *J. med. Chem.*, **19**, 825

Nair, M. G., Campbell, P. T., Braverman, E. and Baugh, C. M. (1975). *Tetrahedron Lett.*, 2745

Nair, M. G., Mercer, L. P. and Baugh, C. M. (1974). *J. med. Chem.*, **17**, 1268

Nederbragt, H., Uitendaal, M. P., van der Grint, L., Levya, A. and Pinedo, H. M. (1981). *Cancer Res.*, **41**, 1193

Nichol, C. A. (1977). *Cancer*, **40**, 519

Nichol, C. A., Cavallito, J. C., Woolley, J. L. and Sigel, C. W. (1977). *Cancer Treat. Rep.*, **61**, 559

Nixon, P. and Bertino, J. R. (1972). *New Engl. J. Med.*, **286**, 175

Oleson, J. J. (1950). *Trans. N. Y. Acad. Sci.*, Ser II, **12**, 118

Osborn, M. J., Freeman, M. and Huennekens, F. M. (1958). *Proc. Soc. exp. Biol. Med.*, **97**, 429

Philips, F. S., Sternberg, S. S., Sodergren, J. E. and Videl, P. (1971). *Cancer Chemother.*

Rep., **55**, 35

Piper, J. R. and Montgomery, J. A. (1979). *Devl Biochem.*, **4**, 261

Pogolotti, A. L. Jr. and Santi, D. V. (1974). *Biochemistry*, **13**, 456

Pogolotti, A. L., Jr., and Santi, D. V. (1977). *Bioorganic Chemistry* (ed. E. E. van Tamelen), Academic Press, New York, 277

Price, L. A., Goldie, J. H. and Hill, B. T. (1975). *Br. med. J.*, ii, 20

Price, L. A. and Hill, B. T. (1976). *Chemotherapy-Proc IXth Int. Congr. Chemother.*, (ed. J. D. Williams) Plenum Press, New York, **8**, 481

Rauckman, B. S. and Roth, B. (1980). *J. med. Chem.*, **23**, 384

Rodriguez, V., Richman, S. P., Benjamin, R. S., Burgess, M. A., Murphy, W. K., Valdivieso, M., Banner, R. L., Gutterman, J. U., Bodey, G. P. and Freireich, E. J. (1977). *Cancer Res.*, **37**, 980

Rosowsky, A. (1973). *J. med. Chem.*, **16**, 1190

Rosowsky, A., Beardsley, G. P., Ensminger, W. D., Lazarus, H. and Yu, C-S. (1978). *J. med. Chem.*, **21**, 380

Rosowsky, A., Ensminger, W. D., Lazarus, H. and Yu, C-S. (1977). *J. med. Chem.*, **20**, 925

Rossmann, M. G. and Argos, P. (1976). *J. molec. Biol.*, **105**, 75

Roth, B., Burrows, R. B. and Hitchings, G. H. (1960). *Abstr. 137th Am. Chem. Meeting, Cleveland, Ohio*, 31N

Roth, B., Smith, J. M. Jr. and Hultquist, M. E. (1950). *J. Am. chem. Soc.*, **72**, 1914

Roth, B., Smith, J. M. Jr. and Hultquist, M. E. (1951). *J. Am. chem. Soc.*, **73**, 2864

Roth, B. and Strelitz, J. Z. (1969). *J. org. Chem.*, **34**, 821

Russell, P. B. and Hitchings, G. H. (1951). *J. Am. chem. Soc.*, **73**, 3763

Sandberg, J. S. and Goldin, I. (1970). *Cancer Res.*, **30**, 1276

Santi, D. V. (1980). *J. med. Chem.*, **23**, 103

Santi, D. V. (1981). *Molecular Actions and Targets for Cancer Chemother. Agents*, Academic Press, New York, 285

Scanlon, K. J., Moroson, B. A., Bertino, J. R. and Hynes, J. B. (1979). *Molec. Pharmac.*, **16**, 261

Sedwick, W. D., Birdwell, R. and Laszlo, J. (1979). *Proc. 11th Int. Congr. Chemother.*, Abstr. 1013

Seeger, D. R., Cosulich, D. B., Smith, J. M. Jr and Hultquist, M. E. (1949). *J. Am. chem. Soc.*, **71**, 1753

Seeger, D. R., Smith, J. M. Jr and Hultquist, M. E. (1947). *J. Am. chem. Soc.*, **69**, 2567

Semb, J., Boothe, J. H., Angier, R. B., Waller, C. W., Mowat, J. H., Hutchings, B. L. and SubbaRow, Y. (1949). *J. Am. chem. Soc.*, **71**, 2310

Sirotnak, F. M., Chello, P. L., DeGraw, J. I., Piper, J. R. and Montgomery, J. A. (1981). *Molecular Actions and Targets for Cancer Chemother. Agents*, Academic Press, New York, 349

Sirotnak, F. M., Chello, P. L., Piper, J. R., Montgomery, J. A. and DeGraw, J. I. (1979). *The Chemistry and Biology of Pteridines* (ed. R. L. Kisliuk and G. M. Brown) Elsevier North-Holland, New York, 597

Sirotnak, F. M. and Donsbach, R. C. (1972). *Cancer Res.*, **32**, 2120

Sirotnak, F. M. and Donsbach, R. C. (1976). *Cancer Res.*, **36**, 1151

Sirotnak, F. M., Kurita, S. and Hutchison, D. J. (1968). *Cancer Res.*, **28**, 75

Skeel, R. T., Sawicki, W. L., Cashmore, A. R. and Bertino, J. R. (1973). *Cancer Res.*, **33**, 2972

Smith, J. M. Jr. (1949). *New Jersey J. Pharm.*, p. 15

Smith, S. L., Patrick, P., Stone, D., Phillips, A. W. and Burchall, J. J. (1979). *J. biol. Chem.*, **254**, 11475

Stone, D. (1981). *Biochemical Soc. Trans.*, **9**, 271

Stone, D., Paterson, S. J., Raper, J. H. and Phillips, A. W. (1979). *J. biol. Chem.*, **254**, 480

Stone, D., Phillips, A. W. and Burchall, J. J. (1977). *Eur. J. Biochem.*, **72**, 613

Sugiura, K. (1956). *Ann. N. Y. Acad. Sci.*, **63**, 962

Tattersall, M. H. N., Jackson, R. C., Connors, T. A. and Harrap, K. R. (1973). *Eur. J. Cancer* **9**, 733

Temple, C., Elliott, R. D., Foye, J. L., Montgomery, J. A. (1971). *J. org. Chem.*, **36**, 2818

Temple, C., Kussner, C. L. and Montgomery, J. A. (1975). *J. org. Chem.*, **40**, 2205

Vogler, W. R. and Jacobs, J. (1971). *Cancer*, **28**, 894

Volz, K., Matthews, D., Alden, R., Freer, S., Hansch, C., Kaufman, B. and Kraut, J. (1982). *J. biol. Chem.*, **257**, 2528

Weinstock, L. T., Grabowski, B. F. and Cheng, C. C. (1970). *J. med. Chem.*, **13**, 995

Weinstock, L. T., O'Brien, D. E. and Cheng, C. C. (1968). *J. med. Chem.*, **11**, 1238

Werkheiser, W. C. (1961). *J. biol. Chem.*, **236**, 888

Werkheiser, W. C. (1969). *Fedn Proc.*, **28**, 362

Williams, J. H. (1952). *Blood*, **7**, 100

Wood, R. C., Ferone, R. and Hitchings, G. H. (1961). *Biochem. Pharmac.*, **6**, 113

Wright, W. B. Jr., Cosulich, D. B., Fahrenbach, M. J., Waller, C. W., Smith, J. M. Jr and Hultquist, M. E. (1949). *J. Am. chem. Soc.*, **71**, 3014

Zaharko, D. S., Bruckner, H. and Oliverio, V. T. (1969). *Science*, **166**, 887

Zaharko, D. S. and Oliverio, V. T. (1970). *Biochem. Pharmac.*, **19**, 2923

Zakrzewski, S. F. and Creavan, P. J. (1982). *Proc. Am. Ass. Cancer Res.*, **23**, 130

Zakrzewski, S. F., Dave, C. and Rosen, F. (1978). *J. natn. Cancer Inst.*, **60**, 1029

Zakrzewski, S. F. and Nichol, C. A. (1958). *Biochim. biophys. Acta*, **27**, 425

Footnote

The reader's attention is drawn to recent amendments to the amino acid sequence for bovine liver DHFR. *See* Lai, P.-H., Pan, Y.-C. E., Gleisner, J. M., Peterson, D. L., Williams, K. R. and Blakley, R. L. (1982). *Biochem.*, **21**, 3284

Subject Index